Geometric Gradient

Geometric Series Present Worth:

To Find P $(P/A,g,i,n)$
Given A_1, g When $i = g$ $P = A_1[n(1 + i)^{-1}]$

To Find P $(P/A,g,i,n)$
Given A_1, g When $i \neq g$ $P = A_1\left[\dfrac{1 - (1 + g)^n(1 + i)^{-n}}{i - g}\right]$

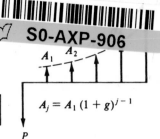

$A_j = A_1(1 + g)^{j-1}$

Continuous Compounding at Nominal Rate r

Single Payment: $F = P[e^{rn}]$ $P = F[e^{-rn}]$

Uniform Series: $A = F\left[\dfrac{e^r - 1}{e^{rn} - 1}\right]$ $A = P\left[\dfrac{e^{rn}(e^r - 1)}{e^{rn} - 1}\right]$

$F = A\left[\dfrac{e^{rn} - 1}{e^r - 1}\right]$ $P = A\left[\dfrac{e^{rn} - 1}{e^{rn}(e^r - 1)}\right]$

Continuous, Uniform Cash Flow (One Period)
With Continuous Compounding at Nominal Rate r

Present Worth:

To Find P
Given $\overline{F}$ $\left[P/\overline{F},r,n\right]$ $P = \overline{F}\left[\dfrac{e^r - 1}{re^{rn}}\right]$

Compound Amount:

To Find F
Given $\overline{P}$ $\left[F/\overline{P},r,n\right]$ $F = \overline{P}\left[\dfrac{(e^r - 1)(e^{rn})}{re^r}\right]$

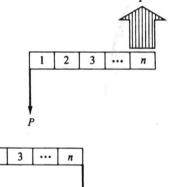

Compound Interest

i = Interest rate per interest period*.

n = Number of interest periods.

P = A present sum of money.

F = A future sum of money. The future sum F is an amount, n interest periods from the present, that is equivalent to P with interest rate i.

A = An end-of-period cash receipt or disbursement in a uniform series continuing for n periods, the entire series equivalent to P or F at interest rate i.

G = Uniform period-by-period increase or decrease in cash receipts or disbursements; the arithmetic gradient.

g = Uniform *rate* of cash flow increase or decrease from period to period; the geometric gradient.

r = Nominal interest rate per interest period*.

m = Number of compounding subperiods per period*.

$\overline{P}, \overline{F}$ = Amount of money flowing continuously and uniformly during one given period.

---

*Normally the interest period is one year, but it could be something else.

Engineering Economic Analysis

Eighth Edition

Donald G. Newnan
Jerome P. Lavelle
Ted G. Eschenbach

New York   Oxford
OXFORD UNIVERSITY PRESS

Oxford University Press   .

Oxford   New York
Athens   Auckland   Bangkok   Bogotá   Buenos Aires   Calcutta
Cape Town   Chennai   Dar es Salaam   Delhi   Florence   Hong Kong   Istanbul
Karachi   Kuala Lumpur   Madrid   Melbourne   Mexico City   Mumbai
Nairobi   Paris   São Paulo   Shanghai   Singapore   Taipei   Tokyo   Toronto   Warsaw

and associated companies in
Berlin   Ibadan

Copyright © 2002 by Oxford University Press, Inc.

Published by Oxford University Press, Inc.
198 Madison Avenue, New York, New York 10016
http://www.oup-usa.org

Oxford is a registered trademark of Oxford University Press

ISBN: 0-19-515152-6

Printing (last digit): 9 8 7 6 5 4 3 2

Printed in the United States of America
on acid-free paper

Contents

Preface

In the first edition of this book we said:

> This book is designed to teach the fundamental concepts of engineering economy to engineers. By limiting the intended audience to engineers it is possible to provide an expanded presentation of engineering economic analysis and do it more concisely than if the book were written for a wider audience.

Our goal was, and still is, to provide an easy to understand and up-to-date presentation of engineering economic analysis. That means the book's writing style must promote the reader's understanding. We most humbly find that our approach has been well received by engineering professors – and more importantly – by engineering students through seven previous editions. This edition has significant improvements. Chapter 1 (*Making Economic Decisions*) has been rewritten to combine Chapters 1 and 2 into a focused unified treatment. In response to adopter suggestions, a new Chapter 2 (*Engineering Costs and Cost Estimating*) has been added. Chapter 10 (*Depreciation*) has been rewritten to focus more on the MACRS depreciation method with a new section on recaptured depreciation and asset disposal. Chapter 11 (*Income Taxes*) has been updated to reflect 1999 tax legislation and rates. Chapter 12 (*Replacement Analysis*) has been updated in the section on after-tax replacement effects.

The most substantial change is in the text's approach to spreadsheets. Rather than relying on spreadsheet templates, the emphasis is on helping students learn to use the enormous capabilities of software that is available on every computer. This approach reinforces the traditional engineering economy factor approach, as the equivalent spreadsheet functions (PMT, PV, RATE, etc.) are used frequently.

For those students who would benefit from a refresher or introduction on how to write good spreadsheets, there is an appendix to introduce spreadsheets. In Chapter 2, spreadsheets are used to draw cash flow diagrams. Then from Chapter 4 to Chapter 15 every chapter has a concluding section on spreadsheet use. Each section is designed to support the other material in the chapter and to add to the student's knowledge of spreadsheets. If spreadsheets are used, the student will be very well prepared to apply this tool to real world problems after graduation.

This approach is designed to support a range of approaches to spreadsheets. Professors and students can rely on the traditional tools of engineering economy and without loss of continuity completely ignore the material on spreadsheets. Or at the other extreme, professors can introduce the concepts and require all computations to be done with spreadsheets. Or a mix of approaches depending on the professor, students, and particular chapter may be taken.

In addition to the specific changes to the various chapters described, over 120 new homework problems have been added at the end of the chapters. Also, many less noticeable changes have been made throughout the book to improve its content and readability.

For instructors and students there is now an expanded set of supplemental materials.

* *Solutions Manual for Engineering Economic Analysis*. This 350-page manual has been revised and checked by the authors for accuracy; all end-of-chapter problems are fully solved by the authors. Available free to adopting professors. (ISBN 1-57645-052-X)

* *Compound Interest Tables*. A separate 32-page pamphlet with the compound interest tables from the textbook. Classroom quantities are free to adopting professors. (ISBN 0-910554-08-0)

* *Exam Files*. Fourteen quizzes prepared by the authors test student knowledge of chapter content. Available free in electronic format to adopting professors. Call 1-800-280-0280 or send an email to college@oup-usa.org.

* *Instructor Lecture Notes and Overhead Transparencies*. Available free in electronic format to adopting professors. Call 1-800-280-0280 or send an email to college@oup-usa.org.

* *Student's Quick Study Guide*: *Engineering Economic Analysis*. This 320-page book features a 32-page summary of engineering economy, followed by 386 problems, each with detailed solutions. (ISBN 1-57645-050-3) Available for purchase only.

Many people have directly or indirectly contributed to the content of the book in its seven editions. We have been influenced by our Stanford and North Carolina State University educations, our university colleagues, and students who have provided invaluable feedback on content and form. We are particularly grateful to Professors:

Dick Bernhard, North Carolina State University
Charles Burford, Texas Tech University
Jeff Douthwaite, University of Washington
Utpal Dutta, University of Detroit, Mercy
Vernon Hillsman, Purdue University
Oscar Lopez, Florida International University
Nic Nigro, Cogswell College North
Ben Nwokolo, Grambling State University
Cecil Peterson, GMI Engineering & Management Institute
Malgorzata Rys, Kansas State University
Robert Seaman, New England College
R. Meenakshi Sundaram, Tennessee Tech University
Roscoe Ward, Miami University
Jan Wolski, New Mexico Institute of Mining and Technology
and particularly Bruce Johnson, U.S. Naval Academy

We would appreciate being informed of errors or receiving other comments about the book. Please write us at Oxford University Press, 198 Madison Avenue, New York, NY 10016, or email the editor: pcg@oup-usa.org.

CHAPTER 1

Making Economic Decisions

This book is about making decisions. ***Decision making*** is a broad topic, for it is a major aspect of everyday human existence. This book will isolate those problems that are commonly faced by engineers and develop the tools to properly analyze and solve them. Even very complex situations can be broken down into components from which sensible solutions are produced. If one understands the decision-making process and has tools for obtaining realistic comparisons between alternatives, one can expect to make better decisions.

Although we will focus on solving problems that confront firms in the marketplace, we will also use examples of how these techniques may be applied to the problems faced in daily life. Since decision making or problem solving is our objective, let us start by looking at some problems.

A Sea Of Problems

A careful look at the world around us clearly demonstrates that we are surrounded by a sea of problems. There does not seem to be any exact way of classifying them, simply because they are so diverse in complexity and "personality." One approach would be to arrange problems by their *difficulty*.

Simple Problems

On the lower end of our classification of problems are simple situations.

- Should I pay cash for an item or use my credit card?
- Do I buy a semester parking pass or use the parking meters?
- Shall we replace a burned-out motor?
- If we use three crates of an item a week, how many crates should we buy at a time?

These are pretty simple problems and do not require much time or effort to come to a good solution.

Intermediate Problems

At this level of complexity we find problems that are primarily economic.

- Shall I buy or lease my next car?
- Which equipment should be selected for a new assembly line?
- Which materials should be used as roofing, siding, and structural support for a new building?
- Shall I buy a 1- or 2-semester parking pass?
- Which printing press should be purchased? A low cost press requiring three operators, or a more expensive one needing only two operators?

Complex Problems

At the upper end of our classification system we discover problems that are indeed complex. They represent a mixture of *economic*, *political*, and *humanistic* elements.

- The decision of Mercedes Benz to build an automobile assembly plant in Tuscaloosa, Alabama illustrates a complex problem. Beside the economic aspects, Mercedes Benz must consider possible reactions in the American auto industry. Will the German government pass legislation to prevent the overseas plant? What about German labor unions?
- The selection of a girlfriend or a boyfriend (who may later become a spouse) is obviously complex. Economic analysis can be of little or no help.
- The annual budget of a corporation is an allocation of resources, but the budget process is heavily influenced by non-economic forces such as power struggles, geographical balancing, and impact on individuals, programs, and profits. For multinational corporations there are even national interests to be considered.

The Role Of Engineering Economic Analysis

Engineering economic analysis is most suitable for intermediate problems and the economic aspects of complex problems. They have these qualities:

1. The problem is *sufficiently important* that we are justified in giving it some serious thought and effort.

2. The problem can't be worked in one's head—that is, a careful analysis *requires that we organize* the problem and all the various consequences, and this is just too much to be done all at once.

3.   The problem has *economic aspects* that are sufficiently important to be a significant component of the analysis leading to a decision.

When problems have these three criteria, engineering economic analysis is an appropriate technique for seeking a solution. Since there are vast numbers of problems that one will encounter in the business world (and in one's personal life) that meet these criteria, engineering economic analysis is often a required tool.

Examples of Engineering Economic Analysis

Engineering economic analysis focuses on costs, revenues, and benefits that occur at different times. For example, when a civil engineer designs a road, a dam, or a building, the construction costs occur in the near future and the benefits to users only begin when construction is finished, but then the benefits continue for a long time.

In fact nearly everything that engineers design has the requirement to spend money to design and build, and after it is built, then revenues or benefits occur–usually for years. Thus the economic analysis of costs, benefits, and revenues occurring over time is called *engineering* economic analysis.

Engineering economic analysis is used to answer many different questions.

- *Which engineering projects are worthwhile?* Has the mining or petroleum engineer shown that the mineral or oil deposit is worth developing?
- *Which engineering projects should have a higher priority?* Has the industrial engineer shown which factory improvement projects should be funded with the available dollars?
- *How the engineering project should be designed?* Has the mechanical or electrical engineer chosen the most economical motor size? Has the civil or mechanical engineer chosen the best thickness for insulation? Has the aeronautical engineer made the best tradeoffs between 1) lighter materials that are expensive to buy but cheaper to fly and 2) heavier materials that are cheap to buy and more expensive to fly?

Engineering economic analysis can also be used to answer questions that are personally important.

- *How to achieve long-term financial goals?* How much should you save each month to buy a house, retire, or fund a trip around the world? Is going to graduate school a good investment–will your additional earnings in later years balance your lost income while in graduate school?
- *How to compare different ways to finance purchases?* Which is a better way to finance your car purchase, use the dealer's low interest rate loan or take the rebate and borrow money from your bank or credit union?
- *How to make short and long-term investment decisions?* Is a higher salary better than stock options? Should you buy a 1- or a 2-semester parking pass?

The Decision Making Process

Decision making may take place by default, that is, without consciously recognizing that an opportunity for decision making exists. This fact leads us to a first element in a definition of decision making. To have a decision-making situation, there must be at least two alternatives available. If only one course of action is available, there can be no decision making, for there is nothing to decide. We would have no alternative but to proceed with the single available course of action. (It is a rather unusual situation when there are no alternative courses of action. More frequently, alternatives simply are not recognized.)

At this point we might conclude that the decision-making process consists of choosing from among alternative courses of action. But this is an inadequate definition. Consider the following:

At a horse race, a bettor was uncertain which of the five horses to bet on in the next race. He closed his eyes and pointed his finger at the list of horses printed in the racing program. Upon opening his eyes, he saw that he was pointing to horse number four. He hurried off to place his bet on that horse.

Does the racehorse selection represent the process of decision making? Yes, it clearly was a process of choosing among alternatives (assuming the bettor had already ruled out the "do-nothing" alternative of placing no bet). But the particular method of deciding seems inadequate and irrational. We want to deal with rational decision making.

Rational Decision Making

Rational decision making is a complex process that contains nine essential elements; which are shown sequentially in Figure 1-1. While these nine steps are shown sequentially, it is common for decision making to repeat steps, take them out of order, and do steps simultaneously. For example, when a new alternative is identified, then more data will be required. Or when the outcomes are summarized it may be clear that the problem needs to be redefined or new goals established.

The value of this sequential diagram is to show all of the steps that are usually required, and to show them in a logical order. Occasionally we will skip a step entirely. For example, a new alternative may be so clearly superior that it is immediately adopted without further analysis.

    1. Recognize problem;

↓

    2. Define the goal or objective;

↓

    3. Assemble relevant data;

↓

    4. Identify feasible alternatives;

↓

    5. Select the criterion to determine the best alternative;

↓

    6. Construct a model;

↓

    7. Predict each alternative's outcomes or consequences;

↓

    8. Choose the best alternative; and

↓

    9. Audit the result.

Figure 1-1 One possible flowchart of the decision process.

The following sections will describe these elements.

1. Recognize the Problem

The starting point in rational decision making is recognizing that a problem exists.

Some years ago, for example, it was discovered that several species of ocean fish contained substantial concentrations of mercury. The decision-making process began with this recognition of a problem, and the rush was on to determine what should be done. Research revealed that fish taken from the ocean decades before, and preserved in laboratories, also contained similar concentrations of mercury. Thus, the problem had existed for a long time but had not been recognized.

In typical situations, recognition is obvious and immediate. An auto accident, an overdrawn check, a burned-out motor, an exhausted supply of parts all produce the recognition of a problem. Once we are aware of the problem, we can solve it as best we can. Many firms establish programs for total quality management (TQM) or continuous improvement (CI) that are designed to identify problems, so that they can be solved.

2. Define the Goal or Objective

Goals or objectives can be a grand, overall goal of a person or a firm. For example, a personal goal could be to lead a pleasant and meaningful life and a firm's goal is usually to operate profitably. The presence of multiple, conflicting goals is often the foundation of complex problems.

But an objective need not be a grand, overall goal of a business or an individual. It may be quite narrow and specific: "I want to pay off the loan on my car by May," or "The plant must produce 300 golf carts in the next two weeks," are more limited objectives. Thus, defining the objective is the act of exactly describing the task or goal.

3. Assemble Relevant Data

To make a good decision, one must first assemble good information. In addition to all the published information, there is a vast quantity of information that is not written down anywhere, but is stored as individual's knowledge and experience. There is also information that remains ungathered. A question like "How many people in Lafayette, Indiana, would be interested in buying a pair of left-handed scissors?" cannot be answered by examining published data or by asking any one person. Market research or other data gathering would be required to obtain the desired information.

From all of this information, which of it is relevant in a specific decision making process? It may be a complex task to decide which data are important and which data are not. The availability of data further complicates this task. Some data are available immediately at little or no cost in published form; other data are available by consulting with specific knowledgeable people; still other data require surveys or research to assemble the information. Some data will be precise and accurate–high quality, while other data may rely on individual judgement for an estimate.

In engineering decision making, an important source of data is a firm's own accounting system. These data must be examined quite carefully. Accounting data focuses on past information, and engineering judgement must often be applied to estimate current and future values. For example, accounting records can show the past cost of buying computers, but engineering judgement is required to estimate the future cost of buying computers.

Financial and cost-accounting is designed to show accounting values and the flow of money—specifically *costs* and *benefits*—in a company's operations. Where costs are directly related to specific operations, there is no difficulty; but there are other costs that are not related to specific operations. These indirect costs, or *overhead,* are usually allocated to a company's operations and products by some arbitrary method. The results are generally satisfactory for cost-accounting purposes but may be unreliable for use in economic analysis.

To create a meaningful economic analysis, we must determine the *true* differences between alternatives, which might require some adjustment of cost-accounting data. The following example illustrates this situation.

EXAMPLE 1-1

The cost-accounting records of a large company show the following average monthly costs for the three-person printing department:

Direct labor and salaries (including employee benefits)	$ 6,000
Materials and supplies consumed	7,000
Allocated overhead costs 200 m² of floor area at $25/m²	5,000
	$18,000

The printing department charges the other departments for its services to recover its $18,000 monthly cost. For example, the charge to run 1000 copies of an announcement is:

Direct labor	$7.60
Materials and supplies	9.80
Overhead costs	9.05
Cost to other departments	$26.45

The shipping department checks with a commercial printer and finds they could have the same 1000 copies printed for $22.95. Although the shipping department only has about 30,000 copies printed a month, they decide to stop using the printing department and have their printing done by the outside printer. The printing department objects to this. As a result, the general manager has asked you to study the situation and recommend what should be done.

Solution:

Much of the printing department's work reveals the company's costs, prices, and other financial information. The company president considers the printing department necessary to prevent disclosing such information to people outside the company.

A review of the cost-accounting charges reveals nothing unusual. The charges made by the printing department cover direct labor, materials and supplies, and overhead. (*Note:* The company's indirect costs—such as heat, electricity, employee insurance, and so forth—must be distributed to its various departments in *some* manner and, like many other firms, it uses *floor space* as the basis for its allocations. The printing department, in turn, must distribute its costs into the charges for the work that it does. The allocation of indirect costs is a customary procedure in cost accounting systems, but it is potentially misleading for decision making.)

| | Printing department | | Outside printer | |
	1000 copies	30,000 copies	1000 copies	30,000 copies
Direct labor	$ 7.60	$228.00		
Materials & supplies	9.80	294.00	$22.95	$688.50
Overhead costs	9.05	271.50		
	$26.45	$793.50	$22.95	$688.50

The shipping department would reduce its cost from $793.50 to $688.50 by using the outside printer. In that case, how much would the printing department's costs decline? We will examine each of the cost components:

1. *Direct labor.* If the printing department had been working overtime, then the overtime could be reduced or eliminated. But, assuming no overtime, how much would the saving be? It seems unlikely that a printer could be fired or even put on less than a 40-hour work week. Thus, although there might be a $228.00 saving, it is much more likely that there will be no reduction in direct labor.
2. *Materials and supplies.* There would be a $294.00 saving in materials and supplies.
3. *Allocated overhead costs.* There will be no reduction in the printing department's monthly $5000 overhead, for there will be no reduction in department floor space. (Actually, of course, there may be a slight reduction in the firm's power costs if the printing department does less work.)

The firm will save $294.00 in materials and supplies and may or may not save $228.00 in direct labor if the printing department no longer does the shipping department work. The maximum saving would be $294.00 + 228.00 = $522.00. But if the shipping department is permitted to obtain its printing from the outside printer, the firm must pay $688.50 a month. The saving from not doing the shipping department work in the printing department would not exceed $522.00, and it probably would be only $294.00. The result would be a net increase in cost to the firm. For this reason, the shipping department should be discouraged from sending its printing to the outside commercial printer. ■

Gathering cost data presents other difficulties. One way to look at the financial consequences—costs and benefits—of various alternatives is as follows.

■ *Market Consequences.* These consequences have an established price in the marketplace. We can quickly determine raw material prices, machinery costs, labor costs, and so forth.

■ *Extra-Market Consequences.* There are other items that are not directly priced in the marketplace. But by indirect means, a price may be assigned to these items.

(Economists call these prices **shadow prices**.) Examples might be the cost of an employee injury or the value to employees of going from a five-day to a four-day, forty-hour week.

■ *Intangible Consequences.* Numerical economic analysis probably never fully describes the real differences between alternatives. The tendency to leave out those consequences that do not have a significant impact on the analysis itself, or on the conversion of the final decision into actual money, is difficult to resolve or eliminate. How does one evaluate the potential loss of workers' jobs due to automation? What is the value of landscaping around a factory? These and a variety of other consequences may be left out of the numerical calculations, but they should be considered in conjunction with the numerical results in reaching a decision.

4. Identify Feasible Alternatives

One must keep in mind that unless the best alternative is considered, the result will always be suboptimal[1]. Two types of alternatives are sometimes ignored. First, in many situations a do-nothing alternative is feasible. This may be the "let's keep doing what we are now doing", or the "let's not spend any money on that problem" alternative. Second, there are often feasible (but unglamourous) alternatives, such as "patch it up and keep it running for another year before replacing it."

There is no way to ensure that the best alternative *is* among the alternatives being considered. One should try to be certain that all conventional alternatives have been listed, and then make a serious effort to suggest innovative solutions. Sometimes a group of people considering alternatives in an innovative atmosphere—***brainstorming***—can be helpful. Even impractical alternatives may lead to a better possibility. The payoff from a new, innovative alternative can far exceed the value of carefully selecting between the existing alternatives.

Any good listing of alternatives will produce both practical and impractical alternatives. It would be of little use, however, to seriously consider an alternative that cannot be adopted. An alternative may be infeasible for a variety of reasons, such as, it violates fundamental laws of science, or it requires resources or materials that cannot be obtained, or it cannot be available in the time specified in the definition of the goal. Only the feasible alternatives are retained for further analysis.

5. Select the Criterion to Determine the Best Alternative

The central task of decision making is choosing from among alternatives. How is the choice made? Logically, one wants to choose the best alternative. This requires that we define what we mean by *best*. There must be a **criterion**, or set of **criteria**, to judge which alternative is best. Now, we recognize that *best* is a relative adjective on one end of this spectrum:

---

[1] A group of techniques called value analysis is sometimes used to examine past decisions. Where the decision made was somehow inadequate, value analysis re-examines the entire decision-making process with the goal of identifying a better solution and, hence, improving decision making.

Worst	Bad	Fair	Good	Better	Best

relative subjective judgement spectrum

Since we are dealing in *relative terms,* rather than *absolute values,* the selection will be the alternative that is relatively the most desirable. Consider a driver found guilty of speeding and given the alternatives of a $175 fine or three days in jail. In absolute terms, neither alternative is good. But on a relative basis, one simply "makes the best of a bad situation."

There may be an unlimited number of ways that one might judge the various alternatives. Several possible criteria are:

- Create the least disturbance to the ecology;
- Improve the distribution of wealth among people;
- Minimize the expenditure of money;
- Ensure that the benefits to those who gain from the decision are greater than the losses of those who are harmed by the decision;[2]
- Minimize the time to accomplish the goal or objective;
- Minimize unemployment;
- Use money in an economically efficient way.

Selecting the criterion for choosing the best alternative may not be easy, because different groups may support different criteria and desire different alternatives. The criteria may conflict. For example, minimizing unemployment may require increasing the expenditure of money. Or minimizing ecological disturbance may conflict with minimizing time to complete the project. The disagreement between management and labor in collective bargaining (concerning wages and conditions of employment) reflects a disagreement over the objective and the criterion for selecting the best alternative.

The last criterion above–use money in an economically efficient way–is the one normally selected in engineering decision making. Using this criterion, all problems fall into one of three categories:

A. *Fixed input.* The amount of money or other input resources (like labor, materials, or equipment) are fixed. The objective is to effectively utilize them.

Examples:

- A project engineer has a budget of $350,000 to overhaul a portion of a petroleum refinery.

- You have $300 to buy clothes for the start of school.

   For economic efficiency, the appropriate criterion is to maximize the benefits or other outputs.

---

[2] Kaldor Criterion.

B. *Fixed output.* There is a fixed task (or other output objectives or results) to be accomplished.

> *Examples:*

- A civil engineering firm has been given the job to survey a tract of land and prepare a "Record of Survey" map.

- You wish to purchase a new car with no optional equipment.

> The economically efficient criterion for a situation of fixed output is to minimize the costs or other inputs.

C. *Neither input not output fixed.* The third category is the general situation where neither the amount of money or other inputs, nor the amount of benefits or other outputs are fixed.

> *Examples:*

- A consulting engineering firm has more work available than it can handle. It is considering paying the staff for working evenings to increase the amount of design work it can perform.

- One might wish to invest in the stock market, but neither the total cost of the investment nor the benefits are fixed.

- An automobile battery is needed. Batteries are available at different prices, and although each will provide the energy to start the vehicle, their useful lives are different.

What should be the criterion in this category? Obviously, we want to be as economically efficient as possible. This will occur when we maximize the difference between the return from the investment (benefits) and the cost of the investment. Since the difference between the benefits and the costs is simply profit, a businessperson would define this criterion as *maximizing profit.*

For the three categories, the proper economic criteria are:

Category	*Economic criterion*
Fixed input	**Maximize the benefits or other outputs.**
Fixed output	**Minimize the costs or other inputs.**
Neither input nor output fixed	**Maximize (benefits or other outputs minus costs or other inputs) or, stated another way, maximize profit.**

6. Constructing the Model

At some point in the decision-making process, the various elements must be brought together. The *objective, relevant data, feasible alternatives,* and *selection criterion* must be merged. For example, if one were considering borrowing money to pay for an automobile, there is a mathematical relationship between the following variables for the loan: amount, interest rate, duration, and monthly payment.

Constructing the interrelationships between the decision-making elements is frequently called **model building** or **constructing the model**. To an engineer, modeling may be of two forms: a scaled *physical representation* of the real thing or system; or a *mathematical equation,* or set of equations, that describe the desired interrelationships. In a laboratory there may be a physical model, but in economic decision making, the model is usually mathematical.

In modeling, it is helpful to represent only that part of the real system that is important to the problem at hand. Thus, the mathematical model of the student capacity of a classroom might be,

$$\text{Capacity} = \frac{lw}{k}, \quad \text{where} \qquad l = \text{length of classroom in meters,}$$

$$w = \text{width of classroom in meters, and}$$
$$k = \text{classroom arrangement factor.}$$

The equation for student capacity of a classroom is a very simple model; yet it may be adequate for the problem being solved.

7. Predicting the Outcomes for Each Alternative

A model and the data are used to predict the outcomes for each feasible alternative. As was suggested earlier, each alternative might produce a variety of outcomes. Selecting a motorcycle, rather than a bicycle, for example, may make the fuel supplier happy, the neighbors unhappy, the environment more polluted, and one's savings account smaller. But, to avoid unnecessary complications, we assume that decision making is based on a single criterion for measuring the relative attractiveness of the various alternatives. If necessary, one could devise a single composite criterion that is the weighted average of several different choice criteria.

To choose the best alternative, the outcomes for each alternative must be stated in a *comparable* way. Usually the consequences of each alternative are stated in terms of money, that is, in the form of costs and benefits. This **resolution of consequences** is done with all monetary and non-monetary consequences. The consequences can also be categorized as follows:

Market consequences—where there are established market prices available;

Extra-market consequences—no direct market prices, so priced indirectly;

Intangible consequences—valued by judgement not monetary prices.

In the initial problems we will examine, the costs and benefits occur over a short time period and can be considered as occurring at the same time. In other situations the various costs and benefits take place in a longer time period. The result may be costs at one point in time followed by periodic benefits. We will resolve these in the next chapter into a *cash flow diagram* to show the timing of the various costs and benefits.

For these longer term problems, the most common error is to assume that the current situation will be unchanged for the do nothing alternative. For example, current profits will shrink or vanish due to the actions of competitors and the expectations of customers. As another example, traffic congestion normally increases over the years as the number of vehicles increases–doing nothing does not imply the situation is unchanged.

8. Choosing the Best Alternative

Earlier we indicated that choosing the best alternative may be simply a matter of determining which alternative best meets the selection criterion. But the solutions to most economics problems have market consequences, extra-market consequences, and intangible consequences. Since the intangible consequences of possible alternatives are left out of the numerical calculations, they should be introduced into the decision-making process at this point. The alternative to be chosen is the one that best meets the choice criterion after looking at both the numerical consequences and the consequences not included in the monetary analysis.

During the decision-making process there are feasible alternatives that are eliminated. These alternatives are dominated by other better alternatives. For example, buying a computer on-line may allow you to buy a custom configured computer for less money than a stock computer in a local store. Buying at the local store is feasible, but dominated. While eliminating dominated alternatives makes the decision-making process more efficient, there are dangers.

Having examined the structure of the decision-making process, it is appropriate to ask, "When is a decision made and who makes it?" If one person performs *all* the steps in decision making, then he is the decision maker. *When* he makes the decision is less clear. The selection of the feasible alternatives may be the key item, with the rest of the analysis a methodical process leading to the inevitable decision. We can see that the decision may be drastically affected, or even predetermined, by the way in which the decision-making process is carried out. This is illustrated by the following example.

Liz, a young engineer, was assigned to make an analysis of what additional equipment to add to the machine shop. The criterion for selection was that the equipment selected should be the most economical, considering both initial costs and future operating costs. A little investigation by Liz revealed three practical alternatives:

1.   A new specialized lathe;
2.   A new general-purpose lathe;
3.   A rebuilt lathe available from a used equipment dealer.

A preliminary analysis indicated that the rebuilt lathe would be the most economical. Liz did not like the idea of buying a rebuilt lathe so she decided to discard that alternative. She prepared a two-alternative analysis which showed the general-purpose lathe was more economical than the specialized lathe. She presented his completed analysis to her manager. The manager assumed that the two alternatives presented were the best of all feasible alternatives, and he approved Liz's recommendation.

At this point we should ask: who was the decision maker, Liz or her manager? Although the manager signed his name at the bottom of the economic analysis worksheets to authorize purchasing the general-purpose lathe, he was merely authorizing what already had been made inevitable, and thus he was not the decision maker. Rather,Liz had made the key decision when she decided to discard the most economical alternative from further consideration. The result was a decision to buy the better of the two *less economically desirable* alternatives.

9. Audit the Results

An audit of the results is a check of what happened as compared with predictions. Do the results of a decision analysis reasonably agree with its projections? If a new machine tool was purchased to save labor and improve quality, did it? If so, the economic analysis seems to be accurate. If the savings are not being obtained, what was overlooked. The audit may help ensure that projected operating advantages are ultimately obtained. On the other hand, the economic analysis projections may have been unduly optimistic. We want to know this, too, so that these mistakes are not repeated. Finally, an effective way to promote *realistic* economic analysis calculations is for all people involved to know that there *will* be an audit of the results!

Engineering Decision Making

Some of the easiest forms of engineering decision making deal with problems of alternate *designs, methods,* or *materials.* Since results of the decision occur in a very short period of time, one can quickly add up the costs and benefits for each alternative. Then, using the suitable economic criterion, the best alternative can be identified. Three example problems illustrate these situations.

EXAMPLE 1-2

A concrete aggregate mix is required to contain at least 31% sand by volume for proper batching. One source of material, which has 25% sand and 75% coarse aggregate, sells for $3 per cubic meter. Another source, which has 40% sand and 60% coarse aggregate, sells for $4.40 per cubic meter. Determine the least cost per cubic meter of blended aggregates.

Solution: The least cost of blended aggregates will result from maximum use of the lower cost material. The higher cost material will be used to increase the proportion of sand up to the minimum level (31%) specified.

Let x = Portion of blended aggregates from $3.00/m^3 source

$1 - x$ = Portion of blended aggregates from $4.40/m^3 source

Sand balance:

$$x\,(0.25) + (1 - x)\,(0.40) = 0.31$$

$$0.25x + 0.40 - 0.40x = 0.31$$

$$x = \frac{0.31 - 0.40}{0.25 - 0.40} = \frac{-0.09}{-0.15}$$

$$= 0.60$$

Thus the blended aggregates will contain:

60% of $3.00/m^3 material

40% of $4.40/m^3 material

The least cost per cubic meter of blended aggregates:

$$= 0.60(\$3.00) + 0.40(\$4.40) = 1.80 + 1.76$$
$$= \$3.56/m^3 \quad ■$$

EXAMPLE 1-3

A machine part is manufactured at a unit cost of 40¢ for material and 15¢ for direct labor. An investment of $500,000 in tooling is required. The order calls for three million pieces. Half-way through the order, a new method of manufacture can be put into effect which will reduce the unit costs to 34¢ for material and 10¢ for direct labor—but it will require $100,000 for additional tooling. What, if anything, should be done?

Solution:

Our problem only concerns the second half of the order, as there is only one alternative for the first 1.5 million pieces.

Alternative A: Continue with present method.

Material cost	1,500,000 pieces x 0.40 =	$600,000
Direct labor cost	1,500,000 pieces x 0.15 =	225,000
Other costs	2.50 x Direct labor cost =	562,500
Cost for remaining	1,500,000 pieces	$1,387,500

Alternative B: Change the manufacturing method.

Additional tooling cost	=	$100,000
Material cost	1,500,000 pieces x 0.34 =	510,000
Direct labor cost	1,500,000 pieces x 0.10 =	150,000
Other costs	2.50 x Direct labor cost =	375,000
Cost for remaining	1,500,000 pieces	$1,135,000

Before making a final decision, one should closely examine the *Other costs* to see that they do, in fact, vary as the *Direct labor cost* varies. Assuming they do, the decision would be to change the manufacturing method. ■

EXAMPLE 1-4

In the design of a cold-storage warehouse, the specifications call for a maximum heat transfer through the warehouse walls of 30,000 joules/hr/sq meter of wall when there is a 30°C temperature difference between the inside surface and the outside surface of the insulation. The two insulation materials being considered are as follows:

Insulation material	Cost/cubic meter	Conductivity $J\text{-}m/m^2\text{-}°C\text{-}hr$
Rock wool	$12.50	140
Foamed insulation	14.00	110

The basic equation for heat conduction through a wall is:

$$Q = \frac{K(\Delta T)}{L} \quad \text{where } Q = \text{Heat transfer in J/hr/m}^2 \text{ of wall}$$

$$K = \text{Conductivity in J-m/m}^2\text{-}°\text{C-hr}$$

$$\Delta T = \text{Difference in temperature between the two surfaces in } °\text{C}$$

$$L = \text{Thickness of insulating material in meters}$$

Which insulation material should be selected?

Solution:

There are two steps required to solve the problem. First, the required thickness of each of the alternate materials must be calculated. Then, since the problem is one of providing a fixed output (heat transfer through the wall limited to a fixed maximum amount), the criterion is to minimize the input (cost).

Required insulation thickness:

Rock wool $\qquad 30,000 = \dfrac{140(30)}{L} \qquad L = 0.14 \text{ m}$

Foamed insulation $\qquad 30,000 = \dfrac{110(30)}{L} \qquad L = 0.11 \text{ m}$

Cost of insulation per square meter of wall:

$$\text{Unit cost} = \text{Cost/m}^3 \text{ x Insulation thickness in meters}$$

Rock wool: $\quad$ Unit cost = \$12.50 x 0.14 m = \$1.75/m^2

Foamed insulation: $\quad$ Unit cost = \$14.00 x 0.11 m = \$1.54/m^2

The foamed insulation is the lesser cost alternative. However, there is an intangible constraint that must be considered. How thick is the available wall space? ■

Summary

Classifying Problems

Many problems are simple and easy to solve. Others are of intermediate difficulty and need considerable thought and/or calculation to properly evaluate. These intermediate problems tend to have a substantial economic component, hence are good candidates for economic analysis. Complex problems, on the other hand, often contain people elements, along with political and economic components. Economic analysis is still very important, but the best alternative must be selected considering all criteria–not just economics.

The Decision Making Process

Rational decision making uses a logical method of analysis to select the best alternative from among the feasible alternatives. The following nine steps can be followed sequentially, but there are often steps that are repeated, undertaken simultaneously, and even skipped.

1. Recognize the problem.
2. Define the goal or objective: what is the task?
3. Assemble relevant data: what are the facts? Is more data needed and is it worth more than the cost to obtain it?
4. Identify feasible alternatives.
5. Select the criterion for choosing the best alternative: possible criteria include political, economic, ecological, and humanitarian. The single criterion may be a composite of several different criteria.
6. *Mathematically model* the various interrelationships.
7. Predict the outcomes for each alternative.
8. Choose the best alternative to achieve the objective.
9. Audit the results.

Engineering decision making refers to solving substantial engineering problems where economic aspects dominate and economic efficiency is the criterion for choosing from possible alternatives. It is a particular case of the general decision-making process. Some of the unusual aspects of engineering decision making are as follows:

1. Cost-accounting systems, while an important source of cost data, contain allocations of indirect costs that may be inappropriate for use in economic analysis.
2. The various consequences—costs and benefits—of an alternative may be of three types:

   a. Market consequences—there are established market prices;
   b. Extra-market consequences—there are no direct market prices, but prices can be

assigned by indirect means;
   c. Intangible consequences—valued by judgement not by monetary prices.
3. The economic criteria for judging alternatives can be reduced to three cases:
   a. For fixed input: maximize benefits or other outputs.
   b. For fixed output: minimize costs or other inputs.
   c. When neither input nor output is fixed: maximize the difference between benefits and costs or, more simply stated, maximize profit.
   The third case states the general rule from which both the first and second cases may be derived.
4. To choose among the alternatives, the market consequences and extra-market consequences are organized into a cash flow diagram. We will see in Chapter 3 that differing cash flows can be compared with engineering economic calculations. These outcomes are compared against the selection criterion. From this comparison *plus* the consequences not included in the monetary analysis, the best alternative is selected.
5. An essential part of engineering decision making is the post audit of results. This step helps to ensure that projected benefits are obtained and to encourage realistic estimates in analyses.

Problems

1-1 Think back to your first hour after awakening this morning. List fifteen decision-making opportunities that existed during that one hour. After you have done that, mark the decision-making opportunities that you actually recognized this morning and upon which you made a conscious decision.

1-2 Some of the problems listed below would be suitable for solution by engineering economic analysis. Which ones are they?
   a. Would it be better to buy an automobile with a diesel engine or a gasoline engine?
   b. Should an automatic machine be purchased to replace three workers now doing a task by hand?
   c. Would it be wise to enroll for an early morning class so you could avoid traveling during the morning traffic rush hours?
   d. Would you be better off if you changed your major?
   e. One of the people you might marry has a job that pays very little money, while another one has a professional job with an excellent salary. Which one should you marry?

1-3 Which one of the following problems is *most* suitable for analysis by engineering economic analysis?

a. Some 45¢ candy bars are on sale for twelve bars for $3.00. Sandy eats a couple of candy bars a week, and must decide whether or not to buy a dozen.

b. A woman has $150,000 in a bank checking account that pays no interest. She can either invest it immediately at a desirable interest rate, or wait one week and know that she will be able to obtain an interest rate that is 0.15% higher.

c. Joe backed his car into a tree, damaging the fender. He has automobile insurance that will pay for the fender repair. But if he files a claim for payment, they may change his "good driver" rating downward, and charge him more for car insurance in the future.

1-4 If you have $300 and could make the right decisions, how long would it take you to become a millionaire? Explain briefly what you would do.

1-5 Many people write books explaining how to make money in the stock market. Apparently the authors plan to make *their* money selling books telling other people how to profit from the stock market. Why don't these authors forget about the books, and make their money in the stock market?

1-6 The owner of a small machine shop has just lost one of his larger customers. The solution to his problem, he says, is to fire three machinists to balance his workforce with his current level of business. The owner says it is a simple problem with a simple solution. The three machinists disagree. Why?

1-7 Every college student had the problem of selecting the college or university to attend. Was this a simple, intermediate, or complex problem for you? Explain.

1-8 Recently the U. S. Government wanted to save money by closing a small portion of all its military installations throughout the United States. While many people agreed it was a desirable goal, areas potentially affected by selection to close soon reacted negatively. The Congress finally selected a panel of people whose task was to develop a list of installations to close, with the legislation specifying that Congress could not alter the list. Since the goal was to save money, why was this problem so hard to solve?

1-9 The college bookstore has put pads of engineering computation paper on sale at half price. What is the minimum and maximum number of pads you might buy during the sale? Explain.

1-10 Consider the seven situations described. Which one situation seems most suitable for solution by engineering economic analysis?

a. Jane has met two college students that interest her. Bill is a music major who is lots of fun to be with. Alex, on the other hand, is a fellow engineering student, but does not like to dance. Jane wonders what to do.

b. You drive periodically to the post office to pick up your mail. The parking meters require 10¢ for six minutes–about twice the time required to get from your car to the post office and back. If parking tickets cost $8.00, do you put money in the meter or not?

c. At the local market, candy bars are 45¢ each or three for $1.00.

d. The cost of automobile insurance varies widely from insurance company to insurance company. Should you check with several companies when your insurance comes up for renewal?

e. There is a special local sales tax ("sin tax") on a variety of things that the town council would like to remove from local distribution. As a result a store has opened up just outside the town and offers an abundance of these specific items at prices about 30% less than is charged in town.

f. Your mother reminds you she wants you to attend the annual family picnic. That same Saturday you already have a date with a person you have been trying to date for months.

g. One of your professors mentioned that you have a poor attendance record in his class. You wonder whether to drop the course now or wait to see how you do on the first midterm exam. Unfortunately, the course is required for graduation.

1-11 An automobile manufacturer is considering locating an automobile assembly plant in Tennessee. List two simple, two intermediate, and two complex problems associated with this proposal.

1-12 Consider the three situations below. Which ones appear to represent rational decision making? Explain.

a. Joe's best friend has decided to become a civil engineer, so Joe has decided that he, too, will become a civil engineer.

b. Jill needs to get to the university from her home. She bought a car and now drives to the university each day. When Jim asks her why she didn't buy a bicycle instead, she replies, "Gee, I never thought of that."

c. Don needed a wrench to replace the spark plugs in his car. He went to the local automobile supply store and bought the cheapest one they had. It broke before he finished replacing all the spark plugs in his car.

1-13 Identify possible objectives for NASA? For your favorite of these, how should alternative plans to achieve the objective be evaluated?

1-14 Suppose you have just two hours to answer the question, "How many people in your home town would be interested in buying a pair of left-handed scissors?" Give a step-by-step outline of how you would seek to answer this question within two hours.

1-15 A college student determines that he will have only $50 per month available for his housing for the coming year. He is determined to continue in the university, so he has decided to list all feasible alternatives for his housing. To help him, list five feasible alternatives.

1-16 Describe a situation where a poor alternative was selected, because there was a poor search for better alternatives.

1-17 Consider a situation where there are only two alternatives available and both are unpleasant and undesirable. What should you do?

1-18 The three economic criteria for choosing the best alternative are: minimize input; maximize output; or maximize the difference between output and input. For each of the following situations, what is the appropriate economic criterion?

    a. A manufacturer of plastic drafting triangles can sell all the triangles he can produce at a fixed price. His unit costs increase as he increases production due to overtime pay, and so forth. The manufacturer's criterion should be _____.

    b. An architectural and engineering firm has been awarded the contract to design a wharf for a petroleum company for a fixed sum of money. The engineering firm's criterion should be _____.

    c. A book publisher is about to set the list price (retail price) on a textbook. If they choose a low list price, they plan on less advertising than if they select a higher list price. The amount of advertising will affect the number of copies sold. The publisher's criterion should be _____.

    d. At an auction of antiques, a bidder for a particular porcelain statue would be trying to _____.

1-19 See Problem **1-18**. For each of the following situations, what is the appropriate economic criterion?

    a. The engineering school held a raffle of an automobile with tickets selling for 50¢ each or three for $1. When the students were selling tickets, they noted that many people were undecided whether to buy one or three tickets. This indicates the buyers' criterion was _____.

    b. A student organization bought a soft-drink machine for use in a student area. There was considerable discussion as to whether they should set the machine to charge 30¢, 35¢, or 40¢ per drink. The organization recognized that the number of soft drinks sold would depend on the price charged. Eventually the decision was made to charge 35¢. Their criterion was _____.

    c. In many cities, grocery stores find that their sales are much greater on days when they have advertised their special bargains. The advertised special prices do not appear to increase the total physical volume of groceries sold by a store. This leads us to conclude that many shoppers' criterion is _____.

    d. A recently graduated engineer has decided to return to school in the evenings to obtain a Master's degree. He feels it should be accomplished in a manner that will allow him the maximum amount of time for his regular day job plus time for recreation. In working for the degree, he will _____.

1-20 Seven criteria are given in the chapter for judging which is the best alternative. After studying the list, devise three additional criteria that might be used.

1-21   Suppose you are assigned the task of determining the route of a new highway through an older section of town. The highway will require that many older homes must be either relocated or torn down. Two possible criteria that might be used in deciding exactly where to locate the highway are:
1. Ensure that there are benefits to those who gain from the decision and no one is harmed by the decision.
2. Ensure that the benefits to those who gain from the decision are greater than the losses of those who are harmed by the decision.

Which criterion will you select to use in determining the route of the highway? Explain.

1-22   Identify benefits and costs for problem **1-21**.

1-23   In the Fall, Jay Thompson decided to live in a university dormitory. He signed a dorm contract under which he was obligated to pay the room rent for the full college year. One clause stated that if he moved out during the year, he could sell his dorm contract to another student who would move into the dormitory as his replacement. The dorm cost was $600 for the two semesters, which Jay already has paid.

A month after he moved into the dorm, he decided he would prefer to live in an apartment. That week, after some searching for a replacement to fulfill his dorm contract, Jay had two offers. One student offered to move in immediately and to pay Jay $30 per month for the eight remaining months of the school year. A second student offered to move in the second semester and pay $190 to Jay.

Jay now has $1050 left (after paying the $600 dorm bill and food for a month) which must provide for all his room and board expenses for the balance of the year. He estimates his food cost per month is $120 if he lives in the dorm and $100 if he lives in an apartment with three other students. His share of the apartment rent and utilities will be $80 per month. Assume each semester is 4½ months long. Disregard the small differences in the timing of the disbursements or receipts.

    a.   What are the three alternatives available to Jay?

    b.   Evaluate the cost for each of the alternatives.

    c.   What do you recommend that Jay do?

1-24   In decision making we talk about the construction of a model. What kind of model is meant?

1-25   An electric motor on a conveyor burned out. The foreman told the plant manager that the motor had to be replaced. The foreman indicated that "there are no alternatives," and asked for authorization to order the replacement. In this situation, is there any decision making taking place? By whom?

1-26   Bill Jones' parents insisted that Bill buy himself a new sport shirt. Bill's father gave specific instructions, saying the shirt must be in "good taste," that is, neither too wildly colored nor too extreme in tailoring. Bill found in the local department store there were three types of sport shirts available:

- rather somber shirts that Bill's father would want him to buy;

- good looking shirts that appealed to Bill; and

- weird shirts that were even too much for Bill.

He wanted a good looking shirt but wondered how to convince his father to let him keep it. The clerk suggested that Bill take home two shirts for his father to see and return the one he did not like. Bill selected a good looking blue shirt he liked, and also a weird lavender shirt. His father took one look and insisted that Bill keep the blue shirt and return the lavender one. Bill did as his father instructed. What was the key decision in this decision process, and who made it?

1-27   A farmer must decide what combination of seed, water, fertilizer, and pest control will be most profitable for the coming year. The local agricultural college did a study of this farmer's situation and prepared the following table.

Plan	Cost/acre	Income/acre
A	$ 600	$ 800
B	1500	1900
C	1800	2250
D	2100	2500

The last page of the college's study was torn off, and hence the farmer is not sure which plan the agricultural college recommends. Which plan should the farmer adopt? Explain.

1-28   Identify the alternatives, outcomes, criteria, and process for the selection of your college major? Did you make the best choice for you?

1-29   Describe a major problem you must address in the next two years. Use the techniques of this chapter to structure the problem and recommend a decision.

1-30   One strategy for solving a complex problem is to break the problem into a group of less complex problems, and then find solutions to the less complex problems. The result is the solution of the complex problem.   Give an example where this strategy will work. Then give another example where this strategy will not work.

1-31 On his first engineering job, Joy Hayes was given the responsibility of determining the production rate for a new product. He has assembled data as indicated on the two graphs:

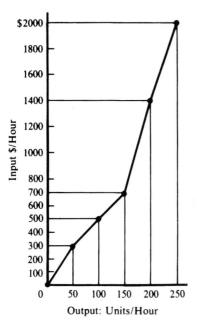 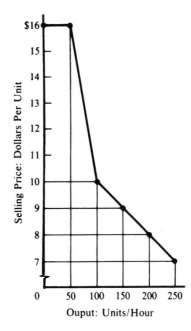

a. Select an appropriate economic criterion and estimate the production rate based upon it.

b. Joy's boss told Joy: "I want you to maximize output with minimum input." Joy wonders if it is possible to achieve her boss's criterion. She asks your advice. What would you tell her?

1-32 Willie Lohmann travels from city to city in the conduct of his business. Every other year he buys a new car for about $12,000. The auto dealer allows about $8000 as a trade-in allowance with the result that the salesman spends $4000 every other year for a car. Willie keeps accurate records which show that all other expenses on his car amount to 12.3¢ per mile for each mile he drives. Willie's employer has two plans by which salesmen are reimbursed for their car expenses:

1. Willie will receive all his operating expenses, and in addition will receive $2000 each year for the decline in value of the automobile.

2. Instead of Plan 1, Willie will receive 22¢ per mile.

If Willie travels 18,000 miles per year, which method of computation gives him the larger reimbursement? At what annual mileage do the two methods give the same reimbursement?

1-33 Maria, a college student, is getting ready for three final examinations at the end of the school year. Between now and the start of exams, she has 15 hours of study time available. She would like to get as high a grade average as possible in her Math, Physics, and Engineering Economy classes.

She feels she must study at least two hours for each course and, if necessary, will settle for the low grade that the limited study would yield. How much time should Maria devote to each class if she estimates her grade in each subject as follows:

Mathematics		Physics		Engineering Economy	
Study hours	Grade	Study hours	Grade	Study hours	Grade
2	25	2	35	2	50
3	35	3	41	3	61
4	44	4	49	4	71
5	52	5	59	5	79
6	59	6	68	6	86
7	65	7	77	7	92
8	70	8	85	8	96

1-34 Two manufacturing companies, located in cities 90 miles apart, have discovered that they both send their trucks four times a week to the other city full of cargo and return empty. Each company pays its driver $185.00 a day (the round trip takes all day) and have truck operating costs (excluding the driver) of 60 cents per mile. How much could each company save each week if they shared the task, with each sending their truck twice a week and hauling the other company's cargo on the return trip?

1-35 A city is in need of increasing its rubbish disposal facilities. There is a choice of two rubbish disposal areas, as follows:

Area A: A gravel pit with a capacity of 16 million cubic meters. Due to the possibility of high ground water, however, the Regional Water Pollution Control Board has restricted the lower 2 million cubic meters of fill to inert material only; for example, earth, concrete, asphalt, paving, brick, and so forth. The inert material, principally clean earth, must be purchased and hauled to this area for the bottom fill.

Area B: Capacity is 14 million cubic meters. The entire capacity may be used for general rubbish disposal. This area will require an average increase in a round-trip haul of five miles for 60% of the city, a decreased haul of two miles for 20% of the city. For the remaining 20% of the city, the haul is the same distance as for Area *A*.

Assume the following conditions:

■ Cost of inert material placed in Area *A* will be $2.35 per cubic meter.

■ Average speed of trucks from last pickup to disposal site is 15 miles per hour.

■ The rubbish truck and a two-man crew will cost $35 per hour.

■ Truck capacity of 4½ tons per load or 20 cubic meters.

■ Sufficient cover material is available at all areas; however, inert material for the bottom fill in Area *A* must be hauled in.

Which of the sites do you recommend? (*Answer:* Area *B*)

1-36   An oil company is considering adding an additional grade of fuel at its service stations. To do this, an additional 3000-gallon tank must be buried at each station. Discussions with tank fabricators indicate that the least expensive tank would be cylindrical with minimum surface area. What size tank should be ordered?

(Answer: 8-ft diameter by 8-ft length.)

1-37   The vegetable buyer for a group of grocery stores has decided to sell packages of sprouted grain in the vegetable section of the stores. The product is perishable and any remaining unsold after one week in the store is discarded. The supplier will deliver the packages to the stores, arrange them in the display space, and remove and dispose of any old packages. The price the supplier will charge the stores depends on the size of the total weekly order for all the stores.

Weekly order	Price per package
Less than 1000 packages	35¢
1000–1499	28
1500–1999	25
2000 or more	20

The vegetable buyer estimates the quantity that can be sold per week, at various selling prices, as follows:

Selling price	Quantity sold per week
60¢	300 packages
45	600
40	1200
33	1700
26	2300

The sprouted grain will be sold at the same price in all the grocery stores. How many packages should be purchased per week, and at which of the five prices listed above should they be sold?

1-38   Cathy Gwynn, a recently graduated engineer, decided to invest some of her money in a "Quick Shop" grocery store. The store emphasizes quick service, a limited assortment of grocery items, and rather high prices. Cathy wants to study the business to see if the store hours (currently 0600 to 0100) can be changed to make the store more profitable. Cathy assembled the following information.

	Daily sales in
Time period	*the time period*
0600–0700	$ 20
0700–0800	40
0800–0900	60
0900–1200	200
1200–1500	180
1500–1800	300
1800–2100	400
2100–2200	100
2200–2300	30
2300–2400	60
2400–0100	20

The cost of the groceries sold averages 70% of sales. The incremental cost to keep the store open, including the clerk's wage and other incremental operating costs, is $10 per hour. To maximize profit, when should the store be opened, and when should it be closed?

1-39   Jim Jones, a motel owner, noticed that just down the street the "Motel 36" advertises their $36-per-night room rental rate on their sign. As a result, they rent all of their eighty rooms every day by late afternoon. Jim, on the other hand, does not advertise his rate, which is $54 per night, and averages only a 68% occupancy of his fifty rooms.

There are a lot of other motels nearby and, except for Motel 36, none of the others advertises their rate on their sign. (Their rates vary from $48 to $80 per night.) Jim estimates that his actual incremental cost per night for each room rented, rather than remaining vacant, is $12. This $12 pays for all the cleaning, laundering, maintenance, utilities, and so on. Jim believes his eight alternatives are:

Alternative	*Advertise and charge*	*Resulting occupancy rate*
1	$35 per night	100%
2	42 per night	94
3	48 per night	80
4	54 per night	66

Do not advertise and charge

5	$48 per night	70%
6	54 per night	68
7	62 per night	66
8	68 per night	56

What should Jim do? Show how you reached your conclusion.

1-40   A firm is planning to manufacture a new product. The sales department estimates that the quantity that can be sold depends on the selling price. As the selling price is increased, the quantity that can be sold decreases. Numerically they estimate:

$P = \$35.00 - 0.02Q$

> where P = Selling price per unit

> Q = Quantity sold per year

On the other hand, the management estimates that the average cost of manufacturing and selling the product will decrease as the quantity sold increases. They estimate

$C = \$4.00Q + \$8000$

> where C = Cost to produce and sell Q per year

The firm's management wishes to produce and sell the product at the rate that will maximize profit, that is, where income minus cost is a maximum. What quantity should they plan to produce and sell each year? (*Answer:* 775 units)

1-41   A manufacturing firm has received a contract to assemble 1000 units of test equipment in the next year. The firm must decide how to organize its assembly operation. Skilled workers, at $22.00 per hour each, could be assigned to individually assemble the test equipment. Each worker would do all the assembly steps and the task would take 2.6 hours per unit. An alternate approach would be to set up teams of four less skilled workers (at $13.00 per hour each) and organize the assembly tasks so each worker does his share of the assembly. The four-man team would be able to assemble a unit in one hour. Which approach would result in more economical assembly?

1-42   A grower estimates that if he picks his apple crop now, he will obtain 1000 boxes of apples, which he can sell at $3 per box. However, he thinks his crop will increase an additional 120 boxes of apples for each week he delays picking, but that the price will drop at a rate of 15¢ per box per week; in addition, he estimates approximately 20 boxes per week will spoil for each week he delays picking. When should he pick his crop to obtain the largest total cash return? How much will he receive for his crop at that time?

CHAPTER 2

Engineering Costs
and Cost Estimating

In this chapter fundamental cost concepts are defined. These include fixed and variable costs, marginal and average costs, sunk and opportunity costs, recurring and non-recurring costs, incremental cash costs, book costs, and life-cycle costs. We then describe the various types of estimates and difficulties sometimes encountered. The models that are described include unit factor, segmenting, cost indexes, power sizing, triangulation, and learning curves. The chapter discusses estimating benefits, developing cash flow diagrams, and drawing these diagrams with spreadsheets.

This chapter addresses the important question of "where do the numbers come from." Understanding engineering costs is fundamental to the engineering economic analysis process.

Engineering Costs

Evaluating a set of feasible alternatives requires that many costs be analyzed. Examples include costs for: initial investment, new construction, facility modification, general labor, parts and material, inspection and quality, contractor and subcontractor labor, training, computer hardware and software, material handling, fixtures and tooling, data management, technical support, as well as general support costs (overhead). In this section we describe several concepts for classifying and understanding these costs.

Fixed, Variable, Marginal, and Average Costs

Fixed costs are constant or unchanging regardless of the level of output or activity. In contrast, *variable* costs depend on the level of output or activity. A *marginal* cost is the variable cost for one more unit, while the *average* cost is the total cost divided by the number of units.

For example, in a production environment a fixed cost, such as costs for factory floor-space and equipment, remains the same even though the production quantity, number of employees, or level of work-in-process are varying. Labor costs are classified as a *variable cost* because they depend on the number of employees in the factory. Thus *fixed costs* are level or constant regardless of output or activity, and *variable costs* are changing and related to the level of output or activity.

As another example, many universities charge full-time students a fixed cost for 12 to 18 hours, and a cost per credit hour for each credit hour over 18. Thus for full-time students who are taking an overload (>18 hours), there is a variable cost that depends on the level of activity.

This example can also be used to distinguish between *marginal* and *average* costs. A marginal cost is the cost of one more unit. This will depend on how many credit hours the student is taking. If currently enrolled for 12 to 17 hours, adding one more is free. The marginal cost of an additional credit hour is $0. However, if the student is taking 18 or more hours, then the marginal cost equals the variable cost of one more hour.

To illustrate average costs, the fixed and variable costs need to be specified. Suppose the cost of 12 to 18 hour is $1800 per term and overload credits are $120/hour. If a student takes 12 hours, the *average* cost is $1800/12 = $150 per credit hour. If the student were to take 18 hours the *average* cost decreases to $1800/18 = $100 per credit hour. If the student takes 21 hours the *average* cost is ($1800 + 3·$120)/21 = $102.86 per credit hour. Average cost is thus calculated by dividing the total cost for all units by the total number of units. Decision makers use *average cost* to attain an overall cost picture of the investment on a per unit basis.

Marginal cost is used for decision-making on whether or not the additional unit should be made, purchased, or enrolled in. For the full-time student at our example university, the marginal cost of another credit is $0 or $120 depending on how many credits the student is already enrolled in.

EXAMPLE 2-1

An entrepreneur named DK was considering the money-making potential of chartering a bus to take people from his hometown to an event in a larger city. DK planned to provide transportation, tickets to the event, and refreshments on the bus for those who signed up. He gathered data and categorized these expenses as either fixed or variable:

DK's Fixed Costs		DK's Variable Costs	
• Bus Rental	$80	• Event Ticket	$12.50 per person
• Gas Expense	$75	• Refreshments	$7.50 per person
• Other Fuels	$20		
• Bus Driver	$50		

Develop an expression of DK's total fixed and total variable costs for chartering this trip.

Solution
DK's fixed costs will be incurred regardless of how many people sign up for the trip (even if only one person signs up!). These costs include bus rental, gas and fuel expense, and the cost to hire a driver. The total fixed cost would be:

$$\text{Total fixed costs} = 80 + 75 + 20 + 50 = \$225.$$

DK's variable costs depend on how many people sign up for the charter which is the level of activity. Total variable costs for event tickets and refreshments would be:

$$\text{Total variable costs} = 12.50 + 7.50 = \$20 \text{ per person.} \quad \blacksquare$$

From the above example we see how it is possible to calculate total fixed and total variable costs. Furthermore, these values can be combined into a single *total cost* equation as follows:

$$\textit{Total cost} = \textit{Total fixed cost} + \textit{Total variable cost} \qquad (2\text{-}1)$$

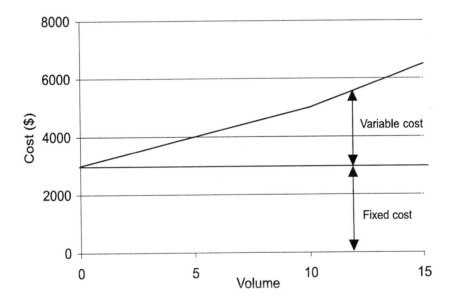

Figure 2-1 Fixed, variable, and total costs

The relationship between total cost and fixed and variable costs are shown in Figure 2-1. The fixed cost portion of $3000 is the same across the entire range of the output variable x. Often, the variable costs are *linear* (straight line between x and y); however, the variable costs can be non-linear. For example, employees are often paid at 150% of their hourly rate for overtime hours, so that production levels requiring overtime have higher variable costs. Total cost in Figure 2-1 is a fixed cost of $3000 plus a variable cost of $200 per unit for straight-time production of up to 10 units and $300 per unit for overtime production of up to 5 more units.

Figure 2-1 can also be used to illustrate marginal and average costs. At a volume of 5 units the marginal cost is $200 per unit, while at a volume of 12 units the marginal cost is $300 per unit. The respective average costs are $800 per unit [= (3000 + 200·5)/5] and $467 per unit [= (3000 + 200·10 + 300·2)/12].

EXAMPLE 2-2

In Example 2-1, DK developed an overall total cost equation for his business expenses. Now in this example DK wants to evaluate the potential to make money from this chartered bus trip.

Solution
Using Equation 2-1 DK's total cost equation would be:

$$\text{Total Cost} = \text{Total Fixed Cost} + \text{Total Variable Cost}$$
$$= \$225 + (\$20)(\text{the number of people on the trip})$$

Let x = number of people on the trip   Thus,   Total Cost = $225 + 20x$

Using this relationship, DK can calculate the total cost for any number of people – up to the capacity of the bus. What he lacks is a *revenue equation* to offset his costs. DK's total revenue from this trip can be expressed as:

Total Revenue = (Charter ticket price)(number of people on the trip) = (ticket price)(x)

DK believes that he could attract 30 people at a charter ticket price of $35. The total profit is:

$$\text{Total Profit} = [\text{Total Revenue}] - [\text{Total Costs}] = [35x] - [225 + 20x] = 15x - 225$$

At $x = 30$ Total Profit $= [(35)(30)] - [225 + (20)(30)] = \$225$

Thus, if 30 people take the charter, then DK would net a profit of $225. This somewhat simplistic analysis ignores the value of DK's time – in this case he would have to "pay himself" out of his $225 profit. ∎

In Examples 2-1 and 2-2 DK developed *total cost* and *total revenue* equations to describe the charter bus proposal. These equations can be used to create what is called a *profit-loss breakeven chart* (see Figure 2-2). Both the *costs* and *revenues* associated with various levels of output (activity) are placed on the same set of *x-y* axes. This allows one to illustrate a *breakeven point* (in terms of costs and revenue) and regions of *profit* and *loss* for some business activity. These terms can be defined as follows.

Breakeven point: The level of business activity at which the total costs to provide the product, good, or service are *equal to* the revenue (or savings) generated by providing the service. This is the level at which one "just breaks even."

Profit region: The output level of the variable *x* above the breakeven point, where total revenue is greater than total costs.

Loss region: The output level of the variable *x* below the breakeven point, where total costs are greater than total revenue.

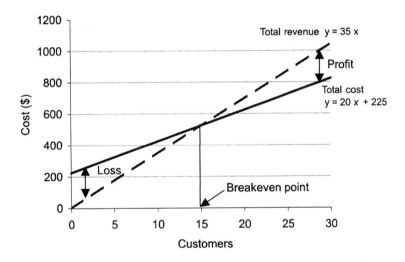

Figure 2-2 Profit-loss breakeven chart for Examples 2-1 and 2-2

Notice in Figure 2-2 that the *breakeven* point for the number of persons on the charter trip is 15 people. For more than 15 people, DK will make a profit. If less than 15 sign up there will be a net loss. At the breakeven level the total cost to provide the charter equals the revenue received from the 15 passengers. We can solve for the *breakeven point* by setting the *total costs* and *total revenue* expressions equal to each other and solving for the unknown value of *x*. From Examples 2-1 and 2-2:

Total Cost = Total Revenue
$225 + 20x = 35x$
thus $x = 15$ people

Sunk Costs

A *sunk cost* is money already spent due to a *past* decision. Sunk costs should be disregarded in our engineering economic analysis, as current decisions cannot change the past. For example, dollars spent last year to purchase new production machinery is money that is *sunk*. In this case, the money to purchase the production machinery has already been spent — there is nothing that can be done now to change that action. As engineering economists we deal with present and future opportunities.

Many times it is difficult not to be influenced by sunk costs. Consider 100 shares of stock in XYZ, Inc. that were purchased for $15 per share last year. The share price has steadily declined over the past 12 months to a price of $10 per share today. Current decisions must focus on the $10 per share that could be attained today (as well as future price potential) not the $15 per share that was paid last year. The $15 per share paid last year is a *sunk cost* and has no influence on present opportunities.

As another example, Regina purchased a newest-generation laptop from the college bookstore for $2000, when she was a sophomore. By the time she graduated the most anyone would pay her for it was $400 because the newest models were faster and had more capabilities. For Regina the original purchase price was a *sunk cost* that has no influence on her present opportunity to sell the laptop at its current market value ($400).

Opportunity Costs

An *opportunity cost* is associated with using a resource in one activity instead of another. Every time we use a business resource (equipment, dollars, manpower, and so on) in one activity, we give up the opportunity to use those same resources at that time in some other activity.

Every day businesses use resources to accomplish various tasks—forklifts are used to transport materials, engineers are used to design products and processes, assembly lines are used to make product A, and parking lots are used to provide parking for employees' vehicles. Each of these resources cost the company money to maintain for those intended purposes. However, that cost is not just made up of the dollar cost, it also includes the opportunity cost. Each resource that a firm owns can feasibly be used in several alternative ways. For instance, the assembly line could produce product B, and the parking lot could be rented out, used as a building site, or converted into a small airstrip. Each of these alternative uses would provide some benefit to the company.

When the firm chooses to use the resource in one way, they are giving up the benefits that would be derived from using it in those other ways. The benefit that would be derived by using the resource in this "other activity" is the *opportunity cost* for using it in the chosen activity. Opportunity cost may also be considered a *foregone opportunity cost* because we are foregoing the benefit that could have been realized. A formal definition of opportunity cost might be:

> An *opportunity cost* is the benefit that is *foregone* by engaging a business resource in a chosen activity instead of engaging that same resource in the foregone activity.

As an example, suppose that a college student has been invited to travel through Europe over the summer break by a group of friends. In considering the offer the student might calculate all of the *out-of-pocket* cash costs that will be incurred. Cost estimates might be made for items such as: air travel, lodging, meals, entertainment, train passes, etc. Suppose this amounts to $3000 for a 10 week period. After checking his bank account the student reports that indeed he can afford the $3000 trip. However, the *true* cost to the student includes not only his *out-of-pocket* cash costs but also his *opportunity cost*. By taking the trip the student is giving up the *opportunity* to earn $5000 as a summer intern at a local business. The student's total cost will include the $3000 cash cost as well as the $5000 opportunity cost (wages foregone) — the total cost to our traveler is thus $8000.

EXAMPLE 2-3

A distributor of electric pumps must decide what to do with a "lot" of old electric pumps that was purchased 3 years ago. Soon after the distributor purchased the lot, technology advances were made. These advances made the old pumps less desirable to customers. The pumps are becoming more obsolescent as they sit in inventory. The pricing manager has the following information.

Distributor's purchase price 3 years ago	= $7,000
Distributor's storage costs to-date	= 1,000
Distributor's list price 3 years ago	= 9,500
Current list price of a "lot" of new-pumps	= 12,000
Amount offered for the old pumps from a buyer 2 years ago	= 5,000
Current price that the lot of old pumps it could be sold for	= 3,000

Looking at the data the pricing manager has concluded that the price should be set at $8000. This is the money that the firm has "tied up" in the lot of old pumps ($7000 purchase and $1000 storage) — it was reasoned that the company should at least recover this cost. Furthermore, the pricing manager has argued that an $8000 price would be $1500 less than the list price from 3 years ago, and it would be $4000 less than what a lot of new pumps would cost ($12,000 - $8000). What would be your advice on price?

Solution

Let's look more closely at each of the data items.

Distributor's purchase price 3 years ago: This is a sunk cost that should not be considered in setting the price today.

Distributor's storage costs to-date: The storage costs for keeping the pumps in inventory are sunk costs that have been previously paid. Since these are sunk costs, they should not influence the pricing decision.

Distributor's list price 3 years ago: If there have been no willing buyers in the past 3 years at this price, it is unlikely that a buyer will emerge in the future. This past list price should have no influence on the current pricing decision.

Current list price of a lot of new-technology pumps: Newer pumps now include technology and features that have made the older pumps less valuable. Directly comparing the older pumps to those with new technology is misleading. However, the price of the new pumps and the value of the new features help determine the market value of the old pumps.

Amount offered from a buyer two years ago: This is a foregone opportunity. At the time of the offer the company chose to keep the lot and thus the $5000 offered became an opportunity cost for keeping the pumps. This amount should not influence the current pricing decision.

Current price that the lot could be sold for: The price that a willing buyer in the marketplace offers is called the asset's *market value.* The lot of old pumps in question is believed to have a current market value of $3000.

From this analysis, it is easy to see the flaw in the pricing manager's reasoning. In an engineering economist analysis we   deal only with *today's* and prospective *future* opportunities. It is impossible to go back in time and change decisions that have been made. Thus, the pricing manager should recommend to the distributor that the price be set at the current value that a buyer is willing to pay, $3000. ■

Recurring and Non-Recurring Costs

Recurring costs refer to any expense that is known, anticipated, and occurs at regular intervals. *Non-recurring costs* are one-of-a-kind and occur at irregular intervals, and thus are sometimes difficult to plan for or anticipate from a budgeting perspective.

Example recurring costs include those for resurfacing a highway and re-shingling a roof. Annual expenses for maintenance and operation are also recurring expenses. Examples of non-recurring costs include: the cost to install a new machine (including any facility modifications required), the cost to augment older-technology equipment to make fit for use, emergency maintenance expenses, or the disposal or close-down costs associated with ending operations.

In engineering economic analyses *recurring costs* are modeled as cash flows that occur at regular intervals (such as every year or every 5 years.) Their magnitude can be estimated and they can be included in the overall analysis. *Non-recurring costs* can be handled easily in our analysis if we are able to anticipate their timing and size. However, this is not always so easy to do.

Incremental Costs

One of the fundamental principles in engineering economic analysis is that in making a choice among a set of competing alternatives, focus should be placed on the *differences* between those alternatives. This is the concept of *incremental costs*. For instance, one may be interested in comparing two options to lease a vehicle for personal use. The two lease options may have several specifics where costs are the same. However, there may be *incremental costs* associated with one option not required or stipulated by the other. In comparing the two leases, the focus should be on the differences between the alternatives, not on the costs that are the same.

EXAMPLE 2-4

Philip is choosing between model *A* (a budget model) and model *B* (with more features and a higher purchase price). What *incremental costs* would Philip incur if he chose model *B* instead of the less expensive model *A*?

Cost Items	*Model A*	*Model B*
Purchase price	$ 10,000	$ 17,500
Installation costs	3,500	5,000
Annual maintenance costs	2,500	750
Annual utility expenses	1,200	2,000
Disposal costs after useful life	700	500

Solution

We are interested in the incremental or *extra* costs that are associated with choosing Model B instead of Model A. To obtain these we subtract Model A costs from Model B costs for each category (cost item) with the following results.

Cost Items	(Model B cost - A cost)	Incremental Cost of B
Purchase price	17,500 - 10,000	$7500
Installation costs	5,000 - 3,500	1500
Annual maintenance costs	750 - 2,500	- 1750/yr
Annual utility expenses	2,000 - 1,200	800/yr
Disposal costs after useful life	500 - 700	-200
		$7850

Notice that for the cost categories given, the *incremental costs* of model B are both positive and negative. Positive incremental costs mean that model B costs more than model A, and negative incremental costs indicate that there would be a *savings* (reduction in cost) if model B where chosen instead.

Because model B has more features, a decision would also have to consider the incremental benefits offered by that model. ∎

Cash Costs versus Book Costs

A *cash cost* requires the cash transaction of dollars "out of one person's pocket" into the "pocket of someone else." When you buy dinner for your friends or make your monthly automobile payment you are incurring a *cash cost* or *cash flow*. Cash costs and cash flows are the basis for engineering economic analysis.

Book costs do not require the transaction of dollars "from one pocket to another." Rather, *book costs* are cost effects from past decisions that are recorded "in the books" (accounting books) of a firm. A common book cost is asset depreciation (which we discuss in Chapter 10), where the expense paid for a particular business asset is "written off" on a company's accounting system over a number of periods. Book costs do not ordinarily represent cash flows and thus are not included in engineering economic analysis. One exception to this is the impact of asset depreciation on tax payments – which are cash flows and included in after-tax analyses.

Life-cycle Costs

The products, goods, and services designed by engineers all progress through a *life-cycle*. This life-cycle is very much like the human life-cycle. People are conceived, go through a growth phase, reach our peak during our maturity, and then gradually decline and expire. The same general pattern holds for products, goods, and services. Like humans, the duration of the different phases, the height of the peak at maturity, and the time of the onset of decline and termination all vary depending upon the individual product, good, or service. Figure 2-3 illustrates the typical phases that a product, good or service progresses through over its life-cycle.

Beginning ——————————— Time ——————————————→ End					
NEEDS ASSESSMENT AND JUSTIFICATION PHASE	**CONCEPTUAL OR PRELIMINARY DESIGN PHASE**	**DETAILED DESIGN PHASE**	**PRODUCTION OR CONSTRUCTION PHASE**	**OPERATIONAL USE PHASE**	**DECLINE AND RETIREMENT PHASE**
REQUIREMENTS	IMPACT ANALYSIS	ALLOCATION OF RESOURCES	PRODUCT, GOODS & SERVICES BUILT	OPERATIONAL USE	DECLINING USE
OVERALL FEASIBILITY	PROOF OF CONCEPT	DETAILED SPECIFICATIONS	ALL SUPPORTING FACILITIES BUILT	USE BY ULTIMATE CUSTOMER	PHASE OUT
CONCEPTUAL DESIGN PLANNING	PROTOTYPE/ BREADBOARD	COMPONENT AND SUPPLIER SELECTION	OPERATIONAL USE PLANNING	MAINTENANCE AND SUPPORT	RETIREMENT
	DEVELOPMENT AND TESTING	PRODUCTION OR CONSTRUCTION PHASE		PROCESSES, MATERIALS AND METHODS USE	RESPONSIBLE DISPOSAL
	DETAILED DESIGN PLANNING			DECLINE AND RETIREMENT PLANNING	

Figure 2-3: Typical life-cycle for products, goods and services.

Life-cycle costing refers to the concept of designing products, goods, and services with a full and explicit recognition of the *associated costs* over the various phases of their life-cycles. Two key concepts in life-cycle costing are that the later design changes are made in the life-cycle the higher the costs, and that decisions made early in the life-cycle tend to "lock in" costs that are incurred later in the life cycle. Figure 2-4 illustrates how costs are committed early in the product life cycle — nearly 70-90% of all costs are set during the design phases. At the same time, as the figure shows, only 10-30% of cumulative life-cycle costs have been spent.

Figure 2-5 reinforces these concepts by illustrating that downstream product changes are more costly and that upstream changes are easier (and less costly) to make. It can be very costly to try and save money at an early design stage. Often a poorer design results, and then there are much more expensive change orders during construction and prototype development.

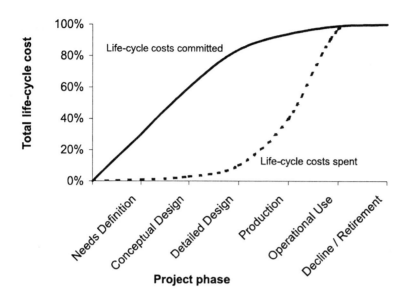

Figure 2-4: Cumulative life-cycle costs committed and dollars spent.

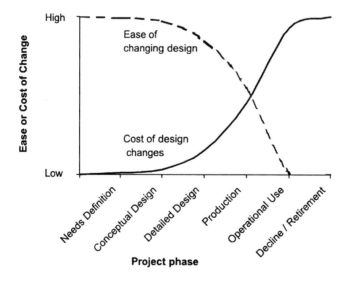

Figure 2-5: Life-cycle design change costs and ease of change

From Figures 2-4 and 2-5 we see that the time to consider all life-cycle effects, and make design changes, is during the needs and conceptual/preliminary design phases — before a lot of dollars are committed. Some of the life-cycle effects that engineers should consider at design time include product costs for: liability, production, material, testing and quality assurance, and maintenance and warranty. Other life-cycle effects include product features based on customer input and product disposal effects on the environment. The key point is that engineers should consider all life-cycle costs when designing products and the systems that produce them.

Cost Estimating

Engineering economic analysis focuses on the future consequences of current decisions. Because these consequences are in the future, usually they must be estimated and they are not known with certainty. Examples of the estimates that may be needed in engineering economic analysis include: purchase costs, annual revenue, yearly maintenance, interest rates for investments, annual labor and insurance costs, equipment salvage values, and tax rates as well as many others.

Estimating is the foundation of economic analysis. As is the case in any analysis procedure, the outcome is only as good as the quality of the numbers used to reach the decision. As an example, consider a person who is interested in estimating her federal income taxes for a given year. The person could do a very detailed analysis, including: social security deductions, retirement savings deductions, itemized personal deductions, exemption calculations, and estimates of likely changes to the tax code. However, this very technical and detailed analysis could be grossly inaccurate if poor data are used to predict the person's income. Thus, it is very important to make careful estimates so that our analysis is a reasonable evaluation of future events.

Types of Estimates

The American poet and novelist Gertrude Stein wrote in *The Making of Americans* in 1925 that "a rose is a rose is a rose is a rose." However, what holds for roses does not necessarily hold for estimates because "an estimate is not an estimate." Ms. Stein's point was a bit deeper than that all roses are the same, but it *is* true that all estimates *are not* the same. Rather, we can define three general types of estimates whose purposes, accuracies, and underlying methods are quite different.

Rough estimates: are order-of-magnitude type estimates used for high-level planning, macro-feasibility, and in a project's initial planning and evaluation phases. Rough estimates tend to involve back-of-the-envelope numbers with little detail or accuracy. The intent is to quantify and consider the order of magnitude of the numbers involved. These estimates require minimum resources to develop, and their accuracy is generally -30% to +60%.

Notice the non-symmetry in the estimating error – this is because there is a tendency by decision makers to underestimate the magnitude of costs (negative economic effects). Also due to Murphy's Law, there seem to be more ways for results to be worse than expected than there are for the results to be better than expected.

Semi-detailed estimates: are used for budgeting purposes at a project's conceptual or preliminary design stages. These estimates are more detailed, and they require additional time and resources to develop. Greater sophistication is used in developing semi-detailed estimates than the rough-order type, and their accuracy is generally -15 to +20%.

Detailed estimates: are used during a project's detailed design and contract bidding phases. These estimates are made from detailed quantitative models, blueprints, product specification sheets, and vendor quotes. Detailed estimates involve the most time and resources to develop and thus are much more accurate than rough or semi-detailed estimates. The accuracy of these estimates is generally -3 to +5%.

In considering the three types of estimates it is important to recognize that each type has its unique purpose, place, and function in a project's life. Rough estimates are used for general feasibility activities, semi-detailed estimates support budgeting and preliminary design decisions, and detailed estimates are used for establishing design details and contracts. As one moves from rough to detailed design, one moves from less to much more accurate estimates.

However, this increased accuracy requires added time and resources. Figure 2-6 illustrates the accuracy versus cost tradeoff. In engineering economic analysis, the resources spent must be justified by the need for detail in the estimate. As an illustration, during the project feasibility stages we would not want to use our resources (people, time, and money) to develop detailed estimates for alternatives that are not feasible and will be quickly eliminated from further consideration. However, regardless of how accurate an estimate is assumed to be, it is an estimate of what the future will be. There will be some error even if ample resources and sophisticated methods are used.

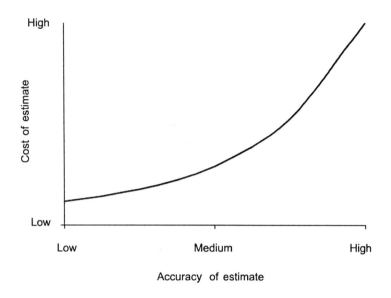

Figure 2-6: Accuracy versus cost tradeoff in estimating.

Difficulties in Estimation

Estimating is difficult because the future is unknown. With few exceptions (such as with legal contracts) it is difficult to anticipate future economic consequences exactly. In this section we discuss several aspects of estimating that "make it a difficult task."

ONE-OF-A-KIND ESTIMATES

Estimated parameters can be for one-of-a-kind or first-run projects. The first time something is done, it is difficult to estimate costs required to design, produce, and maintain a product over its life-cycle. Consider the projected cost estimates that were developed for the first space missions at NASA. The U.S. space program initially had no experience with human flight in outer space, thus the development of the cost estimates for design, production, launch, and recovery of the astronauts, flight hardware, and payloads was a "first time experience." The same is true for any new endeavor in which there is no local or global historical cost data. New products or processes that are unique and fundamentally different make estimating costs difficult.

The good news is that there are very few one-of-a-kind estimates to be made in engineering design and analysis. Nearly all new technologies, products, and processes have "close cousins" that have lead to their development. The concept of *estimation by analogy* allows one to use knowledge about well understood activities to anticipate costs for new activities. In the 1950s at the start of the military missile program, aircraft companies drew on their in-depth knowledge from designing and producing aircrafts when they bid on missile contracts. As another example, consider the problem of estimating the production labor requirements for a brand new product *X*. A company may use its labor knowledge about product *Y*, a similar type product, to build up the estimate for *X*. Thus, although "first run" estimates are difficult to make, *estimation by analogy* can be an effective tool.

TIME AND EFFORT AVAILABLE

Our ability to develop engineering estimates is constrained by time and person-power availability. In an ideal world, it would *cost nothing* to use *unlimited resources* over an *extended period of time*. However, reality requires the use of limited resources in fixed intervals of time. Thus for a rough estimate only limited effort is used.

Constraints on time and person-power can make the overall estimating task more difficult. If the purpose of the estimate does not require as much detail (such as in rough estimating), then time and personnel constraints may not be a factor. In cases where detail is necessary and critical (such as in legal contracts), then anticipating the requirements and planning the resource use is needed.

ESTIMATOR EXPERTISE

Consider two common phrases: *The past is our greatest teacher* and *Knowledge is power*. These simple axioms hold true for much of what we encounter during life, and they are true in engineering estimating as well. The more experienced and knowledgeable the engineering estimator is the less difficult the estimating process will be, the more accurate the estimate will be, the less likely a major error will occur, and the more likely the estimate will be of high quality.

How is experience acquired in industry? One approach is to assign inexperienced engineers relatively smaller jobs to create expertise and build familiarity with products and processes. Another strategy used is to pair inexperienced engineers with mentors who have vast technical experience. Technical boards and review meetings conducted to "justify the numbers" also are used to build knowledge and experience. Finally, many firms maintain databases of their past estimates and the costs that were actually incurred.

Estimating Models

This section develops several estimating models that can be used at the rough, semi-detailed, or detailed-design levels. For rough estimates the models are used with rough data, likewise for detailed-design estimates they are used with detailed data. The level of detail will depend upon the accuracy of the model's data.

Per Unit Model

The *per unit model* uses a "per unit" factor, such as cost per square foot, to develop the estimate desired. This is a very simplistic yet useful technique especially for developing rough or order-of-magnitude type estimates. The per unit model is commonly used in the construction industry. As an example, you may be interested in a new home that is constructed with a certain type of material and has a specific construction style. Based on this information a contractor may quote a cost of $65 per square foot for your home. If you are interested in a 2000 square foot floor plan your cost would thus be: 2000 x 65 = $130,000. Other examples where per unit factors are utilized include:

- service cost per customer
- safety cost per employee
- gasoline cost per mile
- cost of defects per batch

- maintenance cost per window
- mileage cost per vehicle
- utility cost per square foot of floor space
- housing cost per student

It is important to note that the *per unit model* does not make allowances for economies of scale (the fact that higher quantities usually cost less on a per unit basis). However, the model can be effective at getting the decision maker "in the ballpark" of likely costs in most cases, and can be very accurate if accurate data are used.

EXAMPLE 2-5

Use the per unit model to estimate the cost per student that you will incur for hosting 24 foreign exchange students at a local island campground for 10 days. During camp you are planning the following activities:

- 2 days of canoeing
- 3 campsite sponsored day hikes

- 3 days at the lake beach (swimming, volleyball, etc.)
- nightly entertainment

After calling the campground and collecting other information you have accumulated the following data:

- Van rental from your city to the camp (one way) is $50 per 15 person van plus gas.
- Camp is 50 miles away, the van gets 10 miles per gallon, and gas is $1 per gallon.
- Cabins at the camp hold 4 campers, rent is $10 per day per cabin.
- Meals are $10 per day per camper, no outside food is allowed.
- Boat transportation to the island is $2 per camper (one way).
- Insurance/grounds fee/overhead is $1 per day per camper.
- Canoe rentals are $5 per day per canoe, canoes hold 3 campers.
- Day hikes are $2.50 per camper (plus the cost for meals).
- Beach rental is $25 per group per ½ day.
- Nightly entertainment is free.

Solution

You are asked to use the per unit factor to estimate the cost per student on this trip. For planning purposes we assume that there will be 100% participation in all activities. We will break the total cost down into categories of transportation, living, and entertainment.

Transportation Costs:

    1. Van travel to and from camp: 2 vans x 2 trips x ($50/van + 50 miles x 1 gal/10 miles
        x $1/gal) = $220

    2. Boat travel to and from island: 2 trips x $2/camper x 24 campers = $96

    Transportation Costs: = 220 + 96 = $ 316

Living Costs:

    1. Meals for the 10 day period: 24 campers x $10/camper/day x 10 days = $2400

    2. Cabin rental for the 10 day period: 24 campers x 1 cabin/4 campers x $10/day/cabin
        x 10 days = $600

    3. Insurance/Overhead expense for the 10 day period: 24 campers x $1/day/camper x
        10 days = $240

    Living Costs: = 2400 + 600 + 240 = $3240

Entertainment Costs:

    1. Canoe rental costs: 2 canoe days x 24 campers x 1canoe/3 campers x $5/day/canoe
        = $80

    2. Beach rental costs: 3 days x 2 half-days/day x $25/half-day = $150

    3. Day hike costs: 24 campers x 3 day hikes x $2.50/camper/day hike = $180

    4. Nightly entertainment: This is free! Can you believe it?

    Entertainment Costs: = 80 + 150 + 180 + 0 = $410

Total Cost for 10-day period   =
        Transportation costs + Living costs + Entertainment costs
           = 316 + 3240 + 410 = $3966

Thus, the cost per student would be $3966/24 = $165.25

Thus, it would cost you $165.25 per student to host the students at the island campground for the 10 day period. In this case the per unit model gives you a very detailed cost estimate (although it depends upon the accuracy of your data and assumptions you've made). ■

Segmenting Model

This model can be described as "divide and conquer." An estimate is decomposed into its individual components, estimates are made at those lower levels, and then the estimates are aggregated (added) back together. It is much easier to estimate at the lower levels because they are more readily understood. This approach is common in engineering estimating in many applications and for any level of accuracy needed. Consider the camp trip planned in Example 2-5 above, the overall estimate was *segmented* into the cost for travel, living, and entertainment. The example illustrates the segmenting model (division of the overall estimate into the various categories) together with the unit factor model to make the sub-estimates for each category. Example 2-6 provides another example of the segmenting approach.

EXAMPLE 2-6

Clean Lawn Corp. a manufacturer of yard equipment is planning to introduce a new high-end industrial-use lawn mower called the *Grass Grabber*. The *Grass Grabber* is designed as a walk-behind self-propelled mower. Clean Lawn engineers have been asked by the accounting department to estimate the material costs that will make up the new mower. The material cost estimate will be used along with estimates for labor and overhead to evaluate the potential of this new model.

Solution
The engineers decide to decompose the design specifications for the *Grass Grabber* into its sub-components, estimate the material costs for each of the sub-components and then sum these costs up to obtain their overall estimate. The engineers are using a *segmenting* approach to build up their estimate. After careful consideration the engineers have divided the mower into the following major sub-systems: chassis, drive train, controls, and cutting/collection system. Each of these were further divided as appropriate and unit material cost estimates were made at this lowest of levels as follows:

Cost Item	Unit Material Cost Estimate	Cost Item	Unit Material Cost Estimate
A. Chassis		**C. Controls**	
A.1 Deck	$ 7.40	C.1 Handle Assembly	$ 3.85
A.2 Wheels	10.20	C.2 Engine Linkage	8.55
A.3 Axles	4.85	C.3 Blade Linkage	4.70
	$22.45	C.4 Speed Control Linkage	21.50
		C.5 Drive Control Assembly	6.70
B. Drive Train		C.6 Cutting Height Adjuster	7.40
B.1 Engine	$38.50		$52.70
B.2 Starter Assembly	5.90		
B.3 Transmission	5.45	**D. Cutting/Collection System**	
B.4 Drive Disc Assembly	10.00	D.1 Blade Assembly	$10.80
B.5 Clutch Linkage	5.15	D.2 Side Chute	7.05
B.6 Belt Assemblies	7.70	D.3 Grass Bag & Adapter	7.75
	$72.70		$25.60

The total material cost estimate of $173.45 was calculated by summing up the estimates for each of the four major sub-system levels (chassis, drive train, controls, and cutting/collection system). It should be noted that this cost represents only the material portion of the overall cost to produce the mowers. Other costs would include labor and overhead items. ■

In Example 2-6 the engineers at Clean Lawn Corp. decomposed the cost estimation problem into logical elements. The scheme they used of decomposing and numbering (A, A.1, A.2, etc.) the material components is known as a *work breakdown structure*. This technique is commonly used in engineering cost estimating and project management when working with large products, processes, or projects. A work breakdown structure decomposes a large "work package" into its constituent parts which can then be estimated or managed individually. In Example 2-6 the work breakdown structure of the *Grass Grabber* has three levels. At the top level is the product itself, at the second level are the four major sub-systems, and at the third level are the individual cost items. Imagine what the product work breakdown structure for a Boeing 777 looks like. Then imagine trying to manage the 777's design, engineering, construction, and costing without a tool like the work breakdown structure.

Cost Indexes

Cost indexes are numerical values that reflect historical change in engineering (and other) costs. The cost index numbers are dimension-less, and reflect relative price change in either individual cost items (labor, material, utilities) or groups of costs (consumer prices, producer prices). Indexes can be used to update historical costs with the basic ratio relationship given in Equation 2-2.

$$\frac{\text{Cost at time A}}{\text{Cost at time B}} = \frac{\text{Index Value at time A}}{\text{Index Value at time B}} \qquad (2\text{-}2)$$

Equation 2-2 states that the ratio of the cost index numbers at two points in time (*A* and *B*) is equivalent to the dollar cost ratio of the item at the same times (see Example 2-7).

EXAMPLE 2-7

Miriam is interested in estimating the annual labor and material costs for a new production facility. She was able to obtain the following labor and material cost data:

Labor Costs: • Labor Cost Index value was at 124 ten years ago and is 188 today.
            • Annual labor costs for a similar facility were $575,500 ten years ago.

Material Costs: • Material Cost Index value was at 544 three years ago and is 715 today.
            • Annual material costs for a similar facility were $2,455,000 three years ago.

Solution

Miriam will use Equation 2-2 to develop her cost estimates for annual labor and material costs.

Labor:

$$\frac{\text{Annual Cost today}}{\text{Annual Cost 10 years ago}} = \frac{\text{Index Value today}}{\text{Index value 10 years ago}}$$

$$\text{Annual Cost today} = \frac{188}{124} \times \$575,500 = \$871,800$$

Material:

$$\frac{\text{Annual Cost today}}{\text{Annual Cost 3 years ago}} = \frac{\text{Index Value today}}{\text{Index value 3 years ago}}$$

$$\text{Annual Cost today} = \frac{715}{544} \times \$2,455,000 = \$3,227,000 \quad ■$$

Cost index data are collected and published by several private and public sources in the United States (and world). In the U.S. the federal government publishes data through the *Department of Commerce Bureau of Labor Statistics*. The *Statistical Abstract of the United States* publishes cost indexes for labor, construction, and materials. Another useful source for engineering cost index data is the *Engineering News Record*.

Power Sizing Model

The *power sizing model* is used to estimate the costs of industrial plants and equipment. The model "scales up" or "scales down" previously known costs, thereby accounting for "economies of scale" that are common in industrial plant and equipment costs. Consider the cost to build a refinery. Would the cost to build this same refinery with double the capacity be twice as much? It is unlikely. The *power sizing model* uses the exponent (x) in equation 2-3 below, called the *power sizing exponent*, to reflect the effect of economies of scale in the size or capacity.

$$\frac{\text{Cost of equipment A}}{\text{Cost of equipment B}} = \left(\frac{\text{Size(capacity) of equipment A}}{\text{Size(capacity) of B}} \right)^x \tag{2-3}$$

Where: x is the power sizing exponent.
             Cost of A and B are at the same point in time (same dollar basis).
             Size or capacity of A and B are in the same physical units.

The power sizing exponent (x) can be equal to one (indicating a linear cost versus size/capacity relationship) or greater than 1.0 (indicating *dis*economies of scale), but it is usually less than 1.0 (indicating economies of scale). Exponent values for many types of plants and equipment may be found in several sources, including: industry reference books, research reports, and technical journals. Titles where such exponent values may be found include: the *Perry's Chemical Engineers' Handbook, Plant Design and Economics for Chemical Engineers,* and *Preliminary Plant Design in Chemical Engineering.* Table 2-1 gives power sizing exponent values for several types of industrial facilities and equipment. The exponent given applies only to equipment within the size range specified.

Table 2-1 EXAMPLE POWER SIZING EXPONENT VALUES

Equipment/Facility	*Size Range (units)*	*Power Sizing Exponent*
Blower, centrifugal	10,000-100,000 (cubic ft./min.)	0.59
Compressor	200-2100 (hp)	0.32
Crystalizer, vacuum batch	500-7000 (square ft.)	0.37
Dryer, drum, single atmospheric	10-100 (square ft.)	0.40
Fan, centrifugal	20,000-70,000 (sq. ft./min.)	1.17
Filter, vacuum rotary drum	10-1500 (sq. ft.)	0.48
Lagoon, aerated	0.05-20 (million gal. per day)	1.13
Motor	5-20 (hp)	0.69
Reactor, 300 psi	100-1000 (gal.)	0.56
Tank, atmospheric, horizontal	100-40,000 (gal.)	0.57
Tray, bubble cup	3-10 (ft. diameter)	1.20

In Equation 2-3 equipment costs for A and B both occur at the same point in time. This equation is useful for scaling equipment costs but *not* for updating those costs. In cases where the time of the desired cost estimate is different than the time in which the scaling occurs (per Equation 2-3) cost indexes accomplish the time updating. Thus, in cases like Example 2-8 involving both scaling and updating we use the power sizing model together with cost indexes.

EXAMPLE 2-8

Based on her work in Example 2-7, Miriam has been asked to estimate the cost today of a 2500 ft^2 heat exchange system for the new plant being analyzed. She has the following data.

- Her company paid $50,000 for a 1000 ft^2 heat exchanger 5 years ago.
- Heat exchangers within this range of capacity have a power sizing exponent (x) = 0.55.
- The Heat Exchanger Cost Index (HECI) was 1306 five years ago and is 1487 today.

Solution

Miriam will first use Equation 2-3 to scale up the cost of the 1000 ft² exchanger to one that is 2500 ft² using the 0.55 power sizing exponent.

$$\frac{\text{Cost of 2500ft}^2 \text{ equipment}}{\text{Cost of 1000ft}^2 \text{ equipment}} = \left(\frac{2500\text{ft}^2 \text{ equipment}}{1000\text{ft}^2 \text{ equipment}} \right)^{0.55}$$

$$\text{Cost of 2500 ft}^2 \text{ equipment} = \left(\frac{2500}{1000} \right)^{0.55} \times 50{,}000 = \$82{,}800$$

Miriam knows that the $82,800 only reflects the scaling up of the cost of the 1000 ft² model to a 2500 ft² model. Now Miriam will use Equation 2-2 and the HECI data to estimate the cost of a 2500 ft² exchanger today. Miriam's cost estimate would be:

$$\frac{\text{Equipment Cost today}}{\text{Equipment Cost 5 years ago}} = \frac{\text{Index value today}}{\text{Index value 5 years ago}}$$

$$\text{Equipment cost today} = \frac{1487}{1306} \times \$82{,}800 = \$94{,}300$$

Triangulation

Triangulation is used in engineering surveying. A geographical area is divided into triangles from which the surveyor is able to map points within that region. Using triangulation a surveyor uses three fixed points and horizontal angular distances to locate fixed points of interest (like property line reference points). Since any point can be located with two lines, the third line represents an extra perspective and check. We will not use trigonometry to arrive at our cost estimates, but we can utilize the concept of *triangulation*. We should approach our economic estimate from *different perspectives* because such varied perspectives add richness, confidence, and quality to the estimate. Triangulation in cost estimating might involve using different sources of data or using different quantitative models to arrive at the value being estimated. As decision makers we should always seek out varied perspectives.

Improvement and the Learning Curve

One common phenomenon observed, regardless of the task being performed, is as the number of repetitions increases the faster and more accurate performance becomes. This is the concept of *learning* and *improvement* in the activities that people perform. From our own experience we all know that our fiftieth repetition is much faster than the time required to complete a task the first time.

The *learning curve* captures the relationship between task performance and task repetition. In general, as output *doubles* the unit production time will be reduced to some fixed percentage, *the learning curve percentage.* For example it may take 300 minutes to produce the third unit in a production run involving a task with a 95% learning time curve. In this case the sixth (2x3) unit will take $300(0.95) = 285$ minutes to produce. Sometimes the *learning curve* is also known as the *progress curve, improvement curve, experience curve,* or *manufacturing progress function.*

Equation 2-4 gives an expression that can be used for time estimating in repetitive tasks.

$$T_N = T_{initial} \times N^b \tag{2-4}$$

Where: T_N = time requirement for the Nth unit of production
$T_{initial}$ = time requirement for the first (initial) unit of production
N = number of completed units (cumulative production)
b = learning curve exponent (slope of the learning curve)

As given above, a learning curve is often referred to by its percentage learning slope. Thus, a curve with $b = -0.074$ is a 95% learning curve because $2^{-0.074} = 0.95$. This equation uses 2, because the learning curve percentage applies for doubling cumulative production. The learning curve exponent is calculated using Equation 2-5.

$$b = \frac{\log (\text{learning curve expressed as a decimal})}{\log 2.0} \tag{2-5}$$

EXAMPLE 2-9

Calculate the time required to produce the 100th unit of a production run if the first unit took 32.0 minutes to produce and the learning curve rate for production is 80%.

Solution

$$T_{100} = T_1 \bullet 100^{\log (0.80) / \log 2.0}$$
$$T_{100} = 32.0 \bullet 100^{-0.3219}$$
$$T_{100} = 7.27 \text{ minutes} \quad \blacksquare$$

It is particularly important to account for the learning curve effect if the production run involves a small number of units instead of a large number. When thousands or even millions of units are being produced, early inefficiencies tend to be "averaged out" because of the larger batch sizes. However, in the short run, those same inefficiencies can lead to rather poor estimates of production time requirements, and thus production cost estimates may be understated. Consider Example 2-10 and the results that might be observed if the learning curve effect is ignored. Notice in this example that a "steady state" time is given. Steady state is the time at which "no more learning or improvement" can take place due to the physical constraints of performing the task.

EXAMPLE 2-10

Estimate the overall labor cost portion due to a task that has a learning curve rate of 85% and reaches a steady state value after 16 units of 5.0 minutes per unit. Labor and benefits are $22.00 per hour and the task requires two skilled workers. The overall production run is 20 units.

Solution
Because we know the time required for the 16th unit we can use Equation 2-4 to calculate the time required to produce the first unit.

$$T_{16} = T_1 \bullet 16^{\ \log(0.85)\,/\,\log 2.0}$$
$$5.0 = T_1 \bullet 16^{\ -0.2345}$$
$$T_1 = 9.6 \text{ minutes}$$

Now we use Equation 2-4 to calculate the time requirements for each unit in the production run as well as the total production time required.

$$T_N = 9.6 \bullet N^{-0.2345}$$

Unit Number (N)	Time (min) to produce (Nth)	Cumulative Time from 1 to N	Unit Number (N)	Time (min) to produce (Nth)	Cumulative Time from 1 to N
1	9.6	9.6	11	5.5	74.0
2	8.2	17.8	12	5.4	79.2
3	7.4	24.2	13	5.3	84.5
4	6.9	32.1	14	5.2	89.7
5	6.6	38.7	15	5.1	94.8
6	6.3	45.0	16	5.0	99.8
7	6.1	51.1	17	5.0	104.8
8	5.9	57.0	18	5.0	109.8
9	5.7	62.7	19	5.0	114.8
10	5.6	68.3	20	5.0	119.8

The total cumulative time of the production run is 119.8 minutes (2.0 hours.) Thus the total labor cost estimate would be:

$$2.0 \text{ hours} \times \$22/\text{hour per worker} \times 2 \text{ workers} = \$88$$

If we ignore the learning curve effect and calculate the labor cost portion based only on the steady state labor rate, the estimate would be:

$$0.083 \text{ hours/unit} \times 20 \text{ units} \times 22\$/\text{hour per worker} \times 2 \text{ workers} = \$73.04$$

This estimate is about 20% understated from what the true cost would be.   ■

Estimating Benefits

This chapter has focused on cost terms and cost estimating. However, engineering economists must often also estimate benefits. Example benefits include sales of products, revenues from bridge tolls and electric power sales, cost reductions from reduced material or labor costs, reduced time spent in traffic jams, and reduced risk of flooding. These benefits are the reasons that many engineering projects are undertaken.

The cost concepts and cost estimating models can also be applied to economic benefits. Fixed and variable benefits, recurring and non-recurring benefits, incremental benefits, and life-cycle benefits all have meaning. Also, issues regarding the type of estimate (rough, semi-detailed, and detailed) as well as difficulties in estimation (one of a kind, time and effort, and estimator expertise) all apply directly to estimating benefits. Lastly, per unit, segmented, and indexed models are used to estimate benefits. The concept of triangulation is particularly important for estimating benefits.

The uncertainty in benefit estimates is also typically asymmetric, with a broader limit for negative outcomes. Benefits are more likely to be over-estimated than under-estimated, so an example set of limits might be (-50%, + 20%). One difference between cost and benefit estimation is that many costs of engineering projects occur in the near future (for design and construction), but the benefits are further in the future. Because benefits are often further in the future, they are more difficult to estimate accurately and more uncertainty is typical.

The estimation of economic benefits for inclusion in our analysis is an important step that should not be overlooked. Many of the models, concepts, and issues that apply in the estimation of costs also apply in the estimation of economic benefits.

Cash Flow Diagrams

The costs and benefits of engineering projects occur over time and are summarized on a *Cash Flow Diagram* (CFD). Specifically, a CFD illustrates the size, sign, and timing of individual cash flows. In this way the CFD is the basis for engineering economic analysis.

A *Cash Flow Diagram* is created by first drawing a segmented time-based horizontal line, divided into appropriate time units . The time units on the CFD can be years, months, quarters or any other consistent time unit. Then at each time when there is a cash flow, a vertical arrow is added – pointing down for costs and up for revenues or benefits. These cash flows are drawn to relative scale. Consider Figure 2-7, the CFD for a specific investment opportunity whose cash flows are described as follows:

Timing of Cash Flow	**Size of Cash Flow**
At time zero (now or today)	A positive cash flow of $100
1 time period from today	A negative cash flow of $100
2 time periods from today	A positive cash flow of $100
3 time periods from today	A negative cash flow of $150
4 time periods from today	A negative cash flow of $150
5 time periods from today	A positive cash flow of $150

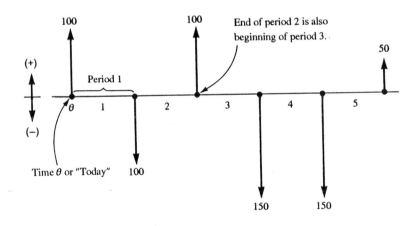

Figure 2-7 An example Cash Flow Diagram (CFD)

Categories of Cash Flows

The expenses and receipts due to engineering projects usually fall into one of the following categories.

First cost ≡ expense to build or to buy and install

Operations and maintenance (O&M) ≡ annual expense, such as electricity, labor, and minor repairs

Salvage value ≡ receipt at project termination for sale or transfer of the equipment (can be a salvage cost)

Revenues ≡ annual receipts due to sale of products or services

Overhaul ≡ major capital expenditure that occurs during the asset's life

Individual projects will often have specific costs, revenues, or user benefits. For example, annual operations and maintenance (O&M) expenses on an assembly line might be divided into direct labor, power, and other. Similarly, a public-sector dam project might have its annual benefits divided into flood control, agricultural irrigation, and recreation.

Drawing a Cash Flow Diagram

The cash flow diagram shows when all cash flows occur. Look at Figure 2-7 and the $100 positive cash flow at the end of period 2. From the time line one can see that this cash flow can also be described as occurring at the *beginning* of period 3. Thus, in a CFD the end of *period t* is the same time as the beginning of *period t* + 1. Beginning of period cash flows (such as rent, lease, and insurance payments) are thus easy to handle, just draw your CFD and put them in where they occur. Thus O&M, salvages, revenues, and overhauls are assumed to be end-of-period cash flows.

The choice of time 0 is arbitrary. For example, it can be 1) when a project is analyzed, 2) when funding is approved, or 3) when construction begins. Usually construction periods are assumed to be short, and first costs are assumed to occur at time 0 and the first annual revenues and costs start at the end of the first period.

Perspective is also important when drawing a CFD. Consider the simple transaction of paying $5000 for some equipment. To the firm buying the equipment, the cash flow is a cost and negative in sign. To the firm selling the equipment, the cash flow is a revenue and positive in sign. This simple example shows that a consistent perspective is required when modeling the cash flows of a problem using a CFD. One person's cash outflow is another person's inflow.

Often two or more cash flows occur in the same year, such as an overhaul and an O&M expense or the salvage value and the last year's O&M expense. Combining these into one total cash flow per year would simplify the cash flow diagram. However, it is better to show each individually so there is a clear connection from the problem statement to each cash flow in the diagram.

Drawing Cash Flow Diagrams with a Spreadsheet

One simple way to draw cash flow diagrams with "arrows" proportional to the size of the cash flows is to use a spreadsheet to draw a stacked bar chart. The data for the cash flows is entered, as shown in the table part of Figure 2-8. To make a quick graph, select cells B1 to D8, which are the three columns of the cash flow. Then select the graph menu and choose column chart and select the stack option. Except for labeling axes (using the cells for year 0 to year 6), choosing the scale for the y-axis, and adding titles – the cash flow diagram is done. Refer to the appendix for a review of basic spreadsheet use.

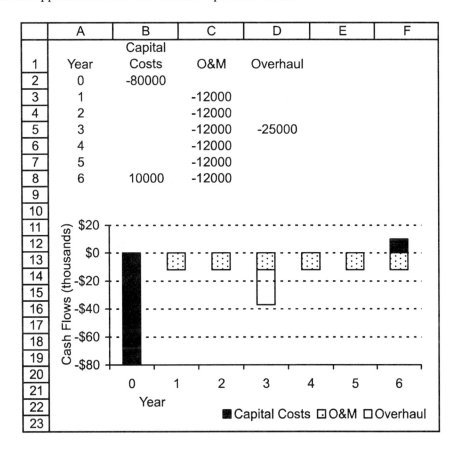

	A	B	C	D	E	F
1	Year	Capital Costs	O&M	Overhaul		
2	0	-80000				
3	1		-12000			
4	2		-12000			
5	3		-12000	-25000		
6	4		-12000			
7	5		-12000			
8	6	10000	-12000			
9						
10						

Figure 2-8 Example of cash flow diagram in spreadsheets.

Summary

This chapter has introduced the following cost concepts: fixed and variable, marginal and average, sunk, opportunity, recurring and nonrecurring, incremental, cash and book, and life-cycle. *Fixed costs* are constant and unchanging as volumes change, while *variable costs* change as output changes. Fixed and variable costs are used to find the breakeven value between costs and revenues, as well as the regions of net profit and loss. A *marginal cost* is for one more unit, while the *average cost* is the total cost divided by the number of units.

Sunk costs result from past decisions and should not influence our attitude toward current and future opportunities. Remember, "sunk costs are sunk." *Opportunity costs* involve the benefit that is foregone when we chose to use a resource in one activity instead of another. *Recurring costs* can be planned and anticipated expenses; *non-recurring costs* are one-of-a-kind costs that are often more difficult to anticipate.

Incremental costs are economic consequences associated with the differences between two choices of action. *Cash costs* are also known as *out of pocket costs* that represent actual cash flows. *Book costs* do not result in the exchange of money, but rather are costs listed in a firm's accounting books. *Life-cycle costs* are those costs that are incurred over the life of a product, process, or service. Thus engineering designers must consider life-cycle costs when choosing materials and components, tolerances, processes, testing, safety, service and warranty, and disposal.

Cost estimating is the process of "developing the numbers" for engineering economic analysis. Unlike a textbook, the real world does not present its challenges with neat problem statements that provide all the data. The types of engineering estimates include *rough estimates* which give us order-of-magnitude numbers and are useful for high-level and initial planning as well as judging the feasibility of alternatives. *Semi-detailed estimates* are more accurate than rough-order estimates, thus requiring more resources (people, time, and money) to develop. These estimates are used in preliminary design and budgeting activities. *Detailed estimates* generally have an accuracy of $\pm$ 3-5%. They are used during the detail design and contract bidding phases of a project.

Difficulties are common in developing estimates. *One-of-a-kind estimates* do not have a previous basis, but this can be addressed through *estimation by analogy*. Lack of time available is best addressed by planning and by matching the estimate's detail to the purpose — one should not spend money developing a detailed estimate when only a rough estimate is needed. *Estimator expertise* must be developed through work experiences and mentors.

Several general models and techniques for developing cost estimates were discussed. The *per unit* and *segmenting models* use different levels of detail and costs per square foot or other unit. *Cost index* data are useful for updating historical costs to formulate current estimates. The *power sizing model* is useful for scaling up or down a previously known cost quantity to account for economies-of-scale with different power sizing exponents for different types of industrial plants and equipment. *Triangulation* suggests that one should seek varying perspectives when developing cost estimates. Different information sources, data bases, and analytical models can all be used to create unique perspectives. As the number of task repetitions increase, efficiency improves because of *learning* or *improvement*. This is summarized in the *learning curve percentage*, where doubling the cumulative production reduces the time to complete the task so it equals the learning curve percentage times the current production time.

Cash flow estimation must include project benefits. These include labor cost savings, avoided quality costs, direct revenue from sales, reduced catastrophic risks, improved traffic flow, and cheaper power supplies. *Cash Flow Diagrams* are used to model the positive and negative cash flows of potential investment opportunities. These diagrams provide a consistent view of the problem (and the alternatives) to support economic analysis.

Problems

2-1   Bob Johnson decided to purchase a new home. After looking at tracts of new homes, he decided a custom-built home was preferable. He hired an architect to prepare the drawings. In due time, the architect completed the drawings and submitted them. Bob liked the plans; he was less pleased that he had to pay the architect a fee of $4000 to design the house. Bob asked a building contractor to provide a bid to construct the home on a lot Bob already owned. While the contractor was working to assemble the bid, Bob came across a book of standard house plans. In the book was a home that he and his wife liked better than the one designed for them by the architect. Bob paid $75 and obtained a complete set of plans for this other house. Bob then asked the contractor to provide a bid to construct this "stock plan" home. In this way Bob felt he could compare the costs and make a decision. The building contractor submitted the following bids:

Custom designed home	$128,000
Stock-plan home	128,500

Both Bob and his wife decided they were willing to pay the extra $500 for it. Bob's wife, however, told Bob they would have to go ahead with the custom-designed home, for, as she put it, "We can't afford to throw away a set of plans that cost $4000." Bob agrees, but he dislikes the thought of building a home that is less desirable than the stock plan home. He asks your advice. Which house would you advise him to build? Explain.

2-2   Venus Computer Co. can produce 23,000 personal computers a year on its daytime shift. The fixed manufacturing costs per year are $2,000,000 and the total labor cost is $9,109,000. To increase its production to 46,000 computers per year, Venus is considering adding a second shift. The unit labor cost for the second shift would be 25% higher than the day shift, but the total fixed manufacturing costs would increase only to $2,400,000 from $2,000,000.

    a. Compute the unit manufacturing cost for the daytime shift.

    b. Would adding a second shift increase or decrease the unit manufacturing cost at the plant?

2-3   A small machine shop, with thirty horsepower of connected load, purchases electricity under the following monthly rates (assume any demand charge is included in this schedule):

First 50 kw-hr per HP of connected load at 8.6¢ per kw-hr;
Next 50 kw-hr per HP of connected load at 6.6¢ per kw-hr;
Next 150 kw-hr per HP of connected load at 4.0¢ per kw-hr;
All electricity over 250 kw-hr per HP of connected load at 3.7¢ per kw-hr.

The shop uses 2800 kw-hr per month.

> **a.** Calculate the monthly bill for this shop. What are the marginal and average costs per kw-hr?
>
> **b.** Suppose Jennifer, the proprietor of the shop, has the chance to secure additional business that will require her to operate her existing equipment more hours per day. This will use an extra 1200 kw-hr per month. What is the lowest figure that she might reasonably consider to be the "cost" of this additional energy? What is this per kw-hr?
>
> **c.** She contemplates installing certain new machines that will reduce the labor time required on certain operations. These will increase the connected load by 10 HP but, as they will operate only on certain special jobs, will add only 100 kw-hr per month. In a study to determine the economy of installing these new machines, what should be considered as the "cost" of this energy? What is this per kw-hr?

2-4   Two automatic roadside-map-dispenser-systems are being compared by the State Highway Department at a new state crossing. A breakeven chart of the comparison of these systems (System I versus System II) is given below. The chart below shows total yearly costs for the number of maps dispensed per year for both alternatives.   Answer the following questions.

> **a.** What is the fixed cost for System I ?
>
> **b.** What is the fixed cost for System II ?
>
> **c.** What is the variable cost per map dispensed for System I ?
>
> **d.** What is the variable cost per map dispensed for System II ?
>
> **e.** What is the breakeven point in terms of maps dispensed at which the two systems have equal annual costs ?
>
> **f.** For what range of annual number of maps dispensed is System I recommended ?
>
> **g.** For what range of annual number of maps dispensed is System II recommended ?
>
> **h.** At 3000 maps per year, what are the marginal and average map costs for each system?

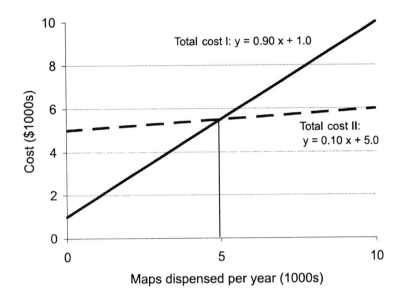

Figure: Breakeven chart for problem 2-4.

2-5   Mr. Sam Spade, the president of Ajax, recently read in a report that a competitor named Bendix has the following cost–production quantity relationship:

$$C = \$3,000,000 - \$18,000Q + \$75Q^2$$

where C = Total manufacturing cost per year, and Q = Number of units produced per year.

A newly hired employee, who previously worked for Bendix, told Mr. Spade that Bendix is now producing 110 units per year. If the selling price remains unchanged, Sam wonders if Bendix is likely to increase the number of units produced per year, in the near future. He asks you to look at the information and tell him what you are able to deduce from it.

2-6   A small computer system is to be purchased for the sales department. There are four alternatives available.

Computer	Pet	Pear	Pal	Pearl
Initial cost	$6000	$8000	$9000	$10,000
Total annual operating cost	3500	3200	2800	2,650

Each computer system will perform the desired tasks, but the more expensive models will take less effort to use and, hence, will have a lower total annual operating cost.

The company requires a 20% Annual Percentage Rate (APR) on all investments or non-essential increments of investment.

$$APR = \frac{\text{Annual benefit}}{\text{Initial cost}} \times 100$$

Which computer system should the company purchase? Show your computations.

2-7   A privately owned summer camp for youngsters has the following data for a 12-week session:

Charge per camper	$120 per week
Fixed costs	$48,000 per session
Variable cost per camper	$80 per week
Capacity	200 campers

   a. Develop the mathematical relationships for total cost and total revenue.
   b. What is the total number of campers that will allow the camp to *just break even*?
   c. What is the profit or loss for the 12-week season if the camp operates at 80% capacity?

2-8   Two new amusement rides are being compared by a local park in terms of their annual operating costs. The two rides are assumed to be able to generate the same level of revenue (and thus the focus on costs). The Tummy Tugger has fixed costs of $10,000 per year, and variable costs of $2.50 per visitor. The Head Buzzer has fixed costs of $4000 per year, and variable costs of $4.00 per visitor. Provide answers to the following questions so the amusement park can make their comparison.

   a. Mathematically determine the breakeven number of visitors per year for the two rides to have equal annual costs.
   b. Develop a graph that illustrates the following: (NOTE: Put visitors per year on the horizontal axis and costs on the vertical axis.)
   • Accurate total cost lines for the two alternatives (show line, slopes, and equations).
   • The breakeven point for the two rides in terms of number of visitors.
   • The ranges of visitors per year where each alternative is preferred.

2-9   Consider the breakeven graph given below for an investment, and answer questions *a-d* given below as they pertain to the graph:

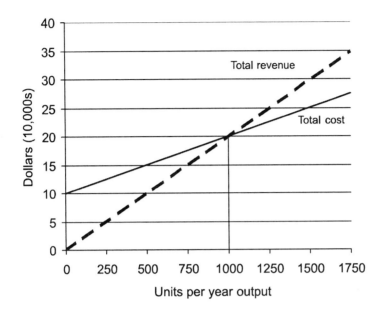

Figure: Breakeven chart for problem 2-9.

a. Give the equation to describe total revenue for x units per year.
b. Give the equation to describe total costs for x units per year.
c. What is the "breakeven" level of x in terms of costs and revenues?
d. If you sell 1,500 units this year, will you have a profit or loss? How much?

2-10 Quatro Hermanas, Inc. is investigating implementing some new production machinery as part of its operations. Three alternatives have been identified and they have fixed and variables costs as given below. Determine the ranges of production (units produced per year) over which each alternative would be recommended for implementation by Quatro Hermanas, Inc. Be exact.

NOTE: Consider the range of production to be from (0-to-30,000) units per year.

Alternative	Annual Fixed Costs	Annual Variable Costs
A	$ 100,000	$ 20.00 per unit
B	200,000	5.00 per unit
C	150,000	7.50 per unit

2-11   Three alternative designs have been created by Snakisco engineers for a new machine that spreads cheese between the crackers in a Snakisco snack. Each machine design has unique total costs (fixed and variable) based on the annual production rate of boxes of these crackers. For the three designs the costs are given below (where x is the annual production rate of boxes of cheese crackers).

	Fixed Cost	*Variable Cost ($ per x)*
Design *A*	$ 100,000	20.5x
Design *B*	350,000	10.5x
Design *C*	600,000	8.0x

You are asked to do the following:

a. Mathematically determine which of the machine designs would be recommended for different levels of annual production of boxes of snack crackers. Management is interested in the production interval of 0 to 150,000 boxes of crackers per year. Over what production volume would each design (*A* or *B* or *C*) be chosen?

b. Depict your solution from part **a** graphically for management to see more easily, include: (NOTE: Put x per year on the horizontal axis and $ on the vertical axis.)

  (1) Accurate total cost lines for each alternative (show line, slopes, and line equations).
  (2) Any relevant breakeven, or cross-over points in terms of costs between the alternatives.
  (3) Ranges of annual production where each alternative is preferred.
  (4) Clearly label your axes and include a *title* for the graph.

2-12   A small company manufactures a certain product. Variable costs are $20 per unit and fixed costs are $10,875. The price-demand relationship for this product is $P = -0.25D+250$. Where P is the unit sales price of the product and D is the annual demand. Answer the questions below, given the data (and helpful hints) that follow:

• Total cost = Fixed cost + Variable cost
• Revenue = Demand x Price
• Profit = Revenue – Total cost
• Set up your graph with:
        - on the y *axis* put Dollars ($), between 0 and $70,000
        - on the x *axis* put Demand [D] (units produced or sold), between 0 and 1,000 units

a. Develop the equations for total cost and total revenue.
b. Find the breakeven quantity (in terms of profit, loss) for the product.
c. What profit would the company obtain by maximizing its total revenue?
d. What is the company's maximum possible profit ?
e. Neatly graph the solutions from **a, b, c** and **d** above.

2-13   A painting operation is performed by a production worker at a labor cost of $1.40 per unit. A robot spray-painting machine, costing $15,000, would reduce the labor cost to $0.20 per unit. If the paint machine would be valueless at the end of three years, what would be the minimum number of units that would have to be painted each year to justify the purchase of the paint machine?

2-14   Company *A* has fixed expenses of $15,000 per year and each unit of product has a $0.002 variable cost. Company *B* has fixed expenses of $5,000 per year and can produce the same product at a $0.05 variable cost. At what number of units of annual production will Company *A* have the same overall cost as Company *B*?

2-15   A firm believes the sales volume (*S*) of its product depends on its unit selling price (*P*), and can be determined from the equation $P = \$100 - S$. The cost (*C*) of producing the product is $1000 + 10S$.

   a.   Draw a graph with the sales volume (*S*) from 0 to 100 on the *x*-axis, and Total Cost and Total Income from 0 to 2500 on the *y*-axis. On the graph draw the line $C = \$1000 + 10S$. Then plot the curve of Total Income [which is sales volume (*S*) x Unit Selling Price ($100 - S$)]. Mark the breakeven points on the graph.
   b.   Determine the breakeven point (lowest sales volume where total sales income just equals total production cost). *Hint:* This may be done by trial and error, or by using the quadratic equation to locate the point where profit is zero.
   c.   Determine the sales volume (*S*) at which the firm's profit is a maximum. *Hint:* Write an equation for profit and solve it by trial and error, or as a minima-maxima calculus problem.

2-16   The Northern Tundra Telephone (NTT) company has received a contract to install emergency phones along a new 100 mile section of the Snow-Moose Turnpike. Fifty emergency phone systems will be installed about 2 miles apart. The material cost of a unit is $125. NTT will need to run underground communication lines which cost NTT $7 500 per mile (including labor) to install. There will also be a one time cost of $10,000 to network these phones into NTT's current communication system. You are asked to develop a cost estimate of the project from NTT's perspective. If NTT adds a profit margin of 35% to its costs how much will it cost the State for this project?

2-17   You and your spouse are planning a second honeymoon to the Cayman Islands this summer and would like to have your house painted while you are away. Estimate the total cost of the paint job from the information given below, where:

$$COST_{TOTAL} = COST_{PAINT} + COST_{LABOR} + COST_{FIXED}$$

PAINT INFORMATION: Your house has a surface area of 6000 square feet. One can of paint can cover 300 square feet. You are estimating the cost to put on two coats of paint for the entire house. The cost per can is given in the table below. Note the incremental decrease in unit cost per can as you purchase more-and-more cans.

Number of Cans Purchased	Cost Per Can For Those Cans
First 10 Cans Purchased	$ 15.00 Per Can
Second 15 Cans Purchased	$ 10.00 Per Can
Up To Next 50 Cans Purchased	$ 7.50 Per Can

LABOR INFORMATION: You plan to hire 5 painters who will paint for 10 hours per day each. You estimate that the job will require 4 and 1/2 days of their painting time. The painter's labor rate is $8.75 per hour.

FIXED COST INFORMATION: There is a fixed cost of $200 per job that the painting company charges to cover travel expenses, clothing, cloths, thinner, administration, etc.

2-18   You are interested in having a mountain cabin built for weekend trips, vacations, to host family, and perhaps eventually to retire in. After further discussing the project with a local contractor you receive an estimate that the total construction cost of your 2000 ft^2 lodge will be $150,000. Costs within each category include labor, material, and overhead items. The percentage of costs for each of several items (categories) is broken down as follows:

Cost Items	Percentage of Total Costs
1. Construction permits, legal, title fees	0.08
2. Roadway, site clearing, preparation	0.15
3. Foundation, concrete, masonry	0.13
4. Wallboard, flooring, carpentry	0.12
5. Heating, ventilation, air conditioning	0.13
6. Electric, plumbing, communications	0.10
7. Roofing, flooring	0.12
8. Painting, finishing	0.17
	1.00

a. What is the cost per square foot of the 2000 ft^2 lodge ?

b. If you are also considering a 4000 ft^2 layout option, estimate your construction costs if:

1. all cost items (in the table) change proportionately to the size increase.

2. cost items 1 and 2 do not change at all, all others are proportionate.

2-19 SungSam Inc. is currently designing a new digital camcorder that is projected to have the following per unit costs to manufacture:

Cost Categories	Unit Costs
Materials costs	$ 112.00
Labor costs	85.00
Overhead costs	213.00
Total Unit Cost	$ 410.00

SungSam adds 30% to its manufacturing cost for corporate profit. Answer the following questions:

a. What unit profit would SungSam realize on each camcorder?

b. What is the overall cost to produce a batch of 10,000 camcorders?

c. What would SungSam's profit be on the batch of 10,000 if historical data shows:

   3% of product will go unsold

   1% of product will be scrapped in manufacturing

   2% of *sold* product will be returned for refund

d. How much can SungSam afford to pay for a contract that would lock in a 50% reduction in the unit material cost previously given? If SungSam does sign the contract the sales price will remain the same as previously.

2-20 Grandma Bell had purchased a set of gold plated silverware 50 years ago for $55.00, which you inherited last year. Unfortunately a house fire at your home destroyed the set. Your insurance company is at a loss to define the replacement cost and has asked your help. You do some research and find that the Aurum Flatware Cost Index (AFCI) for gold plated silverware was 112 when grandma Bell bought her set, and is at 2050 today. Use the AFCI to update the cost of Bell's set to today's cost to show to the insurance company.

2-21   Your boss is the director of reporting for the Athens County Construction Agency (ACCA). It has been his job to track the cost of construction in Athens County. Twenty five years ago he created the ACCA Cost Index to track these costs. Costs during the first year of the index were $12 per square foot of constructed space (the index value was set at 100 for that first year). This past year a survey of contractors revealed that costs were $72 per square foot. What index number will your boss publish in his report for this year? If the index value was 525 last year what was the cost per square foot last year?

2-22   An antique refinisher named Constance has been so successful with her small business that she is planning to expand her shop with all new equipment. She is going to start enlarging her shop by purchasing the equipment given below. What would be the *net* cost to Constance to obtain this equipment — assume that she can trade the old equipment in for 15% of its original cost. Assume also that the relative price to purchase the equipment has not changed over time (that is, there has been no inflation in equipment prices).

Equipment	Original Size	Cost of Original Equipment	Power-Sizing Exponent	New Equipment Size
Varnish Bath	50 gallon	$ 3500	0.80	75 gallon
Power Scraper	¾ hp	$ 250	0.22	1.5 hp
Paint Booth	3 cubic feet	$3000	0.6	12 cubic feet

2-23   Refer to the previous problem and now assume the prices for the equipment that Constance wants to replace have not been constant. Use the cost index data for each piece of equipment to update the costs to the price that would be paid today. Develop the overall cost for Constance, again assuming the 15% trade-in allowance for the old equipment. Use any necessary data from the previous problem.

Original Equipment	Cost Index When Originally Purchased	Cost Index Today
Varnish Bath	154	171
Power Scraper	780	900
Paint Booth	49	76

2-24   A nuclear centrifuge cost $40,000 5 years ago when the relevant cost index was 120. The centrifuge had a capacity of separating 1,500 gallons of ionized solution per hour. Today, it is desired to build a centrifuge with 4,500 gallons per hour capacity, but the cost index now is 300. Assuming a power sizing exponent to reflect economies of scale, *x*, of 0.75, what will be the approximate cost (expressed in today's dollars) of the new reactor using the power sizing model?

2-25 Padre works for a trade magazine that publishes lists of *Power Sizing Exponents (PSE)* that reflect economies-of-scale for developing engineering estimates of various equipment. Padre has been unable to find any published data on the VMIC machine and wants to list its *PSE* value in his next issue. Given the data below (your staff was able to find data regarding costs and sizes of the VMIC machine) calculate the *PSE* value that Padre should publish. NOTE: The VMIC-100 can handle twice the volume of a VMIC-50.

Cost of VMIC-100 today = $100,000   VMIC Equipment Index today = 214

Cost of VMIC-50 5 years ago = $45,000   VMIC Equipment Index 5 years ago = 151

2-26 Develop an estimate for each situation given below:

   a. The cost of a 500 mile automobile trip, if: gasoline is $1.00 per gallon, vehicle wear and tear is $0.08 per mile, and our vehicle gets 20 miles per gallon.

   b. The total number of hours in the average human life, if: the average life is 75 years.

   c. The number of days it takes to travel around the equator using a hot air balloon, if: the balloon averages 100 miles per day, the diameter of the earth is ~4,000 miles [note: circumference = π (diameter)]

   d. The total area in square miles of the United States of America, if: Kansas is an average sized state, Kansas is 390 miles x 200 miles in shape.

2-27 If 200 labor hours were required to produce the 1st unit in a production run, and 60 labor hours were required to produce the 7th unit, what was the *learning curve rate* during production?

2-28 Rose is a project manager at the civil engineering consulting firm Sands, Gravel, Concrete and Waters, Inc. She has been collecting data on a project where concrete pillars were being constructed, however not all of the data are available. She has been able to find out that the 10th pillar required 260 person-hours to construct, and that there was a 75% learning curve that applied. She is interested in calculating the time required to construct the 1st and 20th pillars. Compute the values for her.

2-29 Sally Statistics is implementing a system of statistical process control (SPC) charts in her factory in an effort to reduce the overall cost of scrapped product. The current cost of scrap is $X per month. If an 80% learning curve is expected in the use of the SPC charts to reduce the cost of scrap, what would the *percentage reduction* in monthly scrap cost be after the charts have been implemented for 12 months? *Hint: model each month as a unit of production.*

2-30   Randy Duckout has been asked to develop an estimate of the *per unit selling price* (the price that each unit will be sold for) of a new line of hand-crafted booklets that offer excuses for missed appointments.   His assistant Doc Duckout has collected information that Randy will need in developing his estimate:

Cost of direct labor	$20.00 per hour
Cost of materials	$43.75 per batch of 25 booklets
Cost of overhead items	50% of direct labor cost
Desired profit	20% of total manufacturing cost

Doc also finds out that: (1) they should use a 75% learning curve for estimating the cost of direct labor, (2) the time to complete the 1st booklet is estimated at 0.60 hours, and (3) the estimated time to complete the 25th booklet should be used as their standard time for the purpose of determining the *unit selling price*. What would Randy and Doc's estimate be for the *unit selling price* ?

2-31   On December 1st, Al Smith purchased a car for $18,500. He paid $5000 immediately and agreed to pay three additional payments of $6000 each (which includes principal and interest) at the end of one, two, and three years. Maintenance for the car is projected at $1000 at the end of the first year, and $2000 at the end of each subsequent year. Al expects to sell the car at the end of the fourth year (after paying for the maintenance work) for $7000. Using these facts, prepare a table of cash flows.

2-32   Bonka Toys is considering a robot that will cost $20,000 to buy. After 7 years its salvage value will be $2000. An overhaul costing $5000 will be needed in year 4. O&M costs will be $2500 per year. Draw the cash flow diagram.

2-33   Pine Village needs some additional recreation fields. Construction will cost $225,000 and annual O&M expenses are $85,000. The city council estimates that the value of added youth leagues is about $190,000 annually. In year 6 another $75,000 will be needed to refurbish the fields. The salvage value is estimated to be $100,000 after 10 years. Draw the cash flow diagram.

2-34   Identify your major cash flows for the current school term as first costs, O&M expenses, salvage values, revenues, overhauls, etc. Using a week as the time period, draw the cash flow diagram.

Selected References:

•   Peters, M.S. and Timmerhaus, K.D., *Plant Design and Economics for Chemical Engineers, 4th*, McGraw Hill, 1991.

•   Perry, H.P. and Green, D.W., *Perry's Chemical Engineers' Handbook, 7th*, McGraw Hill, 1997.

CHAPTER **3**

Interest And Equivalence

In the last chapter we saw the full range of the engineering decision-making process. Part of that process includes the prediction of outcomes for each alternative. For many of the situations we examined, the economic consequences of an alternative were immediate, that is, took place either right away or in a very short period of time, as in Example 2-2 (the decision on the design of a concrete aggregate mix) or Ex. 2-3 (the change of manufacturing method). In such relatively simple situations, we total the various positive and negative aspects and quickly reach a decision. But can we do the same if the economic consequences occur over a considerable period of time?

Computing Cash Flows

The installation of an expensive piece of machinery in a plant obviously has economic consequences that occur over an extended period of time. If the machinery were bought on credit, then the simple process of paying for it is one that may take several years. What about the usefulness of the machinery? Certainly it must have been purchased because it would be a beneficial addition to the plant. These favorable consequences may last as long as the equipment performs its useful function. In these circumstances, we do not add up the various consequences; instead, we describe each alternative as cash *receipts* or *disbursements* at different points in *time*. In this way, each alternative is resolved into a set of *cash flows*. This is illustrated by Examples 3-1 and 3-2.

EXAMPLE 3-1
The manager has decided to purchase a new $30,000 mixing machine. The machine may be paid for by one of two ways:

1. Pay the full price now *minus* a 3% discount.

2. Pay $5000 now; at the end of one year, pay $8000; at the end of four subsequent years, pay $6000 per year.

List the alternatives in the form of a table of cash flows.

Solution: In this problem the two alternatives represent different ways to pay for the mixing machine. While the first plan represents a lump sum of $29,100 now, the second one calls for payments continuing until the end of the fifth year. The problem is to convert an alternative into cash receipts or disbursements and show the timing of each receipt or disbursement. The result is called a ***cash flow table*** or, more simply, a set of *cash flows*.

The cash flows for both the alternatives in this problem are very simple. The cash flow table, with disbursements given negative signs, is as follows:

End of Year	Pay in full now	Pay over 5 years
0 (now)	−$29,100	−$5000
1	0	−8000
2	0	−6000
3	0	−6000
4	0	−6000
5	0	−6000

EXAMPLE 3-2

A man borrowed $1000 from a bank at 8% interest. He agreed to repay the loan in two end-of-year payments. At the end of the first year, he will repay half of the $1000 principal amount plus the interest that is due. At the end of the second year, he will repay the remaining half of the principal amount plus the interest for the second year. Compute the borrower's cash flow.

Solution: In engineering economic analysis, we normally refer to the beginning of the first year as "Time 0." At this point the man receives $1000 from the bank. (A positive sign represents a receipt of money and a negative sign, a disbursement.) Thus, at Time 0, the cash flow is + $1000.

At the end of the first year, the man pays 8% interest for the use of $1000 for one year. The interest is 0.08 × $1000 = $80. In addition, he repays half the $1000 loan, or $500. Therefore, the end-of-Year 1 cash flow is −$580.

At the end of the second year, the payment is 8% for the use of the balance of the principal ($500) for the one-year period, or 0.08 × 500 = $40. The $500 principal is also repaid for a total end-of-Year 2 cash flow of −$540. The cash flow is:

End of year	Cash flow
0 (now)	+$1000
1	−580
2	−540

In this chapter, we will demonstrate techniques to compare the value of money at different dates, an ability that is essential to engineering economic analysis. We must be able to compare, for example, a low-cost motor with a higher-cost motor. If there were no other consequences, we would obviously prefer the low-cost one. But if the higher-cost motor were more efficient and thereby reduced the annual electric power cost, we would be faced with the question of whether to spend more money now on the motor to reduce power costs in the future. This chapter will provide the methods for comparing the alternatives to determine which motor is preferred.

Time Value Of Money

We often find that the money consequences of any alternative occur over a substantial period of time—say, a year or more. When money consequences occur in a short period of time, we simply add up the various sums of money and obtain a net result. But can we treat money this same way when the time span is greater?

Which would you prefer, $100 cash today or the assurance of receiving $100 a year from now? You might decide you would prefer the $100 now because that is one way to be certain of receiving it. But suppose you were convinced that you would receive the $100 one year hence. Now what would be your answer? A little thought should convince you that it *still* would be more desirable to receive the $100 now. If you had the money now, rather than a year hence, you would have the use of it for an extra year. And if you had no current use for $100, you could let someone else use it.

Money is quite a valuable asset—so valuable that people are willing to pay to have money available for their use. Money can be rented in roughly the same way one rents an apartment, only with money, the charge for its use is called **interest** instead of rent. The importance of interest is demonstrated by banks and savings institutions continuously offering to pay for the use of people's money, to pay interest.

If the current interest rate is 9% per year, and you put $100 into the bank for one year, how much will you receive back at the end of the year? You will receive your original $100 together with $9 interest, for a total of $109. This example demonstrates the time preference for money: we would rather have $100 today than the assured promise of $100 one year hence; but we might well consider leaving the $100 in a bank if we knew it would be worth $109 one year hence. This is because there is a **time value of money** in the form of the willingness of banks, businesses, and people to pay interest for the use of money.

Simple Interest

Simple interest is interest that is computed on the original sum. Thus if you were to loan a present sum of money P to someone at a simple annual interest rate i (stated as a decimal) for a period of n years, the amount of interest you would receive from the loan would be:

Total interest earned $= P \times i \times n = P i n$ (3-1)

At the end of n years the amount of money due you F would equal the amount of the loan P plus the total interest earned or

   Amount of Money due at the end of the loan

$$F = P + P i n$$

$$F = P(1 + i n)$$ (3-2)

Example 3-3

You have agreed to loan a friend $5000 for five years at a simple interest rate of 8% per year. How much interest will you receive from the loan? How much will your friend pay you at the end of five years?

Solution: Total interest earned $= P i n = (\$5000)(0.08)(5 \text{ yrs}) = \$2000$
Amount due at end of loan $= P + P i n = 5000 + 2000 = \$7000$ ■

In Example 3-3 the interest earned at the end of the first year is $(5000)(0.08)(1) = \$400$, but this money is not paid to the lender until the end of the fifth year. As a result the borrower has the use of the $400 for four years without paying any interest on it. This is how simple interest works, and explains why lenders seldom agree to make simple interest loans.

In actual practice interest is almost always computed by the *compound interest* method. In this method any interest owed, but not paid at the end of the year, is added to the unpaid debt and interest is charged in the next year on both the unpaid debt and unpaid interest. This is what distinguishes compound interest from simple interest. In the next section (and the rest of the book) all the interest computations are based on compound interest.

Repaying a Debt

To better understand the mechanics of interest, let us consider a situation where $5000 is owed and is to be repaid in five years, together with 8% annual interest. There are a great many ways in which debts are repaid; for simplicity, we have selected four specific ways for our example. Table 3-1 tabulates the four plans.

In Plan 1, $1000 will be paid at the end of each year plus the interest due at the end of the year for the use of money to that point. Thus, at the end of the first year, we will have had the use of $5000. The interest owed is 8% × $5000 = $400. The end-of-year payment is, therefore, $1000 principal *plus* $400 interest, for a total payment of $1400. At the end of the second year, another $1000 principal plus interest will be repaid on the money owed during the year. This time the amount owed has declined from $5000 to $4000 because of the $1000 principal payment at the end of the first year. The interest payment is 8% × $4000 = $320, making the end-of-year payment a total of $1320. As indicated in Table 3-1, the series of payments continues each year until the loan is fully repaid at the end of the fifth year.

Plan 2 is another way to repay $5000 in five years with interest at 8%. This time the end-of-year payment is limited to the interest due, with no principal payment. Instead, the $5000 owed is repaid in a lump sum at the end of the fifth year. The end-of-year payment in each of the first four years of Plan 2 is 8% × $5000 = $400. The fifth year, the payment is $400 interest *plus* the $5000 principal, for a total of $5400.

Plan 3 calls for five equal end-of-year payments of $1252 each. At this point, we have not shown how the figure of $1252 was computed (see Ex. 4-3). However, it is clear that there is some equal end-of-year amount that would repay the loan. By following the computations in Table 3-1, we see that this series of five payments of $1252 repays a $5000 debt in five years with interest at 8%.

Plan 4 is still another method of repaying the $5000 debt. In this plan, no payment is made until the end of the fifth year when the loan is completely repaid. Note what happens at the end of the first year: the interest due for the first year—8% × $5000 = $400—is not paid; instead, it is added to the debt. At the second year, then, the debt has increased to $5400. The second year interest is thus 8% × $5400 = $432. This amount, again unpaid, is added to the debt, increasing it further to $5832. At the end of the fifth year, the total sum due has grown to $7347 and is paid at that time.

Note that when the $400 interest was not paid at the end of the first year, it was added to the debt and, in the second year, there was interest charged on this unpaid interest. That is, the $400 of unpaid interest resulted in 8%× $400 = $32 of additional interest charge in the second year. That $32, together with 8%× $5000 = $400 interest on the $5000 original debt, brought the total interest charge at the end of the second year to $432. Charging interest on unpaid interest is called ***compound interest***. We will deal extensively with compound interest calculations later in this chapter.

Table 3-1 FOUR PLANS FOR REPAYMENT OF $5000 IN FIVE YEARS WITH INTEREST AT 8%

(a) Year	(b) Amount owed at beginning of year	(c) Interest owed for that year [8% x (b)]	(d) Total owed at end of year [(b) + (c)]	(e) Principal payment	(f) Total end-of-year payment

Plan 1: At end of each year pay $1000 principal *plus* interest due.

1	$5000	$400	$5400	$1000	$1400
2	4000	320	4320	1000	1320
3	3000	240	3240	1000	1240
4	2000	160	2160	1000	1160
5	1000	80	1080	1000	1080
		$1200		$5000	$6200

Plan 2: Pay interest due at end of each year and principal at end of five years.

1	$5000	$ 400	$5400	$ 0	$400
2	5000	400	5400	0	400
3	5000	400	5400	0	400
4	5000	400	5400	0	400
5	5000	400	5400	5000	5400
		$2000		$5000	$7000

Plan 3: Pay in five equal end-of-year payments.

1	$5000	$400	$5400	$852	$1252*
2	4148	331	4479	921	1252
3	3227	258	3485	994	1252
4	2233	178	2411	1074	1252
5	1159	93	1252	1159	1252
		$1260		$5000	$6260

Plan 4: Pay principal and interest in one payment at end of five years.

1	$5000	$ 400	$5400	$ 0	$ 0
2	5400	432	5832	0	0
3	5832	467	6299	0	0
4	6299	504	6803	0	0
5	6803	544	7347	5000	7347
		$2347		$5000	$7347

*The exact value is $1252.28, which has been rounded to an even dollar amount.

With Table 3-1 we have illustrated four different ways of accomplishing the same task, that is, to repay a debt of $5000 in five years with interest at 8%. Having described the alternatives, we will now use them to present the important concept of *equivalence*.

Equivalence

When we are indifferent as to whether we have a quantity of money now or the assurance of some other sum of money in the future, or series of future sums of money, we say that the present sum of money is ***equivalent*** to the future sum or series of future sums.

If an industrial firm believed 8% was an appropriate interest rate, it would have no particular preference whether it received $5000 now or was repaid by Plan 1 of Table 3-1. Thus $5000 today is equivalent to the series of five end of-year payments. In the same fashion, the industrial firm would accept repayment Plan 2 as equivalent to $5000 now. Logic tells us that if Plan 1 is equivalent to $5000 now and Plan 2 is also equivalent to $5000 now, it must follow that Plan 1 is equivalent to Plan 2. In fact, *all four repayment plans must be equivalent to each other and to $5000 now.*

Equivalence is an essential factor in engineering economic analysis. In Chapter 2, we saw how an alternative could be represented by a cash flow table. How might two alternatives with different cash flows be compared? For example, consider the cash flows for Plans 1 and 2:

Year	Plan 1	Plan 2
1	-$1400	-$400
2	-1320	-400
3	-1240	-400
4	-1160	-400
5	-1080	-5400
	-$6200	-$7000

If you were given your choice between the two alternatives, which one would you choose? Obviously the two plans have cash flows that are different. Plan 1 requires that there be larger payments in the first four years, but the total payments are smaller than the sum of Plan 2's payments. To make a decision, the cash flows must be altered so that they can be compared. The ***technique of equivalence*** is the way we accomplish this.

Using mathematical manipulation, we can determine an equivalent value at some point in time for Plan 1 and a ***comparable equivalent value*** for Plan 2, based on a selected interest rate. Then we can judge the relative attractiveness of the two alternatives, not from their cash flows, but from comparable equivalent values. Since Plan 1 and Plan 2 each repay a *present* sum of $5000 with interest at 8%, they both are equivalent to $5000 *now*; therefore, the alternatives are equally attractive. This cannot be deduced from the given cash flows alone.

It is necessary to learn this by determining the equivalent values for each alternative at some point in time, which in this case is "the present."

Difference in Repayment Plans

The four plans computed in Table 3-1 are equivalent in nature but different in structure. Table 3-2 repeats the end-of-year payment schedule from the previous table. In addition, each plan is graphed to show the debt still owed at any point in time. Since $5000 was borrowed at the beginning of the first year, all the graphs begin at that point. We see, however, that the four plans result in quite different situations on the amount of money owed at any other point in time. In Plans 1 and 3, the money owed declines as time passes. With Plan 2 the debt remains constant, while Plan 4 increases the debt until the end of the fifth year. These graphs show an important difference among the repayment plans—the areas under the curves differ greatly. Since the axes are *Money Owed* and *Time in Years*, the area is their product: Money owed *times* Time in years.

In the discussion of the time value of money, we saw that the use of money over a time period was valuable, that people are willing to pay interest to have the use of money for periods of time. When people borrow money, they are acquiring the use of money as represented by the area under the Money owed vs. Time in years curve. It follows that, at a given interest rate, the amount of interest to be paid will be proportional to the area under the curve. Since in each case the $5000 loan is repaid, the interest for each plan is the Total *minus* the $5000 principal:

Plan	Total interest paid
1	$1200
2	2000
3	1260
4	2347

Using Table 3-2 and the data from Table 3-1, we can compute the area under each of the four curves, that is, the area bounded by the abscissa, the ordinate, and the curve itself. We multiply the ordinate (Money owed) *times* the abscissa (1 year) for each of the five years, then *add*:

Area under curve = (Money owed in Year 1)(1 year)
+ (Money owed in Year 2)(1 year)
+ ...
+ (Money owed in Year 5)(1 year)

or,

Area under curve [(Money owed)(Time)] = ***Dollar-Years***

Table 3-2 END-OF-YEAR PAYMENT SCHEDULES AND THEIR GRAPHS

From Table 3-1: *"Four Plans for Repayment of $5000 in Five Years at 8% Interest"*

Plan 1: At end of each year pay $1000 principal
plus interest due.

Year	End-of-year payment
1	$1400
2	1320
3	1240
4	1160
5	1080
	$6200

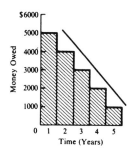

Plan 2: Pay interest due at end of each year and
principal at end of five years.

Year	End-of-year payment
1	$ 400
2	400
3	400
4	400
5	5400
	$7000

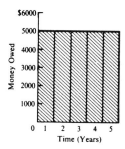

Plan 3: Pay in five equal end-of-year payments.

Year	End-of-year payment
1	$1252
2	1252
3	1252
4	1252
5	1252
	$6260

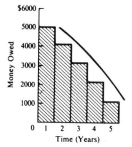

Plan 4: Pay principal and interest in one payment
at end of five years.

Year	End-of-year payment
1	$ 0
2	0
3	0
4	0
5	7347
	$7347

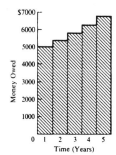

The dollar-years for the four plans would be as follows:

	Dollar-Years			
	Plan 1	*Plan 2*	*Plan 3*	*Plan 4*
(Money owed in Year 1)(1 year)	$ 5,000	$ 5,000	$ 5,000	$ 5,000
(Money owed in Year 2)(1 year)	4,000	5,000	4,148	5,400
(Money owed in Year 3)(1 year)	3,000	5,000	3,227	5,832
(Money owed in Year 4)(1 year)	2,000	5,000	2,233	6,299
(Money owed in Year 5)(1 year)	1,000	5,000	1,159	6,803
Total Dollar-Years	$15,000	$25,000	$15,767	$29,334

With the area under each curve computed in dollar-years, the ratio of total interest paid to area under the curve may be obtained:

Plan	*Total interest paid, dollars*	*Area under curve, dollar-years*	*Ratio:*	$\frac{\textit{Total interest paid}}{\textit{Area under curve}}$
1	$1200	$15,000		0.08
2	2000	25,000		0.08
3	1260	15,767		0.08
4	2347	29,334		0.08

We see that the ratio of total interest paid to area under the curve is constant and equal to 8%. Stated another way, the total interest paid equals the interest rate *times* the area under the curve.

From our calculations, we more easily see why the repayment plans require the payment of different total sums of money, yet are actually equivalent to each other. The key factor is that the four repayment plans provide the borrower with different quantities of dollar-years. Since dollar-years *times* interest rate equals the interest charge, the four plans result in different total interest charges.

Equivalence is Dependent on Interest Rate

In the example of Plans 1–4, all calculations were made at an 8% interest rate. At this interest rate, it has been shown that all four plans are equivalent to a present sum of $5000. But what would happen if we were to change the problem by changing the interest rate?

If the interest rate were increased to 9%, we know that the required interest payment for each plan would increase, and the calculated repayment schedules—Table 3-1, Column (f)—could no longer repay the $5000 debt with the higher interest. Instead, each plan would repay a sum *less* than the principal of $5000, because more money would have to be used to repay the higher interest rate. By some calculations (that will be explained in this chapter and the next), the equivalent present sum that each plan will repay at 9% interest is:

Plan	Repay a present sum of
1	$4877
2	4806
3	4870
4	4775

As predicted, at the higher 9% interest the repayment plans of Table 3-1 each repay a present sum less than $5000. But they do not repay the *same* present sum. Plan 1 would repay $4877 with 9% interest, while Plan 2 would repay $4806. Thus, with interest at 9%, Plans 1 and 2 are no longer equivalent, for they will not repay the same present sum. The two series of payments (Plan 1 and Plan 2) were equivalent at 8%, but not at 9%. This leads to the conclusion ***that equivalence is dependent on the interest rate***. Changing the interest rate destroys the equivalence between two series of payments.

Could we create revised repayment schemes that would be equivalent to $5000 now with interest at 9%? Yes, of course we could: to revise Plan 1 of Table 3-1, we need to increase the total end-of-year payment in order to pay 9% interest on the outstanding debt.

Year	Amount owed at beginning of year	9% interest for year	Total end-of-year payment ($1000 plus interest)
1	$5000	$450	$1450
2	4000	360	1360
3	3000	270	1270
4	2000	180	1180
5	1000	90	1090

Plan 2 of Table 3-1 is revised for 9% interest by increasing the first four payments to 9% × $5000 = $450 and the final payment to $5450. Two plans that repay $5000 in five years with interest at 9% are:

	Revised end-of-year payments	
Year	Plan 1	Plan 2
1	$1450	$ 450
2	1360	450
3	1270	450
4	1180	450
5	1090	5450

We have determined that Revised Plan 1 is equivalent to a present sum of $5000 and Revised Plan 2 is equivalent to $5000 now; it follows that at 9% interest, Revised Plan 1 is equivalent to Revised Plan 2.

Application of Equivalence Calculations

To understand the usefulness of equivalence calculations, consider the following:

Year	*Alternative A:* *lower initial cost,* *higher operating cost*	*Alternative B:* *higher initial cost,* *lower operating cost*
0 (now)	-$600	-$850
1	-115	-80
2	-115	-80
3	-115	-80
.	.	.
.	.	.
.	.	.
10	-115	-80

Is the least cost alternative the one that has the lower initial cost and higher operating costs or the one with higher initial cost and lower continuing costs? Because of the time value of money, one cannot add up sums of money at different points in time directly. This means that a comparison between alternatives cannot be made in actual dollars at different points in time, but must be made in some equivalent comparable sums of money.

It is not sufficient to compare the initial $600 against $850. Instead, we must compute a value that represents the entire stream of payments. In other words, we want to determine a sum that is equivalent to Alternative *A*'s cash flow; similarly, we need to compute the equivalent present sum for Alternative *B*. By computing equivalent sums at the same point in time ("now"), we will have values that may be validly compared. The methods for accomplishing this will be presented later in this chapter and chapter 4.

Thus far we have discussed computing equivalent present sums for a cash flow. But the technique of equivalence is not limited to a present computation. Instead, we could compute the equivalent sum for a cash flow at any point in time. We could compare alternatives in "Equivalent Year 10" dollars rather than "now" (Year 0) dollars. Further, the equivalence need not be a single sum, but could be a series of payments or receipts. In Plan 3 of Table 3-1, we had a situation where the series of equal payments was equivalent to $5000 now. But the equivalency works both ways: if we ask the question, "What is the equivalent equal annual payment continuing for five years, given a present sum of $5000 and interest at 8%?" The answer is "$1252."

Single Payment Compound Interest Formulas

To facilitate equivalence computations, a series of **interest formulas** will be derived. To simplify the presentation, we'll use the following notation:

i = *Interest rate per interest period.* In the equations the interest rate is stated as a decimal (that is, 9% interest is 0.09).

n = *Number of interest periods.*

P = *A present sum of money.*

F = *A future sum of money.* The future sum F is an amount, n interest periods from the present, that is equivalent to P with interest rate i.

Suppose a present sum of money P is invested for one year[1] at interest rate i. At the end of the year, we should receive back our initial investment P, together with interest equal to iP, or a total amount $P + iP$. Factoring P, the sum at the end of one year is $P(1 + i)$.

Let us assume that, instead of removing our investment at the end of one year, we agree to let it remain for another year. How much would our investment be worth at the end of the second year? The end-of-first-year sum $P(1 + i)$ will draw interest in the second year of $iP(1 + i)$. This means that, at the end of the second year. the total investment will become

$$P(1 + i) + iP(1 + i)$$

This may be rearranged by factoring $P(1 + i)$, which gives us

$$P(1 + i)(1 + i)$$
$$\text{or } P(1 + i)^2.$$

If the process is continued for a third year, the end of the third year total amount will be $P(1 + i)^3$; at the end of n years, it will be $P(1 + i)^n$. The progression looks like this:

	Amount at beginning of interest period	+	*Interest for period*	= *Amount at end of interest period*
First year	P		$+ iP$	$= P(1 + i)$
Second year	$P(1 + i)$		$+ iP(1 + i)$	$= P(1 + i)^2$
Third year	$P(1 + i)^2$		$+ iP(1 + i)^2$	$= P(1 + i)^3$
nth year	$P(i + i)^{n-1}$		$+ iP(1 + i)^{n-1}$	$= P(1 + i)^n$

---

[1] A more general statement is to specify "one interest period" rather than "one year." It is easier to visualize one year so the derivation will assume that one year is the interest period.

In other words, a present sum P increases in n periods to $P(1 + i)^n$. We therefore have a relationship between a present sum P and its equivalent future sum, F.

   Future sum = (present sum)$(1 + i)^n$

$$F = P(1 + i)^n \qquad\qquad (3\text{-}3)$$

This is the *single payment compound amount formula* and is written in functional notation as

$$F = P(F/P, i, n) \qquad\qquad (3\text{-}4)$$

The notation within the parenthesis (*F/P,i,n*) can be read as:
   To find a future sum *F*, given a present sum, *P*, at an interest rate *i* per interest period, and *n* interest periods hence.

Functional notation is designed so that the compound interest factors may be written in an equation in an algebraically correct form. In the equation above, for example, the functional notation is interpreted as

$$F = \cancel{P}\left(\frac{F}{\cancel{P}}\right)$$

which is dimensionally correct. Without proceeding further, we can see that when we derive a compound interest factor to find a present sum P, given a future sum F, the factor will be $(P/F, i, n)$; so, the resulting equation would be

$$P = F(P/F, i, n)$$

which is dimensionally correct.

EXAMPLE 3-4

If $500 were deposited in a bank savings account, how much would be in the account three years hence if the bank paid 6% interest compounded annually?

   We can draw a diagram of the problem. Note: To have a consistent notation, we will represent *receipts* by upward arrows (and positive signs), and *disbursements* (or payments) will have downward arrows (and negative signs).

Solution: From the viewpoint of the person depositing the $500, the diagram is:

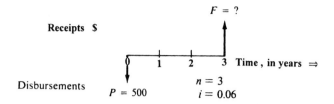

We need to identify the various elements of the equation. The present sum P is $500. The interest rate per interest period is 6%, and in three years there are three interest periods. The future sum F is to be computed.

$P = \$500$ $\qquad$ $i = 0.06$ $\qquad$ $n = 3$ $\qquad$ $F =$ unknown

$F = P(1 + i)^n = 500(1 + 0.06)^3 = \$595.50$

If we deposit $500 in the bank now at 6% interest, there will be $595.50 in the account in three years.

Alternate Solution: The equation $F = P(1 + i)^n$ need not be solved with a hand calculator. Instead, *the single payment compound amount factor,* $(1 + i)^n$, is readily determined from computed tables. The factor is written in convenient notation as

$(1 + i)^n = (F/P, i, n)$

and in functional notation as

$(F/P, 6\%, 3)$

Knowing $n = 3$, locate the proper row in the 6% table. (*Note:* **Compound Interest Tables** appear in the tinted pages in this volume. Each table is computed for a particular value of i.) Read in the first column, which is headed "Single Payment, Compound Amount Factor," 1.191. Thus,

$F = 500(F/P, 6\%, 3) = 500(1.191) = \$595.50$

Before leaving this problem, let's draw another diagram of it, this time from the bank's point of view.

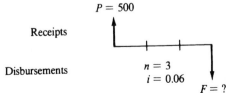

This indicates the bank receives $500 now and must make a disbursement of F at the end of three years. The computation, from the bank's point of view, is

$F = 500(F/P,6\%,3) = 500(1.191) = \$595.50$

This is exactly the same as was computed from the depositor's viewpoint, since this is just the other side of the same transaction. The bank's future disbursement equals the depositor's future receipt. ∎

If we take $F = P(1 + i)^n$ and solve for P, then

$$P = F \frac{1}{(1+i)^n} = F(1+i)^{-n}$$

This is the **single payment present worth formula.** The equation

$$P = F(1 + i)^{-n} \qquad\qquad (3\text{-}5)$$

in our notation becomes

$$P = F(P/F, i, n) \qquad\qquad (3\text{-}6)$$

EXAMPLE 3-5

If you wished to have $800 in a savings account at the end of four years, and 5% interest was paid annually, how much should you put into the savings account now?

Solution: $F = \$800$ $i = 0.05$ $n = 4$ $P = $ unknown

$P = F(1 + i)^{-n} = 800(1 + 0.05)^{-4} = 800(0.8227) = \$658.16$

To have $800 in the savings account at the end of four years, we must deposit $658.16 now.

Alternate Solution:

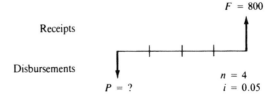

Receipts

Disbursements

$P = ?$

$F = 800$

$n = 4$
$i = 0.05$

$P = F(P/F, i, n) = \$800\ (P/F, 5\%, 4)$

From the Compound Interest Tables,

$(P/F, 5\%, 4) = 0.8227$

$P = \$800(0.8227) = \$658.16$ ■

Here the problem has an exact answer. In many situations, however, the answer is rounded off, recognizing that it can only be as accurate as the input information upon which it is based.

EXAMPLE 3-6

Suppose the bank changed their interest policy in Ex. 3-4 to "6% interest, compounded quarterly." For this situation a $500 deposit now would result in how much money in the account at the end of three years?

Solution: First, we must be certain to understand the meaning of *6% interest, compounded quarterly.* There are two elements:

1. *6% interest:* Unless otherwise described, it is customary to assume the stated interest is for a one-year period. *If the stated interest is for other than a one-year period, then it must be explained.*

2. *Compounded quarterly:* This indicates there are four interest periods per year; that is, an interest period is three months long.

We know that the 6% interest is an annual rate, because if it were anything different, it would have been stated. Since we are dealing with four interest periods per year, it follows that the interest rate per interest period is 1½ %. For the total three-year duration, there are twelve interest periods.

$P = \$500 \qquad i = 0.015 \qquad n = (4 \times 3) = 12 \qquad F = \text{unknown}$

$F = P(1 + i)^n = P(F/P, i, n)$

$\qquad = \$500(1 + 0.015)^{12} = \$500(F/P, 1½\%, 12)$

$\qquad = \$500(1.196) = \$598.00$

A $500 deposit now would yield $598.00 in three years. ■

EXAMPLE 3-7

Consider the following situation:

Year	Cashflow
0	+P
1	0
2	0
3	-400
4	0
5	-600

Solve for *P* assuming a 12% interest rate and using the Compound Interest Tables. Recall that receipts have a plus sign and disbursements or payments have a negative sign. Thus, the diagram is:

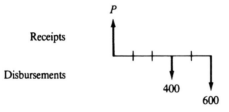

Solution:

$$P = 400(P/F,12\%,3) + 600(P/F,12\%,5)$$

$$= 400(0.7118) + 600(0.5674)$$

$$= \$625.16$$

It is important to understand just what the solution, $625.16, represents. We can say that $625.16 is the amount of money that would need to be invested at 12% annual interest to allow for the withdrawal of $400 at the end of three years and $600 at the end of five years.

Let's examine the computations further.

If $625.16 is invested for one year at 12% interest, it will increase to $[625.16 + 0.12(625.16)] = \$700.18$. If for the second year the $700.18 is invested at 12%, it will increase to $[700.18 + 0.12(700.18)] = \$784.20$. And if this is repeated for another year, $[784.20 + 0.12(784.20)] = \$878.30$.

We are now at the end of Year 3. The original $625.16 has increased through the addition of interest to $878.30. It is at this point that the $400 is paid out. Deducting $400 from $878.30 leaves $478.30.

The $478.30 can be invested at 12% for the fourth year and will increase to $[478.30 + 0.12(478.30)] = \$535.70$. And if left at interest for another year, it will increase to $[535.70 + 0.12(535.70)] = \$600$. We are now at the end of Year 5; with a $600 payout, there is no money remaining in the account.

In other words, the $625.16 was just enough money, at a 12% interest rate, to exactly provide for a $400 disbursement at the end of Year 3 and also a $600 disbursement at the end of Year 5. We neither end up short of money, nor with money left over: this is an illustration of equivalence. The initial $625.16 is *equivalent* to the combination of a $400 disbursement at the end of Year 3 and a $600 disbursement at the end of Year 5.

Alternate Formation of Example 3-7: There is another way to see what the $625.16 value of P represents.

Suppose at Year 0 you were offered a piece of paper that guaranteed you would be paid $400 at the end of three years and $600 at the end of five years. How much would you be willing to pay for this piece of paper if you wanted your money to produce a 12% interest rate?

This alternate statement of the problem changes the signs in the cash flow and the diagram:

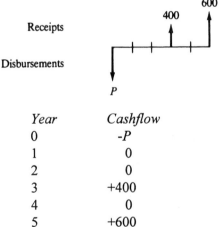

Year	Cashflow
0	-P
1	0
2	0
3	+400
4	0
5	+600

Since the goal is to recover our initial investment P together with 12% interest per year, we can see that P must be *less* than the total amount to be received in the future (that is, $400 + 600 = \$1000$). We must calculate the present sum P that is *equivalent,* at 12% interest, to an aggregate of $400 three years hence and $600 five years hence.

Since we previously derived the relationship
$$P = (1 + i)^{-n}$$
then $P = 400(1 + 0.12)^{-3} + 600(1 + 0.12)^{-5}$
$$= \$625.17$$

This is virtually the same figure as was computed from the first statement of this example. [The slight difference is due to the rounding in the Compound Interest Tables. For example, $(1+0.12)^{-5} = 0.567427$, but the Compound Interest Table shows 0.5674.] ∎

Both problems in Example 3-7 have been solved by computing the value of P that is equivalent to $400 at the end of Year 3 and $600 at the end of Year 5. In the first problem, we received $+P$ at Year 0 and were obligated to pay out the $400 and $600 in later years. In the second (alternate formation) problem, the reverse was true. We paid $-P$ at Year 0 and would receive the $400 and $600 sums in later years. In fact, the two problems could represent the buyer and seller of the same piece of paper. The seller would receive $+P$ at Year 0 while the buyer would pay $-P$. Thus, while the problems looked different, they could have been one situation examined from the viewpoint first of the seller and subsequently of that of the buyer. Either way, the solution is based on an equivalence computation.

EXAMPLE 3-8

One of the cash flows in Example 3-7 was

Year	Cashflow
0	-P
1	0
2	0
3	+400
4	0
5	+600

At a 12% interest rate, P was computed to be $625.17. Suppose the interest rate is increased to 15%. Will the value of P be larger or smaller?

Solution: One can consider P as sum of money invested at 15% from which one is to obtain $400 at the end of 3 years and $600 at the end of 5 years. At 12%, the required P is $627.17. At 15%, P will earn more interest each year, indicating that we can begin with a *smaller P* and still accumulate enough money for the subsequent cash flows. The computation is:

$$P = 400(P/F,15\%,3) + 600(P/F,15\%,5)$$

$$= 400(0.6575) + 600(0.4972)$$

$$= \$561.32$$

The value of P is smaller at 15% than at 12% interest. ∎

Summary

In this chapter, cash flow tables, time value of money and equivalence are described in detail. Also, the single payment compound interest formulas are derived. It is essential that these concepts and the use of the interest formulas be carefully understood, as the remainder of this book is based upon them.

Time Value of Money. The continuing offer of banks to pay interest for the temporary use of other people's money is ample proof that there is a time value of money. Thus, we would always choose to receive $100 today rather than the promise of $100 to be paid at a future date.

Equivalence. What sum would a person be willing to accept a year hence instead of $100 today? If a 9% interest rate is considered to be appropriate, he would require $109 a year hence. If $100 today and $109 a year hence are considered equally desirable, we say the two sums of money are equivalent. But, if on further consideration, we decided that a 12% interest rate is applicable, then $109 a year hence would no longer be equivalent to $100 today. This illustrates that equivalence is dependent on the interest rate.

Single Payment Formulas.

Compound amount:

$$F = P (1+ i)^n = P (F/P,i,n)$$

Present worth:

$$P = F (1+ i)^{-n} = F (P/F,i,n)$$

where:

i = Interest rate per interest period (stated as a decimal).

n = Number of interest periods.

P = A present sum of money.

F = A future sum of money. The future sum F is an amount, n interest periods from the present, that is equivalent to P with interest rate i.

Problems

3-1 A woman borrowed $2000 and agreed to repay it at the end of three years, together with 10% simple interest per year. How much will she pay three years hence?

3-2 A $5000 loan was to be repaid with 8% simple annual interest. A total of $5350 was paid. How long was the loan outstanding?

3-3 Solve the diagram below for the unknown Q assuming a 10% interest rate.

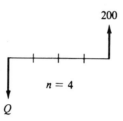

(Answer: Q = $136.60)

3-4 The following series of payments will repay a present sum of $5000 at an 8% interest rate. Using single payment factors, what present sum is equivalent to this series of payments at a 10% interest rate?

Year	End-of-year payment
1	$1400
2	1320
3	1240
4	1160
5	1080

3-5 A man went to his bank and borrowed $750. He agreed to repay the sum at the end of three years, together with the interest at 8% per year, compounded annually. How much will he owe the bank at the end of three years? (*Answer:* $945)

3-6 What sum of money now is equivalent to $8250 two years hence, if interest is 8% per annum, compounded semi-annually? (*Answer:* $7052)

3-7 The local bank offers to pay 5% interest, compounded annually, on savings deposits. In a nearby town, the bank pays 5% interest, compounded quarterly. A man who has $3000 to put in a savings account wonders if the increased interest paid in the nearby town justifies driving his car there to make the deposit. Assuming he will leave all money in the account for two years, how much additional interest would he obtain from the out-of-town bank over the local bank?

3-8 A sum of money invested at 4% interest, compounded semi-annually, will double in amount in approximately how many years? (*Answer:* 17½ years)

3-9 The Apex Company sold a water softener to Marty Smith. The price of the unit was $350. Marty asked for a deferred payment plan, and a contract was written. Under the contract, the buyer could delay paying for the water softener provided that he purchased the coarse salt for re-charging the softener from Apex. At the end of two years, the buyer was to pay for the unit in a lump sum, including 6% interest, compounded quarterly. The contract provided that, if the customer ceased buying salt from Apex at any time prior to two years, the full payment due at the end of two years would automatically become due.

Six months later, Marty decided to buy salt elsewhere and stopped buying from Apex, who thereupon asked for the full payment that was to have been due 18 months hence. Marty was unhappy about this, so Apex offered as an alternative to accept the $350 with interest at 20% per annum compounded semiannually for the six months that Marty had the water softener. Which of these alternatives should Marty accept? Explain.

3-10 Linda Dunlop will deposit $1500 in a bank savings account that pays 10% interest per year, compounded daily. How much will Linda have in her account at the end of two and one-half years?

3-11 For the following cash flow, compute the interest rate at which the $100 cost is equivalent to the subsequent benefits.

Year	Cash Flow
0	-$100
1	+25
2	+45
3	+45
4	+30

3-12 The United States recently purchased $1 billion of 30-year zero-coupon bonds from a struggling foreign nation. The bonds yield 4½% per year interest. The zero coupon bonds pay no interest during their 30-year life. Instead, at the end of 30 years, the U. S. Government is to receive back its $1 billion together with interest at 4½% per year. A U. S. Senator objected to the purchase, claiming that the correct interest rate for bonds like this is 5¼%. The result, he said, was a multimillion dollar gift to the foreign country without the approval of the U. S. Congress. Assuming the Senator's 5¼% interest rate is correct, how much will the foreign country have saved in interest when they repay the bonds at the end of 30 years?

3-13 One thousand dollars is borrowed for one year at an interest rate of 1% per month compounded monthly. If this same sum of money could be borrowed for the same period at an interest rate of 12% per year compounded annually, how much could be saved in interest charges?

3-14 Given a sum of money Q that will be received six years from now. At 5% annual interest the present worth now of Q is $60. At this same interest rate, what would be the value of Q ten years from now?

3-15 In 1995 an anonymous private collector purchased a painting by Picasso entitled *Angel Fernandez de Soto* for $29,152,000. The picture depicts Picasso's friend deSoto seated in a Barcelona cafe drinking absinthe. The painting was done in 1903 and valued then at $600. If the painting was owned by the same family until its sale in 1995, what rate of return did they receive on the $600 investment?

3-16 *a.* If $100 at time "0" will be worth $110 a year hence and was $90 a year ago, compute the interest rate for the past year and the interest rate next year.

      b. Assume $90 invested a year ago will return $110 a year from now. What is the annual interest rate in this situation?

3-17 How much must you invest now at 7.9% interest to accumulate $175,000 in 63 years?

3-18 We know that a certain piece of equipment will cost $150,000 in five years. How much will it cost today using 10% interest?

3-19 The local garbage company charges $6.00 a month for garbage collection. It had been their practice to send out bills to their 100,000 customers at the end of each two-month period. Thus, they would send a bill to each customer for $12 at the end of February for garbage collection during January and February.

Recently the firm changed its billing date so now they send out the two-month bills after one month's service has been performed. Bills for January–February, for example, are sent out at the end of January. The local newspaper claims the firm is receiving half their money before they do the garbage collection. This unearned money, the newspaper says, could be temporarily invested for one month at 1% per month interest by the garbage company to earn extra income.

Compute how much extra income the garbage company could earn each year if it invests the money as described by the newspaper. (*Answer:* $36,000)

3-20 Sally Stanford is buying an automobile that costs $12,000. She will pay $2000 immediately and the remaining $10,000 in four annual end-of-year principal payments of $2500 each. In addition to the $2500, she must pay 15% interest on the unpaid balance of the loan each year. Prepare a cash flow table to represent this situation.

CHAPTER 4

More Interest Formulas

In the last chapter the fundamental components of engineering economic analysis were presented, including formulas to compute equivalent single sums of money at different points in time. Most problems we will encounter are much more complex. Thus this chapter develops formulas for payments that are a uniform series, or are increasing on an arithmetic or geometric gradient. Later in the chapter nominal and effective interest are discussed. Finally, equations are derived for situations where interest is continuously compounded.

Uniform Series Compound Interest Formulas

Many times we will find situations where there are a uniform series of receipts or disbursements. Automobile loans, house payments, and many other loans are based on a *uniform payment series*. It will often be convenient to use tables based on a uniform series of receipts or disbursements. The series A is defined:

> A = An end-of-period[1] cash receipt or disbursement in a uniform series, continuing for n periods, the entire series equivalent to P or F at interest rate i.

The horizontal line in Figure 4-1 is a representation of time with four interest periods illustrated. Uniform payments A have been placed at the end of each interest period, and there are as many A's as there are interest periods n. (Both of these conditions are specified in the definition of A.)

---

[1] In textbooks on economic analysis, it is customary to define A as an end-of-period event rather than beginning-of-period or, possibly, middle-of-period. The derivations that follow are based on this end-of-period assumption. One could, of course, derive other equations based on beginning-of-period or mid-period assumptions.

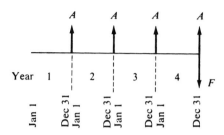

Figure 4-1 The general relationship between A and F.

In the previous section on single payment formulas, we saw that a sum P at one point in time would increase to a sum F in n periods, according to the equation

$$F = P(1 + i)^n$$

We will use this relationship in our uniform series derivation.

Looking at Fig. 4-1, we see that if an amount A is invested at the end of each year for n years, the total amount F at the end of n years will be the sum of the compound amounts of the individual investments.

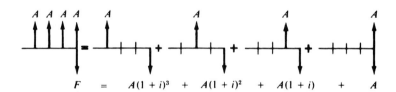

In the general case for n years,

$$F = A(1 + i)^{n-1} + \ldots + A(1 + i)^3 + A(1 + i)^2 + A(1 + i) + A \tag{4-1}$$

Multiplying Eq. 4-1 by $(1 + i)$,

$$(1 + i)F = A(1 + i)^n + \ldots + A(1 + i)^4$$
$$+ A(1 + i)^3 + A(1 + i)^2 + A(1 + i) \tag{4-2}$$

Factoring out A and subtracting Eq. 4-1 gives

$$(1+i)F = A\left[(1+i)^n + \ldots + (1+i)^4 + (1+i)^3 + (1+i)^2 + (1+i)\right]$$

(4-3)

$$- \quad F = A\left[(1+i)^{n-1} + \ldots + (1+i)^3 + (1+i)^2 + (1+i) + 1\right]$$

$$iF = A\left[(1+i)^n - 1\right]$$

(4-4)

Solving Equation 4-4 for F,

$$F = A\left[\frac{(1+i)^n - 1}{i}\right] = A(F/A, i\%, n)$$

(4-5)

Thus we have an equation for F when A is known. The term within the brackets

$$\left[\frac{(1+i)^n - 1}{i}\right]$$

is called the **uniform series compound amount factor** and has the notation **(F/A,i,n)**

EXAMPLE 4-1

A man deposits $500 in a credit union at the end of each year for five years. The credit union pays 5% interest, compounded annually. At the end of five years, immediately following his fifth deposit, how much will he have in his account?

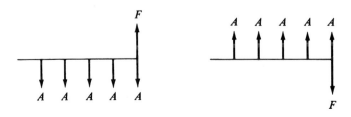

Solution: The diagram on the left shows the situation from the man's point of view; the one on the right, from the credit union's point of view. Either way, the diagram of the five deposits and the desired computation of the future sum F duplicates the situation for the uniform series compound amount formula.

$$F = A\left[\frac{(1+i)^n - 1}{i}\right] = A(F/A, i\%, n)$$

where $A = \$500$, $n = 5$, $i = 0.05$, $F =$ unknown. Filling in the known variables,

$$F = \$500(F/A, 5\%, 5) = \$500(5.526) = \$2763$$

He will have $2763 in his account following the fifth deposit.   ■

If Eq. 4-5 is solved for A, we have

$$A = F\left[\frac{i}{(1+i)^n - 1}\right] = F(A/F, i\%, n)$$

(4-6)

where

$$\left[\frac{i}{(1+i)^n - 1}\right]$$

is called the ***uniform series sinking fund*** [2] factor and is written as *(A/F,i,n)*.

EXAMPLE 4-2

Jim Hayes read that in the western United States, a ten-acre parcel of land could be purchased for $1000 cash. Jim decided to save a uniform amount at the end of each month so that he would have the required $1000 at the end of one year. The local credit union pays 6% interest, compounded monthly. How much would Jim have to deposit each month?

Solution: In this example,

$$F = \$1000 \qquad n = 12 \qquad i = \tfrac{1}{2}\% \qquad A = \text{unknown}$$

$$A = 1000(A/F, \tfrac{1}{2}\%, 12) = 1000(0.0811) = \$81.10$$

Jim would have to deposit $81.10 each month.   ■

---

[2] A *sinking fund* is a separate fund into which one makes a uniform series of money deposits (A) with the goal of accumulating some desired future sum (F) at a given future point in time.

If we use the sinking fund formula (Eq. 4-6) and substitute for F the single payment compound amount formula (Eq. 3-3), we obtain

$$A = F\left[\frac{i}{(1+i)^n - 1}\right] = P(1+i)^n\left[\frac{i}{(1+i)^n - 1}\right]$$

$$A = P\left[\frac{i(1+i)^n}{(1+i)^n - 1}\right] = P(A/P, i\%, n) \tag{4-7}$$

We now have an equation for determining the value of a series of end-of-period payments—or disbursements—A when the present sum P is known.

The portion within the brackets,

$$\left[\frac{i(1+i)^n}{(1+i)^n - 1}\right]$$

is called the **uniform series capital recovery factor** and has the notation **(A/P,i,n)**.

EXAMPLE 4-3

On January 1 a man deposits $5000 in a credit union that pays 8% interest, compounded annually. He wishes to withdraw all the money in five equal end-of-year sums, beginning December 31st of the first year. How much should he withdraw each year?

Solution:

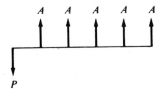

$$P = \$5000 \qquad n = 5 \qquad i = 8\% \qquad A = \text{unknown}$$

$$A = P(A/P, 8\%, 5) = 5000(0.2505) = \$1252$$

The annual withdrawal is $1252. ■

In the example, with interest at 8%, a present sum of $5000 is equivalent to five equal end-of-period disbursements of $1252. This is another way of stating Plan 3 of Table 3-1. The method for determining the annual payment that would repay $5000 in five years with 8% interest has now been explained. The calculation is simply

$$A = 5000(A/P,8\%,5) = 5000(0.2505) = \$1252$$

If the capital recovery formula (Eq. 4-7) is solved for the Present sum P, we obtain the uniform series present worth formula.

$$P = A\left[\frac{(1+i)^n - 1}{i(1+i)^n}\right] = A(P/A, i\%, n) \tag{4-8}$$

and

$$(P/A, i\%, n) = \left[\frac{(1+i)^n - 1}{i(1+i)^n}\right]$$

which is the ***uniform series present worth factor***.

EXAMPLE 4-4

An investor holds a time payment purchase contract on some machine tools. The contract calls for the payment of $140 at the end of each month for a five year period. The first payment is due in one month. He offers to sell you the contract for $6800 cash today. If you otherwise can make 1% per month on your money, would you accept or reject the investor's offer?

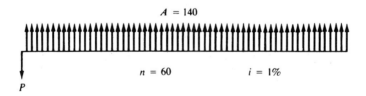

$$A = 140$$

$$n = 60 \qquad i = 1\%$$

$$P$$

Solution: In this problem we are being offered a contract that will pay $140 per month for 60 months. We must determine whether the contract is worth $6800, if we consider 1% per month to be a suitable interest rate. Using the uniform series present worth formula, we will compute the present worth of the contract.

$$P = A(P/A,i,n) = 140(P/A,1\%,60) = 140(44.955)$$

$$= \$6293.70$$

It is clear that if we pay the $6800 asking-price for the contract, we will receive something less than the 1% per month interest we desire. We will, therefore, reject the investor's offer. ∎

EXAMPLE 4-5

Suppose we decided to pay the $6800 for the time purchase contract in Example 4-4. What monthly rate of return would we obtain on our investment?

Solution: In this situation, we know P, A, and n, but we do not know i. The problem may be solved using either the uniform series present worth formula,

$$P = A(P/A,i,n)$$

or the uniform series capital recovery formula,

$$A = P(A/P,i,n)$$

Either way, we have one equation with one unknown.

$$P = \$6800 \qquad A = \$140 \qquad n = 60 \qquad i = \text{unknown}$$

$$P = A(P/A,i,n)$$

$$\$6800 = \$140(P/A,i,60)$$

$$(P/A,i,60) = \frac{6800}{140} = 48.571$$

We know the value of the uniform series present worth factor, but we do not know the interest rate i. As a result, we need to look through several Compound Interest Tables and compute the rate of return i by interpolation. From the tables in the back of this book, we find

Interest rate	$(P/A,i,60)$
½%	51.726
¾%	48.174
1 %	44.955

The rate of return is between ½% and ¾%, and may be computed by a linear interpolation. The interest formulas are not linear, so a linear interpolation will not give an exact solution. To minimize the error, the interpolation should be computed using interest rates as close to the correct answer as possible.

$$\text{Rate of return } i^* = 0.50\% + 0.25\%\left(\frac{51.726 - 48.571}{51.726 - 48.174}\right)$$

$$= 0.50\% + 0.25\%\left(\frac{3.155}{3.552}\right) = 0.50\% + 0.22\%$$

$$= 0.72\% \text{ per month}$$

The monthly rate of return on our investment would be 0.72% per month. ■

EXAMPLE 4-6

Using a 15% interest rate, compute the value of F in the following cash flow:

Year	Cash flow
1	+100
2	+100
3	+100
4	0
5	−F

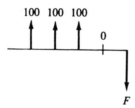

Solution: We see that the cash flow diagram is not the same as the sinking fund factor diagram (see Ex. 4-1):

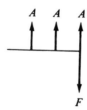

Since the diagrams do not agree, the problem is more difficult than the previous ones we've discussed. The general approach to use in this situation is to convert the cash flow from its present form into standard forms, for which we have compound interest factors and Compound Interest Tables.

One way to solve this problem is to consider the cash flow as a series of single payments P and then to compute their sum F. In other words, the cash flow is broken into three parts, each one of which we can solve.

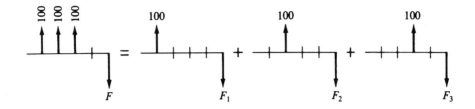

$F = F_1 + F_2 + F_3 = 100(F/P,15\%,4) + 100(F/P,15\%,3) + 100(F/P, 15\%,2)$
$\qquad\qquad\qquad = 100(1.749) + 100(1.521) + 100(1.322)$
$\qquad\qquad\qquad = \$459.20$

The value of F in the illustrated cash flow is \$459.20.

Alternate Solution:

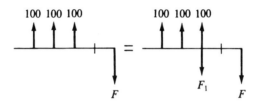

Looked at this way, we first solve for F_1.

$F_1 = 100(F/A,15\%,3) = 100(3.472) = \$347.20$

Now F_1 can be considered a present sum P in the diagram:

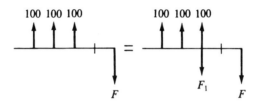

So, $F = F_1 (F/P,15\%,2)$

$\qquad = 347.20(1.322)$

$\qquad = \$459.00$

The slightly different value from the previous computation is due to rounding in the

Compound Interest Tables.

This has been a two-step solution:

$$F_1 = 100(F/A,15\%,3)$$

$$F = F_1(F/P,15\%,2)$$

One could substitute the value of F_1 from the first equation into the second equation and solve for F, without computing F_1.

$$F = 100(F/A,15\%,3)(F/P,15\%,2)$$

$$= 100(3.472)(1.322)$$

$$= \$459.00 \quad ■$$

EXAMPLE 4-7

Consider the following situation:

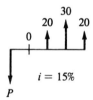

The diagram is not in a standard form, indicating there will be a multiple-step solution. There are at least three different ways of computing the answer. (It is important that you understand how the three computations are made, so please study all three solutions.)

Ex. 4-7, Solution One:

$$P = P_1 + P_2 + P_3$$

$$= 20(P/F,15\%,2) + 30(P/F,15\%,3) + 20(P/F,15\%,4)$$

$$= 20(0.7561) + 30(0.6575) + 20(0.5718)$$

$$= \$46.28$$

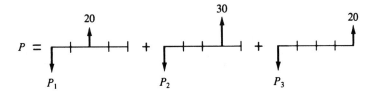

Ex. 4-7, Solution Two:

The relationship between P and F in the diagram is

$$P = F(P/F,15\%,4)$$

Next we compute the future sums of the three payments, as follows:

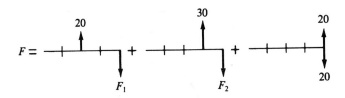

$$F = F_1 + F_2 + 20$$
$$= 20(F/P,15\%,2) + 30(F/P,15\%,1) + 20$$

Combining the two equations,

$$P = [F_1 + F_2 + 20](P/F,15\%,4)$$
$$= [20(F/P,15\%,2) + 30(F/P,15\%,1) + 20](P/F,15\%,4)$$
$$= [20(1.322) + 30(1.150) + 20](0.5718)$$
$$= \$46.28$$

Ex. 4-7, Solution Three:

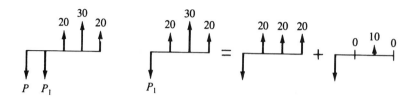

$P = P_1(P/F,15\%,1)$ $P_1 = 20(P/A,15\%,3) + 10(P/F,15\%,2)$

Combining,

$$P = [20(P/A,15\%,3) + 10(P/F,15\%,2)](P/F,15\%,1)$$

$$= [20(2.283) + 10(0.7561)](0.8696)$$

$$= \$46.28 \quad \blacksquare$$

Relationships Between Compound Interest Factors

From the derivations, we see there are several simple relationships between the compound interest factors. They are summarized here.

Single Payment

$$\text{Compound amount factor} = \frac{1}{\text{Present worth factor}}$$

$$(F/P,i,n) = \frac{1}{(P/F,i,n)} \tag{4-9}$$

Uniform Series

$$\text{Capital recovery factor} = \frac{1}{\text{Present worth factor}}$$

$$(A/P,i,n) = \frac{1}{(P/A,i,n)} \tag{4-10}$$

$$\text{Compound amount factor} = \frac{1}{\text{Sinking fund factor}}$$

$$(F/A,i,n) = \frac{1}{(A/F,i,n)} \tag{4-11}$$

The uniform series present worth factor is simply the sum of the N terms of the single payment present worth factor

$$(P/A,i,n) = \sum_{J=1}^{N} (P/F,i,J) \tag{4-12}$$

For example:

$(P/A,5\%,4) = (P/F,5\%,1) + (P/F,5\%,2) + (P/F,5\%,3) + (P/F,5\%,4)$
$3.546 = 0.9524 + 0.9070 + 0.8638 + 0.8227$

The uniform series compound amount factor equals 1 *plus* the sum of $(N-1)$ terms of the single payment compound amount factor

$$(F/A,i,n) = 1 + \sum_{J=1}^{N-1} (F/P,i,J) \tag{4-13}$$

For example,

$(F/A,5\%,4) = 1 + (F/P,5\%,1) + (F/P,5\%,2) + (F/P,5\%,3)$
$4.310 = 1 + 1.050 + 1.102 + 1.158$

The uniform series capital recovery factor equals the uniform series sinking fund factor *plus i.*

$$(A/P,i,n) = (A/F,i,n) + i \tag{4-14}$$

For example,

$$(A/P,5\%,4) = (A/F,5\%,4) + 0.05$$

$$0.2820 = 0.2320 + 0.05$$

This may be proved as follows:

$$(A/P,i,n) = (A/F,i,n) + i$$

$$\left[\frac{i(1+i)^n}{(1+i)^n - 1}\right] = \left[\frac{1}{(1+i)^n - 1}\right] + i$$

Multiply by $(1 + i)^n - 1$ to get

$$i(1 + i)^n = i + i(1 + i)^n - i = i(1 + i)^n.$$

Arithmetic Gradient

We frequently encounter the situation where the cash flow series is not of constant amount A. Instead, there is a uniformly increasing series as shown:

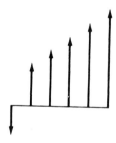

Cash flows of this form may be resolved into two components:

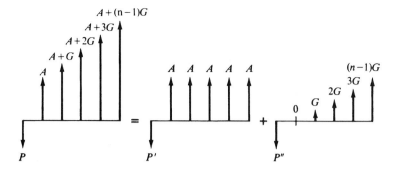

Note that by resolving the problem in this manner, it makes the first cash flow in the arithmetic gradient series equal to zero. We already have an equation for P', and we need to derive an equation for P''. In this way, we will be able to write

$$P = P' + P'' = A(P/A,i,n) + G(P/G,i,n)$$

Derivation of Arithmetic Gradient Factors

The arithmetic gradient is a series of increasing cash flows as follows:

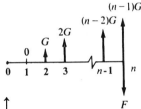

Note: Time = 0 here ↑

The arithmetic gradient series may be thought of as a series of individual cash flows:

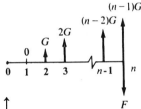

The value of F for the sum of the cash flows $= F^I + F^{II} + \dots F^{III} + F^{IV}$

or $F = G(1 + i)^{n-2} + 2G(1 + i)^{n-3} + \dots + (n - 2)(G)(1 + i)^1 + (n - 1)G$ (4-15)

Multiply Eq. 4-15 by $(1 + i)$ and factor out G, or

$$(1 + i)F = G[(1 + i)^{n-1} + 2(1 + i)^{n-2} + \ldots + (n - 2)(1 + i)^2 + (n - 1)(1 + i)^1] \qquad (4\text{-}16)$$

Rewrite Eq. 4-15 to show other terms in the series,

$$F = G[(1 + i)^{n-2} + \ldots + (n - 3)(1 + i)^2 + (n - 2)(1 + i)^1 + n - 1] \qquad (4\text{-}17)$$

Subtracting Eq. 4-17 from Eq. 4-16, we obtain

$$F + iF - F = G[(1 + i)^{n-1} + (1 + i)^{n-2} + \ldots + (1 + i)^2 + (1 + i)^1 + 1] - nG \qquad (4\text{-}18)$$

In the derivation of Eq. 4-5, the terms within the brackets of Eq. 4-18 were shown to equal the series compound amount factor:

$$\left[(1+i)^{n-1} + (1+i)^{n-2} + \cdots + (1+i)^2 + (1+i)^1 + 1\right] = \frac{(1+i)^n - 1}{i}$$

Thus, Equation 4-18 becomes

$$iF = G\left[\frac{(1+i)^n - 1}{i}\right] - nG$$

Rearranging and solving for F,

$$F = \frac{G}{i}\left[\frac{(1+i)^n - 1}{i} - n\right]$$

$$(4\text{-}19)$$

Multiplying Eq. 4-19 by the single payment present worth factor,

$$P = \frac{G}{i}\left[\frac{(1+i)^n - 1}{i} - n\right]\left[\frac{1}{(1+i)^n}\right]$$

$$= G\left[\frac{(1+i)^n - in - 1}{i^2(1+i)^n}\right]$$

$$(P/G, i, n) = \left[\frac{(1+i)^n - in - 1}{i^2(1+i)^n}\right] \qquad (4\text{-}20)$$

Equation 4-20 is the *arithmetic gradient present worth factor*. Multiplying Eq. 4-19 by the sinking fund factor,

$$A = \frac{G}{i} \left[\frac{(1+i)^n - 1}{i} - n \right] \left[\frac{i}{(1+i)^n - 1} \right]$$

$$= G \left[\frac{(1+i)^n - in - 1}{i(1+i)^n - i} \right]$$

$$(A/G, i, n) = \left[\frac{(1+i)^n - in - 1}{i(1+i)^n - i} \right] = \left[\frac{1}{i} - \frac{n}{(1+i)^n - 1} \right] \qquad (4\text{-}21)$$

Equation 4-21 is the **arithmetic gradient uniform series factor**.

EXAMPLE 4-8

A man purchased a new automobile. He wishes to set aside enough money in a bank account to pay the maintenance on the car for the first five years. It has been estimated that the maintenance cost of an automobile is as follows:

Year	Maintenance cost
1	$120
2	150
3	180
4	210
5	240

Assume the maintenance costs occur at the end of each year and that the bank pays 5% interest. How much should he deposit in the bank now?

Solution:

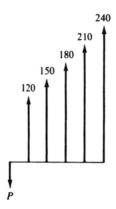

The cash flow may be broken into its two components:

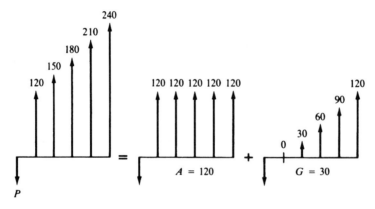

Both of the components represent cash flows for which compound interest factors have been derived. The first is uniform series present worth, and the second is arithmetic gradient series present worth.

$$P = A(P/A,5\%,5) + G(P/G,5\%,5)$$

Note that the value of n in the gradient factor is 5, not 4. In deriving the gradient factor, we had $(n-1)$ terms containing G. Here there are 4 terms containing G.

Thus, $(n-1) = 4,$ so $n = 5$.

$$P = 120(P/A,5\%,5) + 30(P/G,5\%,5)$$

$$= 120(4.329) + 30(8.237) = 519 + 247$$

$$= \$766$$

He should deposit $766 in the bank now. ■

EXAMPLE 4-9

On a certain piece of machinery, it is estimated that the maintenance expense will be as follows:

Year	Maintenance
1	$100
2	200
3	300
4	400

What is the equivalent uniform annual maintenance cost for the machinery if 6% interest is used?

Solution:

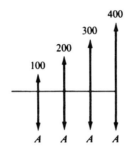

The first cash flow in the arithmetic gradient series is zero, hence the diagram above is *not* in proper form for the arithmetic gradient equation. The cash flow must be resolved into two components as is done in Example 4-8.

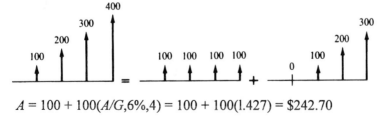

$$A = 100 + 100(A/G,6\%,4) = 100 + 100(1.427) = \$242.70$$

The equivalent uniform annual maintenance cost is $242.70. ■

EXAMPLE 4-10

A textile mill in India installed a number of new looms. It is expected that initial maintenance and repairs will be high, but that they will then decline for several years. The projected cost is:

Year	Maintenance and repair cost
1	24,000 rupees
2	18,000
3	12,000
4	6,000

What is the projected equivalent annual maintenance and repair cost if interest is 10%?

Solution:

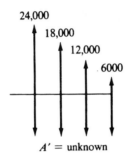

A' = unknown

The projected cash flow is not in the form of the arithmetic gradient factors. Both factors were derived for an increasing gradient over time. The factors cannot be used directly for a declining gradient. Instead, we will subtract an increasing gradient from an assumed uniform series of payments.

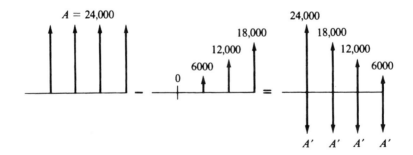

$$A' = 24{,}000 - 6000(A/G, 10\%, 4) = 24{,}000 - 6000(1.381)$$

$$= 15{,}714 \text{ rupees}$$

The projected equivalent uniform maintenance and repair cost is 15,714 rupees per year. ■

EXAMPLE 4-11

Compute the value of *P* in the diagram below. Use a 10% interest rate.

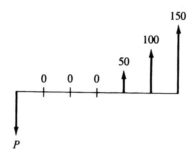

Solution: With the arithmetic gradient series present worth factor, we can compute a present sum *J*.

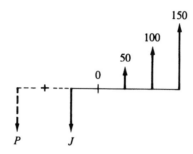

It is important that you closely examine the location of *J*. Based on the way the factor was derived, there will be one zero value in the gradient series to the right of *J*. (If this seems strange or incorrect, review the beginning of this Arithmetic Gradient section.)

$$J = G(P/G, i, n)$$

$$= 50(P/G, 10\%, 4) \qquad (\textit{Note:} \ 3 \ \text{would be incorrect.})$$

$$= 50(4.378) = 218.90$$

Then $P = J(P/F, 10\%, 2)$

To obtain the present worth of the future sum *J*, use the $(P/F, i, n)$ factor. Combining,

$$P = 50(P/G, 10\%, 4)(P/F, 10\%, 2)$$

$$= 50(4.378)(0.8264) = \$180.90$$

The value of *P* is $180.90. ■

Geometric Gradient

In the previous section, we saw that the arithmetic gradient is applicable where the period-by-period change in a cash receipt or payment is a uniform amount. There are other situations where the period-by-period change is a ***uniform rate, g***. An example of this would be where the maintenance costs for an automobile are $100 the first year and increasing at a uniform rate, g, of 10% per year. For the first five years, the cash flow would be:

Year			Cash flow
1	100.00	=	$100.00
2	$100.00 + 10\%(100.00 = 100(1 + 0.10)^1$	=	110.00
3	$110.00 + 10\%(110.00 = 100(1 + 0.10)^2$	=	121.00
4	$121.00 + 10\%(121.00 = 100(1 + 0.10)^3$	=	133.10
5	$133.10 + 10\%(133.10 = 100(1 + 0.10)^4$	=	146.41

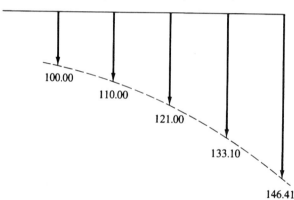

From the table, we can see that the maintenance cost in any year is

$$\$100(1 + g)^{n-1}$$

Stated in a more general form,

$$A_n = A_1(1 + g)^{n-1} \tag{4-22}$$

where

g = Uniform *rate* of cash flow increase/decrease from period to period, that is, the geometric gradient.

A_1 = Value of cash flow at Year 1 ($100 in the example).

A_n = Value of cash flow at any Year n.

Since the present worth P_n of any cash flow A_n at interest rate i is

$$P_n = A_n(1 + i)^{-n} \tag{4-23}$$

we can substitute Eq. 4-22 into Eq. 4-23 to get

$$P_n = A_1(1 + g)^{n-1}(1 + i)^{-n}$$

This may be rewritten as

$$P = A_1(1 + i)^{-1} \sum_{x=1}^{n} \left(\frac{1 + g}{1 + i} \right)^{x-1} \tag{4-24}$$

The present worth of the entire gradient series of cash flows may be obtained by expanding Eq. 4-24:

$$P = A_1(1 + i)^{-1} \sum_{x=1}^{n} \left(\frac{1 + g}{1 + i} \right)^{x-1} \tag{4-25}$$

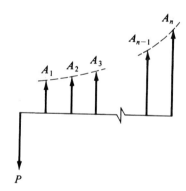

In the general case where $i \neq g$, Eq. 4-24 may be written out as

$$P = A_1(1 + i)^{n-1} + A_1(1 + i)^{-1}\left(\frac{1 + g}{1 + i} \right) + A_1(1 + i)^{-1}\left(\frac{1 + g}{1 + i} \right)^2$$
$$+ \cdots + A_1(1 + i)^{-1}\left(\frac{1 + g}{1 + i} \right)^{n-1} \tag{4-26}$$

Let $a = A_1(1 + i)^{-1}$ and $b = \left(\frac{1 + g}{1 + i} \right)$. Equation 4-26 becomes

$$P = a + ab + ab^2 + \cdots + ab^{n-1} \tag{4-27}$$

Multiply Equation 4-27 by b:

$$bP = ab + ab^2 + ab^3 + \cdots + ab^{n-1} + ab^n \tag{4-28}$$

Subtract Eq. 4-28 from Eq. 4-27:

$$P - bP = a - ab^n$$

$$P(1 - b) = a(1 - b^n)$$

$$P = \frac{a(1 - b^n)}{1 - b}$$

Replacing the original values for a and b, we obtain:

$$P = A_1 (1 + i)^{-1} \left[\frac{1 - \left(\dfrac{1+g}{1+i}\right)^n}{1 - \left(\dfrac{1+g}{1+i}\right)} \right]$$

$$= A_1 \left[\frac{1 - \left(\dfrac{1+g}{1+i}\right)^n}{(1+i) - \left(\dfrac{1+g}{1+i}\right)(1+i)} \right]$$

$$= A_1 \left[\frac{1 - (1+g)^n (1+i)^{-n}}{1 + i - 1 - g} \right]$$

$$P = A_1 \left[\frac{1 - (1+g)^n (1+i)^{-n}}{i - g} \right] \qquad \text{where } i \neq g \tag{4-29}$$

The expression in the brackets of Equation 4-29 is the *geometric series present worth factor* where $i \neq g$.

$$(P/A,g,i,n) = \left[\frac{1-(1+g)^n(1+i)^{-n}}{i-g}\right] \quad \text{where } i \neq g \tag{4-30}$$

In the special case where $i = g$, Eq. 4-29 becomes

$$P = A_1 n(1+i)^{-1} \text{ where } i = g$$

$$(P/A,g,i,n) = [n(1+i)^{-1}] \quad \text{where } i = g \tag{4-31}$$

EXAMPLE 4-12

The first year maintenance cost for a new automobile is estimated to be $100, and it increases at a uniform rate of 10% per year. What is the present worth of cost of the first five years of maintenance in this situation, using an 8% interest rate?

Step-by-Step Solution:

Year n		Maintenance cost		(P/F,8%,n)	PW of maintenance
1	100.00	= 100.00	X	0.9259 =	$ 92.59
2	100.00 +10%(100.00)	= 110.00	X	0.8573 =	94.30
3	110.00 +10%(110.00)	= 121.00	X	0.7938 =	96.05
4	121.00 +10%(121.00)	= 133.10	X	0.7350 =	97.83
5	133.10 +10%(133.10)	= 146.41	X	0.6806 =	99.65
					$480.42

Solution Using Geometric Series Present Worth Factor:

$$P = A_1\left[\frac{1-(1+g)^n(1+i)^{-n}}{i-g}\right] \quad \text{where } i \neq g$$

$$= 100.00\left[\frac{1-(1.10)^5(1.08)^{-5}}{-0.02}\right] = \$480.42$$

The present worth of cost of maintenance for the first five years is $480.42. ■

Nominal And Effective Interest

EXAMPLE 4-13

Consider the situation of a person depositing $100 into a bank that pays 5% interest, compounded semi-annually. How much would be in the savings account at the end of one year?

Solution: Five percent interest, compounded semi-annually, means that the bank pays $2\frac{1}{2}\%$ every six months. Thus, the initial amount P = $100 would be credited with $0.025(100) =$ $2.50 interest at the end of six months, or

$P \longrightarrow P + Pi = 100 + 100(0.025) = 100 + 2.50 = \$102.50$
The $102.50 is left in the savings account; at the end of the second six-month period, the interest earned is $0.025(102.50) = \$2.56$, for a total in the account at the end of one year of $102.50 + 2.56 = \$105.06$, or

$$(P + Pi) \longrightarrow (P + Pi) + i(P + Pi) = P(1 + i)^2 = 100(1 + 0.025)^2$$
$$= \$105.06 \quad \blacksquare$$

> *Nominal interest rate* **per year, r, is the annual interest rate without considering the effect of any compounding.**

In the example, the bank pays $2\frac{1}{2}\%$ interest every six months. The nominal interest rate per year, r, therefore, is $2 \times 2\frac{1}{2}\% = 5\%$.

> *Effective interest rate* **per year, i_a, is the annual interest rate taking into account the effect of any compounding during the year.**

In Example 4-13 we saw that $100 left in the savings account for one year increased to $105.06, so the interest paid was $5.06. The effective interest rate per year, i_a, is $5.06/\$100.00 = 0.0506 = 5.06\%$.

r = Nominal interest rate per interest period (usually one year).

i = Effective interest rate per interest period.

i_a = Effective interest rate per year (annum)

m = Number of compounding subperiods per time period.

Using the method presented in Ex. 4-13, we can derive the equation for the effective interest rate. If a $1 deposit were made to an account that compounded interest m times per year and paid a nominal interest rate per year, r, the *interest rate per compounding subperiod* would be r/m, and the total in the account at the end of one year would be

$$\$1\left(1+\frac{r}{m}\right)^{m} \quad \text{or simply} \quad \left(1+\frac{r}{m}\right)^{m}$$

If we deduct the $1 principal sum, the expression would be

$$\left(1+\frac{r}{m}\right)^{m}-1$$

Therefore,

Effective interest rate per year,

$$i_{a} = \left(1+\frac{r}{m}\right)^{m}-1 \qquad (4\text{-}32)$$

where r = Nominal interest rate per year

m = Number of compounding subperiods per year

Or, substituting the effective interest rate per compounding subperiod, $i = (r/m)$,

Effective interest rate per year, $\qquad i_{a} = (1 + i)^{m} - 1 \qquad (4\text{-}33)$

where i = Effective interest rate per compounding subperiod

m = Number of compounding subperiods per year

Either Eq. 4-32 or 4-33 may be used to compute an effective interest rate per year.

One should note that i was described in chapter three simply as the interest rate per interest period. We were describing the effective interest rate without making any fuss about it. A more precise definition, we now know, is that i is the *effective* interest rate per interest period. Although it seems more complicated, we are describing the same exact situation, but with more care.

The nominal interest rate r is often given for a one-year period (but it could be given for either a shorter or a longer time period). In the special case where the nominal interest rate is given per compounding subperiod, then the effective interest rate per compounding subperiod, i, equals the nominal interest rate per subperiod, r.

In the typical effective interest computation, there are multiple compounding subperiods $(m > 1)$. The resulting effective interest rate is either the solution to the problem, or an intermediate solution, which allows us to use standard compound interest factors to proceed to solve the problem.

For ***continuous compounding*** (which is described in the next section),

---

Effective interest rate per year, $i_a = e^r - 1$ (4-34)

EXAMPLE 4-14

If a savings bank pays 1½% interest every three months, what are the nominal and effective interest rates per year?

Solution:

Nominal interest rate per year, $r = 4 \times 1\frac{1}{2}\% = 6\%$

$$\text{Effective interest rate per year, } i_a = \left(1 + \frac{r}{m}\right)^m - 1$$

$$= \left(1 + \frac{0.06}{4}\right)^4 - 1 = 0.061$$

$$= 6.1\%$$

Alternately,

$$\text{Effective interest rate per year } i_a = (1 + i)^m - 1$$
$$= (1 + 0.015)^4 - 1 = 0.061$$
$$= 6.1\% \quad \blacksquare$$

Table 4-1 tabulates the effective interest rate for a range of compounding frequencies and nominal interest rates. It should be noted that when a nominal interest rate is compounded annually, the nominal interest rate equals the effective interest rate. Also, it will be noted that increasing the frequency of compounding (for example, from monthly to continuously) has only a small impact on the effective interest rate. But if the amount of money is large, even small differences in the effective interest rate can be significant.

Table 4-1 NOMINAL AND EFFECTIVE INTEREST

r	*Yearly*	*Semi-annually*	*Monthly*	*Daily*	*Continuously*
Nominal interest rate per year	*Effective interest rate per year, i_a, when nominal rate is compounded*				
1%	1.0000%	1.0025%	1.0046%	1.0050%	1.0050%
2	2.0000	2.0100	2.0184	2.0201	2.0201
3	3.0000	3.0225	3.0416	3.0453	3.0455
4	4.0000	4.0400	4.0742	4.0809	4.0811
5	5.0000	5.0625	5.1162	5.1268	5.1271
6	6.0000	6.0900	6.1678	6.1831	6.1837
8	8.0000	8.1600	8.3000	8.3278	8.3287
10	10.000	10.2500	10.4713	10.5156	10.5171
15	15.000	15.5625	16.0755	16.1798	16.1834
25	25.000	26.5625	28.0732	28.3916	28.4025

EXAMPLE 4-15

A loan shark lends money on the following terms:
"If I give you $50 on Monday, you owe me $60 on the following Monday."

a. What nominal interest rate per year (*r*) is the loan shark charging?

b. What effective interest rate per year (i_a) is he charging?

c. If the loan shark started with $50 and was able to keep it, as well as all the money he received, out in loans at all times, how much money would he have at the end of one year?

Solution to Ex. 4-15a:

$$F = P(F/P, i, n)$$

$$60 = 50(F/P, i, 1)$$

$$(F/P, i, 1) = 1.2$$

Therefore, $i = 20\%$ per week

Nominal interest rate per year = 52 weeks x 0.20 = 10.40 = 1040%

Solution to Ex. 4-15b:

$$\text{Effective interest rate per year, } i_a = \left(1 + \frac{r}{m}\right)^m - 1$$

$$= \left(1 + \frac{10.40}{52}\right)^{52} - 1 = 13{,}105 - 1$$

$$= 13{,}104 = 1{,}310{,}400\%$$

Or,   *Effective interest rate per year* $i_a = (1 + i)^m - 1$

$$= (1 + 0.20)^{52} - 1 = 13{,}104$$

$$= 1{,}310{,}400\%$$

Solution to Ex. 4-15c:

$$F = P(1 + i)^n = 50(1 + 0.20)^{52}$$

$$= \$655{,}200$$

With a nominal interest rate of 1040% per year and effective interest rate of 1,310,400% per year, if he started with $50, the loan shark would have $655,200 at the end of one year. ■

When the various time periods in a problem match each other, we generally can solve the problem using simple calculations. In Ex. 4-3, for example, there is $5000 in an account paying 8% interest, compounded annually. The desired five equal end-of-year withdrawals are simply computed as

$$A = P(A/P,8\%,5) = 5000(0.2505) = \$1252$$

Consider how this simple problem becomes more difficult if the compounding period is changed to no longer match the annual withdrawals.

EXAMPLE 4-16

On January 1st, a woman deposits $5000 in a credit union that pays 8% nominal annual interest, compounded quarterly. She wishes to withdraw all the money in five equal yearly sums, beginning December 31st of the first year. How much should she withdraw each year?

Solution: Since the 8% nominal annual interest rate r is compounded quarterly, we know that the effective interest rate per interest period, i, is 2%; and there are a total of $4 \times 5 = 20$ interest periods in five years. For the equation $A = P(A/P,i,n)$ to be used, there must be as many periodic withdrawals as there are interest periods, n. In this example we have five withdrawals and twenty interest periods.

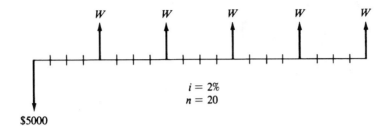

To solve the problem, we must adjust it so that it is in one of the standard forms for which we have compound interest factors. This means we must first either compute an equivalent *A* for each three-month interest period, or an effective *i* for each time period between withdrawals. Let's solve the problem both ways.

Ex. 4-16, Solution One:

Compute an equivalent *A* for each three-month time period.

If we had been required to compute the amount that could be withdrawn quarterly, the diagram would be as follows:

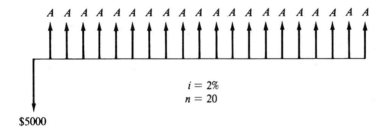

$$A = P(A/P,i,n) = 5000(A/P,2\%,20) = 5000(0.0612) = \$306$$

Now, since we know *A*, we can construct the diagram that relates it to our desired equivalent annual withdrawal, *W*:

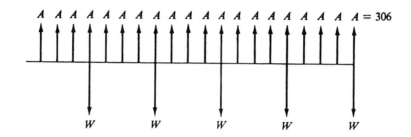

Looking at a one-year period,

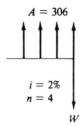

$W = A(F/A,i,n) = 306(F/A,2\%,4) = 306(4.122)$

$\quad = \$1260$

Ex. 4-16, Solution Two:

Compute an effective *i* for the time period between withdrawals.

Between withdrawals, *W*, there are four interest periods, hence *m* = 4 compounding subperiods per year. Since the nominal interest rate per year, *r*, is 8%, we can proceed to compute the effective interest rate per year.

$$\text{Effective interest rate per year, } i_a = \left(1 + \frac{r}{m}\right)^m - 1 = \left(1 + \frac{0.08}{4}\right)^4 - 1$$

$$= 0.0824 = 8.24\% \text{ per year}$$

Now the problem may be redrawn as:

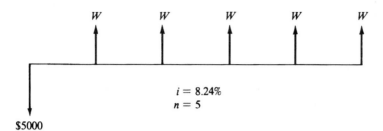

$$i = 8.24\%$$
$$n = 5$$

$5000

This diagram may be directly solved to determine the annual withdrawal W using the capital recovery factor:

$$W = P(A/P,i,n) = 5000(A/P,8.24\%,5)$$

$$= P\left[\frac{i(1+i)^n}{(1+i)^n - 1}\right] = 5000\left[\frac{0.0824(1+0.0824)^5}{(1+0.0824)^5 - 1}\right]$$

$$= 5000(0.2520) = \$1260$$

She should withdraw $1260 per year. ■

Continuous Compounding

Two variables we have introduced are:

   r = Nominal interest rate per interest period

   m = Number of compounding subperiods per time period

Since the interest period is normally one year, the definitions become:

   r = Nominal interest rate per year

   m = Number of compounding subperiods per year

   $\dfrac{r}{m}$ = Interest rate per interest period

   mn = Number of compounding subperiods in n years

Single Payment Interest Factors–Continuous Compounding

The single payment compound amount formula (Equation 3-3)

$$F = P(1 + i)^n$$

may be rewritten as

$$F = P\left(1 + \frac{r}{m}\right)^{mn}$$

If we increase m, the number of compounding subperiods per year, without limit, m becomes very large and approaches infinity, and r/m becomes very small and approaches zero.

This is the condition of **continuous compounding**, that is, where the duration of the interest period decreases from some finite duration Δt to an infinitely small duration dt, and the number of interest periods per year becomes infinite. In this situation of continuous compounding:

$$F = P \lim_{m \to \infty}\left(1 + \frac{r}{m}\right)^{mn}$$

(4-35)

An important limit in calculus is:

$$\lim_{x \to 0} (1 + x)^{1/x} = 2.71828 = e$$ (4-36)

If we set $x = r/m$, then mn may be written as $(1/x)(rn)$. As m becomes infinite, x becomes 0. Equation 4-35 becomes

$$F = P[\lim_{x \to 0} (1 + x)^{1/x}]^{rn}$$

Equation 4-36 tells us the quantity within the brackets equals e, so Eq. 3-3
$F = P(1 + i)^n$ becomes $F = Pe^{rn}$ (4-37)
and $P = F(1 + i)^{-n}$ becomes $P = Fe^{-rn}$ (4-38)
We see that for continuous compounding,
  $(1 + i) = e^r$

or, **Effective interest rate per year, $i_a = e^r - 1$** (4-34)

Single Payment—Continuous Compounding

  Compound Amount $F = P(e^{rn}) = P[F/P,r,n]$ (4-39)

  Present Worth $P = F(e^{-rn}) = F[P/F,r,n]$ (4-40)

Note that square brackets around the notation are used to distinguish continuous compounding. If your hand calculator does not have e^x, use the table of e^{rn} and e^{-rn}, provided at the end of the Compound Interest Tables.

EXAMPLE 4-17

If you were to deposit $2000 in a bank that pays 5% nominal interest, compounded continuously, how much would be in the account at the end of two years?

Solution: The single payment compound amount equation for continuous compounding is

$F = Pe^{rn}$ where r = nominal interest rate = 0.05

n = number of years = 2

$F = 2000e^{(0.05 \times 2)} = 2000(1.1052) = \$2210.40$

There would be $2210.40 in the account at the end of two years. ■

EXAMPLE 4-18

A bank offers to sell savings certificates that will pay the purchaser $5000 at the end of ten years but will pay nothing to the purchaser in the meantime. If interest is computed at 6%, compounded continuously, at what price is the bank selling the certificates?

Solution:

$P = Fe^{-rn}$ where $F = \$5000$; $r = 0.06$; $n = 10$ years

$P = 5000e^{-(0.06 \times 10)}$ 5000(0.5488) = \$2744

Therefore, the bank is selling the $5000 certificates for $2744. ■

EXAMPLE 4-19

How long will it take for money to double at 10% nominal interest, compounded continuously?

$$F = Pe^{rn}$$

$$2 = 1e^{(0.10)n}$$

$$e^{0.10n} = 2$$

$$or, \ 0.10n = \ln 2 = 0.693$$

$$n = 6.93 \text{ years}$$

It will take 6.93 years for money to double at 10% nominal interest, compounded continuously. ■

EXAMPLE 4-20

If the savings bank in Example 4-14 changed its interest policy to 6% interest, compounded continuously, what are the nominal and the effective interest rates?

Solution: The nominal interest rate remains at 6% per year.

$$\text{Effective interest rate} = e^r - 1$$

$$= e^{0.06} - 1 = 0.0618$$

$$= 6.18\% \quad \blacksquare$$

Uniform Payment Series-Continuous Compounding at Nominal Rate *r* per Period

If we substitute the equation $i = e^r - 1$ into the equations for periodic compounding, we get:

Continuous Compounding Sinking Fund:

$$[A/F,r,n] = \frac{e^r - 1}{e^{rn} - 1} \tag{4-41}$$

Continuous Compounding Capital Recovery:

$$[A/P,r,n] = \frac{e^{rn}(e^r - 1)}{e^{rn} - 1} \tag{4-42}$$

Continuous Compounding Series Compound Amount:

$$[F/A,r,n] = \frac{e^{rn} - 1}{e^r - 1} \tag{4-43}$$

Continuous Compounding Series Present Worth:

$$[P/A,r,n] = \frac{e^{rn} - 1}{e^{rn}(e^r - 1)} \tag{4-44}$$

EXAMPLE 4-21

In Example 4-1, a man deposited $500 per year into a credit union that paid 5% interest, compounded annually. At the end of five years, he had $2763 in the credit union. How much would he have if they paid 5% nominal interest, compounded continuously?

Solution: $A = \$500 \quad r = 0.05 \quad n = 5 \text{ years}$

$$F = A[F/A,r,n] = A\left(\frac{e^{rn} - 1}{e^r - 1}\right) = 500\left(\frac{e^{0.05(5)} - 1}{e^{0.05} - 1}\right)$$

$$= \$2769.84$$

He would have $2769.84. ■

EXAMPLE 4-22

In Ex. 4-2, Jim Hayes wished to save a uniform amount each month so he would have $1000 at the end of one year. Based on 6% nominal interest, compounded monthly, he had to deposit $81.10 per month. How much would he have to deposit if his credit union paid 6% nominal interest, compounded continuously?

Solution: The deposits are made monthly; hence, there are twelve compounding subperiods in the one-year time period.

$$F = \$1000 \quad r = \text{nominal interest rate / interest period} = \frac{0.06}{12} = 0.005$$

$n = 12$ compounding subperiods in the one-year period of the problem

$$A = F[A/F,r,n] = F\left(\frac{e^r - 1}{e^{rn} - 1}\right) = 1000\left(\frac{e^{0.005} - 1}{e^{0.005(12)} - 1}\right)$$

$$= 1000\left(\frac{0.005013}{0.061837}\right) = \$81.07$$

He would have to deposit $81.07 per month. Note that the difference between monthly and continuous compounding is just three cents per month. ■

Continuous, Uniform Cash Flow (One Period) With Continuous Compounding at Nominal Interest Rate *r*

Equations for a continuous, uniform cash flow during one period only, with continuous compounding, can be derived as follows. Let the continuous, uniform cash flow totaling $\overline{P}$ be distributed over *m* subperiods within one period ($n = 1$). Thus $\overline{P}/m$ is the cash flow at the end of each subperiod. Since the nominal interest rate per period is *r*, the effective interest rate per subperiod is *r/m*. Substituting these values into the uniform series compound amount equation (Eq. 4-5) gives:

$$F = \frac{\overline{P}}{m}\left[\frac{[1+(r/m)]^m - 1}{r/m}\right] \tag{4-45}$$

Setting $x = r/m$, we obtain:

$$F = \frac{\overline{P}}{m}\left[\frac{[1+x]^{r/x} - 1}{r/m}\right] = \overline{P}\left[\frac{\left[(1+x)^{1/x}\right]^r - 1}{r}\right] \tag{4-46}$$

As *m* increases, *x* approaches zero. Equation 4-36 says

$$\lim_{x \to 0}(1+x)^{1/x} = e$$

hence Eq. 4-46 for one period becomes

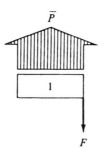

$$F = \overline{P}\left(\frac{e^r - 1}{r}\right) \tag{4-47}$$

Multiplying Eq. 4-43 by the single payment–continuous compounding factors:

For Any Future Time:

$$F = \overline{P}\left(\frac{e^r - 1}{r}\right)\left(e^{r(n-1)}\right) = \overline{P}\left[\frac{(e^r - 1)(e^{rn})}{re^r}\right]$$

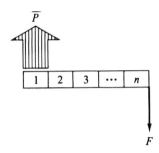

$$\text{Compound Amount}\left[F/\overline{P}, r, n\right] = \left[\frac{(e^r - 1)(e^{rn})}{re^r}\right] \tag{4-48}$$

For Any Present Time:

$$P = \overline{F}\left(\frac{e^r - 1}{r}\right)\left(e^{-rn}\right) = \overline{F}\left[\frac{e^r - 1}{re^{rn}}\right]$$

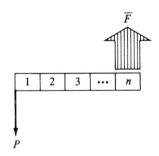

$$\text{Present Worth}\left[P/\overline{F}, r, n\right] = \left[\frac{e^r - 1}{re^{rn}}\right] \tag{4-49}$$

EXAMPLE 4-28

A self-service gasoline station has been equipped with an automatic teller machine (ATM). Customers may obtain gasoline simply by inserting their ATM bank card into the machine and filling their car with gasoline. When they have finished, the ATM unit automatically deducts the gasoline purchase from the customer's bank account and credits it to the gas station's bank account. The gas station receives $40,000 per month in this manner with the cash flowing uniformly throughout the month. If the bank pays 9% nominal interest, compounded continuously, how much will be in the gasoline station bank account at the end of the month?

Solution:

The example problem may be solved by either Eq. 4-47 or Eq.4-48. Here the general equation (Eq. 4-48) is used.

$$r = \text{Nominal interest rate per month} = \frac{0.09}{12} = 0.0075$$

$$F = \bar{P}\left[\frac{(e^r - 1)(e^{rn})}{re^r}\right] = 40,000\left[\frac{(e^{0.0075} - 1)(e^{0.0075(1)})}{0.0075e^{0.0075}}\right]$$

$$= 40,000\left[\frac{(0.0075282)(1.0075282)}{0.00755646}\right] = \$40,150.40$$

There will be $40,150.40 in the station's bank account. ■

Spreadsheets for Economic Analysis

Spreadsheets are used in most real world applications of engineering economy. Common tasks include:

1. Constructing tables of cash flows.

2. Using annuity functions to calculate a P, F, A, N, or i.

3. Using a block function to find the present worth or internal rate of return for a table of cash flows.

4. Making graphs for analysis and convincing presentations.

5. Calculating "what-if" for different assumed values of problem variables.

Constructing tables of cash flows relies mainly on the spreadsheet basics that are covered in Appendix 1. These basics includes 1) using and naming spreadsheet variables, 2) the difference between absolute and relative addresses when copying a formula, and 3) how to format a cell. Appendix 1 uses the example of the amortization schedule shown in Table 3-1 plan 3. This amortization schedule divides each scheduled loan payment into principal and interest, and includes the outstanding balance for each period.

Because spreadsheet functions can be found through pointing and clicking on menus, those steps are not detailed. In Excel the starting point is the f_x button. Excel functions are used here, but there are only minor syntax differences for most other spreadsheet programs.

Spreadsheet Annuity Functions

In tables of engineering economy factors, i is the table; N is the row; and two of P, F, A, and G define a column. The spreadsheet annuity functions list 4 arguments chosen from N, A, P, F, and i, and solve for the fifth argument. The *Type* argument is optional. If it is omitted or 0, then the A and F values are assumed to be end-of-period cash flows. If the A and F values are beginning-of-period cash flows, then a value of 1 can be entered for the Type variable.

to find the equivalent P	$-$PV(i,N,A,F,Type)
to find the equivalent A	$-$PMT(i,N,P,F,Type)
to find the equivalent F	$-$FV(i,N,A,P,Type)
to find N	NPER(i,A,P,F,Type)
to find i	RATE(N,A,P,F,Type,guess)

The sign convention for the first three functions seems odd to some students. The PV of $200 per period for 10 periods is negative and the PV of -$200 per period is positive. So a minus sign is inserted to find the equivalent P, A, or F. Without this minus sign, the calculated value is not equivalent to the four given values. Instead the present worth equals 0 for the four given values and the calculated value.

EXAMPLE 4-29

A new engineer wants to save money for down payment on a house. The initial deposit is $685, and $375 is deposited at the end of each month. The savings account earns interest at an annual nominal rate of 6% with monthly compounding. How much is on deposit after 48 months?

Solution

Because deposits are made monthly, the nominal annual interest rate of 6% must be converted to ½% per month for the 48 months. Thus the question is: find F if $i = 0.5\% = .005$, $N = 48$, $A = 375$, and $P = 685$. Note both the initial and periodic deposits are positive cash flows for the savings account. The Excel function is multiplied by -1 or $-\text{FV}(.005,48,375,685,0)$ and the result is $21,156.97. ■

EXAMPLE 4-30

A new engineer buys a car with 0% down financing from the dealer. The cost with all taxes, registration, and license fees is $15,732. If each of the 48 monthly payments is $398, what is the monthly interest rate? What is the effective annual interest rate?

Solution

The RATE function can be used to find the monthly interest rate, given that $N = 48$, $A = -398$, $P = 15,732$, and $F = 0$. The Excel function is $\text{RATE}(48,-398,15732,0)$ and the result is 0.822%. The effective annual interest rate is $1.00822^{12} - 1 = 10.33\%$ ■

Spreadsheet Block Functions

Cash flows can be specified period-by-period as a block of values. These cash flows are analyzed by **block functions** that identify the row or column entries for which a present worth or an internal rate of return should be calculated. In Excel the two functions are:

Economic Criteria	Excel Function	Values for periods
Net present value	NPV(i,values)	1 to N
Internal rate of return	IRR(values,guess)	0 to N
	guess argument is optional	

These block functions make different assumptions about the range of years included. NPV(i, values) assumes year 0 is NOT included, while IRR(values, guess) assumes year 0 is included. These functions require that a cash flow be identified for each period . The cash flows for 1 to N are assumed to be end-of-period flows. All periods are assumed to be the same length of time.

Also the NPV functions returns the present worth equivalent to the cash flows, unlike the PV annuity function which returns the negative of the equivalent value.

For cash flows involving only constant values of P, F, and A this block approach seems to be inferior to the annuity functions. However, this is a conceptually easy approach for more complicated cash flows, such as arithmetic gradients. Suppose the years (row 1) and the cash flows (row 2) are specified in columns B through E.

	A	B	C	D	E	F
1	Year	0	1	2	3	4
2	Cash Flow	-25000	6000	8000	10000	12000

If an interest rate of .08 is assumed, then the present worth of the cash flows can be calculated as =B2+NPV(.08,C2:F2), which equals \$4172.95. This is the present worth equivalent to the five cash flows, rather than the negative of the present worth equivalent returned by the PV annuity function. The internal rate of return calculated using IRR(B2:F2) is 14.5%. Notice how the NPV function does not include the year 0 cash flow in B2, while the IRR function does.

Basic Graphing Using Spreadsheets

Often we are interested in the relationship between two variables. Examples include the number and size of payments to repay a loan, the present worth of a M.S. degree and how long until we retire, and the interest rate and present worth for a new machine. This kind of 2-variable relationship is best shown with a graph.

As shown in Example 4-31, the goal is to place one variable on each axis of the graph and to plot the relationship. Modern spreadsheets automate most steps of drawing a graph, so that it is quite easy. However, there are two very similar types of charts, so we must be careful to choose the *xy* chart and not the *line* chart. For both charts the *y* variable is measured, however they treat the *x* variable differently. The *xy* chart measures the *x* variable, thus its *x* value is measured along the *x*-axis. For the *line* chart, each *x* value is placed an equal distance along the *x*-axis. Thus *x* values of 1, 2, 4, 8 would be spaced evenly, rather than doubling each distance. The *line* chart is really designed to plot *y* values for different categories, such as prices for models of cars or enrollments for different universities.

Drawing an *xy* plot with Excel is easiest if the table of data lists the *x* values before the *y* values. This convention makes it easy for Excel to specify one set of *x* values and several sets of *y* values. The block of *xy* values is selected, and then the chart tool is selected. Then the spreadsheet guides the user through the rest of the steps.

EXAMPLE 4-31

Graph the loan payment as a function of the number of payments for a possible new car loan. Let the number of monthly payments vary between 36 and 60. The nominal annual interest rate is 12%, and the amount borrowed is $18,000.

Solution

The spreadsheet table shown in Figure 4-1 is constructed first. Cells A5:B10 are selected, and then the chartwizard icon is selected. The first step is to select an *xy (scatter)* plot with smoothed lines and without markers as the chart type. (The other choices are no lines, straight lines, and adding markers for the data points.) The second step shows us the graph and allows the option of changing the data cells selected. The third step is for chart options. Here we add titles for the two axes and turn off showing the legend. (Since we have only one line in our graph, the legend is not needed. Deleting it leaves more room for the graph.) In the fourth step we choose where the chart is placed.

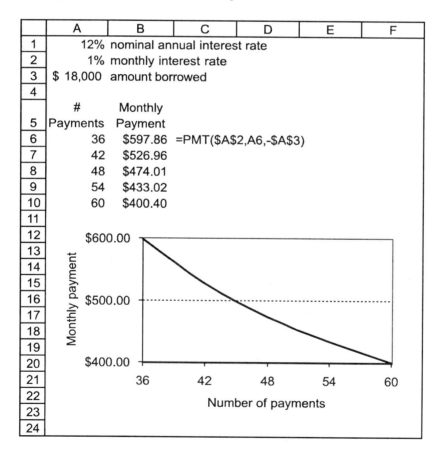

Figure 4-1 Example spreadsheet graph.

Because *xy* plots are normally graphed with the origin set to (0,0), an attractive plot requires that the minimum and maximum values be changed for each axis. This is done by placing the mouse cursor over the axis and "left-clicking." Handles or small black boxes should appear on the axis to show that it has been selected. Right-clicking brings up a menu to select format axis. The scale tab allows us to change the minimum and maximum values.

Summary

The compound interest formulas described in this chapter, along with those in chapter three, will be referred to throughout the rest of this book. It is very important that the reader understand the concepts presented and how these formulas are used.

Compound Interest. The notation used is:

i = Effective interest rate per interest period* (stated as a decimal).

n = Number of interest periods.

P = A present sum of money.

F = A future sum of money. The future sum F is an amount, n interest periods from the present, that is equivalent to P with interest rate i.

A = An end-of-period cash receipt or disbursement in a uniform series continuing for n periods, the entire series equivalent to P or F at interest rate i.

G = Uniform period-by-period increase or decrease in cash receipts or disbursements; the arithmetic gradient.

g = Uniform rate of cash flow increase or decrease from period to period; the geometric gradient.

r = Nominal interest rate per interest period*.

i_a = Effective interest rate per year (annum).

m = Number of compounding subperiods per period*.

$\overline{P},\overline{F}$ = Amount of money flowing continuously and uniformly during one given period.

Single Payment Formulas (derived in chapter 3).

Compound amount:

$$F = P(1 + i)^n = P(F/P,i,n)$$

---

* Normally the interest period is one year, but it could be something else.

Present worth:

$$P = F(1 + i)^{-n} = F(P/F, i, n)$$

Uniform Series Formulas.

Compound amount:

$$F = A\left[\frac{(1 + i)^n - 1}{i}\right] = A(F/A, i, n)$$

Sinking fund:

$$A = F\left[\frac{i}{(1 + i)^n - 1}\right] = F(A/F, i, n)$$

Capital recovery:

$$A = P\left[\frac{i(1 + i)^n}{(1 + i)^n - 1}\right] = P(A/P, i, n)$$

Present worth:

$$P = A\left[\frac{(1 + i)^n - 1}{i(1 + i)^n}\right] = A(P/A, i, n)$$

Arithmetic Gradient Formulas.

Arithmetic gradient present worth:

$$P = G\left[\frac{(1 + i)^n - in - 1}{i^2 (1 + i)^n}\right] = G(P/G, i, n)$$

Arithmetic gradient uniform series:

$$A = G\left[\frac{(1 + i)^n - in - 1}{i(1 + i)^n - i}\right] = G\left[\frac{1}{i} - \frac{n}{(1 + i)^n - 1}\right] = G(A/G, i, n)$$

Geometric Gradient Formulas.

Geometric series present worth, where $i \neq g$:

$$P = A_1 \left[\frac{1 - (1+g)^n (1+i)^{-n}}{i - g} \right] = A_1(P/A, g, i, n)$$

Geometric series present worth, where $i = g$:

$$P = A_1 \left[n(1+i)^{-1} \right] = A_1(P/A, g, i, n)$$

Single Payment Formulas–Continuous Compounding at nominal rate r per period.

Compound amount:

$$F = P(e^{rn}) = P[F/P, r, n]$$

Present worth:

$$P = F(e^{-rn}) = F[P/F, r, n]$$

Note that square brackets around the notation are used to distinguish continuous compounding.

Uniform Payment Series–Continuous Compounding at nominal rate r per period.

Continuous compounding sinking fund:

$$A = F \left[\frac{e^r - 1}{e^{rn} - 1} \right] = F[A/F, r, n]$$

Continuous compounding capital recovery:

$$A = P \left[\frac{e^{rn}(e^r - 1)}{e^{rn} - 1} \right] = P[A/P, r, n]$$

Continuous compounding series compound amount:

$$F = A \left[\frac{e^{rn} - 1}{e^r - 1} \right] = A[F/A, r, n]$$

Continuous compounding series present worth:

$$P = A\left[\frac{e^{rn} - 1}{e^{rn}(e^r - 1)}\right] = A[P/A, r, n]$$

Continuous, Uniform CashFlow (One Period) with Continuous Compounding at Nominal Interest Rate r.

Compound amount:

$$F = \overline{P}\left[\frac{(e^r - 1)(e^{rn})}{re^r}\right] = \overline{P}[F/\overline{P}, r, n]$$

Present worth:

$$P = \overline{F}\left[\frac{e^r - 1}{re^{rn}}\right] = \overline{F}[P/\overline{F}, r, n]$$

Nominal interest rate per year, r:

The annual interest rate without considering the effect of any compounding.

Effective interest rate per year, i_a:

The annual interest rate taking into account the effect of any compounding during the year.

Effective interest rate per year (periodic compounding):

$$i_a = \left(1 + \frac{r}{m}\right)^m - 1 \quad \text{or,}$$

$$i_a = (1 + i)^m - 1$$

Effective interest rate per year (continuous compounding):

$$i_a = e^r - 1$$

Problems

4-1 Solve Diagrams (a)–(c) below for the unknowns R, S, and T, assuming a 10% interest rate.

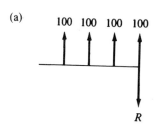

(a)

100 100 100 100

R

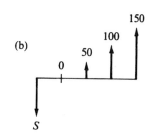

(b)

150

100

50

0

S

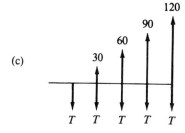

(c)

120

90

60

30

T T T T T

(*Answers:* R = $464.10; S= $218.90; T = $54.30)

4-2 A man borrowed $500 from a bank on October 15th. He must repay the loan in 16 equal monthly payments, due on the 15th of each month, beginning November 15th. If interest is computed at 1% per month, how much must he pay each month?
(*Answer:* $33.95)

4-3 A local finance company will loan $10,000 to a homeowner. It is to be repaid in 24 monthly payments of $499.00 each. The first payment is due thirty days after the $10,000 is received. What interest rate per month are they charging? (*Answer:* 1½%)

4-4 For Diagrams (a)–(d) below, compute the unknown values: B, i, V, x, respectively, using the minimum number of compound interest factors.

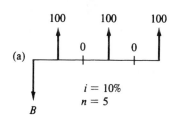

100 100 100

0 0

(a)

i = 10%
n = 5

B

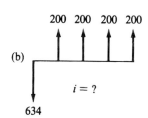

200 200 200 200

(b)

i = ?

634

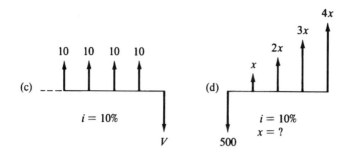

(*Answers:* B= $228.13; i = 10%; V = $51.05; x = $66.24)

4-5   For Diagrams (a)–(d) below, compute the unknown values: *C, i, F, A,* respectively.

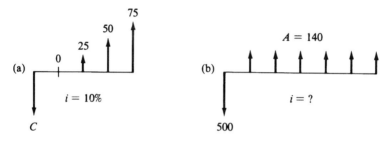

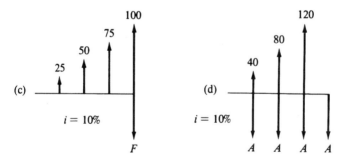

(*Answers:* C = $109.45; i = 17.24%; F = $276.37; A = $60.78)

4-6   For Diagrams (a) through (d) compute the unknown values: *W, X, Y, Z,* respectively.

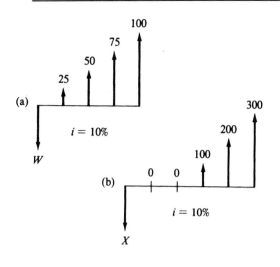

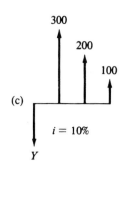

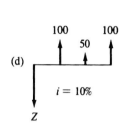

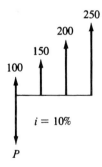

4-7   Using linear interpolation, determine the value of $(P/A, 6\frac{1}{2}\%, 10)$ from the Compound Interest Tables. Compute this same value using the equation. Why do the values differ?

4-8   Four plans have been presented for the repayment of $5000 in five years with interest at 8%. Still another way to repay the $5000 would be to make four annual end-of-year payments of $1000 each, followed by a final payment at the end of the fifth year. How much would the final payment be?

4-9   Compute the value of P in the diagram below:

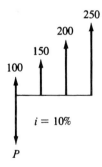

(*Answer:* $589.50)

4-10 Compute the value of *X* in the diagram below using a 10% interest rate.

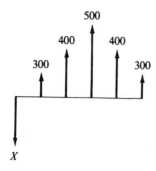

4-11 Compute the value of *P* in the diagram below using a 15% interest rate.

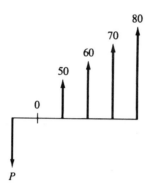

4-12 A local bank will lend a customer $1000 on a two-year car loan as follows:

Money to pay for car	$1000
Two years' interest at 7%: 2 X 0.07 X 1000	140
	$1140

$$24 \text{ monthly payments} = \frac{1140}{24} = \$47.50$$

The first payment must be made in thirty days. What is the nominal annual interest rate the bank is receiving?

4-13 A local lending institution advertises the "51–50 Club." A person may borrow $2000 and repay $51 for the next fifty months, beginning thirty days after receiving the money. Compute the nominal annual interest rate for this loan. What is the effective interest rate?

4-14 A loan company has been advertising on television a plan where one may borrow $1000 and make a payment of $10.87 per month. This payment is for interest only and includes no payment on the principal. What is the nominal annual interest rate that they are charging?

4-15   What effective interest rate per annum corresponds to a nominal rate of 12% compounded monthly? (*Answer:* 12.7%)

4-16   Mr. Sansome withdrew $1000 from a savings account and invested it in common stock. At the end of five years, he sold the stock and received a check for $1307. If Mr. Sansome had left his $1000 in the savings account, he would have received an interest rate of 5%, compounded quarterly. Mr. Sansome would like to compute a comparable interest rate on his common stock investment. Based on quarterly compounding, what nominal annual interest rate did Mr. Sansome receive on his investment in stock? What effective annual interest rate did he receive?

4-17   A woman opened an account in a local store. In the charge account agreement, the store indicated it charges 1½% each month on the unpaid balance. What nominal annual interest rate is being charged? What is the effective interest rate?

4-18   A man buys a car for $3000 with no money down. He pays for the car in thirty equal monthly payments with interest at 12% per annum, compounded monthly. What is his monthly loan payment? (*Answer:* $116.10)

4-19   What amount will be required to purchase, on a man's 40th birthday, an annuity to provide him with thirty equal semi-annual payments of $1000 each, the first to be received on his 50th birthday, if nominal interest is 4% compounded semi-annually?

4-20   Upon the birth of his first child, Dick Jones decided to establish a savings account to partly pay for his son's education. He plans to deposit $20 per month in the account, beginning when the boy is 13 months old. The savings and loan association has a current interest policy of 6% per annum, compounded monthly, paid quarterly. Assuming no change in the interest rate, how much will be in the savings account when Dick's son becomes sixteen years old?

4-21   An engineer borrowed $3000 from the bank, payable in six equal end-of-year payments at 8%. The bank agreed to reduce the interest on the loan if interest rates declined in the United States before the loan was fully repaid. At the end of three years, at the time of the third payment, the bank agreed to reduce the interest rate from 8% to 7% on the remaining debt. What was the amount of the equal annual end-of-year payments for each of the first three years? What was the amount of the equal annual end-of-year payments for each of the last three years?

4-22   On a new car, it is estimated that the maintenance cost will be $40 the first year. Each subsequent year, it is expected to be $10 more than the previous one. How much would you need to set aside when you bought a new car to pay all future maintenance costs if you planned to keep it seven years? Assume interest is 5% per annum.   (*Answer:* $393.76)

4-23   A man decides to deposit $50 in the bank today and make ten additional deposits every six months beginning six months from now, the first of which will be $50 and increasing $10 per deposit after that. A few minutes after he makes the last deposit, he decides to withdraw all the money deposited. If the bank pays 6% nominal interest compounded semi-annually, how much money will he receive?

4-24   A young engineer wishes to become a millionaire by the time he is sixty years old. He believes that by careful investment he can obtain a 15% rate of return. He plans to add a uniform sum of money to his investment program each year, beginning on his 20th birthday and continuing through his 59th birthday. How much money must the engineer set aside in this project each year?

4-25 The council members of a small town have decided that the earth levee that protects the town from a nearby river should be rebuilt and strengthened. The town engineer estimates that the cost of the work at the end of the first year will be $85,000. He estimates that in subsequent years the annual repair costs will decline by $10,000, making the second-year cost $75,000; the third-year $65,000, and so forth. The council members want to know what the equivalent present cost is for the first five years of repair work if interest is 4%. (*Answer:* $292,870)

4-26 A $150 bicycle was purchased on December lst with a $15 down payment. The balance is to be paid at the rate of $10 at the end of each month, with the first payment due on December 31st. The last payment may be some amount less than $10. If interest on the unpaid balance is computed at 1½% per month, how many payments will there be, and what is the amount of the final payment? (*Answers:* 16 payments; final payment: $1.99)

4-27 A company buys a machine for $12,000, which it agrees to pay for in five equal annual payments, beginning one year after the date of purchase, at an interest rate of 4% per annum. Immediately after the second payment, the terms of the agreement are changed to allow the balance due to be paid off in a single payment the next year. What is the final single payment? (*Answer:* $7778)

4-28 An engineering student bought a car at a local used car lot. Including tax and insurance, the total price was $3000. He is to pay for the car in twelve equal monthly payments, beginning with the first payment immediately (in other words, the first payment was the down payment). Nominal interest on the loan is 12%, compounded monthly. After he makes six payments (the down payment plus five additional payments), he decides to sell the car. A buyer agrees to pay a cash amount to pay off the loan in full at the time the next payment is due and also to pay the engineering student $1000. If there are no penalty charges for this early payment of the loan, how much will the car cost the new buyer?

4-29 A bank recently announced an "instant cash" plan for holders of its bank credit cards. A cardholder may receive cash from the bank up to a pre-set limit (about $500). There is a special charge of 4% made at the time the "instant cash" is sent the cardholders The debt may be repaid in monthly installments. Each month the bank charges 1½% on the unpaid balance. The monthly payment, including interest, may be as little as $10. Thus, for $150 of "instant cash," an initial charge of $6 is made and added to the balance due. Assume the cardholder makes a monthly payment of $10 (this includes both principal and interest). How many months are required to repay the debt? If your answer includes a fraction of a month, round up to the next month.

4-30 The treasurer of a firm noted that many invoices were received by his firm with the following terms of payment: "2%–10 days, net 30 days". Thus, if he were to pay the bill within ten days of its date, he could deduct 2%. On the other hand, if he did not promptly pay the bill, the full amount would be due thirty days from the date of the invoice. Assuming a 20-day compounding period, the 2% deduction for prompt payment is equivalent to what effective interest rate per year?

4-31 In 1555, King Henry borrowed money from his bankers on the condition that he pay 5% of the loan at each fair (there were four fairs per year) until he had made forty payments. At that time the loan would be considered repaid. What effective annual interest did King Henry pay?

4-32   A man wants to help provide a college education for his young daughter. He can afford to invest $600/yr for the next four years, beginning on the girl's fourth birthday. He wishes to give his daughter $4000 on her 18th, 19th, 20th, and 21st birthdays, for a total of $16,000. Assuming 5% interest, what uniform annual investment will he have to make on the girl's 8th through 17th birthdays?   (*Answer:* $792.73)

4-33   A man has $5000 on deposit in a bank that pays 5% interest compounded annually. He wonders how much more advantageous it would be to transfer his funds to another bank whose dividend policy is 5% interest, compounded continuously. Compute how much he would have in his savings account at the end of three years under each of these situations.

4-34   A friend was left $50,000 by his uncle. He has decided to put it into a savings account for the next year or so. He finds there are varying interest rates at savings institutions: 4⅜% compounded annually, 4¼% compounded quarterly, and 4⅛% compounded continuously. He wishes to select the savings institution that will give him the highest return on his money. What interest rate should he select?

4-35   One of the local banks indicates that it computes the interest it pays on savings accounts by the continuous compounding method. Suppose you deposited $100 in the bank and they pay 4% per annum, compounded continuously. After five years, how much money will there be in the account?

4-36   A company expects to install smog control equipment on the exhaust of a gasoline engine. The local smog control district has agreed to pay to the firm a lump sum of money to provide for the first cost of the equipment and maintenance during its ten-year useful life. At the end of ten years the equipment, which initially cost $ 10,000, is valueless. The company and smog control district have agreed that the following are reasonable estimates of the end-of-year maintenance costs:

Year 1	$500	Year 6	$200
2	100	7	225
3	125	8	250
4	150	9	275
5	175	10	300

Assuming interest at 6% per year, how much should the smog control district pay to the company now to provide for the first cost of the equipment and its maintenance for ten years?   (*Answer:* $11,693)

4-37   One of the largest automobile dealers in the city advertises a three-year-old car for sale as follows:

Cash price $3575, or a down payment of $375 with 45 monthly payments of $93.41.

Susan DeVaux bought the car and made a down payment of $800. The dealer charged her the same interest rate used in his advertised offer. How much will Susan pay each month for 45 months? What effective interest rate is being charged?   (*Answers:* $81.03; 16.1%)

4-38 At the Central Furniture Company, customers who purchase on credit pay an effective annual interest rate of 16.1 %, based on monthly compounding. What is the nominal annual interest rate that they pay?

4-39 Mary Lavor plans to save money at her bank for use in December. She will deposit $30 a month, beginning on March 1st and continuing through November 1st. She will withdraw all the money on December 1st. If the bank pays ½% interest each month, how much money will she receive on December 1st?

4-40 A man makes an investment every three months at a nominal annual interest rate of 28%, compounded quarterly. His first investment was $100, followed by investments *increasing* $20 each three months. Thus, the second investment was $120, the third investment $140, and so on. If he continues to make this series of investments for a total of twenty years, what will be the value of the investments at the end of that time?

4-41 A debt of $5000 can be repaid, with interest at 8%, by the following payments.

Year	Payment
1	$500
2	1000
3	1500
4	2000
5	X

The payment at the end of the fifth year is shown as X. How much is X?

4-42 Consider the cash flow:

Year	Cash Flow
0	-$100
1	+50
2	+60
3	+70
4	+80
5	+140

Which one of the following is correct for this cash flow?

a. $$100 = 50 + 10(A/G,i,5) + 50(P/F,i,5)$$

b.

$$\frac{50(P/A,i,5) + 10(P/G,i,5) + 50(P/F,i,5)}{100} = 1$$

c. $100(A/P,i,5) = 50 + 10(A/G,i,5)$

d.   None of the equations are correct.

4-43   If $i = 12\%$, what is the value of B in the diagram below?

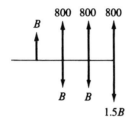

4-44   For a 10% interest rate, compute the value of n for the figure below.

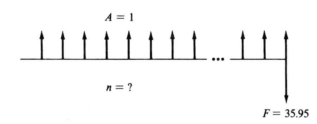

4-45   For the figure below, what is the value of n, based on a 3½% interest rate?

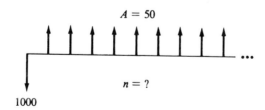

4-46   How many months will it take to pay off a $525 debt, with monthly payments of $15 at the end of each month if the interest rate is 18%, compounded monthly?

(*Answer:* 50 months)

4-47   For the diagram below and a 10% interest rate, compute the value of *J*.

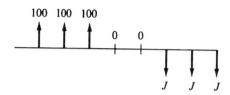

4-48   For the diagram below and a 10% interest rate, compute the value of *C*.

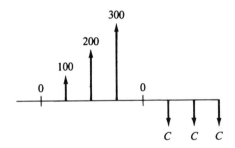

4-49   On January 1st, Frank Jenson bought a used car for $4200 and agreed to pay for it as follows: ⅓ down payment; the balance to be paid in 36 equal monthly payments; the first payment due February 1st; an annual interest rate of 9%, compounded monthly.

    a. What is the amount of Frank's monthly payment?

    b. During the summer, Frank made enough money that he decided to pay off the entire balance due on the car as of October 1st. How much did Frank owe on October 1st?

4-50   If *i* = 12%, compute *G* in the diagram below:

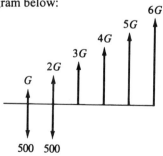

4-51 Compute the value of *D* in the diagram below:

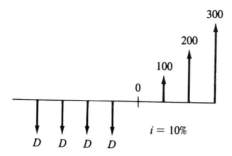

4-52 Compute *E* for the figure below:

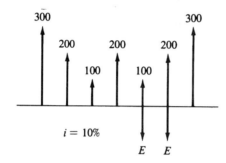

4-53 Compute *B* in the figure below, using a 10% interest rate.

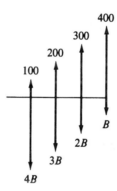

4-54 On January 1st, Laura Brown borrowed $1000 from the Friendly Finance Company. The loan is to be repaid by four equal payments which are due at the end of March, June, September, and December. If the finance company charges 18% interest, compounded quarterly, what is the amount of each payment? What is the effective annual interest rate? (*Answers:* $278.70; 19.3%)

4-55 If you want a 12% rate of return, continuously compounded, on a project that will yield $6000 at the end of 2½ years, how much must you be willing to invest now? (*Answer:* $4444.80)

4-56 What monthly interest rate is equivalent to an effective annual interest rate of 18%?

4-57 A department store charges 1¾% interest per month, compounded continuously, on its customer's charge accounts. What is the nominal annual interest rate? What is the effective interest rate? (*Answers:* 21%; 23.4%)

4-58 A bank is offering to sell six-month certificates of deposit for $9500. At the end of six months, the bank will pay $10,000 to the certificate owner. Based on a six-month interest period, compute the nominal annual interest rate and the effective annual interest rate.

4-59 Two savings banks are located across the street from each other. The West Bank put a sign in the window saying, "We pay 6.50%, compounded daily." The East Bank decided that they would do better, so they put a sign in their window saying, "We pay 6.50%, compounded continuously."

Jean Silva has $10,000 which she will put in the bank for one year. How much additional interest will Jean receive by placing her money in the East Bank rather than the West Bank?

4-60 A bank advertises it pays 7% annual interest, compounded daily, on savings accounts, provided the money is left in the account for four years. What effective annual interest rate do they pay?

4-61 To repay a $1000 loan, a man paid $91.70 at the end of each month for twelve months. Compute the nominal interest rate he paid.

4-62 Sally Struthers wants to have $10,000 in a savings account at the end of six months. The bank pays 8%, compounded continuously. How much should Sally deposit now? (*Answer:* $9608)

4-63 A student bought a $75 used guitar and agreed to pay for it with a single $85 payment at the end of six months. Assuming semi-annual (every six months) compounding, what is the nominal annual interest rate? What is the effective interest rate?

4-64 A firm charges its credit customers 1¾% interest per month. What is the effective interest rate?

4-65 One thousand dollars is invested for seven months at an interest rate of one percent per month. What is the nominal interest rate? What is the effective interest rate? (*Answers:* 12%; 12.7%)

4-66 The *Rule of 78's* is a commonly used method of computing the amount of interest, when the balance of a loan is repaid in advance.

If one adds the numbers representing twelve months,

$$1 + 2 + 3 + 4 + 5 + \cdots + 11 + 12 = 78$$

If a twelve-month loan is repaid at the end of one month, for example, the interest the borrower would be charged is $^{12}/_{78}$ of the year's interest. If the loan is repaid at the end of two months, the total interest charged would be $(12 + 11)/78$, or $^{23}/_{78}$ of the year's interest. After eleven months the interest charge would therefore be $^{77}/_{78}$ of the total year's interest.

Helen Reddy borrowed $10,000 on January 1st at 9% annual interest, compounded monthly. The loan was to be repaid in twelve equal end-of-period payments. Helen made the first two payments and then decided to repay the balance of the loan when she pays the third payment. Thus she will pay the third payment plus an additional sum.

You are to calculate the amount of this additional sum

    a.   Based on the Rule of 78's;

    b.   Based on exact economic analysis methods.

4-67   Consider the following cash flow:

Year	Cash flow
0	−$P
1	+1000
2	+850
3	+700
4	+550
5	+400
6	+400
7	+400
8	+400

Alice was asked to compute the value of P for the cash flow at 8% interest. She wrote the following three equations:

    a. $P = 1000(P/A,8\%,8) - 150(P/G,8\%,8) + 150(P/G,8\%,4)(P/F,8\%,4)$

    b. $P = 400(P/A,8\%,8) + 600(P/A,8\%,5) - 150(P/G,8\%,4)$

    c. $P = 150(P/G,8\%,4) + 850(P/A,8\%,4) + 400(P/A,8\%,4)(P/F,8\%,4)$

Which of the equations is correct?

4-68   Ann Landers deposits $100 at the end of each month into her bank savings account. The bank pays 6% nominal interest, compounded and paid quarterly. No interest is paid on money not in the account for the full three-month period. How much will be in Ann's account at the end of three years? (*Answer:* $3912.30)

4-69   A college professor just won $85,000 in the state lottery. After paying income taxes, about half the money will be left. She and her husband plan to spend her sabbatical year on leave from the university on an around-the-world trip, but she must continue to teach three more years first. She estimates the trip will cost $40,000 and they will spend the money as a continuous flow of funds during their year of travel. She will put enough of her lottery winnings in a bank account now to pay for the trip. The bank pays 7% nominal interest, compounded continuously. She asks you to compute how much she should set aside in the account for their trip.

4-70   Mark Johnson saves a fixed percentage of his salary at the end of each year. This year he saved $1500. For the next five years, he expects his salary to increase at an 8% annual rate, and he plans to increase his savings at the same 8% annual rate. He invests his money in the stock market. Thus there will be six end-of-year investments ($1500 plus five more). Solve the problem using the geometric gradient factor.

a. How much will his investments be worth at the end of six years if they increase in the stock market at a 10% annual rate?

b. How much will Mark have at the end of six years if his stock market investments only increase at an 8% annual rate?

4-71 The *Bawl Street Journal* costs $206, payable now, for a two-year subscription. The newspaper is published 252 days per year (five days per week, except holidays). If a 10% nominal annual interest rate, compounded quarterly, is used:

a. What is the effective annual interest rate in this problem?

b. Compute the equivalent interest rate per $1/252$ of a year.

c. What is a subscriber's cost per copy of the newspaper, taking interest into account?

4-72 Michael Jackson deposited $500,000 into a bank for six months. At the end of that time, he withdrew the money and received $520,000. If the bank paid interest based on continuous compounding:

a. What was the effective annual interest rate?

b. What was the nominal annual interest rate?

4-73 The I've Been Moved Corporation receives a constant flow of funds from its worldwide operations. This money (in the form of checks) is continuously deposited in many banks with the goal of earning as much interest as possible for "IBM." One billion dollars is deposited each month, and the money earns an average of ½% interest per month, compounded continuously. Assume all the money remains in the accounts until the end of the month.

a. How much interest does IBM earn each month?

b. How much interest would IBM earn each month if it held the checks and made deposits to its bank accounts just four times a month?

4-74 A married couple is opening an Individual Retirement Account (IRA) at a bank. Their goal is to accumulate $1,000,000 in the account by the time they retire from work in forty years. The bank manager estimates they may expect to receive 8% nominal annual interest, compounded quarterly, throughout the forty years. The couple believe their income will increase at a 7% annual rate during their working careers. They wish to start with as low a deposit as possible to their IRA now and increase it at a 7% rate each year. Assuming end-of-year deposits, how much should they deposit the first year?

4-75 The Macintosh Co. has an employee savings plan in which an employee may invest up to 5% of his or her annual salary. The money is invested in company common stock with the company guaranteeing the annual return will never be less than 8%. Jill was hired at an annual salary of $52,000. She immediately joined the savings plan investing the full 5% of her salary each year. If Jill's salary increases at an 8% uniform rate, and she continues to invest 5% of it each year. what amount of money is she guaranteed to have at the end of 20 years?

4-76   The football coach at a midwest university was given a five-year employment contract which paid $225,000 the first year, and increased at an 8% uniform rate in each subsequent year. At the end of the first year's football season, the alumni demanded that he be fired. The alumni agreed to buy the coach's remaining four years on the contract by paying him the equivalent present sum, computed using a 12% interest rate. How much will the coach receive?

4-77   A group of ten public-spirited citizens has agreed that they will support the local school hot lunch program. Each year one of the group is to pay the $15,000 years' cost that occurs continuously and uniformly during the year. Each member of the group is to underwrite the cost for one year. Slips of paper numbered Year 1 through Year 10 are put in a hat. As one of the group you draw the slip marked Year 6. Assuming an 8% nominal interest rate per year, how much do you need to set aside now to meet your obligation in Year 6?

4-78   In 1990 Mrs. John Hay Whitney sold her painting by Renoir, *Au Moulin de la Galette*, depicting an open-air Parisian dance hall, for $71 million. The buyer also had to pay the auction house commission of 10%, or a total of $78.1 million. Mrs. Whitney purchased the painting in 1929 for $165,000.

    a. What rate of return did she receive on her investment?

    b. Was the rate of return really as high as you computed in *a*? Explain.

4-79   Derive an equation to find the end of year future sum F that is equivalent to a series of n beginning-of-year payments B at interest rate i. Then use the equation to determine the future sum F equivalent to six B payments of $100 at 8% interest. (Answer: $F = \$792.28$)

4-80   A woman made ten annual end-of-year purchases of $1000 worth of common stock. The stock paid no dividends. Then for four years she held the stock. At the end of the four years she sold all the stock for $28,000. What interest rate did she obtain on her investment?

4-81   For some interest rate i and some number of interest periods n, the uniform series capital recovery factor is 0.1728 and the sinking fund factor is 0.0378. What is the interest rate?

4-82   What interest rate, compounded quarterly, is equivalent to a 9.31% effective interest rate?

4-83   A contractor wishes to set up a special fund by making uniform semiannual end-of-period deposits for 20 years. The fund is to provide $10,000 at the end of each of the last five years of the 20-year period. If interest is 8%, compounded semiannually, what is the required semiannual deposit?

4-84   How long will it take for $10,000, invested at 5% per year, compounded continuously, to triple in value?

4-85   If $200 is deposited in a savings account at the beginning of each of 15 years, and the account draws interest at 7% per year, how much will be in the account at the end of 15 years?

4-86   An automobile may be purchased with a $3000 downpayment now and 60 monthly payments of $280. If the interest rate is 12% compounded monthly, what is the price of the automobile?

4-87   If the nominal annual interest rate is 12% compounded quarterly, what is the effective annual interest rate?

4-88 A man is purchasing a small garden tractor. There will be no maintenance cost the first two years as the tractor is sold with two years free maintenance. The third year the maintenance is estimated at $20. In subsequent years the maintenance cost will increase by $20 per year (that is, fourth year maintenance will be $40; fifth year $60, and so on). How much would need to be set aside now at 8% interest to pay the maintenance costs on the tractor for the first six years of ownership?

4-89 How many months, at an interest rate of one percent per month, does money have to be invested before it will double in value?

4-90 A company deposits $2000 at the end of every year for ten years in a bank. The company makes no deposits during the subsequent five years. If the bank pays 8% interest, how much would be in the account at end of 15 years?

4-91 A bank pays 10% nominal annual interest on special three-year certificates. What is the effective annual interest rate if interest is compounded:

    a.   every three months?

    b.   daily?

    c.   continuously?

4-92 Find the value of P for the cash flow diagram shown below.

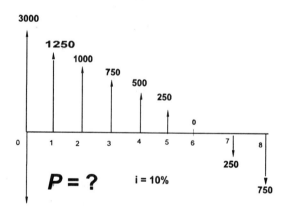

4-93 A bank is offering a loan of $25,000 with a nominal interest rate of 18% compounded monthly, payable in 60 months.

    a. What is the monthly payment?

    b. If a loan origination fee of 2% is charged at the time of the loan, compute the effective interest rate?

    Hint -- The loan origination fee of 2% will be taken out from the loan amount.

4-94 Select the best alternative among the following five alternatives. Assume the investment is for a period of 4 years. $P = $10,000

a. 11.98% interest rate compounded continuously

b. 12.00% interest rate compounded daily

c. 12.01% interest rate compounded monthly

d. 12.02% interest rate compounded quarterly

e. 12.03% interest rate compounded yearly

4-95 Pete Samprass borrows $10,000 to purchase a car. He must repay the loan in 48 equal end-of-period monthly payments. Interest is calculated at 1.25% per month. Determine the following:

a. The nominal annual interest rate.

b. The effective annual interest rate.

c. The amount of the monthly payment.

4-96 Picabo Street borrows $1,000. To repay the amount she makes twelve equal monthly payments of $90.30. Determine the following:

a. The effective monthly interest rate.

b. The nominal annual interest rate.

c. The effective annual interest rate.

4-97 Bart Simpson wishes to tour the country with his friends. To do this, he is saving money in order to buy a bus.

a. How much money must Bart deposit in a savings account paying 8% nominal annual interest, compounded continuously, in order to have $8000 in 4-1/2 years?

b. A friend offers to repay Bart $8000 in 4-1/2 years if Bart gives him $5000 now. Assuming continuous compounding, what is the nominal annual interest rate of this offer?

4-98 How much must be deposited now at 5.25% interest to produce $300 at the end of every year for 10 years?

4-99 The following two cash flow transactions are said to be equivalent in terms of economic desirability at an interest rate of 12% compounded annually. Determine the unknown value A.

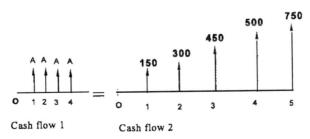

Cash flow 1   Cash flow 2

4-100 A realtor sold a house on August 31, 1997 for $150,000 to a buyer in which a 20% down payment was made. The buyer took a 15-year mortgage on the property with an effective interest rate of 8% per annum. The buyer intends to pay off the mortgage owed in yearly payments starting on August 31, 1998.

> *a.* How much of the mortgage will still be owed after the August 31, 2004 payment will have been paid?
>
> *b.* Solve the same problem by separating the interest and the principal amounts.

4-101 To provide for a college education for his daughter, a man opened an escrow account in 1981 in which equal deposits were made. The first deposit was made on January 1, 1981 and the last deposit was made on January 1, 1998. The yearly college expenses including tuition were estimated to be $8000, for each of the 4 years. Assuming the interest rate to be 5.75%, what was the size of the yearly deposit in the escrow account in order for the daughter to draw $8000 per year for four years beginning January 1, 1998?

4-102 What is the present worth of a series of equal quarterly payments of $3000 that extends over a period of 8 years if the interest rate is 10% compounded monthly?

4-103 You are taking a $2000 loan. You will pay it back in four equal amounts, paid every six months starting three years from now. The interest rate is 6% compounded semiannually. Calculate:

> *a.* The effective interest rate, based on both semiannual and continuous compounding.
>
> *b.* The amount of each semiannual payment.
>
> *c.* The total interest paid.

4-104 Develop a complete amortization table for a loan of $4500, to be paid back in 24 uniform monthly installments, based on an interest rate of 6%. The amortization table must include the following column headings:

> payment number, principal owed (beginning of period), interest owed in each period, total owed (end of each period), principal paid in each payment, uniform monthly payment amount.

You must also show the equations used to calculate each column of the table. You are encouraged to use spreadsheets. *The entire table must be shown.*

4-105 Using the loan and payment plan developed in problem 4-104, determine the month that the final payment is due, and the amount of the final payment, if $500 is paid for payment 8 and $280 is paid for payment 10. This problem requires <u>a separate amortization table</u>.

4-106 The following beginning of month(BOM) and end of month (EOM) amounts are to be deposited in a savings account that pays interest at 9%, compounded monthly:

Today (BOM 1)	$400
EOM 2	270
EOM 6	100
EOM 7	180
BOM 10	200

Set up a spreadsheet to calculate the account balance at the end of the first year (EOM12). The spreadsheet must include the following column headings: month number; deposit BOM; account balance at BOM; interest earned in each month; deposit EOM; account balance at EOM. Also, draw a cash flow diagram of this problem and solve for the account balance at the EOM 12 using the compound interest tables.

4-107 Eighty thousand dollars are borrowed at 7% to purchase a house. The loan is to be repaid in equal monthly payments over a 30 year period. The first payment is paid exactly at the end of the first month. Calculate the interest and principal in the second payment, if the second payment is made 33 days after the first payment.

4-108 A student wants to have $30,000 at graduation four years from now to buy a new car. His grandfather gave him $10,000 as a high school graduation present. How much must the student save each year if he deposits the $10,000 today and can earn 12% on both the $10,000 and his earnings in a mutual fund his grandfather recommends?

4-109 A city engineer knows that she will need $25,000,000 in three years to replace toll booths on a toll road in the city. Traffic on the road is estimated to be 20,000,000 vehicles per year. How much per vehicle should the toll price be to cover the cost of the toll booth replacement project? Interest is 10%. (Simplify your analysis by assuming that the toll receipts are received at the end of each year in a lump sum)

4-110 Traffic at a certain intersection is currently 2000 cars per day. A consultant has told the city that traffic is expected to grow at a continuous rate of 5% per year for the next 4 years. How much traffic will be expected at the end of two years?

4-111 What single amount on April 1, 1998 is equivalent to a series of equal, semiannual cash flows of $1000 that starts with a cash flow on January 1, 1996, and ends with a cash flow on January 1, 2005? The interest rate is 14% and compounding is quarterly.

4-112 A forklift truck costs $29,000. A company agrees to purchase the forklift truck with the understanding that it will make a single payment for the balance due in 3 years. The vendor agrees to the deal and offers two different interest schedules. The first schedule uses an annual effective interest rate of 13%. The second schedule uses 12.75% compounded continuously. *a.* Which schedule should the company accept? *b.* What would be size of the single payment?

4-113 PARC Co. has money to invest in an employee benefit plan and you have been chosen as the plan's trustee. As an employee yourself you want to maximize the interest earned on this investment and have found an account that pays 14% compounded continuously. PARC Co. is providing you $1200 per month to put into your account for seven years. What is the balance in this account after the seven year period?

4-114 Paco's saving account earns 13% compounded weekly and receives quarterly deposits of $38,000. His first deposit occurs on October 1, 1996, and the last deposit occurs on April 1, 2012. Tisha's account earns 13% compounded weekly. Semiannual deposits of $18,000 are made into her account with the first one occurring on July 1, 2006, and the last one occurring on January 1, 2015. What single amount on January 1, 2007, is equivalent to both cash flow series?

4-115 The first of a series of equal semiannual cash flows occurs on July 1, 1997, and the last occurs on January 1, 2010. Each cash flow is equal to $128,000. The nominal interest rate is 12% compounded semiannually. What single amount on July 1, 2001 is equivalent to this cash flow system?

4-116 The first of a series of equal, monthly cash flows of $2000 occurs on April 1, 1998, and the last of the monthly cash flows occurs on February 1, 2000. This series of monthly cash flows is equivalent to a series of semiannual cash flows. The first semiannual cash flow occurs on July 1, 2001, and the last semiannual cash flow occurs on January 1, 2010. What is the amount of each semiannual cash flow? Use a nominal interest rate of 12% with monthly compounding on all accounts.

4-117 What single amount on October 1, 1997, is equal to a series of $1000 quarterly deposits made into an account? The first deposit occurs on October 1, 1997 and the last deposit occurs on January 1, 2011. The account earns 13% compounded continuously.

4-118 Barry, a recent Texas Tech graduate, never took Engineering Economics. When he graduated, he was employed by a prominent architectural firm. The earnings from this job allowed him to deposit $750 each quarter into a savings account. There were two banks offering a savings account in his town (a small town!). The first bank was offering 4.5% interest compounded continuously. The second bank offered 4.6% compounded monthly. Barry decided to deposit in the first bank since it offered continuous compounding. Based on this information, did he make the right decision?

4-119 A series of monthly cash flows is deposited into an account which earns 12% nominal interest compounded monthly. Each monthly deposit is equal to $2100. The first monthly deposit occurs on June 1, 1998 and the last monthly deposit occurs on January 1, 2005. The above account (the series of monthly deposits, 12% nominal interest, and monthly compounding) also has equivalent quarterly withdrawals from it. The first quarterly withdrawal is equal to $5000 and occurs on October 1, 1998. The last $5000 withdrawal occurs on January 1, 2005. How much remains in the account after the last withdrawal?

4-120 Our cat, Fred, wants to purchase a new litter box. The cost is $100 and he'll finance it over 2 years at an annual rate of 18% compounded monthly and to be repaid in 24 monthly payments.

    a. What is his monthly payment?

    b. At the time of the 13th payment, Fred decides to pay off the remainder of the loan.

    Using regular compound interest factors, determine the amount of this last payment.

4-121 Our cat, Fred, has convinced me that I should set up an account such that he will be assured of his "Meow Mix" for the next four years. I will deposit an amount "*P*" today such that he can make end of the month withdrawals of $10 for the next 48 months. Consider an interest rate of 6%

compounded monthly and that the account will be emptied with the last withdrawal.

  a. What is the value of "*P*" that I must deposit today?

  b. What is the account balance immediately after the 24th withdrawal has been made?

4-122 When Jerry Garcia was alive he bought a house for $500,000 and made a $100,000 down payment. He obtained a 30 year loan for the remaining amount. Payments were made monthly. The nominal annual interest rate was 9%. After ten years (120 payments) he decided to pay the remaining balance on the loan.

  a. What was his monthly loan payment?

  b. What must he have paid (in addition to his regular 120th monthly payment) to pay the remaining balance of his loan?

  c. Recompute part (a) using 6% compounded continuously.

4-123 Jim Duggan made an investment of $10,000 in a savings account 10 years ago. This account paid interest of 5-1/2% for the first 4 years and 6-1/2% interest for the remaining 6 years. The interest charges were compounded quarterly.

  a. How much is this investment worth now?

  b. What is the equivalent effective interest rate per year on this investment?

4-124 Net revenues at an older manufacturing plant will be $2M for this year. The net revenue will decrease 15% per year for five years, when the assembly plant will be closed (at the end of year 6). If the firm's interest rate is 10%, calculate the PW of the revenue stream.

4-125 What is the present worth of cash flows that begin at $10,000 and increase at 8% per year for 4 years? The interest rate is 6%.

4-126 What is the present worth of cash flows that begin at $30,000 and decrease at 15% per year for 6 years? The interest rate is 10%.

4-127 Calculate and print out an amortization schedule for a used car loan. The nominal interest is 12% per year, compounded monthly. Payments are made monthly for 3 years. The original loan is for $11,000.

4-128 Calculate and print out an amortization schedule for a new car loan. The nominal interest is 9% per year, compounded monthly. Payments are made monthly for 5 years. The original loan is for $17,000.

4-129 For the used car loan of Problem 4.127, graph the monthly payment.

  a. As a function of the interest rate (5% to 15%).

  b. As a function of the number of payments (24 to 48).

4-130 For the new car loan of Problem 4.128, graph the monthly payment.

a. As a function of the interest rate (4% to 14%).

b. As a function of the number of payments (36 to 84).

🖳 **4-131** Your beginning salary is $50,000. You deposit 10% at the end of each year in a savings account that earns 6% interest. Your salary increases by 5% per year. What value does your savings book show after 40 years?

🖳 **4-132** The market volume for widgets is increasing by 15% per year from current profits of $200,000. Investing in a design change will allow the profit per widget to stay steady, otherwise they will drop 3% per year. What is the present worth of the savings over the next 5 years? Ten years? The interest rate is 10%.

🖳 **4-133** A 30-year mortgage for $120,000 has been issued. The interest rate is 10% and payments are made monthly. Print an amortization schedule.

🖳 **4-134** A homeowner may upgrade a fuel oil based furnace to a natural gas unit. The investment will be $2500 installed. The cost of the natural gas will average $60 per month over the year, instead of the $145 per month that the fuel oil costs. If the interest rate is 9% per year, how long will it take to recover the initial investment?

🖳 **4-135** Develop a general purpose spreadsheet to calculate an amortization schedule for a loan. The user inputs to the spreadsheet will be the loan amount, the number of payments per year, the number of years payments are made, and the nominal interest rate. Submit printouts of your analysis of a loan in the amount of $15,000 @ 8.9% nominal rate for 36 months and for 60 months of payments.

🖳 **4-136** Use the spreadsheet developed for Problem 4-135 to analyze 180-month and 360-month house loan payments. Analyze a $100,000 mortgage loan at a nominal interest rate of 7.5% and submit a graph of the interest and principal paid over time. You need not submit the printout of the 360 payments because it will not fit on one page.

CHAPTER 5

Present Worth Analysis

In the previous chapters two important tasks were accomplished. First, the concept of equivalence was presented. We are powerless to compare series of cash flows unless we can resolve them into some equivalent arrangement. Second, equivalence, with alteration of cash flows from one series to an equivalent sum or series of cash flows, created the need for compound interest factors. A whole series of compound interest factors were derived—some for periodic compounding and some for continuous compounding. This background sets the stage for the chapters that follow.

Economic Criteria

We have shown how to manipulate cash flows in a variety of ways, and in so doing we can now solve many kinds of compound interest problems. But engineering economic analysis is more than simply solving interest problems. The decision process (see Figure 2-2) requires that the outcomes of feasible alternatives be arranged so that they may be judged for *economic efficiency* in terms of the selection criterion. Depending on the situation, the economic criterion will be one of the following:[1]

---

[1]This short table summarizes the discussion on selection of criteria, early in Chapter 3.

Situation	*Criterion*
For fixed input	Maximize output
For fixed output	Minimize input
Neither input nor output fixed	Maximize (output – input)

We will now examine ways to resolve engineering problems, so that criteria for economic efficiency can be applied.

Equivalence provides the logic by which we may adjust the cash flow for a given alternative into some equivalent sum or series. To apply the selection criterion to the outcomes of the feasible alternatives, we must first resolve them into comparable units. The question is, how should they be compared? In this chapter we'll learn how analysis can resolve alternatives into *equivalent present consequences,* referred to simply as **present worth analysis.** Chapter 6 will show how given alternatives are converted into an *equivalent uniform annual cash flow,* and Chapter 7 solves for the interest rate at which favorable consequences—that is, *benefits*—are equivalent to unfavorable consequences—or *costs.*

As a general rule, any economic analysis problem may be solved by the methods presented in this and in the two following chapters. This is true because *present worth, annual cash flow,* and *rate of return* are exact methods that will always yield the same solution in selecting the best alternative from among a set of mutually exclusive alternatives.[2] Some problems, however, may be more easily solved by one method than another. For this reason, we now focus on the kinds of problems that are most readily solved by present worth analysis.

Applying Present Worth Techniques

One of the easiest ways to compare mutually exclusive alternatives is to resolve their consequences to the present time. The three criteria for economic efficiency are restated in terms of present worth analysis in Table 5-1.

Present worth analysis is most frequently used to determine the present value of future money receipts and disbursements. It would help us, for example, to determine a present worth of income-producing property, like an oil well or an apartment house. If the future income and costs are known, then using a suitable interest rate, the present worth of the property may be calculated. This should provide a good estimate of the price at which the property could be

---

[2]Mutually exclusive is where selecting one alternative precludes selecting any other alternative. An example of mutually exclusive alternatives would be deciding between constructing a gas station or a drive-in restaurant on a particular piece of vacant land.

bought or sold. Another application might be determining the valuation of stocks or bonds based on the anticipated future benefits from owning them.

In present worth analysis, careful consideration must be given to the time period covered by the analysis. Usually the task to be accomplished has a time period associated with it. In that case, the consequences of each alternative must be considered for this period of time which is usually called the **analysis period,** or sometimes the **planning horizon**.

Table 5-1 PRESENT WORTH ANALYSIS

	Situation	*Criterion*
Fixed input	Amount of money or other input resources are fixed	Maximize present worth of benefits or other outputs
Fixed output	There is a fixed task, benefit, or other output to be accomplished	Minimize present worth of costs or other inputs
Neither input nor output is fixed	Neither amount of money, or other inputs, nor amount of benefits, or other output, is fixed	Maximize (present worth of benefits *minus* present worth of costs), that is, maximize net present worth

There are three different analysis-period situations that are encountered in economic analysis problems:

1. The useful life of each alternative equals the analysis period.
2. The alternatives have useful lives different from the analysis period.
3. There is an infinite analysis period, $n = \infty$

1. Useful Lives Equal the Analysis Period

Since different lives and an infinite analysis period present some complications, we will begin with four examples where the useful life of each alternative equals the analysis period.

EXAMPLE 5-1

A firm is considering which of two mechanical devices to install to reduce costs in a particular situation. Both devices cost $1000 and have useful lives of five years and no salvage value. Device *A* can be expected to result in $300 savings annually. Device *B* will provide cost savings of $400 the first year but will decline $50 annually, making the second year savings $350, the third year savings $300, and so forth. With interest at 7%, which device should the firm purchase?

Solution: The analysis period can conveniently be selected as the useful life of the devices, or five years. Since both devices cost $1000, we have a situation where, in choosing either A or B, there is a fixed input (cost) of $1000. The appropriate decision criterion is to choose the alternative that maximizes the present worth of benefits.

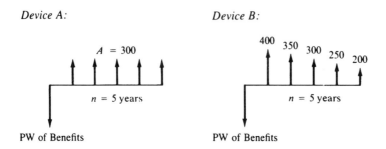

PW of benefits $A = 300 \ (P/A,7\%,5) = 300(4.100) = \$1230$
PW of benefits $B = 400 \ (P/A,7\%,5) - 50(P/G,7\%,5)$
$$= 400(4.100) - 50(7.647) = \$1257.65$$

Device B has the larger present worth of benefits and is, therefore, the preferred alternative. It is worth noting that, if we ignore the time value of money, both alternatives provide $1500 worth of benefits over the five-year period. Device B provides greater benefits in the first two years and smaller benefits in the last two years. This more rapid flow of benefits from B, although the total magnitude equals that of A, results in a greater present worth of benefits. ■

EXAMPLE 5-2

Wayne County will build an aqueduct to bring water in from the upper part of the state. It can be built at a reduced size now for $300 million and be enlarged 25 years hence for an additional $350 million. An alternative is to construct the full-sized aqueduct now for $400 million.

Both alternatives would provide the needed capacity for the 50-year analysis period. Maintenance costs are small and may be ignored. At 6% interest, which alternative should be selected?

Solution: This problem illustrates stage construction. The aqueduct may be built in a single stage, or in a smaller first stage followed many years later by a second stage to provide the additional capacity when needed.

For the two-stage construction:
$$\text{PW of cost} = \$300 \text{ million} + 350 \text{ million} \ (P/F,6\%,25)$$
$$= \$300 \text{ million} + 81.6 \text{ million} = \$381.6 \text{ million}$$

For the single-stage construction:

PW of cost = $400 million
The two-stage construction has a smaller present worth of cost and is the preferred construction plan. ■

EXAMPLE 5-3

A purchasing agent is considering the purchase of some new equipment for the mailroom. Two different manufacturers have provided quotations. An analysis of the quotations indicates the following:

Manufacturer	Cost	Useful life, in years	End-of-useful-life salvage value
Speedy	$1500	5	$200
Allied	1600	5	325

The equipment of both manufacturers is expected to perform at the desired level of (fixed) output. For a five-year analysis period, which manufacturer's equipment should be selected? Assume 7% interest and equal maintenance costs.

Solution: For fixed output, the criterion is to minimize the present worth of cost.
Speedy:
PW of cost $= 1500 - 200(P/F,7\%,5) = 1500 - 200(0.7130)$
$= 1500 - 143 = \$1357$

Allied:
PW of cost $= 1600 - 325(P/F,7\%,5) = 1600 - 325(0.7130)$
$= 1600 - 232 = \$1368$

Since it is only the *differences between alternatives* that are relevant, maintenance costs may be left out of the economic analysis. Although the PW of cost for each of the alternatives is nearly identical, we would, nevertheless, choose the one with minimum present worth of cost unless there were other tangible or intangible differences that would change the decision. Buy the Speedy equipment. ■

EXAMPLE 5-4

A firm is trying to decide which of two alternate weighing scales it should install to check a package filling operation in the plant. The scale would allow better control of the filling operation and result in less overfilling. If both scales have lives equal to the six-year analysis period, which one should be selected? Assume an 8% interest rate.

Alternatives	Cost	Uniform annual benefit	End-of-useful-life salvage value
Atlas scale	$2000	$450	$100
Tom Thumb scale	3000	600	700

Solution:

Atlas scale:

PW of benefits – PW of cost = $450(P/A,8\%,6) + 100(P/F,8\%,6) - 2000$
$$= 450(4.623) + 100(0.6302) - 2000$$
$$= 2080 + 63 - 2000 = \$143$$

Tom Thumb scale:

PW of benefits – PW of cost = $600(P/A,8\%,6) + 700(P/F,8\%,6) - 3000$
$$= 600(4.623) + 700(0.6302) - 3000$$
$$= 2774 + 441 - 3000 = \$215$$

The salvage value of the scale, it should be noted, is simply treated as another benefit of the alternative. Since the criterion is to maximize the present worth of benefits minus the present worth of cost, the preferred alternative is the Tom Thumb scale. ∎

Net Present Worth

In Example 5-4, we compared two alternatives and selected the one where present worth of benefits *minus* present worth of cost was a maximum. The criterion is called the *net present worth criterion* and written simply as **NPW**:

Net present worth = Present worth of benefits – Present worth of cost
$$NPW = PW \text{ of benefits } - PW \text{ of cost}$$

2.   Useful Lives Different from the Analysis Period

In present worth analysis, there always must be an identified analysis period. It follows, then, that each alternative must be considered for the entire period. In the previous Examples, the useful life of each alternative was equal to the analysis period. Often we can arrange it this way, but there will be many more situations where the alternatives have useful lives different from the analysis period. This section examines the problem and describes how to overcome this difficulty.

In Ex. 5-3, suppose that the Allied equipment was expected to have a ten year useful life, or twice that of the Speedy equipment. Assuming the Allied salvage value would still be $325 ten years hence, which equipment should now be purchased? We will recompute the present worth of cost of the Allied equipment.

Allied:

PW of cost = $1600 - 325(P/F,7\%,10) = 1600 - 325(0.5083)$
$$= 1600 - 165 = \$1435$$

The present worth of cost has increased. This is due, of course, to the more distant recovery of the salvage value. More importantly, we now find ourselves attempting to compare Speedy equipment, with its five-year life, against the Allied equipment with a ten-year life. This variation in the useful life of the equipment means we no longer have a situation of *fixed output*. Speedy equipment in the mailroom for five years is certainly not the same as ten years of service with Allied equipment. For present worth calculations, it is important that we select an analysis period and judge the consequences of each of the alternatives during the selected analysis period.

The analysis period for an economy study should be determined from the situation. In some industries with rapidly changing technologies, a rather short analysis period or planning horizon might be in order. Industries with more stable technologies (like steel making) might use a longer period (say, ten to twenty years), while government agencies frequently use analysis periods extending to fifty years or more.

Not only is the firm and its economic environment important in selecting an analysis period, but also the specific situation being analyzed is important. If the Allied equipment (Ex. 5-3) has a useful life of ten years, and the Speedy equipment will last five years, one method is to select an analysis period which is the **least common multiple** of their useful lives. Thus we would compare the ten-year life of Allied equipment against an initial purchase of Speedy equipment *plus* its replacement with new Speedy equipment in five years. The result is to judge the alternatives on the basis of a ten-year requirement in the mailroom. On this basis the economic analysis is as follows:

Speedy: Assuming the replacement Speedy equipment will also cost $1500 five years hence,

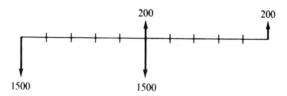

$$\text{PW of cost} = 1500 + (1500 - 200)(P/F,7\%,5) - 200(P/F,7\%,10)$$
$$= 1500 + 1300(0.7130) - 200(0.5083)$$
$$= 1500 + 927 - 102 = \$2325$$

Allied:

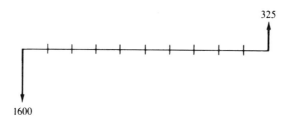

$$\text{PW of cost} = 1600 - 325(P/F,7\%,10) = 1600 - 325(0.5083) = \$1435$$

For the fixed output of ten years of service in the mailroom, the Allied equipment, with its smaller present worth of cost, is preferred. ■

We have seen that setting the analysis period equal to the least common multiple of the lives of the two alternatives seems reasonable in the revised Ex. 5-3. What would one do, however, if in another situation the alternatives had useful lives of 7 and 13 years, respectively? Here the least common multiple of lives is 91 years. An analysis period of 91 years hardly seems realistic. Instead, a suitable analysis period should be based on how long the equipment is likely to be needed. This may require that terminal values be estimated for the alternatives at some point prior to the end of their useful lives.

Figure 5-1 graphically represents this concept. As Fig. 5-1 indicates, it is not necessary for the analysis period to equal the useful life of an alternative or some multiple of the useful life. To properly reflect the situation at the end of the analysis period, an estimate is required of the market value of the equipment at that time. The calculations might be easier if everything came out even, but it is not essential.

3. Infinite Analysis Period—Capitalized Cost

Another difficulty in present worth analysis arises when we encounter an infinite analysis period ($n = \infty$). In governmental analyses, at times there are circumstances where a service or condition is to be maintained for an infinite period. The need for roads, dams, pipelines, or whatever is sometimes considered permanent. In these situations a present worth of cost analysis would have an infinite analysis period. We call this particular analysis *capitalized cost*.

Capitalized cost is the present sum of money that would need to be set aside now, at some interest rate, to yield the funds required to provide the service (or whatever) indefinitely. To accomplish this, the money set aside for future expenditures must not decline. The interest received on the money set aside can be spent, but not the principal. When one stops to think about an infinite analysis period (as opposed to something relatively short, like one hundred years), we see that an undiminished principal sum is essential, otherwise one will of necessity run out of money prior to infinity.

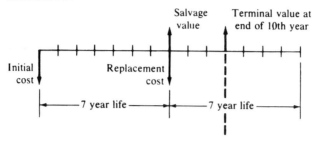

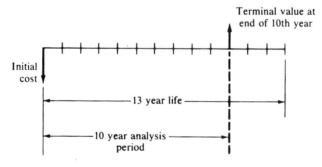

Figure 5-1 Superimposing an analysis period on 7- and 13-year alternatives.

In Chapter 4 we saw:

Principal sum + Interest for the period = Amount at end of period
$$P \quad + \quad iP \quad = \quad P + iP$$

If we spend iP, then in the next interest period the principal sum P will again increase to $(P + iP)$. Thus, we can again spend iP.

This concept may be illustrated by a numerical example: suppose you deposited $200 in a bank that paid 4% interest annually. How much money could be withdrawn each year without reducing the balance in the account below the initial $200? At the end of the first year, the $200 would earn 4%($200) = $8 interest. If this interest were withdrawn, the $200 would remain in the account. At the end of the second year, the $200 balance would again earn 4%($200) = $8. This $8 could also be withdrawn and the account would still have $200. This procedure could be continued indefinitely and the bank account would always contain $200.

The year-by-year situation would be depicted like this:

Year 1: $200 initial $P \rightarrow$ 200 + 8 =   208
              Withdrawal iP =   – 8

                       Year 2: $200 $\rightarrow$ 200 + 8 =   208
                              Withdrawal iP =   –   8
                                              $200
                                          and so on.

Thus, for any initial present sum P, there can be an end-of-period withdrawal of A equal to iP each period, and these withdrawals may continue forever without diminishing the initial sum P. This gives us the basic relationship:

For $n = \infty$, $A = Pi$

This relationship is the key to capitalized cost calculations. We previously defined capitalized cost as the present sum of money that would need to be set aside at some interest rate to yield the funds to provide the desired task or service forever. Capitalized cost is therefore the P in the equation $A = iP$. It follows that:

Capitalized cost $P = \dfrac{A}{i}$

If we can resolve the desired task or service into an equivalent A, the capitalized cost may be computed. The following examples illustrate such computations.

EXAMPLE 5-5

How much should one set aside to pay $50 per year for maintenance on a gravesite if interest is assumed to be 4%? For perpetual maintenance, the principal sum must remain undiminished after making the annual disbursement.

Solution:

$$\text{Capitalized cost } P = \frac{\text{Annual disbursement } A}{\text{Interest rate } i}$$

$$P = \frac{50}{0.04} = \$1250.$$

One should set aside $1250. ■

EXAMPLE 5-6

A city plans a pipeline to transport water from a distant watershed area to the city. The pipeline will cost $8 million and have an expected life of seventy years. The city anticipates it will need to keep the water line in service indefinitely. Compute the capitalized cost assuming 7% interest.

Solution: We have the capitalized cost equation:

$$P = \frac{A}{i}$$

that is simple to apply when there are end-of-period disbursements A. Here we have renewals of the pipeline every seventy years. To compute the capitalized cost, it is necessary to first compute an end-of-period disbursement A that is equivalent to $8 million every seventy years.

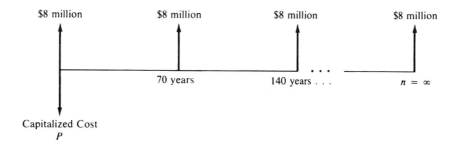

The $8 million disbursement at the end of seventy years may be resolved into an equivalent A.

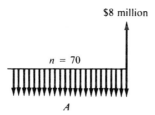

$$A = F(A/Fi,n) = \$8 \text{ million}(A/F,7\%,70)$$
$$= \$8 \text{ million}(0.00062) = \$4960$$

Each seventy-year period is identical to this one and the infinite series is shown in Fig. 5-2.

$$\text{Capitalized cost } P = \$8 \text{ million} + \frac{A}{i} = \$8 \text{ million} + \frac{4960}{0.07}$$
$$= \$8,071,000$$

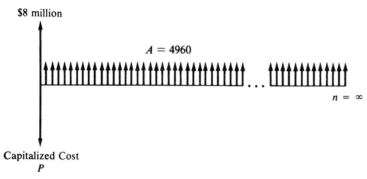

Figure 5-2 Infinite series computed using the sinking fund factor.

Alternate Solution: Instead of solving for an equivalent end-of-period payment A based on a *future* $8 million disbursement, we could find A, given a *present* $8 million disbursement.

$$A = P(A/P,i,n) = \$8 \text{ million}(A/P,7\%,70)$$
$$= \$8 \text{ million}(0.0706) = \$565,000$$

On this basis, the infinite series is shown in Fig. 5-3. Carefully note the difference between this and Fig. 5-2. Now:

$$\text{Capitalized cost } P = \frac{A}{i} = \frac{565,000}{0.07} = \$8,071,000 \qquad ■$$

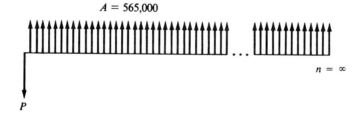

Figure 5-3 Infinite series computed using the capital recovery factor.

Alternate Solution Two:

Another way of solving the problem is to assume the interest is 70 years. Compute an equivalent interest rate for the 70 year period. Then the capitalized cost may be computed. Using Eq. 4-33 for $m = 70$

$$i_{70\,yrs} = \left(1 + i_{1\,yr}\right)^{70} - 1 = \left(1 + 0.07\right)^{70} - 1 = 112.989$$

$$\text{Capitalized cost} = \$8 \text{ million} + \frac{\$8 \text{ million}}{112.989} = \$8,071,000 \quad \blacksquare$$

Multiple Alternatives

So far the discussion has been based on examples with only two alternatives. But multiple-alternative problems may be solved by exactly the same methods employed for problems with two alternatives. (The only reason for avoiding multiple alternatives was to simplify the examples.) Examples 5-7 and 5-8 have multiple alternatives.

EXAMPLE 5-7

A contractor has been awarded the contract to construct a six-miles-long tunnel in the mountains. During the five-year construction period, the contractor will need water from a nearby stream. He will construct a pipeline to convey the water to the main construction yard. An analysis of costs for various pipe sizes is as follows:

	Pipe size			
	2"	3"	4"	6"
Installed cost of pipeline and pump	$22,000	$23,000	$25,000	$30,000
Cost per hour for pumping	$1.20	$0.65	$0.50	$0.40

The pipe and pump will have a salvage value at the end of five years equal to the cost to remove them. The pump will operate 2000 hours per year. The lowest interest rate at which the contractor is willing to invest money is 7%. (The minimum required interest rate for invested money is called the **minimum attractive rate of return**, or MARR.) Select the alternative with the least present worth of cost.

Solution: We can compute the present worth of cost for each alternative. For each pipe size, the Present worth of cost is equal to the Installed cost of the pipeline and pump plus the Present worth of five years of pumping costs.

	Pipe size			
	2"	3"	4"	6"
Installed cost of pipeline and pump	$22,000	$23,000	$25,000	$30,000
1.20 × 2000 hr × (P/A,7%,5)	9,840			
0.65 × 2000 hr × 4.100		5,330		
0.50 × 2000 hr × 4.100			4,100	
0.40 × 2000 hr × 4.100				3,280
Present worth of cost	$31,840	$28,330	$29,100	$33,280

Select the 3" pipe size.　■

EXAMPLE 5-8

An investor paid $8000 to a consulting firm to analyze what he might do with a small parcel of land on the edge of town that can be bought for $30,000. In their report, the consultants suggested four alternatives:

Alternatives	Total investment including land*	Uniform net annual benefit	Terminal value at end of 20 yr
A: Do nothing	$ 0	$ 0	$ 0
B: Vegetable market	50,000	5,100	30,000
C: Gas station	95,000	10,500	30,000
D: Small motel	350,000	36,000	150,000

*Includes the land and structures but does not include the $8000 fee to the consulting firm.

Assuming 10% is the minimum attractive rate of return, what should the investor do?

Solution: Alternative A represents the "do nothing" alternative. Generally, one of the feasible alternatives in any situation is to remain in the present status and do nothing. In this problem, the investor could decide that the most attractive alternative is not to purchase the property and develop it. This is clearly a "donothing" decision.

We note, however, that even if he does nothing, the total venture would not be a very satisfactory one. This is due to the fact that the investor spent $8000 for professional advice on the possible uses of the property. But because the $8000 is a past cost, it is a *sunk cost.* Sunk cost is the name given to past costs. The only relevant costs in an economic analysis are *present* and *future* costs; past events and past costs are gone and cannot be allowed to affect future planning. (The only place where past costs may be relevant is in computing depreciation charges and income taxes.) It should not deter the investor from making the best decision now, regardless of the costs that brought him to this situation and point of time.

This problem is one of neither fixed input nor fixed output, so our criterion will be to maximize the Present worth of benefits *minus* the Present worth of cost, or, simply stated, maximize net present worth.

Alternative A, Do nothing:
NPW = 0

Alternative B, Vegetable market:
NPW $= -50,000 + 5100(P/A,10\%,20) + 30,000(P/F,10\%,20)$
$= -50,000 + 5100(8.514) + 30,000(0.1486)$
$= -50,000 + 43,420 + 4460$
$= -2120$

Alternative C, Gas station:
NPW $= -95,000 + 10,500(P/A,10\%,20) + 30,000(P/F,10\%,20)$
$= -95,000 + 89,400 + 4460$
$= -1140$

Alternative D, Small motel:
NPW $= -350,000 + 36,000(P/A,10\%,20) + 150,000(P/F,10\%,20)$
$= -350,000 + 306,500 + 22,290$
$= -21,210$

The criterion is to maximize net present worth. In this situation, one alternative has NPW equal to zero, and three alternatives have negative values for NPW. We will select the best of the four alternatives, namely, the do-nothing Alt. A with NPW equal to zero. ■

EXAMPLE 5-9

A piece of land may be purchased for $610,000 to be strip-mined for the underlying coal. Annual net income will be $200,000 per year for ten years. At the end of the ten years, the surface of the land will be restored as required by a federal law on strip mining. The cost of reclamation will be $1,500,000 more than the resale value of the land after it is restored. Using a 10% interest rate, determine whether the project is desirable.

Solution: The investment opportunity may be described by the following cash flow:

Year	Cash flow, in thousands
0	-$610
1–10	+200 (per year)
10	-1500

$$NPW = -610 + 200(P/A,10\%,10) - 1500(P/F,10\%,10)$$
$$= -610 + 200(6.145) - 1500(0.3855)$$
$$= -610 + 1229 - 578$$
$$= +41$$

Since NPW is positive, the project is desirable. ■

EXAMPLE 5-10

Two pieces of construction equipment are being analyzed:

Year	Alternative A	Alternative B
0	−$2000	−$1500
1	+1000	+700
2	+850	+300
3	+700	+300
4	+550	+300
5	+400	+300
6	+400	+400
7	+400	+500
8	+400	+600

Based on an 8% interest rate, which alternative should be selected?

Solution:

Alternative A:

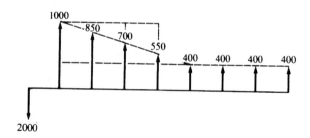

$$\text{PW of benefits} = 400(P/A,8\%,8) + 600(P/A,8\%,4) - 150(P/G,8\%,4)$$
$$= 400(5.747) + 600(3.312) - 150(4.650)$$
$$= 3588.50$$
$$\text{PW of cost} = 2000$$
$$\text{Net present worth} = 3588.50 - 2000 = +1588.50$$

Alternative B:

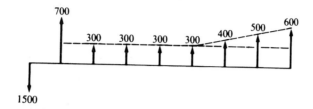

PW of benefits = 300 $(P/A,8\%,8)$ + (700 – 300)$(P/F,8\%,1)$
                 + 100$(P/G,8\%,4)(P/F,8\%,4)$
             = 300(5.747) + 400(0.9259) + 100(4.650)(0.7350)
             = 2436.24
PW of cost   = 1500
Net present worth = 2436.24 – 1500
             = +936.24
To maximize NPW, choose Alt. A. ■

Assumptions In Solving Economic Analysis Problems

One of the difficulties of problem solving is that most problems tend to be very complicated. It becomes apparent that *some* simplifying assumptions are needed to make such problems manageable. The trick, of course, is to solve the simplified problem and still be satisfied that the solution is applicable to the *real* problem! In the paragraphs that follow, we will consider six different items and explain the customary assumptions that are made.

End-of-Year Convention

As we indicated in Chapter 4, economic analysis textbooks follow the end-of-year convention. This makes *"A"* a series of end-of-period receipts or disbursements. (We generally assume in problems that all series of receipts or disbursements occur at the *end* of the interest period. This is, of course, a very self-serving assumption, for it allows us to use values from our Compound Interest Tables without any adjustments.)

A cash flow diagram of *P, A,* and *F* for the end-of-period convention is as follows:

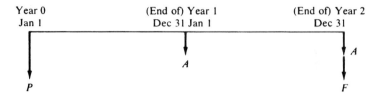

If one were to adopt a middle-of-period convention, the diagram would be:

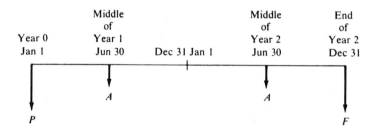

As the diagrams illustrate, only *A* shifts; *P* remains at the beginning-of-period and *F* at the end-of-period, regardless of the convention. The Compound Interest Tables in the Appendix are based on the end-of-period convention.

Viewpoint of Economic Analysis Studies

When we make economic analysis calculations, we must proceed from a point of reference. Generally, we will want to take the point of view of a total firm when doing industrial economic analyses. Example 2-1 vividly illustrated the problem: a firm's shipping department decided it could save money having its printing work done outside rather than by the in-house printing department. An analysis from the viewpoint of the shipping department supported this, as it could get for $688.50 the same printing it was paying $793.50 for in-house. Further analysis showed, however, that its printing department costs would decline *less* than using the commercial printer would save. From the viewpoint of the firm, the net result would be an increase in total cost.

From Ex. 2-1 we see it *is* important that the **viewpoint of the study** be carefully considered. Selecting a narrow viewpoint, like the shipping department, may result in a suboptimal decision from the viewpoint of the firm. For this reason, the viewpoint of the total firm is the normal point of reference in industrial economic analyses.

Sunk Costs

We know that it is the ***differences between alternatives*** that are relevant in economic analysis. Events that have occurred in the past really have no bearing on what we should do in the future. When the judge says, "$200 fine or three days in jail," the events that led to these unhappy alternatives really are unimportant. What *is* important are the current and future differences between the alternatives. Past costs, like past events, have no bearing on deciding between alternatives unless the past costs somehow affect the present or future costs. In general, past costs do not affect the present or the future, so we refer to them as *sunk costs* and disregard them.

Borrowed Money Viewpoint

In most economic analyses, the proposed alternatives inevitably require money to be spent, and so it is natural to ask the source of that money. The source will vary from situation to situation. In fact, there are *two* aspects of money to determine: one is the ***financing***—the obtaining of money—problem; the other is the ***investment***—the spending of money—problem. Experience has shown that these two concerns should be distinguished. When separated, the problems of obtaining money and of spending it are both logical and straightforward. Failure to separate them sometimes produces confusing results and poor decision making.

The conventional assumption in economic analysis is that the money required to finance alternatives/solutions in problem solving is considered to be ***borrowed at interest rate i.***

Effect of Inflation and Deflation

For the present we will assume that ***prices are stable.*** This means that a machine that costs $5000 today can be expected to cost the same amount several years hence. Inflation and deflation is a serious problem in many situations, but we will disregard it for now.

Income Taxes

This aspect of economic analyses, like inflation-deflation, must be considered if a realistic analysis is to be done. We will defer our introduction of income taxes into economic analyses until later.

Spreadsheets and Present Worth

Spreadsheets make it easy to build more accurate models with shorter time periods. When using factors, it is common to assume that costs and revenues are uniform for *N* years. With spreadsheets it is easy to use 120 months instead of 10 years, and the cash flows can be estimated for each month. For example, energy costs for air conditioning peak in the summer and in many areas construction during the winter is limited. Cash flows that depend on population often increase at $x\%$ per year, such as for electric power and transportation costs.

In spreadsheets any interest rate is entered exactly – so no interpolation is needed. This makes it easy to calculate the monthly repayment schedule for a car loan or a house mortgage. Examples 5-11 and 5-12 illustrate using spreadsheets to calculate PW's.

EXAMPLE 5-11

NLE Construction is bidding on a project whose costs are divided into $30,000 for startup and $240,000 for the first year. If the interest rate is 1% per month or 12.68% per year, what is the present worth with monthly compounding?

Solution

Figure 5-4 illustrates the spreadsheet solution with the assumption that costs are distributed evenly throughout the year (-20,000 = -240,000/12).

	A	B	C	D
1	1% *i*			
2	-30,000	initial cash flow		
3	-240,000	annual amount		
4				
5	Month	Cash Flow		
6	0	-30,000		
7	1	-20000		
8	2	-20000		
9	3	-20000		
10	4	-20000		
11	5	-20000		
12	6	-20000		
13	7	-20000		
14	8	-20000		
15	9	-20000		
16	10	-20000		
17	11	-20000		
18	12	-20000		
19	NPV	-$255,102	=NPV(A1,B7:B18)+B6	

Figure 5-4 Spreadsheet with monthly cash flows.

Since the costs are uniform, the factor solution is:

$PW_{annual} = -30,000 - 240,000 \cdot (P/F,12.68\%,1) = -30,000 - 240,000/1.1268 = -\$242,993$

$PW_{mon} = -30,000 - 20,000 \cdot (P/A,1\%,12) = -\$255,102$

These two values differ by more than $12,000, because $20,000 at the end of months 1 through 12 is not the same as $240,000 at the end month 12. The effective interest rates are the same. ■

EXAMPLE 5-12

Regina Industries has a new product whose sales are expected to be 1.2, 3.5, 7, 5, and 3 million units per year over the next 5 years. Production, distribution, and overhead costs are stable at $120 per unit. The price will be $200 per unit for the first 2 years, and then $180, $160, and $140 for the next 3 years. The remaining R&D and production costs are $400 million. If *i* is 15%, what is the present worth of the new product?

Solution

It is easiest to calculate the yearly net revenue per unit before building the spreadsheet shown in Figure 5-5. Those values are the yearly price minus the $120 of costs, which equals $80, $80, $60, $40, and $20.

	A	B	C	D	E
1	12% *i*				
2			Net	Cash	
3	Year	Sales (M)	Revenue	Flow ($M)	
4	0			-300	
5	1	1.2	80	96	
6	2	3.5	80	280	
7	3	7	64	448	
8	4	5	40	200	
9	5	3	20	60	
10		D4+NPV(A1,D5:D9)=		$489 Million	

Figure 5-5 Present worth of a new product.   ■

SUMMARY

Present worth analysis is suitable for almost any economic analysis problem. But it is particularly desirable when we wish to know the present worth of future costs and benefits. And we frequently want to know the value today of such things as income-producing assets, stocks, and bonds.

For present worth analysis, the proper economic criteria are:

Fixed input	Maximize the PW of benefits
Fixed output	Minimize the PW of costs
Neither input nor output is fixed	Maximize (PW of benefits − PW of costs) or, more simply stated: Maximize NPW

To make valid comparisons, we need to analyze each alternative in a problem over the same *analysis period* or *planning horizon*. If the alternatives do not have equal lives, some technique must be used to achieve a common analysis period. One method is to select an analysis period equal to the least common multiple of the alternative lives. Another method is to select an analysis period and then compute end-of-analysis-period salvage values for the alternatives.

Capitalized cost is the present worth of cost for an infinite analysis period ($n = \infty$). When $n = \infty$, the fundamental relationship is $A = iP$. Some form of this equation is used whenever there is a problem with an infinite analysis period.

There are a number of assumptions that are routinely made in solving economic analysis problems. They include the following:

1. Present sums P are beginning of period and all series receipts or disbursements A and future sums F occur at the end of the interest period. The compound interest tables were derived on this basis.

2. In industrial economic analyses, the appropriate point of reference from which to compute the consequences of alternatives is the total firm. Taking a narrower view of the consequences can result in suboptimal solutions.

3. Only the differences between the alternatives are relevant. Past costs are sunk costs and generally do not affect present or future costs. For this reason they are ignored.

4. The investment problem should be isolated from the financing problem. We generally assume that all required money is borrowed at interest rate *i*.

5. For now, stable prices are assumed. The problem of inflation-deflation is deferred to Chapter 13. Similarly, income taxes are deferred to Chapter 11.

Problems

5-1   Compute *P* for the following diagram.

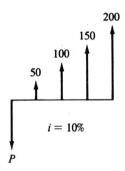

$i = 10\%$

5-2   Compute the value of P that is equivalent to the four cash flows in the diagram below.

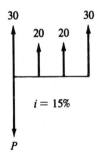

$i = 15\%$

5-3   What is the value of *P* for the situation shown below?

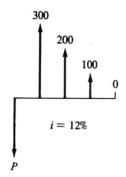

(Answer: P = $498.50)

5-4 Compute the value of Q in the figure below.

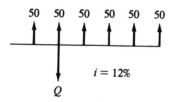

5-5 For the diagram, compute P.

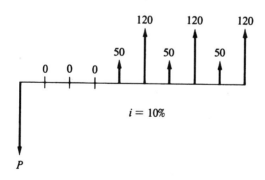

5-6 Compute P for the following diagram.

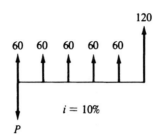

$$120$$

$$60 \quad 60 \quad 60 \quad 60 \quad 60$$

$$i = 10\%$$

$$P$$

(*Answer:* $P = \$324.71$)

5-7   The annual income from a rented house is $12,000. The annual expenses are $3000. If the house can be sold for $145,000 at the end of ten years, how much could you afford to pay for it now, $A = 9,000$ if you considered 18% to be a suitable interest rate?   (*Answer:* $68,155)

5-8   Consider the following cash flow. At a 6% interest rate, what is the value of P, at the end of Year 1, that is equivalent to the benefits at the end of Years 2 through 7?

Year	Cash flow
1	$-P$
2	$+100$
3	$+200$
4	$+300$
5	$+400$
6	$+500$
7	$+600$

5-9   A rather wealthy man decided he would like to arrange for his descendants to be well educated. He would like each child to have $60,000 for his or her education. He plans to set up a perpetual trust fund so that six children will receive this assistance each generation. He estimates that there will be four generations per century, spaced 25 years apart. He expects the trust to be able to obtain a 4% rate of return, and the first recipients to receive the money ten years hence. How much money should he now set aside in the trust? (*Answer:* $389,150)

5-10   How much would the owner of a building be justified in paying for a sprinkler system that will save $750 a year in insurance premiums if the system has to be replaced every twenty years and has a salvage value equal to 10% of its initial cost? Assume money is worth 7%. (*Answer:* $8156)

5-11   A man had to have the muffler replaced on his two-year-old car. The repairman offered two alternatives. For $50 he would install a muffler guaranteed for two years. But for $65 he would install a muffler guaranteed "for as long as you own the car." Assuming the present owner expects to keep the car for about three more years, which muffler would you advise him to have installed if you thought 20% were a suitable interest rate and the less expensive muffler would only last two years?

5-12 A consulting engineer has been engaged to advise a town how best to proceed with the construction of a 200,000 m³ water supply reservoir. Since only 120,000 m³ of storage will be required for the next 25 years, an alternative to building the full capacity now is to build the reservoir in two stages. Initially, the reservoir could be built with 120,000 km³ of capacity and then, 25 years hence, the additional 80,000 m³ of capacity could be added by increasing the height of the reservoir. Estimated costs are as follows:

	Construction cost	Annual maintenance cost
Build in two stages.		
First stage: 120,000 m³ reservoir	$14,200,000	$75,000
Second stage: Add 80,000 m³ of capacity, additional construction and maintenance costs	12,600,000	25,000
Build full capacity now. 200,000 m³ reservoir	$22,400,000	$100,000

If interest is computed at 4%, which construction plan is preferred?

5-13 An engineer has received two bids for an elevator to be installed in a new building. The bids, plus his evaluation of the elevators, are as follows:

	Bids		Engineer's estimates	
Alternatives	Installed cost	Service life, in years	Annual operating cost, including repairs	Salvage value at end of service life
Westinghome	$45,000	10	$2700/yr	$3000
Itis	54,000	15	2850/yr	4500

The engineer will make a present worth analysis using a 10% interest rate. Prepare the analysis and determine which bid should be accepted.

5-14 A railroad branch line is to be constructed to a missile site. It is expected the railroad line will be used for 15 years, after which the missile site will be removed and the land turned back to agricultural use. The railroad track and ties will be removed at that time.

In building the railroad line, either treated or untreated wood ties may be used. Treated ties have an installed cost of $6 and a ten-year life; untreated ties are $4.50 with a six-year life. If at the end of fifteen years the ties then in place have a remaining useful life of four years or more, they will be used by the railroad elsewhere and have an estimated salvage value of $3 each. Anytime ties are removed that are at the end of their service life, or are too close to the end of their service life to be used elsewhere, they are sold for $0.50 each.

Determine the most economical plan for the initial railroad ties and their replacement for the fifteen-year period. Make a present worth analysis assuming 8% interest.

5-15 A weekly business magazine offers a one-year subscription for $58 and a three-year subscription for $116. If you thought you would read the magazine for at least the next three years, and consider 20% as a minimum rate of return, which way would you purchase the magazine, with three one-year subscriptions or a single three-year subscription. (*Answer:* Choose the three-year subscription.)

5-16 A manufacturer is considering purchasing equipment which will have the following financial effects:

Year	Disbursements	Receipts
0	$4400	$0
1	660	880
2	660	1980
3	440	2420
4	220	1760

If money is worth 6%, should he invest in the equipment?

5-17 Jerry Stans, a young industrial engineer, prepared an economic analysis for some equipment to replace one production worker. The analysis showed that the present worth of benefits (of employing one less production worker) just equaled the present worth of the equipment costs, based on a ten-year useful life for the equipment. It was decided not to purchase the equipment.

A short time later, the production workers won a new three-year union contract that granted them an immediate 40¢-per-hour wage increase, plus an additional 25¢-per-hour wage increase in each of the two subsequent years. Assume that in each and every future year, a 25¢-per-hour wage increase will be granted.

Jerry Stans has been asked to revise his earlier economic analysis. The present worth of benefits of replacing one production employee will now increase. Assuming an interest rate of 8%, the justifiable cost of the automation equipment (with a ten-year useful life) will increase by how much? Assume the plant operates a single eight-hour shift, 250 days per year.

5-18 The management of an electronics manufacturing firm believes it is desirable to install some automation equipment in their production facility. They believe the equipment would have a ten-year life with no salvage value at the end of ten years. The plant engineering department has surveyed the plant and suggested there are eight mutually exclusive alternatives available.

Plan	Initial cost, in thousands	Net annual benefit, in thousands
1	$265	$51
2	220	39
3	180	26
4	100	15
5	305	57
6	130	23
7	245	47
8	165	33

If the firm expects a 10% rate of return, which alternative, if any, should they adopt? (*Answer:* Plan 1)

5-19   The president of the E. L. Echo Corporation thought it would be appropriate for his firm to "endow a chair" in the Industrial Engineering Department of the local university; that is, he was considering making a gift to the university of sufficient money to pay the salary of one professor forever. One professor in the department would he designated the E. L. Echo Professor of Industrial Engineering, and his salary would come from the fund established by the Echo Corporation. If the professor will receive $67,000 per year, and the interest received on the endowment fund is expected to remain at 8%, what lump sum of money will the Echo Corporation need to provide to establish the endowment fund? (*Answer:* $837,500)

5-20   A man who likes cherry blossoms very much would like to have an urn full of them put on his grave once each year forever after he dies. In his will, he intends to leave a certain sum of money in the trust of a local bank to pay the florist's annual bill. How much money should be left for this purpose? Make whatever assumptions you feel are justified by the facts presented. State your assumptions, and compute a solution.

5-21   A local symphony association offers memberships as follows:

Continuing membership, per year $ 15
Patron lifetime membership   375

The patron membership has been based on the symphony association's belief that it can obtain a 4% rate of return on its investment. If you believed 4% to be an appropriate rate of return, would you be willing to purchase the patron membership? Explain why or why not.

5-22   A battery manufacturing plant has been ordered to cease discharging acidic waste liquids containing mercury into the city sewer system. As a result, the firm must now adjust the pH and remove the mercury from its waste liquids. Three firms have provided quotations on the necessary equipment. An analysis of the quotations provided the following table of costs.

Bidder	Installed cost	Annual operating cost	Annual income from mercury recovery	Salvage value
Foxhill Instrument	$35,000	$8000	$2000	$20,000
Quicksilver	40,000	7000	2200	0
Almaden	100,000	2000	3500	0

If the installation can be expected to last twenty years and money is worth 7%, which equipment should be purchased? (*Answer:* Almaden)

5-23   A firm is considering three mutually exclusive alternatives as part of a production improvement program. The alternatives are:

	A	B	C
Installed cost	$10,000	$15,000	$20,000
Uniform annual benefit	1,625	1,530	1,890
Useful life, in years	10	20	20

For each alternative, the salvage value at the end of its useful life is zero. At the end of ten years, *A* could be replaced with another *A* with identical cost and benefits. The minimum attractive rate of return is 6%. Which alternative should be selected?

5-24   A steam boiler is needed as part of the design of a new plant. The boiler can be fired either by natural gas, fuel oil, or coal. A decision must be made on which fuel to use. An analysis of the costs shows that the installed cost, with all controls, would be least for natural gas at $30,000; for fuel oil it would be $55,000; and for coal it would be $180,000. If natural gas is used rather than fuel oil, the annual fuel cost will increase by $7500. If coal is used rather than fuel oil, the annual fuel cost will be $15,000 per year less. Assuming 8% interest, a twenty-year analysis period, and no salvage value, which is the most economical installation?

5-25   An investor has carefully studied a number of companies and their common stock. From his analysis, he has decided that the stocks of six firms are the best of the many he has examined. They represent about the same amount of risk and so he would like to determine the one in which to invest. He plans to keep the stock for four years and requires a 10% minimum attractive rate of return.

Common stock	Price per share	Annual end-of-year dividend per share	Estimated price at end of 4 years
Western House	$23¾	$1.25	$32
Fine Foods	45	4.50	45
Mobile Motors	30⅝	0	42
Trojan Products	12	0	20
U.S. Tire	33⅜	2.00	40
Wine Products	52½	3.00	60

Which stock, if any, should the investor consider purchasing? (*Answer:* Trojan Products)

5-26 A home builder must construct a sewage treatment plant and deposit sufficient money in a perpetual trust fund to pay the $5000 per year operating cost and to replace the treatment plant every forty years. The plant will cost $150,000, and future replacement plants will also cost $150,000 each. If the trust fund earns 8% interest, what is the builder's capitalized cost to construct the plant and future replacements, and to pay the operating costs?

5-27 Using an eight-year analysis period and a 10% interest rate, determine which alternative should be selected:

	A	B
First cost	$5300	$10,700
Uniform annual benefit	1800	2,100
Useful life, in years	4	8

5-28 The local botanical society wants to ensure that the gardens in a local park are properly cared for. They just recently spent $100,000 to plant the gardens. They would like to set up a perpetual fund to provide $100,000 for future replantings of the gardens every ten years. If interest is 5%, how much money would be needed to forever pay the cost of replanting?

5-29 An elderly lady decided to distribute most of her considerable wealth to charity and to retain for herself only enough money to provide for her living. She feels that $1000 a month will amply provide for her needs. She will establish a trust fund at a bank which pays 6% interest, compounded monthly. At the end of each month she will withdraw $1000. She has arranged that, upon her death, the balance in the account is to be paid to her niece, Susan. If she opens the trust fund and deposits enough money to pay her $1000 a month forever, how much will Susan receive when her Aunt dies?

5-30 Solve the diagram below for P using a geometric gradient factor.

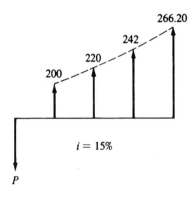

5-31   If $i = 10\%$, what is the value of P?

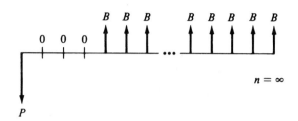

5-32   A stonecutter was carving the headstone for a well-known engineering economist.

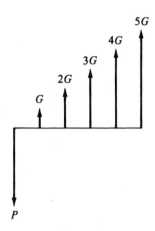

He carved the figure above and then started the equation as follows:
$$P = G(P/G, i, 6)$$
He realized he had made a mistake. The equation should have been
$$P = G(P/G, i, 5) + G(P/A, i, 5)$$
The stonecutter does not want to discard the stone and start over. He asks you to help him with his problem. The right side of Eq. 1 can be multiplied by one compound interest factor and then the equation will be correct for the carved figure. Equation 1 will be of the form:
$$P = G(P/G, i, 6)(\qquad , i, \ )$$
Write the complete equation.

5-33 In a present worth analysis, one alternative has a Net Present Worth of +420, based on a six-year analysis period that equals the useful life of the alternative. A 10% interest rate was used in the computations.

    The alternative is to be replaced at the end of the six years by an identical piece of equipment with the same cost, benefits, and useful life. Based on a 10% interest rate, compute the Net Present Worth of the equipment for the twelve-year analysis period.   (*Answer:* NPW = +657.09)

5-34 A project has a Net Present Worth (NPW) of –140 as of Jan. 1, 2000. If a 10% interest rate is used, what is the project NPW as of Dec. 31, 1997?

5-35 Consider the following four alternatives. Three are "do something" and one is "do nothing."

	Alternative			
	A	*B*	*C*	*D*
Cost	$0	$50	$30	$40
Net annual benefit	0	12	4.5	6
Useful life, in years		5	10	10

At the end of the five-year useful life of *B*, a replacement is not made. If a ten-year analysis period and a 10% interest rate are selected, which is the preferred alternative?

5-36 Six mutually exclusive alternatives are being examined. For an 8% interest rate, which alternative should be selected? Each alternative has a six-year useful life.

	Alternatives					
	A	*B*	*C*	*D*	*E*	*F*
Initial cost	$20.00	$35.00	$55.00	$60.00	$80.00	$100.00
Uniform annual benefit	6.00	9.25	13.38	13.78	24.32	24.32

5-37 A building contractor obtained bids for some asphalt paving, based on a specification. Three paving subcontractors quoted the following prices and terms of payment:

	Price	*Payment schedule*
1. *Quick Paving Co.*	$85,000	50% payable immediately; 25% payable in six months; 25% payable at the end of one year.
2. *Tartan Paving Co.*	$82,000	Payable immediately.
3. *Faultless Paving Co.*	$84,000	25% payable immediately; 75% payable in six months.

The building contractor uses a 12% nominal interest rate, compounded monthly, in this type of bid analysis. Which paving subcontractor should be awarded the paving job?

5-38 A cost analysis is to be made to determine what, if anything, should be done in a situation where there are three "do something" and one "do nothing" alternatives. Estimates of the cost and benefits are as follows:

Alternatives	Cost	Uniform annual benefit	End-of-useful-life salvage value	Useful line, in years
1	$500	$135	$ 0	5
2	600	100	250	5
3	700	100	180	10
4	0	0	0	0

Use a ten-year analysis period for the four mutually exclusive alternatives. At the end of five years, Alternatives 1 and 2 may be replaced with identical alternatives (with the same cost, benefits, salvage value, and useful life).

a. If an 8% interest rate is used, which alternative should be selected?

b. If a 12% interest rate is used, which alternative should be selected?

5-39 Consider five mutually exclusive alternatives:

	Alternatives				
	A	B	C	D	E
Initial cost	$600	$600	$600	$600	$600
Uniform annual benefits for first five years	100	100	100	150	150
Uniform annual benefits for last five years	50	100	110	0	50

The interest rate is 10%. If all the alternatives have a ten-year useful life, and no salvage value, which alternative should be selected?

5-40 On February 1, the Miro Company needs to purchase some office equipment. The company is presently short of cash and expects to be short for several months. The company treasurer has indicated that he could pay for the equipment as follows:

Date	Payment
April 1	$150
June 1	300
Aug. 1	450
Oct. 1	600
Dec. 1	750

A local office supply firm has been contacted, and they will agree to sell the equipment to the firm now and to be paid according to the treasurer's payment schedule. If interest will be charged at 3% every two months, with compounding once every two months, how much office equipment can the Miro Company buy now? (*Answer:* $2020)

5-41 Using 5% nominal interest, compounded continuously, solve for *P*.

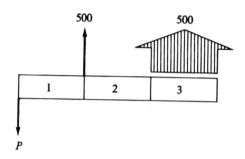

5-42 By installing some elaborate inspection equipment on its assembly line, the Robot Corp. can avoid hiring an extra worker. The worker would have earned $26,000 a year in wages, and Robot would have paid an additional $7500 a year in employee benefits. The inspection equipment has a six-year useful life and no salvage value. Use a nominal 18% interest rate in your calculations. How much can Robot afford to pay for the equipment if the wages and worker benefits are paid:

    a. at the end of each year?
    b. monthly?
    c. continuously?
    d. Explain why the answers in *b* and *c* are larger than in *a*.

Assume the compounding matches the way the wages and benefits are paid, that is, annually, monthly, and continuously, respectively.

5-43 Using capitalized cost, determine which type of road surface is preferred on a particular section of highway. Use 12% interest rate.

	A	B
Initial cost	$500,000	$700,000
Annual maintenance	35,000	25,000
Periodic resurfacing	350,000	450,000
	every 10 years	every 15 years

5-44 What amount of money deposited 50 years ago at 8% interest would provide a perpetual payment of $10,000 a year beginning this year?

5-45 Annual maintenance costs for a particular section of highway pavement are $2000. The placement of a new surface would reduce the annual maintenance cost to $500 per year for the first five years and to $1000 per year for the next five years. After ten years the annual maintenance would again be $2000. If maintenance costs are the only saving, what investment can be justified for the new surface. Assume interest at 4%.

5-46 A small dam was constructed for $2 million. The annual maintenance cost is $15,000. If interest is 5%, compute the capitalized cost of the dam, including maintenance.

5-47 Twenty-five thousand dollars is deposited in a savings account that pays 5% interest, compounded semi-annually. Equal annual withdrawals are to be made from the account, beginning one year from now and continuing forever. What is the maximum equal annual withdrawal?

5-48 Two alternative courses of action have the following schedules of disbursements:

Year	A	B
0	−$1300	
1	0	−$100
2	0	−200
3	0	−300
4	0	−400
5	0	−500
	−$1300	−$1500

Based on a 6% interest rate, which alternative should be selected?

5-49 An investor is considering buying a 20-year corporate bond. The bond has a face value of $ 1000 and pays 6% interest per year in two semi-annual payments. Thus the purchaser of the bond will receive $30 every six months and, in addition, he/she will receive $1000 at the end of 20 years, along with the last $30 interest payment. If the investor thinks he/she should receive 8% interest, compounded semi-annually, how much would the investor be willing to pay for the bond?

5-50 A trust fund is to be established for three purposes: (1) provide $750,000 for the construction and $250,000 for the initial equipment of a small engineering laboratory; (2) pay the $150,000 per year laboratory operating cost; and (3) pay for $100,000 of replacement equipment every four years, beginning four years from now.

At 6% interest, how much money is required in the trust fund to provide for the laboratory and equipment and its perpetual operation and equipment replacement?

5-51 A city has developed a plan which will provide for future municipal water needs. The plan proposes an aqueduct which passes through 500 feet of tunnel in a nearby mountain. Two alternatives are being considered. The first proposes to build a full-capacity tunnel now for $556,000. The second proposes to build a half-capacity tunnel now (cost = $402,000) which should be adequate for 20 years, and then to build a second parallel half-capacity tunnel. The maintenance cost of the tunnel lining for the full-capacity tunnel is $40,000 every 10 years, and for each half-capacity tunnel it is $32,000 every 10 years.

The friction losses in the half-capacity tunnel will be greater than if the full-capacity tunnel were built. The estimated additional pumping costs in the single half-capacity tunnel will be $2000 per year, and for the two half-capacity tunnels it will be $4000 per year. Based on capitalized cost and a 7% interest rate, which alternative should be selected?

5-52   A road building contractor has received a major highway construction contract that will require 50,000 m^3 of crushed stone each year for five years. The needed stone can be obtained from a quarry for $5.80/m^3. As an alternative the contractor has decided to try and purchase the quarry. He believes if he owned the quarry the stone would only cost him $4.30/m^3. He thinks he could resell the quarry at the end of five years for $40,000. If the contractor uses a 10% interest rate, how much would he be willing to pay for the quarry?

5-53   A new office building was constructed five years ago by a consulting engineering firm. At that time the firm obtained a bank loan for $100,000 with a 12% annual interest rate, compounded quarterly. The terms of the loan call for equal quarterly payments to repay the loan in 10 years. The loan also allows for its prepayment at any time without penalty.

Due to internal changes in the firm, it is now proposed to refinance the loan through an insurance company. The new loan would be for a 20-year term with an interest rate of 8% per year, compounded quarterly. The new equal quarterly payments would repay the loan in the 20-year period. The insurance company requires the payment of a 5% loan initiation charge (often described as a "five-point loan fee") which will be added to the new loan.

> *a.*   What is the balance due on the original mortgage if 20 payments have been made in the last five years?

> *b.*   What is the difference between the equal quarterly payments on the present bank loan and the proposed insurance company loan?

5-54 Given the following data, using <u>present worth analysis</u> find the best alternative.

	Alt. A	Alt. B	Alt. C
Initial Cost	$10,000	15,000	$12,000
Annual Benefit	6,000	10,000	5,000
Salvage Value	1,000	-2,000	3,000
Useful Life	2 years	3 years	4 years
MARR	10%	10%	10%

Analysis Period = 12 years $i = 10\%$

5-55 The local Audubon Society has just put a new bird feeder in the park at a cost of $500. The feeder has a useful like of 5 years and an annual maintenance cost of $50. Our cat, Fred, was very impressed with the project. He wants to extablish a fund that will maintin the feeder in perpetuity (that's forever!). Replacement feeders cost $500 every 5 years. If the fund will earn 5% interest, what amount must he raise for its establishment? Note that it will cover both maintenance and replacement costs following the intial investment.

5-56 We want to donate a marble birdbath to the city park as a memorial to our cat, Fred, while he can still enjoy it. We also want to set up a perpetual care fund to cover futute expenses "forever". The initial cost of the bath is $5000. Routine annual operating costs are $200 per year but every fifth year the cost will be $500 to cover major cleaning and maintenance as well as operation.

    a. What is the capitalized cost of this project if the interest rate is 8 percent?

    b. How much is the present worth of this project if it is to be demolished after 75 years?

The final $500 payment in the 75th year will cover the year's operating cost and the site reclamation.

5-57 A corporate bond has a face value of $1000 with maturity date 20 years from today. The bond pays interest semi-annually at a rate of 8% per year based on the face value. The interest rate paid on similar corporate bonds has decreased to a current rate of 6%. Determine the market value of the bond.

5-58 IBP Inc. is considering establishing a new machine to automate a meat packing process. The machine will save $50,000 in labor annually. The machine can be purchased for $200,000 today and will be used for a period of 10 years. It is has a salvage value of $10,000 at the end of its useful life. The new machine will require an annual maintenance cost of $9000. The corporation has a minimum rate of return of 10% Do you recommend automating the process?

5-59 Argentina is considering constructing a bridge across the Rio de La Plata to connect its Northern coast to the Southern coast of Uruguay. If this bridge is constructed, it will reduce the travel time from Buenos Aires,Argentina to Sao Paulo,Brazil by over 10 hours, and has the potential to significantly improve the flow of manufactured goods between the two countries. The cost of the new bridge, which will be the longest bridge in the world and span over 50 miles, will be $700 million. The bridge will require an annual maintenance of $10 million for repairs and upgrades, and is estimated to last 80 years. It is estimated that 550,000 vehicles will use the bridge during the first year of operation, and an additional 50,000 vehicles per year until the 10th year. These data are based on an a toll charge of $90 per vehicle. The annual traffic for the remainder of the life of the bridge life will be 1,000,000 vehicles per year. The government requires a minimum rate of return of 9% in order to proceed with the project.

    a. Does this project provide sufficient revenues to offset its costs?

    b. What other considerations are there besides economics in deciding whether or not to construct the bridge?

5-60 Telefono Mexico is expanding its facitites to serve a new manufacturing plant. The new plant will require 2000 telephone lines this year, and another 2000 lines after expansion in 10 years. The plant will be in operation for 30 years. The telephone company is evaluating two options to serve the demand.

    Option 1:   Provide one cable now with capacity to serve 4000 lines. The cable cost will be $200,000, and will require $15,000 in annual maintenance.

    Option 2:   Provide a cable with capacity to serve 2000 lines now, and a second cable to serve the other 2000 lines in 10 years. The cost of each cable will be $150,000 and each cable will have an annual maintenance of $10,000.

The telephone cables will last at least 30 years, and the cost to remove the cables are offset by the salvage value.

   a. Which alternative should be selected based on a 10% interest rate?

   b. Will your answer to part (a) change if the demand for additional lines is in 5 years instead of 10 years?

5-61 Dr. Fog E. Professor is retiring and wants to endow a chair of engineering economics at his university. It is expected that he will need to cover an annual cost of $100,000 forever. What lump sum must he donate to the university today if the endowment will earn 10% interest?

5-62 Dick Dickerson Construction Inc. has asked to you help them select a new backhoe. You have a choice between a wheel-mounted version which costs $50,000, has an expected life of 5 years and a salvage value of $2000 and a track-mounted one which costs $80,000, has a 5 year life, and an expected salvage value of $10,000. Both machines will achieve the same productivity. Interest is 8%. Which one will you recommend? Use a Present Worth analysis.

5-63 A student has a job which leaves her with $250 per month in disposable income. She decides that she will use the money to buy a car. Before looking for a car, she arranges a 100% loan whose terms are $250 per month for 36 months at 18% annual interest. What is the maximum car purchase price that she can afford with her loan?

5-64 The student in problem 5-63 finds a car she likes and the dealer offers to arrange financing. His terms are 12% interest for 60 months and no down payment. The car's sticker price is $12,000. Can she afford to purchase this car with her $250 monthly disposable income?

5-65 The student in problem 5-64 really wants this particular car. She decides to try and negotiate a different interest rate. What is the highest interest rate that she can accept, given a 60 month term and $250 per month payments?

5-66 Walt Wallace Construction Enterprises is investigating purchasing a new dump truck. Interest is 9%. They have found two models that they like. Their cash flows are shown below:

Model	First Cost	Annual Operating Cost	Annual Income	Salvage Value	Life
A	$50,000	$2000	$9,000	$10,000	10yrs
B	$80,000	$1000	$12,000	$30,000	10yrs

   a. Using Present Worth analysis, which truck should they buy and why?

   b. Before they can close the deal, the dealer sells out of Model B and cannot get anymore. What should they do now and why?

5-67 We know a car costs 60 monthly payments of $199. The car dealer has set us a nominal interest rate of 4.5% compounded daily. What is the purchase price of the car?

5-68 A machine costs $980,000 to purchase and will provide $200,000 a year in benefits. The company plans to use the machine for 13 years and then will sell the machine for scrap, receiving $20,000. The company interest rate is 12%. Should the machine be purchased?

5-69 Two different companies are offering a punch press for sale. Company A charges $250,000 to deliver and install the device. Company A has estimated that the machine will have maintenance and operating costs of $4000 a year and will provide an annual benefit of $89,000. Company B charges $205,000 to deliver and install the device. Company B has estimated that their press will have maintenance and operating costs of $4300 a year and will provide an annual benefit of $86,000. Both machines will last 5 years and can be sold for $15,000 for the scrap metal. Use an interest rate of 12%. Which machine should your company purchase, based on the above data?

5-70 Austin General Hospital is evaluating new office equipment offered by three companies. The equipment have the following characteristics:

	Company A	*Company B*	*Company C*
First cost	$15,000	$25,000	$20,000
Maintenance & operating costs	1,600	400	900
Annual benefit	8,000	13,000	11,000
Salvage value	3,000	6,000	4,500
Useful life, in years	4	4	4

MARR = 15 % Using NPW analysis, from which company should you purchase the equipment?

5-71 Calculate the present worth of a 4.5% $5000 bond with interest paid semiannually. The bond matures in 10 years, and the investor desires to make 8% per year compounded quarterly on the investment.

5-72 The following costs are associated with three tomato-peeling machines being considered for use in a food canning plant.

	Machine A	*Machine B*	*Machine C*
First Cost	$52,000	$63,000	$67,000
Maintenance & operating costs	15,000	9,000	12,000
Annual benefit	38,000	31,000	37,000
Salvage value	13,000	19,000	22,000
Useful life, in years	4	6	12

If the canning company uses a MARR of 12%, which is the best alternative? Use NPW to make your decision. Note: Consider the least common multiple as the study period.

5-73 Find P for the cash flow diagram given below

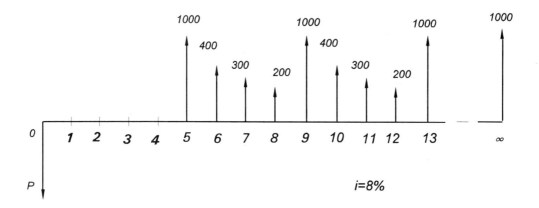

CHAPTER 6

Annual Cash Flow Analysis

This chapter is devoted to annual cash flow analysis—the second of the three major analysis techniques. As we've said, alternatives must be resolved into such a form that they may be compared. This means we must use the equivalence concept to convert from a cash flow representing the alternative into some equivalent sum or equivalent cash flow.

With present worth analysis, we resolved an alternative into an equivalent cash sum. This might have been an equivalent present worth of cost, an equivalent present worth of benefit, or an equivalent net present worth. But instead of computing equivalent present sums, we could compare alternatives based on their equivalent annual cash flows. Depending on the particular situation, we may wish to compute the equivalent uniform annual cost (EUAC), the equivalent uniform annual benefit (EUAB), or their difference (EUAB—EUAC).

To prepare for a discussion of annual cash flow analysis, we will review some annual cash flow calculations, then examine annual cash flow criteria. Following this, we will proceed with annual cash flow analysis.

Annual Cash Flow Calculations

Resolving a Present Cost to an Annual Cost

Equivalence techniques were used in prior chapters to convert money, at one point in time, to some equivalent sum or series. In annual cash flow analysis, the goal is to convert money to an equivalent uniform annual cost or benefit. The simplest case is to convert a present sum P to a series of equivalent uniform end-of-period cash flows. This is illustrated in Example 6-1.

EXAMPLE 6-1

A woman bought $1000 worth of furniture for her home. If she expects it to last ten years, what will be her equivalent uniform annual cost if interest is 7%?

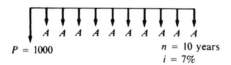

P = 1000   *n* = 10 years
              i = 7%

Solution:

Equivalent uniform annual cost = $P(A/P,i,n)$

$$= 1000(A/P,7\%,10)$$

$$= \$142.40$$

Her equivalent uniform annual cost is $142.40.   ■

Treatment of Salvage Value

In a situation where there is a salvage value, or future value at the end of the useful life of an asset, the result is to decrease the equivalent uniform annual cost.

EXAMPLE 6-2

The woman in Ex. 6-1 now believes she can resell the furniture at the end of ten years for $200. Under these circumstances, what is her equivalent uniform annual cost?

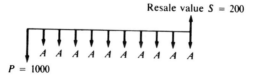

Resale value *S* = 200

P = 1000

Solution: For this situation, the problem may be solved by each of three different calculations, as follows:

Ex. 6-2, Solution One:

$$EUAC = P(A/P,i,n) - S(A/F,i,n) \tag{6-1}$$

$$= 1000(A/P,7\%,10) - 200(A/F,7\%,10)$$

$$= 1000(0.1424) - 200(0.0724)$$

$$= 142.40 - 14.48 = \$127.92$$

This method reflects the Annual cost of the cash disbursement minus the Annual benefit of the future resale value.

Ex. 6-2, Solution Two: Equation 6-1 describes a relationship that may be modified by an identity presented in Chapter 4:

$$(A/P,i,n) = (A/F,i,n) + i \tag{6-2}$$

Substituting this into Eq. 6-1 gives:

$$EUAC = P(A/F,i,n) + Pi - S(A/F,i,n)$$

$$= (P - S)(A/F,i,n) + Pi \tag{6-3}$$

$$= (1000 - 200)(A/F,7\%,10) + 1000(0.07)$$

$$= 800(0.0724) + 70 = 57.92 + 70$$

$$= \$127.92$$

This method computes the equivalent annual cost due to the unrecovered $800 when the furniture is sold, and adds annual interest on the $1000 investment.

Ex. 6-2, Solution Three: If the value for $(A/Fi,n)$ from Eq. 6-2 is substituted into Eq. 6-1, we obtain:

$$EUAC = P(A/P,i,n) - S(A/P,i,n) + Si$$

$$= (P - S)(A/P,i,n) + Si \tag{6-4}$$

$$= (1000 - 200)(A/P,7\%,10) + 200(0.07)$$

$$= 800(0.1424) + 14 = 113.92 + 14 = \$127.92$$

This method computes the annual cost of the $800 decline in value during the ten years, plus interest on the $200 tied up in the furniture as the salvage value.   ■

Example 6-2 illustrates that when there is an initial disbursement P followed by a salvage value S, the annual cost may be computed in three different ways:

1. $EUAC = P(A/P,i,n) - S(A/F,i,n)$ (6-1)

2. $EUAC = (P - S)(A/F,i,n) + Pi$ (6-3)

3. $EUAC = (P - S)(A/P,i,n) + Si$ (6-4)

Each of the three calculations gives the same results. In practice, the first and third methods are most commonly used.

EXAMPLE 6-3

Bill owned a car for five years. One day he wondered what his uniform annual cost for maintenance and repairs had been. He assembled the following data:

Year	Maintenance and repair cost for year
1	$ 45
2	90
3	180
4	135
5	225

Compute the equivalent uniform annual cost (EUAC) assuming 7% interest and end-of-year disbursements.

Solution: The EUAC may be computed for this irregular series of payments in two steps:

1. Compute the present worth of cost for the five years using single payment present worth factors.

2. With the PW of cost known, compute EUAC using the capital recovery factor.

$$PW \text{ of cost} = 45(P/F,7\%,1) + 90(P/F,7\%,2) + 180(P/F,7\%,3)$$
$$+ 135(P/F,7\%,4) + 225(P/F,7\%,5)$$
$$= 45(0.9346) + 90(0.8734) + 180(0.8163)$$
$$+ 135(0.7629) + 225(0.7130)$$
$$= \$531$$

$$EUAC = 531(A/P,7\%,5) = 531(0.2439) = \$130 \quad ∎$$

EXAMPLE 6-4

Bill reexamined his calculations and found that he had reversed the Year 3 and 4 maintenance and repair costs in his table. The correct table is:

Year	Maintenance and repair cost for year
1	$ 45
2	90
3	135
4	180
5	225

Recompute the EUAC.

Solution: This time the schedule of disbursements is an arithmetic gradient series plus a uniform annual cost, as follows:

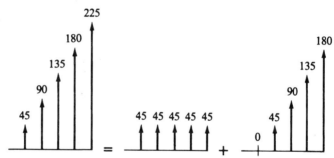

EUAC = 45 + 45(*A/G*,7%,5)

     = 45 + 45(1.865)

     = $129

Since the timing of the Ex. 6-3 and 6-4 expenditures is different, we would not expect to obtain the same EUAC. ■

The examples have shown four essential points concerning cash flow calculations:

1. **There is a direct relationship between the present worth of cost and the equivalent uniform annual cost. It is**

   EUAC = (PW of cost)(*A/P,i,n*)

2. **In a problem, an expenditure of money increases the EUAC, while a receipt of money (like selling something for its salvage value) decreases EUAC.**

3. When there are irregular cash disbursements over the analysis period, a convenient method of solution is to first determine the PW of cost; then, using the equation in Item 1 above, the EUAC may be calculated.

4. Where there is an increasing uniform gradient, EUAC may be rapidly computed using the arithmetic gradient uniform series factor, $(A/G,i,n)$.

Annual Cash Flow Analysis

The criteria for economic efficiency are presented in Table 6-1. One notices immediately that the table is quite similar to Table 5-1. In the case of fixed input, for example, the present worth criterion is *maximize PW of benefits,* and the annual cost criterion is *maximize equivalent uniform annual benefits.* It is apparent that, if you are maximizing the present worth of benefits, simultaneously you must be maximizing the equivalent uniform annual benefits. This is illustrated in Example 6-5.

Table 6-1 ANNUAL CASH FLOW ANALYSIS

	Situation	*Criterion*
Fixed input	**Amount of money or other input resources is fixed**	**Maximize equivalent uniform benefits (maximize EUAB)**
Fixed output	**There is a fixed task, benefit, or other output to be accomplished**	**Minimize equivalent uniform annual cost (minimize EUAC)**
Neither input nor output is fixed	**Neither amount of money, or other inputs, nor amount of benefits, or other outputs, is fixed**	**Maximize (EUAB– EUAC)**

EXAMPLE 6-5

A firm is considering which of two devices to install to reduce costs in a particular situation. Both devices cost $1000 and have useful lives of five years with no salvage value. Device *A* can be expected to result in $300 savings annually. Device *B* will provide cost savings of $400 the first year but will decline $50 annually, making the second year savings $350, the third year savings $300, and so forth. With interest at 7%, which device should the firm purchase?

Solution:

Device A:

   EUAB = $300

Device B:

   EUAB = 400 - 50(A/G,7%,5) = 400 - 50(1.865)

   = $306.75

To maximize EUAB, select device *B*.   ∎

Example 6-5 was previously presented as Ex. 5-1 where we found:

   PW of benefits A = 300(P/A,7%,5) = 300(4.100) = $1230

This is converted to EUAB by multiplying by the capital recovery factor:

   $EUAB_A$ = 1230(A/P,7%,5) = 1230(0.2439) = $300

   PW of benefits B = 400(P/A,7%,5) - 50(P/G,7%,5)

   = 400(4.100) - 50(7.647) = $1257.65

and, hence,

   $EUAB_B$ = 1257.65(A/P,7%,5) = 1257.65(0.2439)

   = $306.75

We see, therefore, that it is easy to convert the present worth analysis results into the annual cash flow analysis results. We could go from annual cash flow to present worth just as easily using the series present worth factor. And, of course, both methods indicate the same device *B* as the preferred alternative.

EXAMPLE 6-6

Three alternatives are being considered for improving an operation on the assembly line along with the do-nothing alternative. The cost of the equipment varies as do their annual benefits compared to the present situation. Each of Plans A, B, and C has a ten-year life and a scrap value equal to 10% of its original cost.

	Plan A	Plan B	Plan C
Installed cost of equipment	$15,000	$25,000	$33,000
Material and labor savings per year	14,000	9,000	14,000
Annual operating expenses	8,000	6,000	6,000
End-of-useful life scrap value	1,500	2,500	3,300

If interest is 8%, which plan, if any, should be adopted?

Solution: Since neither installed cost nor output benefits are fixed, the economic criterion is to maximize (EUAB – EUAC).

	Plan A	Plan B	Plan C	Do nothing
Equivalent uniform annual benefit (EUAB):				
Material and labor per year	$14,000	$9,000	$14,000	$0
Scrap value (*A/F*,8%,10)	104	172	228	0
EUAB =	$14,104	$9,172	$14,228	$0
Equivalent uniform annual cost (EUAC):				
Installed cost (*A/P*,8%,10)	$ 2,235	$3,725	$ 4,917	$0
Annual operating expenses	8,000	6,000	6,000	0
EUAC =	$10,235	$9,725	$10,917	$0
(EUAB – EUAC) =	$ 3,869	-$553	$ 3,311	$0

Based on our criterion of maximizing (EUAB – EUAC), Plan A is the best of the four alternatives. We note, however, that since the do-nothing alternative has (EUAB – EUAC) = 0, it is a more desirable alternative than Plan B. ■

Analysis Period

In the last chapter, we saw that the analysis period was an important consideration in computing present worth comparisons. It was essential that a common analysis period be used for each alternative. In annual cash flow comparisons, we again have the analysis period question. Example 6-7 will help in examining the problem.

EXAMPLE 6-7

Two pumps are being considered for purchase. If interest is 7%, which pump should be bought?

	Pump A	*Pump B*
Initial cost	$7000	$5000
End-of-useful-life salvage value	1500	1000
Useful life, in years	12	6

Solution: The annual cost for twelve years of Pump *A* can be found using Eq. 6-4:

$$EUAC = (P - S)(A/P, \ i,n) + Si$$

$$= (7000 - 1500)(A/P,7\%,12) + 1500(0.07)$$

$$= 5500(0.1259) + 105 = \$797$$

Now compute the annual cost for six years of Pump *B:*

$$EUAC = (5000 - 1000)(A/P,7\%,6) + 1000(0.07)$$

$$= 4000(0.2098) + 70 = \$909$$

For a common analysis period of twelve years, we need to replace Pump *B* at the end of its six-year useful life. If we assume that another Pump *B'* can be obtained that has the same $5000 initial cost, $1000 salvage value and six-year life, the cash flow will be as follows:

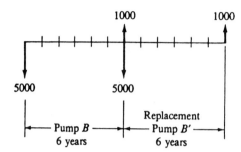

For the twelve-year analysis period, the annual cost for Pump *B:*

$$\text{EUAC} = [5000 - 1000(P/F,7\%,6) + 5000(P/F,7\%,6)$$

$$- 1000(P/F,7\%,12)] \times (A/P,7\%,12)$$

$$= [5000, - 1000(0.6663) + 5000(0.6663) - 1000(0.4440)] \times (0.1259)$$

$$= (5000 - 666 + 3331 - 444)(0.1259)$$

$$= (7211)(0.1259) = \$909$$

The annual cost of *B* for the six-year analysis period is the same as the annual cost for the twelve-year analysis period. This is not a surprising conclusion when one recognizes that the annual cost of the first six-year period is repeated in the second six-year period. Thus the lengthy calculation of EUAC for twelve years·of Pump was not needed. By assuming that the shorter-life equipment is replaced by equipment with identical economic consequences, we have avoided a lot of calculations and the analysis period problem. Select Pump *A.* ■

Analysis Period Equal to Alternative Lives

When the analysis period for an economy study coincides with the useful life for each alternative, we have an ideal situation which causes no difficulties. The economy study is based on this analysis period.

Analysis Period a Common Multiple of Alternative Lives

When the analysis period is a common multiple of the alternative lives (for example, in Ex. 6-7, the analysis period was twelve years with six- and twelve-year alternative lives), a "replacement with an identical item with the same costs, performance, and so forth" is frequently assumed. This means that when an alternative has reached the end of its useful life, it is assumed to be replaced with an identical item. As shown in Ex. 6-7, the result is that the EUAC for Pump *B* with a six-year useful life is equal to the EUAC for the entire analysis period based on Pump *B plus* Replacement Pump *B'*.

Under these circumstances of identical replacement, it is appropriate to compare the annual cash flows computed for alternatives based on their own service lives. In Ex. 6-7, the annual cost for Pump *A*, based on its 12-year service life, was compared with the annual cost for Pump *B*, based on its six-year service life.

Analysis Period For a Continuing Requirement

Many times the economic analysis is to determine how to provide for a more or less continuing requirement. One might need to pump water from a well as a continuing requirement. There is no distinct analysis period. In this situation, the analysis period is assumed to be long but undefined.

If, for example, we had a continuing requirement to pump water and alternative Pumps *A* and *B* had useful lives of seven and eleven years, respectively, what should we do? The customary assumption is that Pump *A's* annual cash flow (based on a seven-year life) may be compared to Pump *B's* annual cash flow (based on an eleven-year life). This is done without much concern that the least common multiple of the seven- and eleven-year lives is 77 years. This comparison of "different-life" alternatives assumes identical replacement (with identical costs, performance, and so forth) when an alternative reaches the end of its useful life. Example 6-8 illustrates the situation.

EXAMPLE 6-8

Pump *B* in Ex. 6-7 is now believed to have a -nine-year useful life. Assuming the same initial cost and salvage value, compare it with Pump *A* using the same 7% interest rate.

Solution: If we assume that the need for *A* or *B* will exist for some continuing period, the comparison of annual costs for the unequal lives is an acceptable technique. For twelve years of Pump *A*:

$$EUAC = (7000 - 1500)(A/P, 7\%, 12) + 1500(0.07) = \$797$$

For nine years of Pump *B*:

$$\text{EUAC} = (5000 - 1000)(A/P, 7\%, 9) + 1000(0.07) = \$684$$

For minimum EUAC, select Pump *B*. ■

Infinite Analysis Period

At times we have an alternative with a limited (finite) useful life in an infinite analysis period situation. The equivalent uniform annual cost may be computed for the limited life. The assumption of identical replacement (replacements have identical costs, performance, and so forth) is often appropriate. Based on this assumption, the same EUAC occurs for each replacement of the limited-life alternative. The EUAC for the infinite analysis period is therefore equal to the EUAC computed for the limited life. With identical replacement,

$$\text{EUAC} _{\text{for infinite analysis period}} = \text{EUAC} _{\text{for limited life } n}$$

A somewhat different situation occurs when there is an alternative with an infinite life in a problem with an infinite analysis period:

$$\text{EUAC} _{\text{for infinite analysis period}} = P(A/P, i, \infty) + \text{any other annual costs}$$

When $n = \infty$, we have $A = Pi$ and, hence, $(A/P, i\infty)$ equals i.

$$\text{EUAC} _{\text{for infinite analysis period}} = Pi + \text{any other annual costs}$$

EXAMPLE 6-9

In the construction of the aqueduct to expand the water supply of a city, there are two alternatives for a particular portion of the aqueduct. Either a tunnel can be constructed through a mountain, or a pipeline can be laid to go around the mountain. If there is a permanent need for the aqueduct, should the tunnel or the pipeline be selected for this particular portion of the aqueduct? Assume a 6% interest rate.

Solution:

	Tunnel through mountain	*Pipeline around mountain*
Initial cost	$5.5 million	$5 million
Maintenance	0	0
Useful life	Permanent	50 years
Salvage value	0	0

Tunnel: For the tunnel, with its permanent life, we want $(A/P,6\%,\infty)$. For an infinite life, the capital recovery is simply interest on the invested capital. So $(A/P,6\%,\infty) = i$,

$$\text{EUAC} = Pi = \$5.5 \text{ million}(0.06)$$

$$= \$330,000$$

Pipeline:

$$\text{EUAC} = \$5 \text{ million}(A/P, 6\%,50)$$

$$= \$5 \text{ million}(0.0634) = \$317,000$$

For fixed output, minimize EUAC. Select the pipeline.   ■

The difference in annual cost between a long life and an infinite life is small unless an unusually low interest rate is used. In Example 6-9 the tunnel is assumed to be permanent. For comparison, compute the annual cost if an 85-year life is assumed for the tunnel?

$$\text{EUAC} = \$5.5 \text{ million}(A/P,6\%,85)$$

$$= \$5.5 \text{ million}(0.0604) = \$332,000$$

The difference in time between 85 years and infinity is great indeed, yet the difference in annual costs in Example 6-9 is very small.

Some Other Analysis Period

The analysis period in a particular problem may be something other than one of the four we have so far described. It may be equal to the life of either the shorter-life alternative, the longer-life alternative, or something entirely different. One must carefully examine the consequences of each alternative throughout the analysis period and, in addition, see what differences there might be in salvage values, and so forth, at the end of the analysis period.

EXAMPLE 6-10

Consider a situation where Alternative 1 has a 7-year life and a salvage value at the end of that time. The replacement cost at the end of 7 years may be more or less than the original cost. If the replacement is retired prior to 7 years, it will have a terminal value that exceeds the end-of-life salvage value. Alternative 2 has a 13-year life and a terminal value whenever it is retired. If the situation indicates that 10 years is the proper analysis period, set up the equations to properly compute the EUAC for each alternatice.

Solution:

Alternative 1:

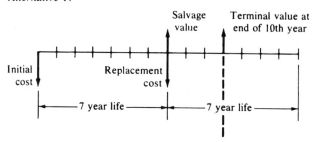

Alternative 2:

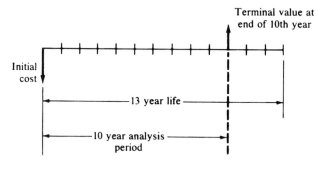

Alternative 1

$$EUAC_1 = [\text{ Initial cost } +(\text{Replacement cost -Salvage value})(P/F,i,7)$$
$$- (\text{Terminal value})(P/F,i,10)] (A/P,i,10)$$

Alternative 2

$$EUAC_2 = [\text{ Initial cost } - (\text{Terminal value})(P/F,i,10)] (A/P,i,10) \quad ■$$

Using Spreadsheets to Analyze Loans

Loan and bond payments are made by firms, agencies, and individual engineers. Usually, the payments in each period are constant. Spreadsheets make it easy to:

>Calculate the loan's amortization schedule
>Decide how a payment is split between principal and interest
>Find the balance due on a loan
>Calculate the number of payments remaining on a loan.

Building an Amortization Schedule

As illustrated in Chapter 4 and Appendix 1, an amortization schedule lists for each payment period: the loan payment, interest paid, principal paid, and remaining balance. For each period the interest paid equals the interest rate times the balance remaining from the period before. Then the principal payment equals the payment minus the interest paid. Finally, this principal payment is applied to the balance remaining from the previous period to calculate the new remaining balance. As a basis for comparison with spreadsheet loan functions, Figure 6-1 shows this calculation for Example 6-11.

EXAMPLE 6-11

An engineer wanted to celebrate graduating and getting a job by buying $2400 of new furniture. Luckily the store was offering 6-month financing at the low interest rate of 6% per year nominal (really ½% per month). Calculate the amortization schedule.

Solution
The first step is to calculate the monthly payment.

$$A = 2400 \ (A/P, \tfrac{1}{2}\%, 6) = 2400 \cdot 0.1696 = \$407.0$$

	A	B	C	D	E
1	2400	Initial balance			
2	0.50%	*i*			
3	6	*N*			
4	$407.03	payment	= -PMT(A2,A3,A1)		
5					
6			principal	ending	
7	month	interest	payment	balance	
8	0			2400.00	=A1
9	1	12.00	395.03	2004.97	=D8-C9
10	2	10.02	397.00	1607.97	
11	3	8.04	398.99	1208.98	
12	4	6.04	400.98	807.99	
13	5	4.04	402.99	405.00	
14	6	2.03	405.00	0.00	
15				=$A$4-B14	
16				= payment - interest	
17			=$A$2*D13		
18			= rate*previous balance		

Figure 6-1 Amortization schedule for furniture loan.

How Much to Interest? How Much to Principal?

For a loan with constant payments, we can answer these questions for any period without the full amortization schedule. For a loan with constant payments, the functions IPMT and PPMT directly answer these questions. For simple problems, both functions have four arguments $(i,t,N,-P)$, where t is the time period being calculated. Both functions have optional arguments that permit adding a balloon payment (an F) and changing from end-of-period payments to beginning-of-period payments.

For example, consider period 4 of Example 6-11. The spreadsheet formulas give the same answer as shown in Figure 6-1.

interest$_4$ = IPMT(0.5%,4,6,-2400) = $6.04
principal payment$_4$ = PPMT(0.5%,4,6,-2400) = $400.98

Finding the Balance Due on a Loan

An amortization schedule can be used to calculate the balance due on a loan. Or more easily the balance due equals the present worth of the remaining payments. Interest is paid in full after each payment, so later payments are simply based on the balance due.

EXAMPLE 6-12

A car is purchased with a 48-month, 9% nominal loan with an initial balance of $15,000. What is the balance due half-way through the four years?

Solution

The first step is to calculate the monthly payment, at a monthly interest rate of ¾%. This equals

$$\text{payment} = 15,000 \cdot (A/P,0.75\%,48) \text{ or } = \text{PMT}(0.75\%,48,-15000)$$
$$= 15,000 \cdot 0.0249 = \$373.50$$
$$\text{or} \qquad = \$373.28$$

The next step will use the spreadsheet answer, because it is more accurate (there are only three significant digits in the tabulated factor).

After 24 payments and with 24 left, the remaining balance equals $(P/A,i,N_{remaining})\cdot\text{payment}$

$$\text{balance} = (P/A,0.75\%,24)\cdot\$373.28 \text{ or } = \text{PV}(0.75\%,24,373.28)$$
$$= 21.889 \cdot 373.28 = \$8170.73$$
$$\text{or} \qquad = \$8170.78$$

Thus half-way through the repayment schedule 54.5% of the original balance is still owed. ■

Payoff Debt Sooner by Increasing Payments

Paying off debt can be a good investment, as the investment earns the rate of interest on the loan. For example, this could be 8% for a mortgage, 10% for a car loan, or 19% for a credit card. When making extra payments on a loan, the common question is: How much sooner will the debt be paid off? Until the debt is paid off, any early payments are essentially locked up, because the same payment amount is owed each month.

The first reason that spreadsheets are convenient is fractional interest rates. For example, an auto loan might be at a nominal rate of 13% with monthly compounding or 1.08333% per month. The second reason is that NPER calculates the number of periods remaining on a loan.

NPER can be used to calculate "How much difference does one extra payment make?" or "How much difference does increasing all payments by $x\%$ make?" Extra payments are applied entirely to principal, so the interest rate, remaining balance, and payment amounts are all known. $N_{remaining}$ equals NPER(i, payment, remaining balance) with optional arguments for beginning of period cash flows and balloon payments. The signs of the payment and the remaining balance must be different.

EXAMPLE 6-13

Maria has a 7.5% mortgage with monthly payments for 30 years. Her original balance was $100,000, and she just made her 12th payment. Each month she also pays into a reserve account, which the bank uses to pay her fire and liability insurance ($900 annually) and property taxes ($1500 annually). How much does she shorten the loan by, if she makes an extra *loan* payment today? If she makes an extra *total* payment? If she increases each total payment to 110% of her current total payment?

Solution
The first step is to calculate her *loan* payment for the 360 months. Rather than calculating a six-significant digit monthly interest rate, it is easier to use 0.075/12 in the spreadsheet formulas.

payment = PMT(0.075/12,360,-100000) = $699.21

The remaining balance after 12 such payments is the present worth of the remaining 348 payments.

balance$_{12}$ = PV(0.075/12,348,699.21) = $99,077.53

(after 12 payments she has paid off $922!)

If she pays an extra $699.21, then the number of periods remaining is

NPER(0.075/12, -699.21, 99077.53-699.21) = 339.5

This is 8.5 payments less than the 348 periods left before the extra payment.
If she makes an extra total payment, then

total payment = 699.21 + 900/12 + 1500/12 = $899.21 / month

NPER(0.075/12, -699.21, 99244 - 899.21) = 337.1 or 2.4 more payments saved.

If she makes an extra 10% payment on the total payment of $899.21, then

NPER(0.075/12,-(1.1*899.21-200), 99077.53) = 246.5 payments or 101.5 payments saved.

Note that $200 of the total payment goes to pay for insurance and taxes. ■

Summary

Annual cash flow analysis is the second of the three major methods of resolving alternatives into comparable values. When an alternative has an initial cost P and salvage value S, there are three ways of computing the equivalent uniform annual cost:

- EUAC = $P(A/P,i,n)$ - $S(A/F,i,n)$

- EUAC = $(P - S)(A/F,i,n) + Pi$

- EUAC = $(P - S)(A/P,i,n) + Si$

All three equations give the same answer.
   The relationship between the present worth of cost and the equivalent uniform annual cost is:

- EUAC = (PW of cost)$(A/P,i,n)$

The three annual cash flow criteria are:

For fixed input	Maximize EUAB
For fixed output	Minimize EUAC
Neither input nor output fixed	Maximize (EUAB–EUAC)

In present worth analysis there must be a common analysis period. Annual cash flow analysis, however, allows some flexibility provided the necessary assumptions are suitable in the situation being studied. The analysis period may be different from the lives of the alternatives, and a valid cash flow analysis made, provided the following two criteria are met:

1. When an alternative has reached the end of its useful life, it is assumed to be replaced by an identical replacement (with the same costs, performance, and so forth).

2. The analysis period is a common multiple of the useful lives of the alternatives, or there is a continuing or perpetual requirement for the selected alternative.

If both these conditions do not apply, then it is necessary to make a detailed study of the consequences of the various alternatives over the entire analysis period with particular attention to the difference between the alternatives at the end of the analysis period.
   There is very little numerical difference between a long-life alternative and a perpetual alternative. As the value of n increases, the capital recovery factor approaches i. At the limit, $(A/P,i,\infty) = i$.

Problems

6-1   On April 1st, $100 is loaned to a man. The loan is to be repaid in three equal semi-annual (every six months) payments. If the annual interest rate is 7% compounded semiannually, how much is each payment? (*Answer:* $35.69)

6-2   Compute the value of *C* for the following diagram, based on a 10% interest rate.

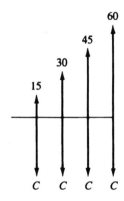

(*Answer: C* = $35.72)

6-3   Compute the value of *B* for the following diagram:

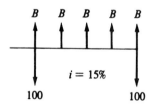

6-4   Compute the value of *E:*

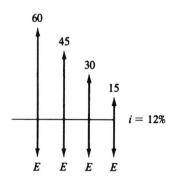

6-5 If *i*= 6%, compute the value of *D* that is equivalent to the two disbursements shown.

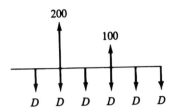

(*Answer:* *D* = $52.31)

6-6 For the diagram, compute the value of *D*:

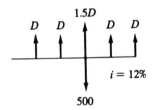

6-7 What is *C* in the figure below?

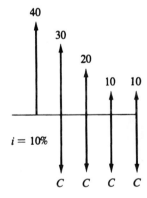

6-8   If interest is 10%, what is *A*?

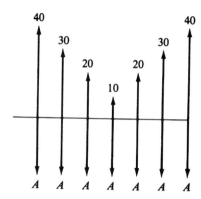

6-9   A certain industrial firm desires an economic analysis to determine which of two different machines should be purchased. Each machine is capable of performing the same task in a given amount of time. Assume the minimum attractive return is 8%. The following data are to be used in this analysis:

	Machine X	Machine Y
First cost	$5000	$8000
Estimated life, in years	5	12
Salvage value	0	$2000
Annual maintenance cost	0	150

Which machine would you choose? Base your answer on annual cost.
(*Answers: X* = $1252; *Y* = $1106)

6-10   An electronics firm invested $60,000 in a precision inspection device. It cost $4000 to operate and maintain in the first year, and $3000 in each of the subsequent years. At the end of four years, the firm changed their inspection procedure, eliminating the need for the device. The purchasing agent was very fortunate in being able to sell the inspection device for the $60,000 that had originally been paid for it. The plant manager asks you to compute the equivalent uniform annual cost of the device during the four years it was used. Assume interest at 10% per year.
(*Answer:* $9287)

6-11   A firm is about to begin pilot plant operation on a process it has developed. One item of optional equipment that could be obtained is a heat exchanger unit. The company finds that one can be obtained now for $30,000, and that this unit can be used in other company operations. It is estimated that the heat exchanger unit will be worth $35,000 at the end of eight years. This seemingly high salvage value is due primarily to the fact that the $30,000 purchase price is really a rare bargain. If the firm believes 15% is an appropriate rate of return, what annual benefit is needed to justify the purchase of the heat exchanger unit? (*Answer:* $4135)

6-12 The maintenance foreman of a plant in reviewing his records found that a large press had the following maintenance cost record for the last five years:

5 years ago:	$ 600
4 years ago:	700
3 years ago:	800
2 years ago:	900
Last year:	1000

After consulting with a lubrication specialist, he changed the preventive maintenance schedule. He believes that this year maintenance will be $900 and will decrease $100 a year in each of the following four years. If his estimate of the future is correct, what will be the equivalent uniform annual maintenance cost for the ten-year period? Assume interest at 8%. (*Answer:* $756)

6-13 A firm purchased some equipment at a very favorable price of $30,000. The equipment resulted in an annual net saving of $1000 per year during the eight years it was used. At the end of eight years, the equipment was sold for $40,000. Assuming interest at 8%, did the equipment purchase prove to be desirable?

6-14 A manufacturer is considering replacing a production machine tool. The new machine would cost $3700, have a life of four years, have no salvage value, and save the firm $500 per year in direct labor costs and $200 per year indirect labor costs. The existing machine tool was purchased four years ago at a cost of $4000. It will last four more years and have no salvage value at the end of that time. It could be sold now for $1000 cash. Assume money is worth 8%, and that the difference in taxes, insurance, and so forth, for the two alternatives is negligible. Determine whether or not the new machine should be purchased.

6-15 Two possible routes for a power line are under study. Data on the routes are as follows:

	Around the lake	Under the lake
Length	15 km	5 km
First cost	$5000/km	$25,000/km
Maintenance	$200/km/yr	$400/km/yr
Useful life, in years	15	15
Salvage value	$3000/km	$5000/km
Yearly power loss	$500/km	$500/km
Annual property taxes	2% of first cost	2% of first cost

If 7% interest is used, should the power line be routed around the lake or under the lake? (*Answer:* Around the lake.)

6-16 Steve Lowe must pay his property taxes in two equal installments on December 1 and April 1. The two payments are for taxes for the fiscal year that begins on July 1 and ends the following

June 30. Steve purchased a home on September 1. He estimates the annual property taxes will be $850 per year. Assuming the annual property taxes remain at $850 per year for the next several years, Steve plans to open a savings account and to make uniform monthly deposits the first of each month. The account is to be used to pay the taxes when they are due.

To begin the account, Steve deposits a lump sum equivalent to the monthly-payments-that-will-not-have-been-made for the first year's taxes. The savings account pays 9% interest, compounded monthly and payable quarterly (March 31, June 30, September 30, and December 31). How much money should Steve put into the account when he opens it on September 1? What uniform monthly deposit should he make from that time on? (A careful *exact* solution is expected.)
(*Answers:* Initial deposit $350.28; Monthly deposit $69.02)

6-17 An oil refinery finds that it is now necessary to process its waste liquids in a costly treating process before discharging them into a nearby stream. The engineering department estimates that the waste liquid processing will cost $30,000 at the end of the first year. By making process and plant alterations, it is estimated that the waste treatment cost will decline $3000 each year. As an alternate, a specialized firm, Hydro-Clean, has offered a contract to process the waste liquids for the ten years for a fixed price of $15,000 per year, payable at the end of each year. Either way, there should be no need for waste treatment after ten years. If the refinery manager considers 8% a suitable interest rate, should he accept the Hydro-Clean offer or not?

6-18 Bill Anderson buys an automobile every two years as follows: initially he pays a downpayment of $6000 on a $15,000 car. The balance is paid in 24 equal monthly payments with annual interest at 12%. When he has made the last payment on the loan, he trades in the two-year old car for $6000 on a new $15,000 car, and the cycle begins over again.

Doug Jones decided on a different purchase plan. He thought he would be better off if he paid $15,000 cash for a new car. Then he would make a monthly deposit in a savings account so that, at the end of two years, he would have $9000 in the account. The $9000 plus the $6000 trade-in value of the car will allow Doug to replace his two-year-old car by paying $9000 for a new one. The bank pays 6% interest, compounded quarterly.

a. What is Bill Anderson's monthly payment to pay off the loan on the car?

b. After he purchased the new car for cash, how much per month should Doug Jones deposit in his savings account to have sufficient money for the next car two years hence?

c. Why is Doug's monthly savings account deposit smaller than Bill's payment?

6-19 Claude James, a salesman, needs a new car for use in his business. He expects to be promoted to a supervisory job at the end of three years, and so his concern now is to have a car for the three years he expects to be "on the road." The company will reimburse their salesmen each month at the rate of 25¢ per mile driven. Claude has decided to drive a low-priced automobile. He finds. however, that there are three different ways of obtaining the automobile:

a. Purchase for cash; the price is $13,000.

b. Lease the car; the monthly charge is $350 on a 36-month lease, payable at the end of each month; at the end of the three-year period, the car is returned to the leasing company.

c. Lease the car with an option to purchase it at the end of the lease; pay $360 a month for 36 months; at the end of that time, Claude could purchase the car, if he chooses, for $3500.

Claude believes he should use a 12% interest rate in determining which alternative to select. If the car could be sold for $4000 at the end of three years, which method should he use to obtain it?

6-20 A college student has been looking for a new tire for his car and has located the following alternatives:

Tire warranty	Price per tire
12 mo.	$39.95
24 mo.	59.95
36 mo.	69.95
48 mo.	90.00

If the student feels that the warranty period is a good estimate of the tire life and that a 10% interest rate is appropriate, which tire should he buy?

6-21 A suburban taxi company is considering buying taxis with diesel engines instead of gasoline engines. The cars average 50,000 kilometers a year, with a useful life of three years for the taxi with the gas engine, and four years for the diesel taxi. Other comparative information is as follows:

	Diesel	Gasoline
Vehicle cost	$13,000	$12,000
Fuel cost per liter	48¢	51¢
Mileage, in km/liter	35	28
Annual repairs	300	200
Annual insurance premium	500	500
End-of-useful-life resale value	2,000	3,000

Determine the more economical choice if interest is 6%.

6-22 When he started work on his 22nd birthday, D. B. Cooper decided to invest money each month with the objective of becoming a millionaire by the time he reaches his 65th birthday. If he expects his investments to yield 18% per annum, compounded monthly, how much should he invest each month? (*Answer*: $6.92 a month.)

6-23 Linda O'Shay deposited $30,000 in a savings account as a perpetual trust. She believes the account will earn 7% annual interest during the first ten years and 5% interest thereafter. The trust is to provide a uniform end-of-year scholarship at the University. What uniform amount could be used for the student scholarship each year, beginning at the end of the first year and continuing forever?

6-24 A motorcycle is for sale for $2600. The motorcycle dealer is willing to sell it on the following terms:

No downpayment; pay $44 at the end of each of the first four months; pay $84 at the end of each month after that, until the motorcycle is paid in full.

Based on these terms and a 12% annual interest rate compounded monthly, how many $84 payments will be required?

6-25 A machine costs $20,000 and has a five-year useful life. At the end of the five years, it can be sold for $4000. If annual interest is 8%, compounded semi-annually, what is the equivalent uniform annual cost of the machine? (An *exact* solution is expected.)

6-26 The average age of engineering students when they graduate is a little over 23 years. This means the working career of most engineers is almost exactly 500 months. How much would an engineer need to save each month to become a millionaire by the end of his working career? Assume a 15% interest rate, compounded monthly.

6-27 As shown in the cash flow diagram, there is an annual disbursement of money that varies from year to year from $100 to $300 in a fixed pattern that repeats forever. If interest is 10%, compute the value of A, also continuing forever, that is equivalent to the fluctuating disbursements.

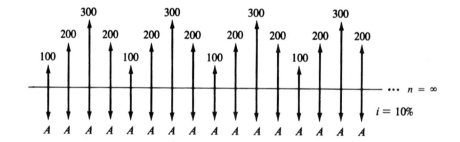

6-28 Alice White has arranged to buy some home recording equipment. She estimates that it will have a five-year useful life and no salvage value. The dealer, who is a friend, has offered Alice two alternative ways to pay for the equipment:

a. Pay $2000 immediately and $500 at the end of one year.

b. Pay nothing until the end of four years when a single payment of $3000 must be made.

If Alice believes 12% is a suitable interest rate, which method of payment should she select? (*Answer:* Select *b*)

6-29 A company must decide whether to buy Machine *A* or Machine *B*:

	Machine A	Machine B
Initial cost	$10,000	$20,000
Useful life, in years	4	10
End-of-useful-life salvage value	$10,000	$10,000
Annual maintenance	1,000	0

At a 10% interest rate, which machine should be installed? (*Answer:* Machine A)

6-30 The Johnson Company pays $200 a month to a trucker to haul wastepaper and cardboard to the city dump. The material could be recycled if the company would buy a $6000 hydraulic press bailer and spend $3000 a year for labor to operate the bailer. The bailer has an estimated useful life of thirty years and no salvage value. Strapping material would cost $200 per year for the estimated 500 bales a year that would be produced. A wastepaper company will pick up the bales at the plant and pay the Johnson Co. $2.30 per bale for them.

a. If interest is 8%, is it economical to install and operate the bailer?

b. Would you recommend that the bailer be installed?

6-31 Consider the following:

| | Alternative | |
	A	B
Cost	$50	$180
Uniform annual benefit	15	60
Useful life, in years	10	5

The analysis period is ten years, but there will be no replacement for Alternative B at the end of five years. Based on a 15% interest rate, determine which alternative should be selected.

6-32 Consider the following two mutually exclusive alternatives:

	Alternative	
	A	B
Cost	$100	$150
Uniform annual benefit	16	24
Useful life, in years	∞	20

Alternative B may be replaced with an identical item every twenty years at the same $150 cost and will have the same $24 uniform annual benefit. Using a 10% interest rate, determine which alternative should be selected.

6-33 Some equipment will be installed in a warehouse that a firm has leased for seven years. There are two alternatives:

	Alternative	
	A	B
Cost	$100	$150
Uniform annual benefit	55	61
Useful life, in years	3	4

At any time after the equipment is installed, it has no salvage value. Assume that Alternatives A and B will be replaced at the end of their useful lives by identical equipment with the same costs and benefits. For a seven-year analysis period and a 10% interest rate, determine which alternative should be selected.

6-34 When he purchased his home, Al Silva borrowed $80,000 at 10% interest to be repaid in 25 equal annual end-of-year payments. Ten years later, after making ten payments, Al found he could refinance the balance due on his loan at 9% interest for the remaining 15 years.

To refinance the loan, Al must pay the original lender the balance due on the loan, plus a penalty charge of 2% of the balance due; to the new lender he also must pay a $ 1000 service charge to obtain the loan. The new loan would be made equal to the balance due on the old loan, plus the 2% penalty charge, and the $1000 service charge. Should Al refinance the loan, assuming that he will keep the house for the next 15 years?

6-35 Consider the following three mutually exclusive alternatives:

	A	B	C
Cost	$100	$150.00	$200.00
Uniform annual benefit	10	17.62	55.48
Useful life, in years	∞	20	5

Assuming that Alternatives B and C are replaced with identical replacements at the end of their useful lives, and an 8% interest rate, which alternative should be selected? (*Answer:* Select C)

6-36 When Sandra began working, she resolved to save $ 1000 a year from her income. After working a couple of months, she realized that it is easier to set a goal than to follow it. Her overall goal is to have saved a "reasonable sum of money" after ten years of work; she decides she can accomplish the same goal in a less painful way by changing her savings pattern: instead of saving $ 1000 a year in a bank account that pays 6% annual interest, she will save an annual sum based on the geometric gradient. She believes her salary will increase 7% each year. Thus, she can save a fixed percentage of her salary each year, and still achieve her goal of saving $ 1000/year, by making a smaller deposit at the end of the first year and increasing the amount of the deposit each year by the same 7% rate her salary increases.

a. What amount does she want to have in the bank at the end of ten years?

b. Following her revised savings plan, how much should she deposit in her savings account at the end of the first year?

6-37 An engineer has a fluctuating future budget for the maintenance of a particular machine. During each of the first five years, $1000 per year will be budgeted. During the second five years, the annual budget will be $1500 per year. In addition, $3500 will be budgeted for an overhaul of the machine at the end of the fourth year, and another $3500 for an overhaul at the end of the eighth year.

The engineer asks you to compute what uniform annual expenditure would be equivalent to these fluctuating amounts, assuming interest at 6% per year.

6-38 An engineer wishes to have five million dollars by the time he retires in 40 years. Assuming 15% nominal interest, compounded continuously, what annual sum must he set aside? (*Answer:* $2011)

6-39 Two mutually exclusive alternatives are being considered.

Year	A	B
0	-$3000	-$5000
1	+845	+1400
2	+845	+1400
3	+845	+1400
4	+845	+1400
5	+845	+1400

One of the alternatives must be selected. Using a 15% nominal interest rate, compounded continuously, determine which one. Solve by annual cash flow analysis.

6-40 A company must decide whether to provide their salesmen with company-owned automobiles, or to pay the salesmen a mileage allowance and have them drive their own automobiles. New automobiles would cost about $18,000 each and could be resold four years later for about $7000 each. Annual operating costs would be $600 per year plus 12¢ per mile. If the salesmen drive their own automobiles, the company probably would pay them 30¢ per mile. Calculate the number of miles each salesman would have to drive each year for it to be economically practical for the company to provide the automobiles. Assume a 10% annual interest rate.

6-41 A pump is required for ten years at a remote location. The pump can be driven by an electric motor if a powerline is extended to the site. Otherwise, a gasoline engine will be used. Based on the following data and a 10% interest rate, how should the pump be powered?

	Gasoline	Electric
First cost	$2400	$6000
Annual operating cost	1200	750
Annual maintenance	300	50
Salvage value	300	600
Life in years	5 yrs	10 yrs

6-42 The town of Dry Gulch needs an additional supply of water from Pine Creek. The town engineer has selected two plans for comparison. *Gravity plan:* Divert water at a point ten miles up Pine Creek and carry it through a pipeline by gravity to the town. *Pumping plan:* Divert water at a point closer to the town and pump it to the town. The pumping plant would be built in two stages, with one-half capacity installed initially and the other half installed ten years later.

An analysis will assume a 40-year life, 10% interest and no salvage value. Costs are as follows:

	Gravity	Pumping
Initial investment	$2,800,000	$1,400,000
Additional investment in 10th year	None	200,000
Operation and maintenance	10,000/year	25,000/year
Power cost		
Average first 10 years	None	50,000/year
Average next 30 years	None	100,000/year

Determine the more economical plan.

6-43 Uncle Elmo needs to replace the family privy. The local sanitary engineering firm has submitted two alternative structural proposals with respective cost estimates as shown below. Which construction should Uncle Elmo choose if his minimum attractive rate of return is 6%. Use both a Present Worth and Annual Cost approach to compare.

	Masonite	Brick
First Cost	$250	$1000
Annual Maintenance	20	10
Service Life	4 yr	20 yr
Salvage Value	$10	$100

6-44 Art Arfons, a K-State educated engineer, has made a considerable fortune. He wishes to start a perpetual scholarship for engineering students at K-State. The scholarship will provide a student with an annual stipend of $2500 for each of four years (freshmen through senior), plus an additional $5000 during the senior year to cover entertainment expenses. Assume that students graduate in four years, a new award is given every four years, and the money is provided at the beginning of each year with the first award at the beginning of year one. The interest rate is 8%.

   a. Determine the equivalent uniform annual cost (EUAC) of providing the scholarship.

   b. How much money must Art donate to K-State?

6-45 Jenny McCarthy is an engineer for a municipal power plant. The plant uses natural gas which is currently provided from an existing pipeline at an annual cost of $10,000 per year. Jenny is considering a project to construct a new pipeline. The initial cost of the new pipeline would be $35,000 but it would reduce the annual cost to $5000 per year. Assume an analysis period of 20 years and no salvage value for either the existing or new pipeline. The interest rate is 6%.

   a. Determine the equivalent uniform annual cost (EUAC) for the new pipeline?

   b. Should the new pipeline be constructed?

6-46 Your company must make a $500,000 balloon payment on a lease 2 years and 9 months from today. You have been directed to deposit an amount of money quarterly, beginning today to provide for the $500,000 payments. The account pays 4% per year, compounded quarterly. What is the required quarterly deposit? Note: Lease payments are beginning of the quarter.

6-47 A machine has a first cost of $150,000, an annual Operation and Maintenance cost of $2500, a life of 10 years, and a salvage value of $30,000. At the end of years 4 and 8, it requires a major service which costs $20,000 and $10,000 respectively. At the end of year 5, it will need to be overhauled at a cost of $45,000. What is the Equivalent Uniform Annual Cost of owning and operating this particular machine?

6-48 Mr. Wiggley wants to buy a new house. It will cost $178,000. The bank will loan us 90% of the purchase price at a nominal interest rate of 10.75% compounded weekly and we will make monthly payments. What is the amount of the monthly payments if Mr. Wiggley intends to pay the house off in 25 years?

6-49 The manager in a canned food processing plant is trying to decide between two labeling machines. Their respective costs and benefits are as follows:

	Machine A	Machine B
First cost	$15,000	$25,000
Maintenance and operating costs	1,600	400
Annual benefit	8,000	13,000
Salvage value	3,000	6,000
Useful life, in years	7	10

Assume an interest rate of 12%. Use annual cash flow analysis to determine which machine should be selected.

6-50 Carp, Inc. wants to evaluate two methods of shipping their products. The following cash flows are associated with the alternatives:

	Alternative	
	A	B
First cost	$700,000	$1,700,000
Maintenance & operating costs	18,000	29,000
+ Cost gradient (begin yr-1)	+900/yr	+750/yr
Annual benefit	154,000	303,000
Salvage value	142,000	210,000
Useful life, in years	10	20

Using a MARR of 15% and annual cash flow analysis, decide which is the most desirable alternative.

6-51 A new car is purchased for $12,000 with a 0% down, 9% loan. The loan's length is 4 years. After making 30 payments the owner desires to pay off the loan's remaining balance. How much is owed?

6-52 A year after buying her car, Anita has been offered a job in Europe. Her car loan is for $15,000 at a 9% nominal interest rate for 60 months. If she can sell the car for $12,000, how much does she get to keep after paying off the loan?

6-53 A $78,000 mortgage has a 30 year term and a 9% nominal interest rate.

a. What is the monthly payment?

b. After the first year of payments, what is the outstanding balance?

c. How much interest is paid in month 13? How much principal?

6-54 A $92,000 mortgage has a 30 year term and a 9% nominal interest rate.

a. What is the monthly payment?

b. After the first year of payments, what fraction of the loan has been repaid?

c. After the first ten years of payments, what is the outstanding balance?

d. How much interest is paid in month 25? How much principal?

🖥 **6-55** A 30 year mortgage for $95,000 is issued at a 9% nominal interest rate.

   a. What is the monthly payment?

   b. How long does it take to pay off the mortgage, if $1000 per month is paid?

   c. How long does it take to pay off the mortgage, if double payments are made?

🖥 **6-56** A 30 year mortgage for $145,000 is issued at a 6% nominal interest rate.

   a. What is the monthly payment?

   b. How long does it take to pay off the mortgage, if $1000 per month is paid?

   c. How long does it take to pay off the mortgage, if 20% extra is paid each month?

🖥 **6-57** Solve Problem 6-15 for the breakeven first cost/km of going under the lake.

🖥 **6-58** Redo Problem 6-21 to calculate the EUAW of the alternatives as a function of miles driven per year to see if there is a crossover point in the decision process. Graph your results.

🖥 **6-59** Set up Problem 6-30 on a spreadsheet to make all the input data variable and determine various scenarios which would make the bailer economical.

🖥 **6-60** Develop a spreadsheet to solve Problem 6-42. What is the breakeven cost of the additional pumping investment in the 10th year?

CHAPTER 7

Rate Of Return Analysis

The third of the three major analysis methods is rate of return. In this chapter we will examine three aspects of rate of return. First, we describe the meaning of "rate of return"; then, the calculation of rate of return is illustrated; finally, rate of return analysis problems will be presented. In an Appendix to this chapter, we describe difficulties sometimes encountered in attempting to compute an interest rate for certain kinds of cash flows.

Internal Rate Of Return

In Chapter 3 we examined four plans to repay $5000 in five years with interest at 8% (Table 3-1). In each of the four plans the amount loaned ($5000) and the loan duration (five years) was the same. Yet the total interest paid to the lender varied from $1200 to $2347, depending on the loan repayment plan. We saw, however, that the lender received 8% interest each year on the amount of money actually owed. And, at the end of five years, the principal and interest payments exactly repaid the $5000 debt with interest at 8%. We say the lender received an "8% rate of return."

> *Internal rate of return* is defined as the interest rate paid on the unpaid balance of a *loan* such that the payment schedule makes the unpaid loan balance equal to zero when the final payment is made.

241

Instead of lending money, we might invest $5000 in a machine tool with a five-year useful life and an equivalent uniform annual benefit of $1252. An appropriate question is, "What rate of return would we receive on this investment?" The cash flow would be as follows:

Year	Cash flow
0	-$5000
1	+1252
2	+1252
3	+1252
4	+1252
5	+1252

We recognize the cash flow as Plan 3 of Table 3-1. We know that five payments of $1252 are equivalent to a present sum of $5000 when interest is 8%. Therefore, the rate of return on this investment is 8%. Stated in terms of an investment, we may define internal rate of return as follows:

Internal rate of return is the interest rate earned on the unrecovered investment such that the payment schedule makes the unrecovered investment equal to zero at the end of the life of the investment.

It must be understood that the 8% rate of return does not mean an annual return of 8% on the $5000 investment, or $400 in each of the five years. Instead, each $1252 payment represents an 8% return on the Unrecovered investment *plus* the Partial return of the investment. This may be tabulated as follows:

Year	Cash flow	Unrecovered investment at beginning of year	8% return on unrecovered investment	Investment repayment at end of year	Unrecovered investment at end of year
0	-$5000				
1	+1252	$5000	$ 400	$ 852	$4148
2	+1252	4148	331	921	3227
3	+1252	3227	258	994	2233
4	+1252	2233	178	1074	1159
5	+1252	1159	93	1159	0
			$1260	$5000	

This cash flow represents a situation where the $5000 investment has benefits that produce an 8% rate of return. But, in the five-year period, the total return is only $1260, far less than $400 per year for five years. The reason, we can see, is because internal rate of return is defined as the interest rate earned on the unrecovered investment.

Although the two definitions of internal rate of return are stated differently, one in terms of a loan and the other in terms of an investment, there is only one fundamental concept being described. It is that ***the internal rate of return is the interest rate at which the benefits are equivalent to the costs.*** Since we are describing situations where funds remain within the investment throughout its life, the resulting rate of return is described as the internal rate of return, i^*.

Calculating Rate Of Return

To calculate a rate of return on an investment, we must convert the various consequences of the investment into a cash flow. Then we will solve the cash flow for the unknown value of i^*, which is the internal rate of return. Five forms of the cash flow equation are:

$$\text{PW of benefits - PW of costs} = 0 \tag{7-1}$$

$$\frac{\text{PW of benefits}}{\text{PW of costs}} = 1 \tag{7-2}$$

$$\text{Net Present Worth} = 0 \tag{7-3}$$

$$\text{EUAB - EUAC} = 0 \tag{7-4}$$

$$\text{PW of costs} = \text{PW of benefits} \tag{7-5}$$

The five equations represent the same concept in different forms. They can relate costs and benefits with rate of return i^* as the only unknown. The calculation of rate of return is illustrated by the following Examples.

EXAMPLE 7-1

An $8200 investment returned $2000 per year over a five-year useful life. What was the rate of return on the investment?

Solution: Using Equation 7-2,

$$\frac{\text{PW of benefits}}{\text{PW of costs}} = 1 \qquad \frac{2000(P/A, i, 5)}{8200} = 1$$

Rewriting the equation, we see that

$$(P/A, i, 5) = \frac{8200}{2000} = 4.1$$

Then look at the Compound Interest Tables for the value of i where $(P/A,i,5) = 4.1$; if no tabulated value of i gives this value, we will then find values on either side of the desired value (4.1) and interpolate to find the rate of return i^*.

From interest tables we find:

i	$(P/A,i,5)$
6%	4.212
7%	4.100
8%	3.993

In this example, no interpolation is needed as the internal rate of return i^*, for this investment is exactly 7%. ■

EXAMPLE 7-2

An investment resulted in the following cash flow. Compute the rate of return.

Year	Cash flow
0	-$700
1	+100
2	+175
3	+250
4	+325

Solution:

$$EUAB - EUAC = 0$$

$$100 + 75(A/G,i,4) - 700(A/P,i,4) = 0$$

In this situation, we have two different interest factors in the equation. We will not be able to solve it as easily as Ex. 7-1. Since there is no convenient direct method of solution, we will solve the equation by trial and error. Try $i^* = 5\%$ first:

$$EUAB - EUAC = 0$$

$$100 + 75(A/G,5\%,4) - 700(A/P,5\%,4) = 0$$

$$100 + 75(1.439) - 700(0.2820) = 0$$

At $i^* = 5\%$, EUAB - EUAC = 208 - 197 = +11

The EUAC is too low. If the interest rate is increased, EUAC will increase. Try $i^* = 8\%$:

$$EUAB - EUAC = 0$$

$$100 + 75(A/G,8\%,4) - 700(A/P,8\%,4) = 0$$

$$100 + 75(1.404) - 700(0.3019) = 0$$

At $i^* = 8\%$, EUAB - EUAC = 205 - 211 = -6

This time the EUAC is too large. We see that the true rate of return is between 5% and 8%. Try $i^* = 7\%$:

$$EUAB - EUAC = 0$$

$$100 + 75(A/G,7\%,4) - 700(A/P,7\%,4) = 0$$

$$100 + 75(1.416) - 700(0.2952) = 0$$

At $i^* = 7\%$, EUAB - EUAC = 206 - 206 = 0

The rate of return i^* is 7% ∎

EXAMPLE 7-3

Given the cash flow below, calculate the rate of return on the investment.

Year	Cash flow
0	-$100
1	+20
2	+30
3	+20
4	+40
5	+40

Solution: Using NPW = 0, try $i = 10\%$:

$$NPW = -100 + 20(P/F,10\%,1) + 30(P/F,10\%,2) + 20(P/F,10\%,3)$$
$$+ 40(P/F,10\%,4) + 40(P/F,10\%,5)$$
$$= -100 + 20(0.9091) + 30(0.8264) + 20(0.7513) + 40(0.6830)$$
$$+ 40(0.6209)$$
$$= -100 + 18.18 + 24.79 + 15.03 + 27.32 + 24.84$$
$$= -100 + 110.16 = +10.16$$

The trial interest rate i is too low. Select a second trial, $i = 15\%$:

$$NPW = -100 + 20(0.8696) + 30(0.7561) + 20(0.6575) + 40(0.5718)$$

$$+ 40(0.4972)$$

$$= -100 + 17.39 + 22.68 + 13.15 + 22.87 + 19.89$$

$$= -100 + 95.98$$

$$= -4.02$$

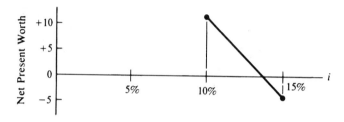

Figure 7-1 Plot of NPW *vs.* interest rate i.

These two points are plotted in Figure 7-1. By linear interpolation we compute the rate of return as follows:

$$i^* = 10\% + (15\% - 10\%)\left(\frac{10.16}{10.16 + 4.02}\right) = 13.5\% \qquad ■$$

We can prove that the rate of return is very close to 13½% by showing that the unrecovered investment is very close to zero at the end of the life of the investment.

Year	Cash flow	Unrecovered investment at beginning of year	13½% return on unrecovered investment	Investment repayment at end of year	Unrecovered investment at end of year
0	-$100				
1	+20	$100.0	$13.5	$ 6.5	$93.5
2	+30	93.5	12.6	17.4	76.1
3	+20	76.1	10.3	9.7	66.4
4	+40	66.4	8.9	31.1	35.3
5	+40	35.3	4.8	35.2	0.1*

*This small unrecovered investment indicates that the rate of return is slightly less than 13½%.

If in Figure 7-1 NPW had been computed for a broader range of values of i, Figure 7-2 would have been obtained. From this figure it is apparent that the error resulting from linear interpolation increases as the interpolation width increases.

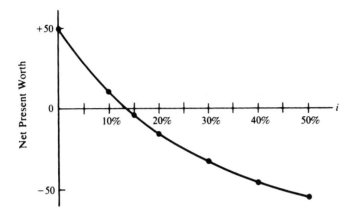

Figure 7-2 Replot of NPW *vs.* interest rate i over a larger range of values.

Plot of NPW *vs.* Interest Rate *i*

Figure 7-2—the plot of NPW *vs.* interest rate i—is an important source of information. A cash flow representing an investment followed by benefits from the investment would have an NPW *vs.* i plot (we will call it an **NPW *plot*** for convenience) in the form of Fig. 7-3.

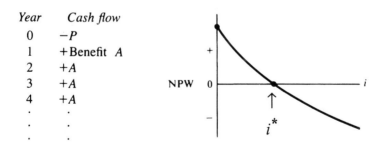

Year	Cash flow
0	$-P$
1	$+$Benefit A
2	$+A$
3	$+A$
4	$+A$
.	.
.	.
.	.

Figure 7-3 Typical NPW plot for an investment.

If, on the other hand, borrowed money was involved, the NPW plot would appear as in Fig. 7-4. This form of cash flow typically results when one is a borrower of money. In such a case, the usual pattern is a receipt of borrowed money early in the time period with a later repayment of an equal sum, plus payment of interest on the borrowed money. In all cases where interest is charged, the NPW at 0% will be negative.

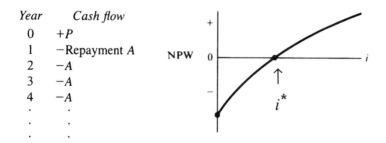

Year	Cash flow
0	$+P$
1	$-$Repayment A
2	$-A$
3	$-A$
4	$-A$
.	.
.	.
.	.

Figure 7-4 Typical NPW plot for borrowed money.

How do we determine the interest rate paid by the borrower in this situation? Typically we would write an equation, such as PW of income $=$ PW of disbursements, and solve for the unknown i^*. Is the resulting i^* positive or negative from the borrower's point of view? If the lender said he was receiving, say, +11% on the debt, it seems reasonable to state that the borrower is faced with −11% interest. Yet this is not the way interest is discussed; rather, interest is referred to in absolute terms without associating a positive or negative sign to it. A banker says he pays 5% interest on savings accounts and charges 11% on personal loans.

Thus, we implicitly recognize interest as a charge for the use of someone else's money and a receipt for letting others use our money. In determining the interest rate in a particular situation, we solve for a single unsigned value of it. We then view this value of i^* in the customary way, that is, either as a charge for borrowing money, or a receipt for lending money.

EXAMPLE 7-4

A new corporate bond was initially sold by a stockbroker to an investor for $1000. The issuing corporation promised to pay the bondholder $40 interest on the $1000 face value of the bond every six months, and to repay the $1000 at the end of ten years. After one year the bond was sold by the original buyer for $950.

    a. What rate of return did the original buyer receive on his investment?

    b. What rate of return can the new buyer (paying $950) expect to receive if he keeps the bond for its remaining nine-year life?

Solution to Ex. 7-4a:

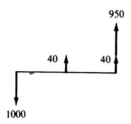

Since $40 is received each six months, we will solve the problem using a six month interest period.

Let PW of cost = PW of benefits

$$1000 = 40(P/A,i,2) + 950(P/F,i,2)$$

Try $i^* = 1\frac{1}{2}\%$:

$$1000 \stackrel{?}{=} 40(1.956) + 950(0.9707) = 78.24 + 922.17$$

$$\stackrel{?}{=} 1000.41$$

The interest rate per six months is very close to $1\frac{1}{2}\%$. This means the nominal (annual) interest rate is $2\times1.5\% = 3\%$. The effective (annual) interest rate $= (1 + 0.015)^2 - 1 = 3.02\%$.

Solution to Ex. 7-4b:

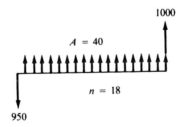

We have the same $40 semi-annual interest payments. For six-month interest periods:

$$950 = 40(P/A,i,18) + 1000(P/F,i,18)$$

Try $i^* = 5\%$:

$$950 \stackrel{?}{=} 40(11.690) + 1000(0.4155) = 467.60 + 415.50$$

$$\stackrel{?}{=} 883.10$$

The PW of benefits is too low. Try a lower interest rate, say, $i^* = 4\%$:

$$950 \doteq 40(12.659) + 1000(0.4936) = 506.36 + 493.60$$

$$\doteq 999.96$$

The value of i is between 4% and 5%. By interpolation,

$$i = 4\% + (1\%)\left(\frac{999.96 - 950.00}{999.96 - 883.10}\right) = 4.43\%$$

The nominal interest rate is $2 \times 4.43\% = 8.86\%$. The effective interest rate is $(1 + 0.0443)^2 - 1 = 9.05\%$. ∎

Rate Of Return Analysis

Rate of return analysis is probably the most frequently used exact analysis technique in industry. Although problems in computing rate of return sometimes occur, its major advantage outweighs the occasional difficulty. The major advantage is that we can compute a single figure of merit that is readily understood.

Consider these statements:

■ The net present worth on the project is $32,000.

■ The equivalent uniform annual net benefit is $2800.

■ The project will produce a 23% rate of return.

While none of these statements tells the complete story, the third one gives a measure of desirability of the project in terms that are widely understood. It is this acceptance by engineers and businessmen alike of rate of return that has promoted its more frequent use than present worth or annual cash flow methods.

There is another advantage to rate of return analysis. In both present worth and annual cash flow calculations, one must select an interest rate for use in the calculations—and this may be a difficult and controversial item. In rate of return analysis, no interest rate is introduced into the calculations (except as described in Chapter 7A). Instead, we compute a rate of return (more accurately called *internal rate of return*) from the cash flow. To decide how to proceed, the calculated rate of return is compared with a preselected *minimum attractive rate of return,* or simply MARR. This is the same value of i used for present worth and annual cash flow analysis.

When there are two alternatives, rate of return analysis is performed by computing the *incremental rate of return*—ΔROR—on the difference between the alternatives. Since we want to look at increments of investment, the cash flow for the difference between the alternatives is computed by taking the higher initial-cost alternative *minus* the lower initial-cost alternative. If the ΔROR is ≥ the MARR, choose the higher-cost alternative. If the, ΔROR is < the MARR, choose the lower-cost alternative.

Two-alternative situation	*Decision*
ΔROR ≥ MARR	Choose the higher-cost alternative
ΔROR < MARR	Choose the lower-cost alternative

Rate of return analysis is illustrated by Examples 7-5 through 7-8.

EXAMPLE 7-5

If an electromagnet is installed on the input conveyor of a coal processing plant, it will pick up scrap metal in the coal. The removal of this metal will save an estimated $1200 per year in machinery damage being caused by metal. The electromagnetic equipment has an estimated useful life of five years and no salvage value. Two suppliers have been contacted: Leaseco will provide the equipment in return for three beginning-of-year annual payments of $1000 each; Saleco will provide the equipment for $2783. If the MARR is 10%, which supplier should be selected?

Solution: Since both suppliers will provide equipment with the same useful life and benefits, this is a fixed-output situation. In rate of return analysis, the method of solution is to examine the differences between the alternatives. By taking (Saleco – Leaseco) we obtain an increment of investment.

Year	Leaseco	Saleco	Difference between alternatives Saleco – Leaseco
0	-$1000	-$2783	-$1783
1	$\begin{cases} -1000 \\ +1200 \end{cases}$	+1200	+1000
2	$\begin{cases} -1000 \\ +1200 \end{cases}$	+1200	+1000
3	+1200	+1200	0
4	+1200	+1200	0
5	+1200	+1200	0

Compute the NPW at various interest rates on the increment of investment represented by the difference between the alternatives.

Year *n*	Cash flow Saleco − Leaseco	PW* at 0%	PW* at 8%	PW* at 20%	PW* at ∞%
0	-$1783	-$1783	-$1783	-$1783	-$1783
1	+1000	+1000	+926	+833	0
2	+1000	+1000	+857	+694	0
3	0	0	0	0	0
4	0	0	0	0	0
5	0	0	0	0	0
NPW =		+217	0	-256	-1783

*Each year the cash flow is multiplied by $(P/F,i,n)$.
  At 0%: $(P/F,0\%,n) = 1$ for all values of n
  At ∞%: $(P/F,\infty\%,0) = 1$
    $(P/F,\infty\%,n) = 0$ for all other values of n

These data are plotted in Figure 7-5. From the figure we see that NPW = 0 at $i = 8\%$.

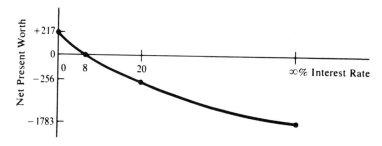

Figure 7-5 NPW plot for Example 7-5.

Thus, the incremental rate of return—ΔROR—of selecting Saleco rather than Leaseco is 8%. This is less than the 10% MARR. Select Leaseco. ■

EXAMPLE 7-6

You are given, the choice of selecting one of two mutually exclusive alternatives. The alternatives are as follows:

Year	Alternative 1	Alternative 2
0	-$10	-$20
1	+15	+28

Any money not invested here may be invested elsewhere at the MARR of 6%. If you can only choose one alternative one time, which one would you select?

Solution: We will select the lesser-cost Alt. 1, unless we find the additional cost of Alt. 2 produces sufficient additional benefits that we would prefer it. If we consider Alt. 2 in relation to Alt. 1, then

$$\begin{bmatrix} \text{Higher - cost} \\ \text{Alt. 2} \end{bmatrix} = \begin{bmatrix} \text{Lower - cost} \\ \text{Alt. 1} \end{bmatrix} + \text{ Differences between Alt. 1 and 2}$$

or

$$\text{Differences between Alt. 1 and 2} = \begin{bmatrix} \text{Higher - cost} \\ \text{Alt. 2} \end{bmatrix} - \begin{bmatrix} \text{Lower - cost} \\ \text{Alt. 1} \end{bmatrix}$$

The choice between the two alternatives reduces to an examination of the differences between them. We can compute the rate of return on the differences between the alternatives. Writing the alternatives again,

Year	Alt. 1	Alt. 2	Alt. 2 – Alt. 1
0	-$10	-$20	-$20 - (-$10) = -$10
1	+15	+28	+28 - (+15) = +13

PW of cost = PW of benefit
$$10 = 13 \ (P/F, i, 1)$$

$$(P/F, i, 1) = \frac{10}{13} = 0.7692$$

One can see that if $10 increases to $13 in one year, the interest rate must be 30%. The Interest Tables confirm this conclusion. The 30% rate of return on the difference between the alternatives is far higher than the 6% MARR. The additional $10 investment to obtain Alt. 2 is superior to investing the $10 elsewhere at 6%. To obtain this desirable increment of investment, with its 30% rate of return, Alt. 2 is selected. ∎

To understand more about Example 7-6, compute the rate of return for each alternative.

Alternative 1:

PW of cost = PW of benefit

$$\$10 = \$15(P/F, i, 1)$$

$$(P/F, i, 1) = \frac{10}{15} = 0.6667$$

From the Interest Tables: rate of return = 50%.

Alternative 2:

PW of cost = PW of benefit

$$\$20 = \$28(P/F, i, 1)$$

$$(P/F, i, 1) = \frac{20}{28} = 0.7143$$

From the Interest Tables: rate of return = 40%.

One is tempted to select Alt. 1, based on these rate of return computations. We have already seen, however, that this is not the correct solution. Solve the problem again, this time using present worth analysis.

Present Worth Analysis:

Alternative 1:

$$NPW = -10 + 15(P/F, 6\%, 1) = -10 + 15(0.9434) = +\$4.15$$

Alternative 2:

$$NPW = -20 + 28(P/F, 6\%, 1) = -20 + 28(0.9434) = +\$6.42$$

Alternative 1 has a 50% rate of return and an NPW (at the 6% MARR) of +$4.15. Alternative 2 has a 40% rate of return on a larger investment, with the result that its NPW (at the 6% MARR) is +$6.42. Our economic criterion is to maximize the return, rather than the rate of return. To maximize NPW, select Alt. 2. This agrees with the rate of return analysis on the differences between the alternatives.

EXAMPLE 7-7

If the computations for Example 7-6 do not convince you, and you still think Alternative 1 looks preferred, try this problem.

You have $20.00 in your wallet and two alternative ways of lending Bill some money.

a. Lend Bill $10.00 with his promise of a 50% return. That is, he will pay you back $15.00 at the agreed time.

b. Lend Bill $20.00 with his promise of a 40% return. He will pay you back $28.00 at the same agreed time.

You can select whether to lend Bill $10.00 or $20.00. This is a one-time situation and any money not lent to Bill will remain in your wallet. Which alternative do you choose?

Solution:

So you see that a 50% return on the smaller sum is less rewarding to you than 40% on the larger sum? Since you would prefer to have $28.00 rather than $25.00 ($15.00 from Bill plus $10.00 remaining in your wallet) after the loan is paid, lend Bill $20.00. ■

EXAMPLE 7-8

Solve Ex. 7-6 again, but this time compute the interest rate on the increment (Alt. 1 – Alt. 2). How do you interpret the results?

Solution: This time the problem is being viewed as:

Alt. 1 = Alt. 2 + [Alt. 1 – Alt. 2]

Year	Alt. 1	Alt. 2	[Alt. 1 - Alt. 2]
0	-$10	-$20	-$10 - (-$20) = +$10
1	+15	+28	+15 - (+28) = -13

We can write one equation in one unknown:

NPW = PW of benefit – PW of cost = 0

$$+ 10 - 13(P/F,i,1) = 0$$

$$(P/F,i,1) = \frac{10}{13} = 0.7692$$

Once again the interest rate is found to be 30%. The critical question is, what does the 30% represent? Looking at the increment again:

Year	Alt. 1 – Alt. 2
0	+$10
1	-13

The cash flow does *not* represent an investment; instead, it represents a loan. It is as if we borrowed $10 in Year 0 (+$10 represents a receipt of money) and repaid it in Year 1 (-$13 represents a disbursement). The 30% interest rate means this is the amount *we would pay* for the use of the $10 borrowed in Year 0 and repaid in Year 1.

Is this a desirable borrowing? Since the MARR on investments is 6%, it is reasonable to assume our maximum interest rate on borrowing would also be 6%. Here the interest rate is 30%, which means the borrowing is undesirable. Since Alt. 1 = Alt. 2 + (Alt. 1 – Alt. 2), and we do not like the (Alt. 1 – Alt. 2) increment, we should reject Alternative 1 as it contains the undesirable increment. This means we should select Alternative 2—the same conclusion reached in Ex. 7-6. ■

This example illustrated that one can analyze either **increments of investment** or **increments of borrowing.** When looking at increments of investment, we accept the increment when the incremental rate of return equals or exceeds the minimum attractive rate of return (ΔROR $\geq$ MARR). When looking at increments of borrowing, we accept the increment when the incremental interest rate is less than or equal to the *minimum* attractive rate of return ($\Delta i \leq$ MARR). One way to avoid much of the possible confusion is to organize the solution to any problem so that one is examining increments of investment. This is illustrated in the next example.

EXAMPLE 7-9

A firm is considering which of two devices to install to reduce costs in a particular situation. Both devices cost $1000, have useful lives of five years and no salvage value. Device *A* can be expected to result in $300 savings annually. Device *B* will provide cost savings of $400 the first year but will decline $50 annually, making the second-year savings $350, the third-year savings $300, and so forth. For a 7% MARR, which device should the firm purchase?

Solution: This problem has been solved by present worth analysis (Ex. 5-1) and annual cost analysis (Ex. 6-5). This time we will use rate of return analysis. The example has fixed input ($1000) and differing outputs (savings).

In determining whether to use an $(A - B)$ or $(B - A)$ difference between the alternatives, we seek an increment of investment. By looking at both $(A - B)$ and $(B - A)$, we find that $(A - B)$ is the one that represents an increment of investment.

Year	Device A	Device B	Difference between alternatives Device A – Device B
0	-$1000	-$1000	$0
1	+300	+400	-100
2	+300	+350	-50
3	+300	+300	0
4	+300	+250	+50
5	+300	+200	+100

For the difference between the alternatives, write a single equation with i as the only unknown.

EUAC = EUAB

$$[100(P/F,i,1) + 50(P/F,i,2)](A/P,i,5) = [50(F/P,i,1) +100](A/F,i,5)$$

The equation is cumbersome, but need not be solved. Instead, we observe that the sum of the costs (-100 and -50) equals the sum of the benefits (+50 and + 100). This indicates that 0% is the ΔROR on the $A - B$ increment of *investment*. This is less than the 7% MARR; therefore, the increment is undesirable. Reject Device *A* and choose Device *B*.

As described in Ex. 7-8, if the increment examined is $(B - A)$, the interest rate would again be 0%, indicating a desirable *borrowing* situation. We would choose Device *B*. ■

Analysis Period

In discussing present worth analysis and annual cash flow analysis, an important consideration is the analysis period. This is also true in rate of return analysis. The method of solution for two alternatives is to examine the differences between the alternatives. The examination must necessarily cover the selected analysis period. An assumption that an alternative can be replaced with one of identical costs and performance appears dubious at best. For now, we can only suggest that the assumptions made should reflect one's perception of the future as accurately as possible.

Example 7-9 is a problem where the analysis period is a common multiple of the alternative service lives, and where identical replacement is assumed. It will illustrate an analysis of the differences between the alternatives over the analysis period.

EXAMPLE 7-10

Two machines are being considered for purchase. If the MARR (here, the minimum required interest rate) is 10%, which machine should be bought?

	Machine X	Machine Y
Initial cost	$200	$700
Uniform annual benefit	95	120
End-of-useful-life salvage value	50	150
Useful life, in years	6	12

Solution:

The solution is based on a twelve-year analysis period and a Replacement Machine *X* that is identical to the present Machine *X*. The cash flow for the differences between the alternatives is as follows:

Year	Machine X	Machine Y	Mach. Y – Mach. X
0	-$200	-$700	-$500
1	+95	+120	+25
2	+95	+120	+25
3	+95	+120	+25
4	+95	+120	+25
5	+95	+120	+25
6	+95 +50 -200	+120	+25 +150
7	+95	+120	+25
8	+95	+120	+25
9	+95	+120	+25
10	+95	+120	+25
11	+95	+120	+25
12	+95 +50	+120 +150	+25 +100

PW of cost = PW of benefits

$$500 = 25(P/A, i, 12) + 150(P/F, i, 6) + 100(P/F, i, 12)$$

The sum of the benefits over the twelve years is $550 which is only a little greater than the $500 additional cost. This indicates that the rate of return is quite low. Try $i^* = 1\%$.

$$500 \stackrel{?}{=} 25(11.255) + 150(0.942) + 100(0.887)$$

$$\stackrel{?}{=} 281 + 141 + 89 = 511$$

The interest rate is too low. Try $i^* = 1\frac{1}{2}\%$:

$$500 \stackrel{?}{=} 25(10.908) + 150(0.914) + 100(0.836)$$

$$\stackrel{?}{=} 273 + 137 + 84 = 494$$

The rate of return on the $Y - X$ increment is about 1.3%, far below the 10% minimum attractive rate of return. The additional investment to obtain Y yields an unsatisfactory rate of return, therefore X is the preferred alternative. ■

Spreadsheets and Rate of Return Analysis

The spreadsheet functions covered in earlier chapters are particularly useful in calculating internal rates of returns (IRRs). If a cash flow diagram can be reduced to at most one P, one A, and/or one F, then the RATE *investment function* can be used. Otherwise the IRR *block function* is used with a cash flow in each period.

The Excel investment function is RATE(N,A,P,F,type,guess). The A, P, and F cannot all be the same sign. The F, type, and guess are optional arguments. The "type" is end or beginning of period cash flows (for A and F), and the "guess" is the starting value in the search for the IRR.

Considering Example 7-1, where $P = -8200$, $A = 2000$, and $N = 6$, the RATE function would be:

RATE(6,2000,−8200)

which gives an answer of 7.00%, which matches that found in Example 7-1.

For Example 7-2, where $P = -700$, $A = 100$, $G = 75$, and $N = 4$, the RATE function cannot be used, since it has no provisions for the arithmetic gradient, G. Suppose the years (row 1) and the cash flows (row 2) are specified in columns B through E. The internal rate of return calculated using IRR(B2:F2) is 6.91%.

	A	B	C	D	E	F
1	Year	0	1	2	3	4
2	Cash Flow	-700	100	175	250	325

Figure 7-6 illustrates using a spreadsheet to graph the present worth of a cash flow series versus the interest rate. The interest rate with present worth equal to 0 is the IRR. The y-axis on this graph has been modified so that the x-axis intersects at a present worth of -5 rather than at 0. To do this click on the y-axis, then right click to bring up the format axis option. Select this and then select the tab for the *scale* of the axis. This has a selection for the intersection of the x-axis. This process ensures that the x-axis labels are outside of the graph.

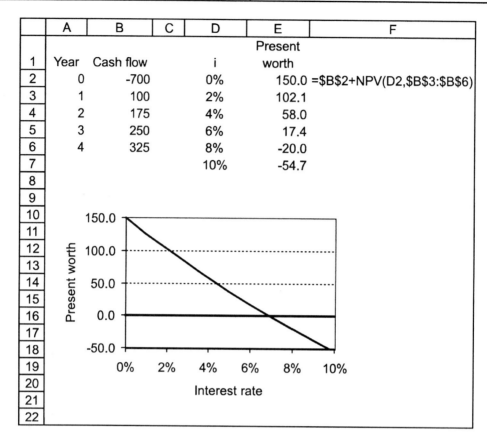

	A	B	C	D	E	F
					Present	
1	Year	Cash flow		i	worth	
2	0	-700		0%	150.0	=$B$2+NPV(D2,$B$3:$B$6)
3	1	100		2%	102.1	
4	2	175		4%	58.0	
5	3	250		6%	17.4	
6	4	325		8%	-20.0	
7				10%	-54.7	
8						
9						
10						
11						
12						
13						
14						
15						
16						
17						
18						
19						
20						
21						
22						

Figure 7-6 Graphing present worth versus *i*.

SUMMARY

Rate of return may be defined as the interest rate paid on the unpaid balance of a loan such that the loan is exactly repaid by the schedule of payments. On an investment, rate of return is the interest rate earned on the unrecovered investment such that the payment schedule makes the unrecovered investment equal to zero at the end of the life of the investment. Although the two definitions of rate of return are stated differently, there is only one fundamental concept being described. It is that the rate of return is the interest rate i^* at which the benefits are equivalent to the costs.

There are a variety of ways of writing the cash flow equation in which the rate of return i^* may be the single unknown. Five of them are:

PW of benefits − PW of costs = 0

$$\frac{PW \text{ of Benefits}}{PW \text{ of Costs}} = 1$$

$$NPW = 0$$

$$EUAB − EUAC = 0$$

PW of costs = PW of benefits

Rate of Return Analysis: Rate of return analysis is the most frequently used method in industry, as the resulting rate of return is readily understood. Also, the difficulties in selecting a suitable interest rate to use in present worth and annual cash flow analysis are avoided.

Criteria:

Two alternatives.

Compute the incremental rate of return—ΔROR—on the increment of *investment* between the alternatives. Then,

■

If ΔROR ≥ MARR, choose the higher-cost alternative; or,

■ If ΔROR < MARR, choose the lower-cost alternative.

When an increment of *borrowing* is examined,

■ If Δi ≤ MARR, the increment is acceptable; or,

■ If Δi > MARR, the increment is not acceptable.

Three or more alternatives.

Incremental analysis is needed, which we move to in Chapter 8.

Rate of return is further described in Chapter 7A and still further in Chapter 18. These two chapters concentrate on the difficulties that occur with some cash flows that yield more than one rate of return.

Problems

7-1 A woman went to the Beneficial Loan Company and borrowed $3000. She must pay $119.67 at the end of each month for the next thirty months.
  a. Calculate the nominal annual interest rate she is paying to within ±0.15%.
  b. What effective annual interest rate is she paying?

7-2 For the diagram shown, compute the interest rate to within ½%.

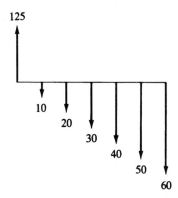

7-3 Helen is buying a $12,375 automobile with a $3000 down payment, followed by 36 monthly payments of $325 each. The down payment is paid immediately, and the monthly payments are due at the end of each month. What nominal annual interest rate is Helen paying? What effective interest rate? (*Answers:* 15%; 16.08%)

7-4 Consider the following cash flow:

Year	Cash flow
0	-$500
1	+200
2	+150
3	+100
4	+50

Compute the rate of return represented by the cash flow.

7-5 For the diagram below, compute the interest rate at which the costs are equivalent to the benefits.

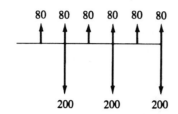

80 80 80 80 80 80

200   200   200

(*Answer:* 50%)

7-6   For the diagram below, compute the rate of return.

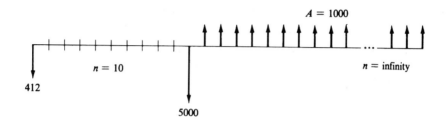

$A = 1000$

$n = 10$

$n = $ infinity

412

5000

7-7   Consider the following cash flow:

Year	Cash flow
0	-$1000
1	0
2	+300
3	+300
4	+300
5	+300

Compute the rate of return on the $1000 investment to within 0.1%. (*Answer:* 5.4%)

7-8   Peter Minuit bought an island from the Manhattoes Indians in 1626 for $24 worth of glass beads and trinkets. The 1991 estimate of the value of land on this island is $12 billion. What rate of return would the Indians have received if they had retained title to the island rather than selling it for $24?

7-9   A man buys a corporate bond from a bond brokerage house for $925. The bond has a face value of $1000 and pays 4% of its face value each year. If the bond will be paid off at the end of ten years, what rate of return will the man receive? (*Answer:* 4.97%) $PWB - PWC = 0$

7-10 A well-known industrial firm has issued $1000 bonds that carry a 4% nominal annual interest, rate paid semi-annually. The bonds mature twenty years from now, at which time the industrial firm will redeem them for $1000 plus the terminal semi-annual interest payment. From the financial pages of your newspaper you learn that the bonds may be purchased for $715 each ($710 for the bond plus a $5 sales commission). What nominal annual rate of return would you receive if you purchased the bond now and held it to maturity twenty years hence? (*Answer:* 6.6%)

7-11 One aspect of obtaining a college education is the prospect of improved future earnings, compared to non-college graduates. Sharon Shay estimates that a college education has a $28,000 equivalent cost at graduation. She believes the benefits of her education will occur throughout forty years of employment. She thinks she will have a $3000-per-year higher income during the first ten years out of college, compared to a non-college graduate. During the subsequent ten years, she projects an annual income that is $6000-per-year higher. During the last twenty years of employment, she estimates an annual salary that is $12,000 above the level of the non-college graduate. Assuming her estimates are correct, what rate of return will she receive as a result of her investment in a college education? $PW \text{ of } B - PW \text{ of } C = 0$

7-12 An investor purchased a one-acre lot on the outskirts of a city for $9000 cash. Each year he paid $80 of property taxes. At the end of four years, he sold the lot. After deducting his selling expenses, the investor received $15,000. What rate of return did he receive on his investment? (*Answer:* 12.92%)

7-13 A popular reader's digest offers a lifetime subscription to the magazine for $200. Such a subscription may be given as a gift to an infant at birth (the parents can read it in those early years), or taken out by an individual for himself. Normally, the magazine costs $12.90 per year. Knowledgeable people say it probably will continue indefinitely at this $12.90 rate. What rate of return would be obtained if a life subscription were purchased for an infant, rather than paying $12.90 per year beginning immediately? You may make any reasonable assumptions, but the compound interest factors must be *correctly* used.

7-14 On April 2, 1988, an engineer buys a $1000 bond of an American airline for $875. The bond pays 6% on its principal amount of $1000, half in each of its April 1 and October 1 semi-annual payments; it will repay the $1000 principal sum on October 1, 2001. What nominal rate of return will the engineer receive from the bond if he holds it to its maturity (on October 1, 2001)? (*Answer:* 7.5%)

7-15 The cash price of a machine tool is $3500. The dealer is willing to accept a $1200 down payment and 24 end-of-month monthly payments of $110 each. At what effective interest rate are these terms equivalent? (*Answer:* 14.4%)

7-16 A local bank makes automobile loans. It charges 4% per year in the following manner: if $3600 is borrowed to be repaid over a three-year period, the bank interest charge is $3600 × 0.04 × 3 years = $432. The bank deducts the $432 of interest from the $3600 loan and gives the customer $3168 in cash. The customer must repay the loan by paying $1/36$ of $3600, or $100, at the end of each month for 36 months. What nominal annual interest rate is the bank actually charging for this loan?

7-17 Upon graduation every engineer must decide whether or not to go on to graduate school. Estimate the costs of going full time to the university to obtain a Master of Science degree. Then estimate the resulting costs and benefits. Combine the various consequences into a cash flow table and compute the rate of return. Non-financial benefits are probably relevant here too.

7-18   In his uncle's will, Frank is to choose one of two alternatives:

Alternative 1: $2000 cash.

Alternative 2: $150 cash now plus $100 per month for twenty months beginning
the first day of next month.

    a.   At what rate of return are the two alternatives equivalent?

    b.   If Frank thinks the rate of return in *a* is too low, which alternative should he select?

7-19   A man buys a table saw at a local store for $175. He may either pay cash for it, or pay $35
now and $12.64 a month for twelve months beginning thirty days hence. If the man chooses the time
payment plan, what is the nominal annual interest rate he will be charged? (*Answer:* 15%)

7-20   An investment of $5000 in Biotech common stock proved to be very profitable. At the end
of three years the stock was sold for $25,000. What was the rate of return on the investment?

7-21   A man owns a corner lot. He must decide which of several alternatives to select in trying to
obtain a desirable return on his investment. After much study and calculation, he decides that the two
best alternatives are:

	Build gas station	*Build soft ice cream stand*
First cost	$80,000	$120,000
Annual property taxes	3,000	5,000
Annual income	11,000	16,000
Life of building, in years	20	20
Salvage value	0	0

If the owner wants a minimum attractive rate of return on his investment of 6%, which of the two
alternatives would you recommend to him?

7-22   Two alternatives are as follows:

Year	A	B
0	-$2000	-$2800
1	+800	+1100
2	+800	+1100
3	+800	+1100

If 5% is considered the minimum attractive rate of return, which alternative should be selected?

7-23   The Southern Guru Copper Company operates a large mine in a South American country.
A legislator in the National Assembly said in a speech that most of the capital for the mining
operation was provided by loans from the World Bank; in fact, Southern Guru has only $500,000 of
its own money actually invested in the property. The cash flow for the mine is:

Year	Cash flow
0	$0.5 million investment
1	3.5 million profit
2	0.9 million profit
3	3.9 million profit
4	8.6 million profit
5	4.3 million profit
6	3.1 million profit
7	6.1 million profit

The legislator divided the $30.4 million total profit by the $0.5 million investment. This produced, he said, a 6080% rate of return on the investment. Southern Guru claims their actual rate of return is much lower. They ask you to compute their rate of return.

7-24 Two alternatives are being considered:

	A	B
First cost	$9200	$5000
Uniform annual benefit	1850	1750
Useful life, in years	8	4

If the minimum attractive rate of return is 7%, which alternative should be selected?

7-25 Jean has decided it is time to purchase a new battery for her car. Her choices are:

	Zappo	Kicko
First cost	$56	$90
Guarantee period, in months	12	24

Jean believes the batteries can be expected to last only for the guarantee period. She does not want to invest extra money in a battery unless she can expect a 50% rate of return. If she plans to keep her present car another two years, which battery should she buy?

7-26 For the diagram below, compute the rate of return on the $3810 investment.

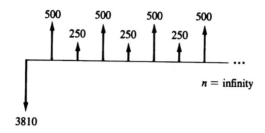

7-27 Consider the following cash flow:

Year	Cash flow
0	-$400
1	0
2	+200
3	+150
4	+100
5	+50

Write one equation, with i as the only unknown, for the cash flow. In the equation you are not to use more than two single payment compound interest factors. (You may use as many other factors as you wish.) Then solve your equation for i.

7-28 Compute the rate of return for the following cash flow to within ½%.

Year	Cash flow
0	-$100
1–10	+27

(*Answer:* 23.9%)

7-29 For the following diagram, compute the rate of return.

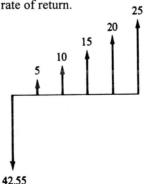

7-30 Solve the cash flow below for the rate of return to within 0.5%.

Year	Cash flow
0	-$500
1	-100
2	+300
3	+300
4	+400
5	+500

7-31 For the cash flow below, compute the rate of return.

Year	Cash flow
1–5	-$233
6–10	+1000

7-32 Compute the rate of return for the following cash flow to within 0.5%.

Year	Cash flow
0	-$640
1	0
2	100
3	200
4	300
5	300

(*Answer:* 9.3%)

7-33 Consider two mutually exclusive alternatives:

Year	X	Y
0	-$100	-$50.0
1	35	16.5
2	35	16.5
3	35	16.5
4	35	16.5

If the minimum attractive rate of return is 10%, which alternative should be selected?

7-34   Consider these two mutually exclusive alternatives:

Year	A	B
0	-$50	-$53
1	17	17
2	17	17
3	17	17
4	17	17

At a MARR of 10%, which alternative should be selected? (*Answer: A*)

7-35   Two mutually exclusive alternatives are being considered. Both have a ten-year useful life. If the MARR is 8%, which alternative is preferred?

	A	B
Initial cost	$100.00	$50.00
Uniform annual benefit	19.93	11.93

7-36   Two alternatives are being considered:

	A	B
Initial cost	$9200	$5000
Uniform annual benefit	1850	1750
Useful life, in years	8	4

Base your computations on a MARR of 7% and an eight-year analysis period. If identical replacement is assumed, which alternative should be selected?

7-37   Two investment opportunities are as follows:

	A	B
First cost	$150	$100
Uniform annual benefit	25	22.25
End-of-useful-life salvage value	20	0
Useful life, in years	15	10

At the end of ten years, Alt. *B* is not replaced. Thus, the comparison is 15 years of *A* vs. 10 years of *B*. If the MARR is 10%, which alternative should be selected?

7-38   An insurance company is offering to sell an annuity for $20,000 cash. In return they will guarantee to pay the purchaser 20 annual end-of-year payments, with the first payment amounting to $1100. Subsequent payments will increase at a uniform 10% rate each year (second payment is $1210; third payment is $1331, and so on). What rate of return will the purchaser receive if he buys the annuity?

7-39   Consider two mutually exclusive alternatives:

Year	X	Y
0	-$5000	-$5000
1	-3000	+2000
2	+4000	+2000
3	+4000	+2000
4	+4000	+2000

If the MARR is 8%, which alternative should be selected?

7-40   A bank proudly announces that they have changed their interest computation method to continuous compounding. Now $2000 left in the bank for nine years will double to $4000.

   a.   What nominal interest rate, compounded continuously, are they paying?

   b.   What effective interest rate are they paying?

7-41   Fifteen families live in Willow Canyon. Although several water wells have been drilled, none has produced water. The residents take turns driving a water truck to a fire hydrant in a nearby town. They fill the truck with water and then haul it to a storage tank in Willow Canyon. Last year it cost $3180 in truck fuel and maintenance costs. This year the residents are seriously considering spending $100,000 to install a pipeline from the nearby town to their storage tank. What rate of return would the Willow Canyon residents receive on their new water supply pipeline if the pipeline is considered to last:

   a.   forever?

   b.   100 years?

   c.   50 years?

   d.   Would you recommend that the pipeline be installed? Explain.

7-42   Compute the rate of return for the following cash flow.

Year	Cash flow
0	-$ 500
1–3	0
4	+4500

7-43   Jan purchased 100 shares of Peach Computer stock for $18 per share, plus a $45 brokerage commission. Every six months she received a 50 cents per share dividend from Peach. At the end of two years, just after receiving the fourth dividend, she sold the stock for $23 per share, and paid a $58 brokerage commission from the proceeds. What annual rate of return did she receive on her investment?

7-44   The Diagonal Stamp Co., which sells used postage stamps to collectors, advertises that their average price has increased from $1 to $5 in the last five years. Thus, they say, investors would have received a 100% rate of return each year if they had purchased stamps from Diagonal.

    a.   To check their calculations, compute the annual rate of return.

    b.   Why is your computed rate of return less than 100%?

7-45   An investor purchased 100 shares of Omega common stock for $9000. He held the stock for nine years. For the first four years he received annual end-of-year dividends of $800. For the next four years he received annual dividends of $400. He received no dividend for the ninth year. At the end of the ninth year he sold his stock for $6000. What rate of return did he receive on his investment?

7-46   You spend $100 and in return receive two payments of $1094.60—one at the end of three years and the other at the end of six years. Calculate the resulting rate of return.

7-47   A mine is for sale for $240,000. It is believed the mine will produce a profit of $65,000 the first year, but the profit will decline $5000 a year after that until it becomes zero and the mine is worthless. What rate of return would this produce for the purchaser of the mine?

7-48   Two mutually exclusive alternatives are being considered.

Year	A	B
0	-$2500	-$6000
1	+746	+1664
2	+746	+1664
3	+746	+1664
4	+746	+1664
5	+746	+1664

If the minimum attractive rate of return is 8%, which alternative should be selected?

Solve the problem by

    a. Present worth analysis

    b. Annual cash flow analysis

    c. Rate of return analysis

7-49 Fred, our cat, just won the local feline lottery to the tune of 3000 cans of "9-Lives" cat food (assorted flavors). A local grocer offers to take the 3000 cans and in return, supply Fred with 30 cans a month for the next 10 years. What rate of return, in terms of nominal annual rate, will Fred realize on this deal? (Compute to nearest .01%)

7-50 The following advertisement appeared in the Wall Street Journal on Thursday, February 9,1995.

> **"There's nothing quiet like the Seville SmartLease. Seville SLS**
>
> **$0 down, $599* a month/36 months."**

*First Month's lease payment of $599 plus $625 refundable security deposit and a consumer down payment of $0 for a Total of $1224 due at Lease Signing. Monthly payment is based on a net capitalized cost of $39,264 for total monthly payments of $21,564. Payment examples based on a 1995 Seville SLS: $43,658 MSRP including destination charge. Tax, license, title fees and insurance extra. Option to purchase at lease end for $27,854. Mileage charge of $0.15 per mile over 36,000 miles.

 a. Set up the cash flows and

 b. Determine the interest rate (nominal and effective) for the lease.

7-51 After 15 years of working for one employer, you transfer to a new job. During these few years your employer contributed (that is, she diverted from your salary) $1500 each year to an account for your retirement (a fringe benefit), and you contributed a matching amount each year. The whole fund was invested at 5% during that time and the value of the account now stands at $30,000. You are now faced with two alternatives. (1) You may leave both contributions in the fund until retirement in thirty five years during which you will get the future value of this amount at 5% interest per year. (2) The other alternative is for you to take out the total value of "your" contributions, which is $15,000 (one-half of the total $30,000). You can do as you wish with the money you take out, but the other half will be lost as far as you are concerned. In other words, you can give up $15,000 today for the sake of getting now the other $15,000. Otherwise, you must wait thirty five years more to get the accumulated value of the entire fund. Which alternative is more attractive? Explain your choice.

7-52 A contractor is considering whether to purchase a new machine for his layout site work or to lease one. Purchasing a new machine will cost $12,000 with a salvage value of $1200 at the end of the machine's useful life of 8 years. On the other hand, leasing one requires an annual lease payment of $3000. Assuming the MARR is 15% and on the basis of an internal rate of return analysis, which alternative should the contractor be advised to accept. The cash flows are as given below.

Year (n)	Alternative A (Purchase)	Alternative B (lease)
0	- $12,000	0
1		- $3000
2		- 3000
3		- 3000
4		- 3000
5		- 3000
6		- 3000
7		- 3000
8	+ 1,200	- 3000

7-53 An apartment building is currently for sale in your neighborhood. This apartment building can be bought for $140,000. The building has four units which are rented at $500 per month each. The tenants have long term leases that expire in 5 years. Maintenance and other expenses for care and upkeep are $8,000 annually. A new university is being built in the vicinity and it is expected that the building could be sold for $160,000 after 5 years.

    a. What is the internal rate of return for this investment?

    b. Should this investment be accepted if the your other options have a rate of return of 12%?

7-54 A finance company is using the "Money by Mail" offer shown below. Calculate the yearly nominal ROR received by the company if a customer chooses the loan of $2000 and accepts the credit insurance (Life and Dis.).

Money by Mail	Non-Negotiable INE
	1/96

For the
Amount of __$3000 or $2000 or $1000_____Dollars
Pay to the Order of I Feel Rich
Limited Time Offer

To Borrow $3000, $2000, or $1000.

For the Amount of $3000 Dollars   APR 18.95%
Pay to the   Finance Charge $1034.29
Order of ____I Feel Rich_____
Total of Payments $4,280.40   Monthly Payment $118.90
Number of Monthly Payments   Credit Line Premium
     36 Months $83.46*
Amount Financed $3,246.25   Credit Disabilit Premium $162.65*

$3000 loan terms

For the Amount of $2000 Dollars   APR 19.95%
Pay to the   Finance Charge $594.25
Order of ____I Feel Rich_____
Total of Payments $2,731.50   Monthly Payment $91.05
Number of Monthly Payments   Credit Line Premium
     30 Months $44.38*
Amount Financed $2,137.25   Credit Disabilit Premium $92.87*

$2000 loan terms

For the Amount of $1000 Dollars   APR 20.95%
Pay to the   Finance Charge $245.54
Order of ____I Feel Rich_____
Total of Payments $1,300.80   Monthly Payment $54.20
Number of Monthly Payments   Credit Line Premium
     24 Months $16.91*
Amount Financed $1055.26   Credit Disabilit Premium $39.02*

$1000 loan terms

*Credit Insurance. If selected, premium will be paid from Amount Financed.
  If not selected, cash advance is total Amount Financed.

7-55 A finance company is using the "Money by Mail" offer shown in problem 7-54 above. Calculate the yearly nominal ROR received by the company if a customer chooses the $3000 loan, but declines the credit insurance.

7-56 A new machine can be purchased today for $300,000. The annual revenue from the machine is calculated to be $67,000 and the equipment will last 10 years. Expect the maintenance and operating costs to be $3000 a year and increase $600 per year. The salvage value of the machine will be $20,000. What is the rate of return for this machine?

7-57 Two hazardous environment facilities are being evaluated, with the projected life of each facility being 10 years. The cash flows for each facility are:

	Alternative	
	A	B
First cost	$615,000	$300,000
Maintenance &	10,000	25,000
Annual benefit	158,000	92,000
Salvage value	65,000	-5,000

The company uses a MARR of 15%. Using rate of return analysis, which alternative should be selected?

7-58 Al Larson asked a bank to lend him money on January 1st, based on the following repayment plan: the first payment would be $2 on February 28th, with subsequent monthly payments increasing by $2 a month on an arithmetic gradient. (The March 31st payment would be $4; the April 30th payment would be $6, and so on.) The payments are to continue for eleven years, making a total of 132 payments.

    a. Compute the total amount of money Al will pay the bank, based on the proposed repayment plan.

    b. If the bank charges interest at 12% nominal per year, compounded monthly, how much would it be willing to lend Al on the proposed repayment plan?

Difficulties Solving For An Interest Rate

Occasionally we encounter a cash flow that cannot be solved for a single positive interest rate. In this appendix to Chapter 7, we examine ways to resolve this difficulty. Example 7A-1 illustrates the situation.

EXAMPLE 7A-1

The Going Aircraft Company has an opportunity to supply a large airplane to Interair, a foreign airline. Interair will pay $19 million when the contract is signed and $10 million one year later. Going estimates its second- and third- year net cash flows at $50 million each when the airplane is being produced. Interair will take delivery of the airplane during Year 4, and agrees to pay $20 million at the end of that year and the $60 million balance at the end of Year 5. Compute the rate of return on this project.

Solution: Computation of NPW at various interest rates, using single payment present worth factors,[1] is presented:

Year	Cash flow	0%	10%	20%	40%	50%
0	+$19	+$19	+$19	+$19	+$19	+$19
1	+10	+10	+9.1	+8.3	+7.1	+6.7
2	-50	-50	-41.3	-34.7	-25.5	-22.2
3	-50	-50	-37.6	-28.9	-18.2	-14.8
4	+20	+20	+13.7	+9.6	+5.2	+4.0
5	+60	+60	+37.3	+24.1	+11.2	+7.9
	NPW =	+$9	+$0.2	-$2.6	-$1.2	+$0.6

---

[1]For example, for Year 2 and $i = 10\%$: PW = $-50(P/F, 10\%, 2) = -50(0.826) = -41.3$

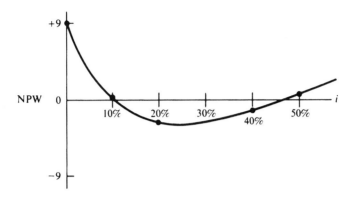

Figure 7A-1 NPW plot.

The NPW plot for this cash flow is represented in Fig. 7A-1. We see this cash flow produces *two* points at which NPW = 0. Thus, there are *two* positive rates of return, one at about 10.1% and the other at about 47%. ■

Example 7A-1 produced unexpected and undesirable results. We want to know when we may expect this kind of result, what it means, and how we may resolve the difficulty. In the next section, we find that the solution of an economic analysis problem is really the solution of a mathematical equation.

Converting a Cash Flow to a Mathematical Equation

Given a simple cash flow,

Year	Cash flow
0	$-P$
1	$+A_1$
2	$+A_2$
.	.
.	.
.	.
n	$+A_n$

Setting NPW = PW of benefits *minus* PW of cost = 0, and using single payment present worth factors, the cash flow may be rewritten as

$$+A_1(1+i)^{-1} + A_2(1+i)^{-2} + \ldots + A_n(1+i)^{-n} - P = 0 \qquad (7A\text{-}1)$$

If we let $X = (1+i)^{-1}$, then Eq. 7A-1 may be written

$$+A_1 X + A_2 X^2 + \ldots + A_n X^n - P = 0 \qquad (7A\text{-}2)$$

Rearranging terms,

$$A_n X^n + \ldots + A_2 X^2 + A_1 X - P = 0 \qquad (7A\text{-}3)$$

Equation 7A-3 is an nth order polynomial to which ***Descartes' Rule*** may be applied. The rule is:

> **If a polynomial with real coefficients has m sign changes, then the number of positive roots will be m - $2k$, where k is a positive integer or zero ($k = 0, 1, 2, 3, \ldots$).**

A sign change is where successive nonzero terms, written according to descending powers of X, have different signs (that is, change from + to - or *vice versa*). Descartes' Rule means that the number of positive roots of the polynomial cannot exceed m, the number of sign changes; the number of positive roots must either be equal to m or less by an even integer.

Descartes' Rule for polynomials gives the following:

Number of sign changes, m	Number of positive values of X
0	0
1	1
2	2 or 0
3	3 or 1
4	4, 2, or 0

But Equation 7A-3's polynomial is not the equation that represents the economic analysis problem. By substituting $(1+i)^{-1}$ in place of X, we return to Eq. 7A-1. We see from the relationship $X = (1+i)^{-1}$ that a positive value of X does not ensure a positive value of i. In fact, whenever X is greater than 1, i is negative. Thus, in a particular situation when m equals two sign changes, there would either be two or zero positive values of X. But this means there could be two, one, or zero positive values of i for the corresponding economic analysis equation with two sign changes.

From this discussion, we form a ***cash flow rule of signs***.

Cash Flow Rule Of Signs

There may be as many positive values of *i* as there are sign changes in the cash flow.

Sign changes are computed the same way as for Descartes' Rule. A sign change is where successive nonzero values in the cash flow have different signs (that is, change from + to -, or *vice versa*). A zero cash flow is ignored. The five cash flows in Table 7A-1 illustrate the counting of the number of sign changes.

Table 7A-1 SIGN CHANGES IN CASH FLOWS

	Cash flow				
Year	A	B	C	D	E
0	+$100	-$100	-$100	+$50	+$50
1	+10	+10	0	+40	-50
2	+50	+50	+50	-100	+50
3	+20	+20	0	+10	-10
4	+40	+40	+80	+10	-30
Sign changes:	0	1	1	2	3

The cash flow rule of signs indicates the following possibilities:

Number of sign changes, *m*	Number of positive values of *i*
0	0
1	1 or 0
2	2, 1, or 0
3	3, 2, 1, or 0

Thus there are three possibilities that need examination: zero sign changes, one sign change, and two or more sign changes.

Zero Sign Changes

There are two situations that produce zero sign changes in a cash flow. Either all terms have a positive sign, representing receipts, or all terms have a negative sign, reflecting a series of disbursements.

The first case would be like walking into a store and finding oneself their millionth customer and, hence, the recipient of money and gifts. This is the utopian something-for-nothing situation. There is no value of *i* that can be computed, for there are no disbursements to offset the receipts.

The second case represents the less-happy situation of disbursements without any compensating receipts. It would be like the periodic purchase of lottery tickets–but never winning anything. There is no value of *i* that would reflect this economic situation.

One Sign Change

Unless there are rather unusual circumstances, one sign change is the normal cash flow pattern. Given one sign change, it is very likely that we have a situation where there will be a single positive value of *i*.

There is no positive value of *i* whenever *in an investment situation* the subsequent benefits do not at least equal the magnitude of the investment. An example would be:

Year	Cash flow
0	-$50
1	+20
2	+20

Also, there is no positive value of *i* whenever in a *borrowing situation* the subsequent repayments do not equal the magnitude of the borrowed money.

Year	Cash flow
0	+$50
1	-20
2	-20

These circumstances can be readily identified and, hence, present no confusion or particular problem.

Two or More Sign Changes

When there are two or more sign changes in the cash flow, we know that there are several possibilities concerning the number of positive values of i. Probably the greatest danger in this situation is to fail to recognize the multiple possibilities and to solve for a value of i. Then one might assume–incorrectly or correctly–that the value of i that is obtained is the only positive value of i. In this multiple sign change situation, one approach is to prepare an NPW plot like Fig. 7A-1. This may be a tedious procedure if done by hand, but is very easy with a spreadsheet.

What the Difficulties Mean

If there is a single positive value of i, we have no problem. On the other hand, a situation with no positive value of i or multiple positive values represents a situation that may be attractive, unattractive, or confusing. Where there are multiple values of i, none of them should be considered a suitable measure of the rate of return or attractiveness of the cash flow.

    We classify a cash flow as an investment situation if we put money into a project and benefits of the project come back to us. The rate of return tells us something about the attractiveness of the project. But what was the situation in Example 7A-1?

Year	Cash flow
0	+$19
1	+10
2	-50
3	-50
4	+20
5	+60

    At the beginning, money is generated and flows out of the project. In Years 2 and 3, money is spent on production, followed by receipts in the final two years. We know that 10.1% is a value of i at which NPW = 0. What does this mean? If the initial outflows from the project are invested in some *external investment* at 10.1% and then their compound amount returned to the project in Year 2, the project, or internal investment, will have i = 10.1%. Similarly, if the initial outflows from the project can be put into an external investment at 47%, then the project or internal investment will also show a 47% interest rate. This sounds somewhat unbelievable, so let's demonstrate both situations.

For *i* = 10.1%:

Year	Cash flow
0	+$19 invest for 2 years in an *external investment* at 10.1%
	$F = +19(F/P,10.1\%,2)$
	$F = +19(1.21) = +23$ ————————
1	+10 invest for 1 year in an *external investment* at 10.1%
	$F = +10(F/P,10.1\%,1)$
	$F = +10(1.101) = +11$ ————————
2	+11 return to internal investment ←
	+23 return to internal investment ←
	−50
3	−50
4	+20
5	+60

When the external investment phase is completed, the cash flow for the internal investment is as follows:

Year	Cash flow
0	$0
1	0
2	-16
3	-50
4	+20
5	+60

We see that for the internal investment there is one sign change, so we can solve for the internal value of *i* directly. It will be 10.1% if, when we use that value, NPW = 0.

$$NPW = -16(P/F,10.1\%,2) - 50(P/F,10.1\%,3) + 20(P/F,10.1\%,4)$$

$$+ 60(P/F, 10.1\%,5)$$

$$= -16(0.825) - 50(0.749) + 20(0.681) + 60(0.618)$$

$$= -13.2 - 37.5 + 13.6 + 37.1$$

$$= 0$$

Similarly, for i = 47%:

Year	Cash flow
0	+$19 invest for 2 years in an *external investment* at 47%
	$F = +19(F/P,47\%,2) = +19(2.161) = +41.1$
1	+10 invest for 1 year in an *external investment* at 47%
	$F = +10(F/P,47\%,1) = +10(1.47) = +14.7$
2	+14.7 return to project ←
	+41.1 return to project ←
	−50
3	−50
4	+20
5	+60

When the cash outflows to the external investment are returned, the transformed cash flow is tabulated below:

Year	Cash flow
0	$0
1	0
2	+5.8
3	-50
4	+20
5	+60

There is the need for external investment of the 5.8 at Year 2 that will be needed in the project at Year 3.

$$F = +5.8(F/P,47\%,1) = 5.8(1.47) = +8.5$$

When the 8.5 is returned to the project in Year 3, the resulting net required investment in Year 3 is -50 + 8.5 = -41.5. Now the internal investment cash flow is,

Year	Cash flow
0	$0
1	0
2	0
3	-41.5
4	+20
5	+60

Solving for the rate of return on the internal investment,

PW of costs = PW of benefits

Thus, at Year 3,

$$41.5 = 20(P/F,i,1) + 60(P/F,i,2)$$

Try $i = 47\%$:

$$41.5 = 20(0.680) + 60(0.464)$$

$$= 13.6 + 27.9$$

$$= 41.5$$

From the computations, we have seen that the two positive interest rates (10.1% and 47%) require that the internal investment and the external investment both earn the same interest rate. Thus, if one were prepared to agree that the appropriate interest rate for external investments is 47%, then we would have to agree that the resulting interest rate on the internal investment is 47%. But if a suitable interest rate on external investments (say, putting the money in a savings account) were only 6%, then what is the rate of return on the internal investment? None of the calculations we have made so far tell us. But, in the next section, we find that this *is* a practical approach for solving the multiple interest rate problem.

External Interest Rate

From the discussion of the meaning of multiple interest rates, we see an important general situation.

> **Solving a cash flow for an unknown interest rate means that money in any required external investment is assumed to earn the same interest rate as money invested in the internal investment.**

This occurs regardless of whether there is only one or more than one positive interest rate.

There can be no particular reason why we would assume that external investments earn the same rate of return as internal investments. The required external investment may be of short duration, like a year or two. And, as will be discussed later, it is very likely that the rate of return available on a capital investment in the business is two or three times the rate of return available on short-duration external deposits, or other external investments of money.

Thus, two interest rates are reasonable, one on the internal investment and one on the temporary external investment.

By separating the interest rates, we have also provided the means for resolving any difficulties that arise from multiple positive rates of return. What we desire is to determine the rate of return on the internal investment assuming a realistic value for the rate of return available on the external investment. With the external interest rate we can compute the effect of any required external investments. The results can be introduced back into the cash flow in the same manner as was done in our detailed examination of the external investment assumed in Ex. 7A-1. Example 7A-1 will now be presented again, this time with a preselected external interest rate of 6%.

EXAMPLE 7A-2

Take the Ex. 7A-1 cash flow and assume that any money held outside of the project earns 6% interest (that is, the external interest rate = 6%).

Year	Cash flow
0	+$19
1	+10
2	-50
3	-50
4	+20
5	+60

Solution:

At both Year 0 and Year 1, there is a flow of money resulting from the advance payments before the aircraft is manufactured. The money will be needed later to help pay the production costs. If the external interest rate is 6%, the +19 (million dollars) will be invested externally for two years and the +10 for one year. Their compound amount at the end of Year 2 will be:

$$\text{Compound amount at end of Year 2} = +19(F/P,6\%,2) + 10(F/P,6\%,1)$$

$$= +19(1.124) + 10(1.06)$$

$$= +21.4 + 10.6$$

$$= +32$$

When this amount is returned to the project, the net cash flow for Year 2 becomes -50 + 32 = -18. The resulting cash flow for the project is,

*Computation of NPW at various
interest rates using single payment
present worth factors*

Year	Cash flow	0%	8%	10%
		0%	8%	10%
0	$0	$0	$0	$0
1	0	0	0	0
2	-18	-18	-15.4	-14.9
3	-50	-50	-39.7	-37.6
4	+20	+20	+14.7	+13.7
5	+60	+60	+40.8	+37.3
		NPW = +$12	+$0.4	-$1.5

The cash flow has one sign change indicating there is either zero or one positive interest rate. We have located a point where NPW = 0 at

$$i = 8\% + 2\%\left(\frac{0.4}{1.5+0.4}\right) = 8\% + 2\%(0.21) = 8.4\%$$

Thus, we have identified the single positive root for the cash flow. Assuming an external interest rate of 6%, the rate of return on the Interair plane contract is 8.4%. ∎

In Ex. 7A-2 we accomplished two tasks:

1. A realistic interest rate was used to find equivalent sums when money must be invested externally. This external interest rate should reflect the rate on external investment opportunities and, therefore, be independent of the rate of return on any particular internal investment.

2. Through the use of an external interest rate, the number of sign changes in the cash flow was reduced to one, ensuring that there will not be multiple positive rates of return.

Resolving Multiple Rate of Return Problems

Example 7A-1 contains a cash flow that produces two positive rates of return. Yet, on closer examination, we saw a cash flow for external investment in the initial part of the problem. The outflow of cash was invested at the external interest rate until it was needed in the project. It was returned to the project at the end of Year 2. In so doing, the number of sign changes in the cash flow was reduced from two to one. A single positive rate of return was then computed. In this way, the multiple rate of return difficulty was resolved.

The general method for handling cash flows with multiple sign changes is: alter the cash flow through the use of an external interest rate to reduce the number of sign changes to one, thereby ensuring no more than one positive rate of return. We must point out, however, that the changes made with the external interest rate affect the project, or internal rate of return. To keep the sensitivity of the internal rate of return to the external interest rate as small as possible, the cash flow adjustments should be kept to a minimum.

A Further Look at the Computation of Rate of Return

The cash flow rule of signs tells us we can have cash flows with multiple sign changes but only one positive rate of return. Using the approach in this subchapter, we may alter a cash flow when no alteration is needed (or be making too great an alteration when one is needed). There is thus more to be said about solving a cash flow for a rate of return, but for now, we continue our broader examination of engineering economic analysis. We will take a further look at the computation of rate of return in Chapter 18.

Summary

In some situations, we find that solution of a cash flow equation results in more than one positive rate of return. This is possible by the cash flow rule of signs. A sign change is where successive nonzero values in the cash flow have different signs (that is, change from + to -, *or vice versa*).

Zero sign changes indicates there is no rate of return, as the cash flow is either all disbursements or all receipts.

One sign change is the usual situation and a single positive rate of return generally results. There will, however, be no rate of return whenever loan repayments are less than the loan or an investment fails to return benefits at least equal to the investment.

Multiple sign changes may result in multiple positive rates of return. The difficulty is not that it will happen, but that the analyst may not recognize that the cash flow has multiple sign changes and may have multiple positive rates of return. When they occur, none of the multiple rates of return are a suitable measure of the economic desirability of the project represented by the cash flow.

Multiple positive rates of return indicate a project that at some time has money invested outside the project. Since investments outside the project may earn interest at a different rate from the internal project rate of return, an external rate should be selected. This approach leaves the rate of return on the money actually invested in the project as the single unknown. The number of sign changes are thereby reduced to one, eliminating the possibility of multiple positive rates of return. This topic is also discussed further in Chapter 18.

Problems

7A-1 The owner of a walnut orchard wished to enjoy some of his wealth and yet not sell the orchard for ten years. He negotiated an agreement with the Omega Insurance Company as follows. Omega would pay the owner $4000 per year in twenty equal annual payments beginning immediately. At the end of the tenth year, the owner is obligated to sell the orchard and with the proceeds pay Omega $75,000 at that time. Omega will, of course, continue the $4000 annual payments to the retired orchardist for nine more years. What interest rate was used in devising the agreement between Omega and the orchardist?

7A-2 A group of businessmen formed a partnership to buy and race an Indianapolis-type racing car. They agreed to pay an individual $50,000 for the car and associated equipment. The payment was to be in a lump sum at the end of the year. In what must have been "beginner's luck," the group won a major race the first week and $80,000. The rest of the first year, however, was not so good: at the end of the first year, the group had to pay out $35,000 for expenses plus the $50,000 for the car and equipment. The second year was a poor one: the group had to pay $70,000 at the end of the second year just to clear up the racing debts. During the third and fourth years, racing income just equalled costs. When the group was approached by a prospective buyer for the car, they readily accepted $80,000 cash, which was paid at the end of the fourth year. What rate of return did the businessmen obtain from their racing venture?

7A-3 A student organization, at the beginning of the Fall quarter, purchased and operated a soft-drink vending machine as a means of helping finance its activities. The vending machine cost $75 and was installed at a gasoline station near the university. The student organization pays $75 every three months to the station owner for the right to keep their vending machine at the station. During the year the student organization owned the machine, they received the following quarterly income from it, before making the $75 quarterly payment to the station owner:

	Income
Fall quarter	$150
Winter quarter	25
Spring quarter	125
Summer quarter	150

At the end of one year, the student group resold the machine for $50. Determine the quarterly cash flow. Then answer *a, b,* and *c* below.

a.   Assume the cash flow has a single positive rate of return. Proceed to compute the nominal annual rate of return the organization received on their investment.

b.   Using a nominal external interest rate of 12% (3% per quarter-year) transform the cash flow to one sign change. Then compute the nominal rate of return.

c.   Why do the answers for *a* and *b* differ? Which one is the "correct" answer?

7A-4 Given the following cash flow:

Year	Cash flow
0	-$500
1	+2000
2	-1200
3	-300

Determine the rate of return on the internal investment. If necessary, assume external investments earn 6% interest. (*Answer:* 20.2%)

7A-5 Given the following cash flow:

Year	Cash flow
0	-$500
1	+200
2	-500
3	+1200

Determine the rate of return on the internal investment. If necessary, assume external investments earn 6% interest. (*Answer:* 19.6%)

7A-6 Given the following cash flow:

Year	Cash flow
0	-$100
1	+360
2	-570
3	+360

Determine the rate of return on the internal investment. If necessary, assume external investments earn 6% interest.

7A-7 Consider the following cash flow:

Year	Cash flow
0	-$110
1	-500
2	+300
3	-100
4	+400
5	+500

Assume external investment of money is at a 10% interest rate. Compute the rate of return on the internal investment.

7A-8 Consider the following cash flow:

Year	Cash flow
0	-$50.0
1	+20.0
2	-40.0
3	+36.8
4	+36.8
5	+36.8

Assume any external investment of money is at a 10% interest rate. Compute the rate of return on the internal investment. (*Answer:* 15%)

7A-9 A firm invested $15,000 in a project that appeared to have excellent potential. Unfortunately, a lengthy labor dispute in Year 3 resulted in costs that exceeded benefits by $8000. The cash flow for the project is as follows:

Year	Cash flow
0	-$15,000
1	+10,000
2	+6,000
3	-8,000
4	+4,000
5	+4,000
6	+4,000

Compute the rate of return for the project, assuming a 12% interest rate on external investments.

7A-10 The textbook tells us that for the following cash flow there is no positive interest rate.

Year	Cash flow
0	-$50
1	+20
2	+20

There is, however, a negative interest rate. Compute its value.
   (*Answer: i = -13.7% or -146%*)

7A-11 For the following cash flow compute the internal rate of return, assuming 15% interest on external investments. Do not transform the cash flow any more than is essential.

Year	Cash flow
0	$0
1	0
2	-20
3	0
4	-10
5	+20
6	-10
7	+100

7A-12 Given the following cash flow:

Year	Cash flow
0	-$800
1	+500
2	+500
3	-300
4	+400
5	+275

If external investments earn 10%, what is the rate of return on the internal investment?
(*Answer: 25%*)

7A-13 Consider the following cash flow.

Year	Cash flow
0	-$100
1	+240
2	-143

a. Solve the cash flow for all positive values of i.

b. Assuming any external investment of money will earn 12% per year, compute the rate of return on the internal investment.

c. If the minimum attractive rate of return is 12%, should the project be undertaken?

7A-14 Refer to the strip-mining project in Example 5-9. Compute the rate of return for the project, assuming if necessary a 10% interest rate on external investments.

7A-15 Consider the following cash flow.

Year	Cash flow
0	-$500
1	+800
2	+170
3	-550

Compute the rate of return on the internal investment. Assume any external investment of money is at 10%.

7A-16 Consider the cash flow: PW of B - PW of C = O

Year	Cash flow
0	-$100
1	+360
2	-428
3	+168

a. Solve the cash flow for all positive rates of return. Make a plot of NPW vs. i.

b. Determine the rate of return on the internal investment. If necessary, assume any external investments earn 10% interest.

7A-17 Consider the cash flow:

Year	Cash flow
0	-$1200
1	+358
2	+358
3	+358
4	+358
5	+358
6	-394

Assume any external investment of money is at a 10% interest rate. Compute the rate of return on the internal investment.

7A-18 Consider cash flow:

Year	Cash flow
0	-$3570
1–3	+1000
4	-3170
5–8	+1500

Determine the rate of return on the internal investment. If necessary, any external investments earn 8% interest.

7A-19 Bill purchased a vacation lot he saw advertised on television for an $800 downpayment and monthly payments of $55. When he visited the lot he had purchased, he found it was not something he wanted to own. After 40 months he was finally able to sell the lot. The new purchaser assumed the balance of the loan on the lot and paid Bill $2500. What rate of return did Bill receive on his investment?

7A-20 Consider the cash flow:

Year	Cash flow
0	-$ 850
1	+600
2–9	+200
10	-1800

a. Compute a rate of return for the cash flow.

b. Compute the Net Present Worth (NPW) of the cash flow at the firm's 10% minimum attractive rate of return.

c. Plot a graph of NPW vs. i for the cash flow.

d. Compute the rate of return on the internal investment. Assume any external investment of money is at 10%.

7A-21 Assume the following cash flows are associated with a project:

Year	Cash
0	-$16,000
1	-8,000
2	11,000
3	13,000
4	-7,000
5	8,950

Use an external interest rate e* of 12% and compute the rate of return for this cash flow.

7A-22 Using an $e*$ of 13%, find the internal rate of return for the following cash flow:

Year	Cash Flow
1	-$15,000
2	10,000
3	-8,000
4	11,000
5	13,000

7A-23 A tomato press has the following annual cost data associated with it:

Year	Cash Flow
0	-$210,000
1	88,000
2	68,000
3	62,000
4	-31,000
5	30,000
6	55,000
7	65,000

The external interest rate, $e*$, is 12%. What is the rate of return associated with this project?

7A-24 A project has been in operation for five years. The following annual cash flows were obtained from the project:

Year	Cash Flow
0	-$103,000
1	102,700
2	-87,000
3	94,500
4	-8,300
5	38,500

Use an $e*$ and MARR of 18%. Calculate the rate of return and state whether it has been an acceptable rate of return.

CHAPTER 8

Incremental Analysis

We now see how to solve problems by each of three major methods, with one exception: for three or more alternatives, no rate of return solution was given. The reason is that under these circumstances, incremental analysis is required and it has not been discussed. This chapter will show how to solve that problem.

Incremental analysis can be defined as the examination of the differences between alternatives. By emphasizing alternatives, we are really deciding whether or not differential costs are justified by differential benefits.

In retrospect, we see that the simplest form of incremental analysis was presented in Chapter 7. We did incremental analysis by the rate of return evaluation of the differences between two alternatives. We recognized that the two alternatives could be related as follows:

$$\begin{array}{ccc} \text{Higher-cost} & = & \text{Lower-cost} & + & \text{Differences} \\ \text{alternative} & & \text{alternative} & & \text{between them} \end{array}$$

We will see that incremental analysis can be examined either graphically or numerically. We will first look at graphical representations of problems, proceed with numerical solutions of rate of return problems, and see that a graphical representation may be useful in examining problems whether using incremental analysis or not.

Graphical Solutions

In the last chapter, we examined problems with two alternatives. Our method of solution represented a form of incremental analysis. A graphical review of that situation will help to introduce incremental analysis.

EXAMPLE 8-1

This is a review of Ex. 7-6. There were two mutually exclusive alternatives:

Year	Alt. 1	Alt. 2
0	−$10	−$20
1	+15	+28

If 6% interest is assumed, which alternative should be selected?

Solution: For this problem, we will plot the two alternatives on a PW of benefits *vs.* PW of cost graph.

Alternative 1:

$$\text{PW of cost} = \$10$$

$$\text{PW of benefit} = \$15(P/F,6\%,1) = 15(0.9434)$$

$$= \$14.15$$

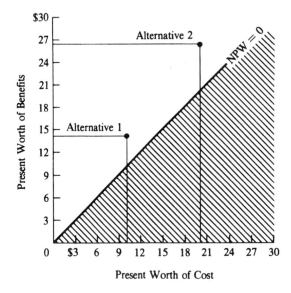

Figure 8-1 PW of benefits *vs.* PW of cost graph.

Alternative 2:

PW of cost = $20

PW of benefit = $28(P/F,6\%,1) = 28(0.9434)$

= $26.42

The alternatives are plotted in Figure 8-1, which looks very simple yet tells us a great deal about the situation.

On a graph of PW of benefits *vs.* PW of cost (for convenience, we will call it a ***benefit–cost graph***), there will be a line where NPW = 0. Where the scales used on the two axes are identical, as in this case, the resulting line will be at a 45° angle. If unequal scales are used, the line will be at some other angle.

For the chosen interest rate (6% in this example), this NPW = 0 line divides the graph into an area of desirable alternatives and undesirable alternatives. To the left (or above) the line is desirable, while to the right (or below) the line is undesirable. We see that to the left of the line, PW of benefits exceeds the PW of cost or, we could say, NPW is positive. To the right of the line, in the shaded area, PW of benefits is less than PW of cost; thus, NPW is negative.

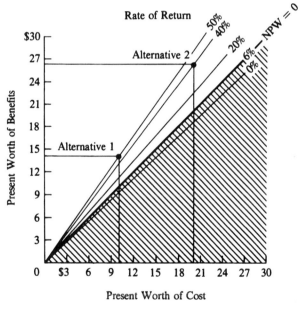

Figure 8-2 Benefit–cost graph for Example 8-1 with a one-year analysis period.

In this example, both alternatives are to the left of the NPW = 0 line. Therefore, both alternatives will have a rate of return greater than the 6% interest rate used in constructing the graph. In fact, other rate of return lines could also be computed and plotted on the graph *for this special case of a one-year analysis period*. We must emphasize at the outset that the additional rate of return lines shown in Fig. 8-2 can be plotted only for this special situation. For analysis periods greater than one year, the NPW = 0 line is the only line that can be accurately drawn. The graphical results in Fig. 8-2 agree with the calculations made in Ex. 7-6, where the rates of return for the two alternatives were 50% and 40%, respectively.

Figure 8-2 shows that the slope of a line on the graph represents a particular rate of return for this special case of a one-year analysis period. Between the origin and Alt. 1, the slope represents a 50% rate of return; while from the origin to Alt. 2, the slope represents a 40% rate of return. Since

$$\underset{\text{Alt. 2}}{\text{Higher-cost}} = \underset{\text{Alt. 1}}{\text{Lower-cost}} + \underset{\text{between}}{\text{Difference}}$$

the differences between the alternatives can be represented by a line shown in Fig. 8-3.

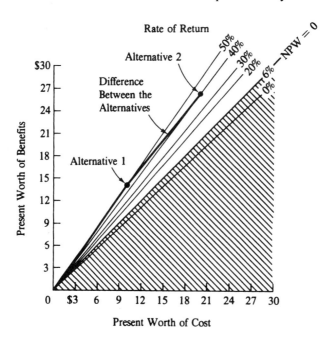

Figure 8-3 Benefit–cost graph for Alternatives 1 and 2 with a one-year analysis period.

Viewed in this manner, we clearly see that Alt. 2 may be considered two separate increments of investment. The first increment is Alt. 1 and the second one is the difference between the alternatives. Thus we will select Alt. 2 if the difference between the alternatives is a desirable increment of investment.

Since the slope of the line represents a rate of return, we see that the increment is desirable if the slope of the increment is greater than the slope of the 6% line that corresponds to NPW = 0. We can see that the slope is greater; hence, the increment of investment is, attractive. In fact, a careful examination shows that the "Difference between the alternatives" line has the same slope as the 30% rate of return line. We can say, therefore, that the incremental rate of return from selecting Alt. 2 rather than Alt. 1 is 30%. This is the same as was computed in Ex. 7-6. We conclude that Alt. 2 is the preferred alternative. ■

EXAMPLE 8-2

Solve Ex. 7-10 by means of a benefit-cost graph. Two machines are being considered for purchase. If the minimum attractive rate of return (MARR) is 10%, which machine should be bought?

	Machine X	Machine Y
Initial cost	$200	$700
Uniform annual benefit	95	120
End-of-useful-life salvage value	50	150
Useful life, in years	6	12

Solution: Using a 12-year analysis period,

Machine X:

$$PW \text{ of cost}^1 = 200 + (200 - 50)(P/F, 10\%, 6) - 50(P/F, 10\%, 12)$$

$$= 200 + 150(0.5645) - 50(0.3186) = 269$$

$$PW \text{ of benefit} = 95(P/A, 10\%, 12) = 95(6.814) = 647$$

Machine Y:

$$PW \text{ of cost}^1 = 700 - 150(P/F, 10\%, 12) = 700 - 150(0.3186) = 652$$

$$PW \text{ of benefit} = 120(P/A, 10\%, 12) = 120(6.814) = 818$$

---

[1] Salvage value is considered a reduction in cost rather than a benefit.

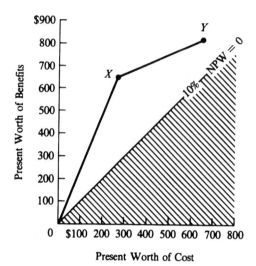

Figure 8-4 Benefit–cost graph.

The two alternatives are plotted in Fig. 8-4; we see that the increment $Y–X$ has a slope much less than the 10% rate of return line. The rate of return on the increment of investment is less than 10%; hence, the increment is undesirable. This means that Machine X should be selected rather than Machine Y. ■

The two example problems show us some aspects of incremental analysis. We now will examine some multiple-alternative problems. We can solve multiple-alternative problems by present worth and annual cash flow analysis without any difficulties. Rate of return analysis requires that, for two alternatives, the differences between them must be examined to see whether or not they are desirable. Now, if we can choose between two alternatives, then by a successive examination we can choose from multiple alternatives. Figure 8-5 illustrates the method:

Three Alternatives:

Two-Alternative Analyses

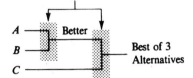

Four Alternatives:

Two Alternative Analyses

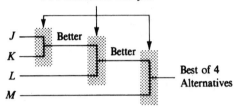

Five Alternatives:

Two-Alternative Analyses

Figure 8-5 Solving multiple-alterative problems by successive two-alternative analyses.

EXAMPLE 8-3

Consider the three mutually exclusive alternatives below.

	A	B	C
Initial cost	$2000	$4000	$5000
Uniform annual benefit	410	639	700

Each alternative has a twenty-year life and no salvage value. If the MARR is 6%, which alternative should be selected?

Solution: At 6%, (PW of benefits) = (Uniform annual benefit) × (Series present worth factor).

PW of benefits = (Uniform annual benefit)$(P/A,6\%,20)$
PW of benefits for Alt. A = $410(11.470) = $4703
Alt. B = $639(11.470) = $7329
Alt. C = $700(11.470) = $8029

Figure 8-6 is a plot of the situation; we see that the slope of the line from the origin to A is greater than the 6% line (NPW = 0). Thus the rate of return for A is greater than 6%. For the increment of additional cost of B over A, the slope of Line B–A is greater than the 6% line. This indicates that the rate of return on the increment of investment also exceeds 6%. But the slope of Increment C–B indicates its rate of return is less than 6%, hence, undesirable. We conclude that the A investment is satisfactory as well as the B–A increment; therefore, B is satisfactory. The C–B increment is unsatisfactory; so C is undesirable compared to B. Our decision is to select Alternative B. ■

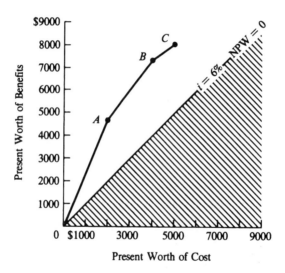

Figure 8-6 Benefit-cost graph for Example 8-3.

EXAMPLE 8-4

Further study of the three alternatives of Ex. 8-3 reveals that Alt. $A's$ uniform annual benefit was overstated. It is now projected to be $122 rather than $410. Replot the benefit–cost graph for this changed situation.

Solution:

Alt. *A'*: PW of benefits = 122(*P/A*,6%,20)

= 122(11.470) = 1399

Figure 8-7 shows the revised plot of the three alternatives. The graph shows that the revised Alt. *A'* is no longer desirable. We see that it has a rate of return less than 6%.

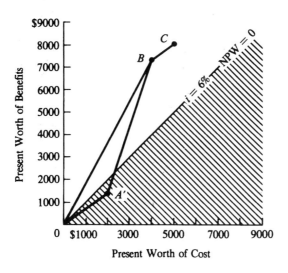

Figure 8-7 Benefit–cost graph for Example 8-4.

Now we wish to examine Alt. *B*. Should we compare it to the do-nothing alternative (which is represented by the origin), or as a *B–A'* increment over *A'*? Graphically, should we examine Line *0–B* or *B–A'*? Since *A'* is an undesirable alternative, it should be discarded and not considered further. Thus ignoring *A'*, we should compare *B* to the do-nothing alternative, which is Line *0–B*. Alternative B is preferred over the do-nothing alternative since it has a rate of return greater than 6%. Then Increment *C–B* is examined and, as we saw previously, it is an undesirable increment of investment. The decision to select *B* has not changed, which should be no surprise: if an inferior *A* has become an even less attractive *A'*, we still would select the superior alternative, *B*. ■

The graphical solution of the four example problems has helped us visualize the mechanics of incremental analysis. While problems could be solved this way, in practice they are solved mathematically rather than graphically. We will now proceed to solve problems mathematically by incremental rate of return analysis.

Incremental Rate Of Return Analysis

To illustrate incremental rate of return analysis, we will solve three of the examples again by mathematical, rather than graphical, methods.

EXAMPLE 8-5

Solve Ex. 8-1 mathematically. With two mutually exclusive alternatives and a 6% MARR, which alternative should be selected?

Year	Alt. 1	Alt. 2
0	−$10	−$20
1	+15	+28

Solution: Examine the differences between the alternatives.

Year	Alt. 2 − Alt. 1
0	−20 − (−10) = −10
1	+28 − (+15) = +13

Incremental rate of return (ΔROR),

$$10 = 13(P/F, i, 1)$$

$$(P/F, i, 1) = \frac{10}{13} = 0.7692$$

$$\Delta\text{ROR} = 30\%$$

The ΔROR is greater than the MARR; hence, we will select the alternative that gives this increment.

$$\begin{bmatrix} \text{Higher - cost} \\ \text{Alt. 2} \end{bmatrix} = \begin{bmatrix} \text{Lower - cost} \\ \text{Alt.1} \end{bmatrix} + \begin{bmatrix} \text{Increment} \\ \text{between them} \end{bmatrix}$$

Select Alternative 2.   ■

EXAMPLE 8-6

Recompute Example 8-3. MARR = 6%. Each alternative has a twenty-year life and no salvage value.

	A	B	C
Initial cost	$2000	$4000	$5000
Uniform annual benefit	410	639	700

Solution: A practical first step is to compute the rate of return for each alternative.

Alternative A:

$$2000 = 410(P/A, i, 20)$$

$$(P/A, i, 20) = \frac{2000}{410} = 4.878 \quad i^* = 20\%$$

Alternative B:

$$4000 = 639(P/A, i, 20)$$

$$(P/A, i, 20) = \frac{4000}{639} = 6.259 \quad i^* = 15\%$$

Alternative C:

$$5000 = 700(P/A, i, 20)$$

$$(P/A, i, 20) = \frac{5000}{700} = 7.143$$

The rate of return is between 12% and 15%:

$$i^* = 12\% + \left(\frac{7.469 - 7.143}{7.469 - 6.259} \right)(3\%) = 12.8\%$$

At this point, we would reject any alternative that fails to meet the MARR criterion of 6%. All three alternatives exceed the MARR in this example.

Next, we arrange the alternatives in order of increasing initial cost. Then we can examine the increments between the alternatives.

	A	B	C
Initial cost	$2000	$4000	$5000
Uniform annual benefit	410	639	700
Rate of return	20%	15%	12.8%

	Increment B–A	Increment C–B
Incremental cost	$2000	$1000
Incremental uniform annual benefit	229	61

Incremental rate of return:

$$2000 = 229(P/A, i, 20)$$

$$(P/A, i, 20) = \frac{2000}{229} \qquad \begin{array}{c} \Delta ROR \\ 9.6\% \end{array}$$

$$1000 = 61(P/A, i, 20)$$

$$(P/A, i, 20) = \frac{1000}{61} \qquad \begin{array}{c} \Delta ROR \\ 2.0\% \end{array}$$

The *B–A* increment is satisfactory; therefore, *B* is preferred over *A*. The *C–B* increment has an unsatisfactory 2% rate of return; therefore, *B* is preferred over *C*. Conclusion: select Alternative *B*. ■

EXAMPLE 8-7

Solve Ex. 8.4 mathematically. Alternative *A* in the previous example was believed to have an overstated benefit. The new situation for *A* (we will again call it *A'*) is a uniform annual benefit of 122. Compute the rate of return for *A'*.

Solution: $2000 = 122(P/A, i, 20)$

$$(P/A, i, 20) = \frac{2000}{122} = 16.39 \quad i^* = 2\%$$

This time Alt. *A'* has a rate of return less than the MARR of 6%. Alternative *A'* is rejected, and the problem now becomes selecting the better of *B* and *C*. In Ex. 8-6 we saw that the increment *C–B* had a ΔROR of 2% and it, too, was undesirable.

Thus, we again select Alternative *B*. ■

EXAMPLE 8-8

The following information is provided for five mutually exclusive alternatives that have twenty-year useful lives. If the minimum attractive rate of return is 6%, which alternative should be selected?

	A	B	C	D	E
Cost	$4000	$2000	$6000	$1000	$9000
Uniform annual benefit	639	410	761	117	785
PW of benefit*	7330	4700	8730	1340	9000
Rate of return	15%	20%	11%	10%	6%

*PW of benefit = (Uniform annual benefit)$(P/A,6\%,20)$ = 11.470(Uniform annual benefit). These values will be used later to plot a PW of cost *vs.* PW of benefit curve.

Solution: We see that the rate of return for each alternative equals or exceeds the MARR, therefore, no alternatives are rejected at this point. Next, we rearrange the alternatives to put them in order of increasing cost:

	D	B	A	C	E
Cost	$1000	$2000	$4000	$6000	$9000
Uniform annual benefit	117	410	639	761	785
Rate of return	10%	20%	15%	11%	6%

	Increment B–D	Increment A–B	Increment C–A
ΔCost	$1000	$2000	$2000
ΔAnnual benefit	293	229	122
ΔRate of return	29%	10%	2%

Beginning with the analysis of Increment B–D, we compute a ΔROR of 29%. Alternative B is thus preferred to Alt. D and D may be discarded at this point. The ΔROR for A–B is also satisfactory, so A is retained and B is now discarded. The C–A increment has a rate of return less than the MARR. Therefore, C is discarded and A continues to be retained.

At this point, we have examined four alternatives—D, B, A, C—and retained A after discarding the other three. Now we must decide whether A or E is the superior alternative. The increment we will examine is E–A. (*Note:* Increment E–C would have no particular meaning for we have already discarded C.)

	Increment E–A
ΔCost	$5000
ΔAnnual benefit	146

Over the twenty-year useful life, the total benefits (20 × 146 = 2920) are less than the cost. There is no rate of return on this increment (or one might say the ΔROR < 0%). This is an unsatisfactory increment, so E is discarded. Alternative A is the best of the five alternatives. ■

The benefit–cost graph (Fig. 8-8) of this example problem illustrates an interesting situation. All five alternatives have rates of return equal to or greater than the MARR of 6%. Yet, on detailed examination, we see that Alternatives *C* and *E* contain increments of investment that are unsatisfactory. Even though *C* has an 11% rate of return, it is unsatisfactory when compared to Alt. *A*.

Also noteworthy is the fact that the project with the greatest rate of return—Alternative *B*-is *not* the best alternative, because the proper economic criterion in this situation is to accept all separable increments of investment that have a rate of return greater than the 6% MARR. We found a desirable *A–B* increment with a 10% ΔROR. A relationship between Alternatives *A* and *B*, and the computed rates of return, are:

Higher-cost Alt. A		Alt. B		Differences between *A* and *B*
	=		+	
15% rate of return		20% rate of return		10% rate of return

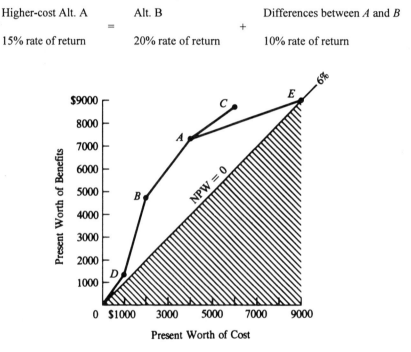

Figure 8-8 Benefit-cost graph for Example 8-8 data.

By selecting *A* we have, in effect, acquired a 20% rate of return on $2000 and a 10% rate of return on an additional $2000. Both of these are desirable. Taken together as *A*, the result is a 15% rate of return on a $4000 investment. This is economically preferable to a 20% rate of return on a $2000 investment, assuming we seek to invest money whenever we can identify an investment opportunity that meets our 6% MARR criterion. This implies that we have sufficient money to accept *all* investment opportunities that come to our attention where the MARR is exceeded. This abundant supply of money is considered appropriate in most industrial analyses, but it is not likely to be valid for individuals. The selection of an appropriate MARR is discussed in Chapter 15.

Elements In
Incremental Rate Of Return Analysis

1. Be sure all the alternatives are identified. In textbook problems the alternatives will be well-defined, but industrial problems may be less clear. Before proceeding, one must have all the mutually exclusive alternatives tabulated, including the "do-nothing" or the "keep doing the same thing" alternative, if appropriate.

2. (Optional) Compute the rate of return for each alternative. If one or more alternatives has a rate of return equal to or greater than the minimum attractive rate of return (ROR ≥ MARR), then any other alternatives with a ROR < MARR may be immediately rejected. This optional step requires calculations that may or may not eliminate alternatives. In an exam, this step probably should be skipped, but in less pressing situations, these are logical computations.

3. Arrange the remaining alternatives in ascending order of investment. The goal is to organize the alternatives so that the incremental analysis will be of separable increments of investment when we analyze the Higher-cost alternative *minus* the Lower-cost alternative. In textbook problems, this is usually easy—but there are lots of potential difficulties in following this simple rule.

The ordering of alternatives is not a critical element in incremental analysis; in fact, the differences between any two alternatives, like X and Y, can be examined either as an X–Y increment or a Y–X increment. If this is done in a random fashion, however, we will be looking sometimes at an increment of borrowing and sometimes at an increment of investment. It can get a little difficult to keep all of this straight; the basic goal is to restrict the incremental analysis, where possible, to increments of investment.

4. Make a two-alternative analysis of the first two alternatives. In the typical we have:

$$\begin{bmatrix} \text{Higher - cost} \\ \text{Alt. } Y \end{bmatrix} = \begin{bmatrix} \text{Lower - cost} \\ \text{Alt. } X \end{bmatrix} + \begin{bmatrix} \text{Differences between} \\ \text{them } (Y - X) \end{bmatrix}$$

so the increment examined is $(Y - X)$, which represents an increment of investment.

Compute the ΔROR on the increment of *investment*. The criterion is:

- If ΔROR ≥ MARR, retain the higher-cost Alt. Y.
- If ΔROR < MARR, retain the lower-cost Alt. X.
- Reject the other alternative used in the analysis.

Sometimes the two alternatives being examined cannot be described as "higher cost" and "lower cost." In Example 7-8, we encountered two alternatives, A and B, with equal investments. There we selected the $(A - B)$ increment because it was an increment of investment.

In other situations one may encounter cash flows of the differences between alternatives that have multiple sign changes (like Cash flows D and E in Table 7A-1). The logical approach is to look at both possible differences between alternatives, for example, $(X - Y)$ and $(Y - X)$, and select the one for analysis where the investment component dominates. In situations where an increment of *borrowing* is examined, the criterion is:

■ If ΔROR $\leq$ MARR, the increment is acceptable.

■ If ΔROR $>$ MARR, the increment is not acceptable.

5. Take the preferred alternative from Step 4, and the next alternative from the list created in Step 3. Proceed with another two-alterative comparison .

6. Continue until all alternatives have been examined and the best of the multiple alternatives has been identified.

Incremental Analysis Where There Are Unlimited Alternatives

There are situations where the possible alternatives are a more or less continuous function. For example, an analysis to determine the economical height of a dam represents a situation where the number of alternatives could be infinite. If the alternatives were limited, however, to heights in even feet, the number of alternatives would still be large and have many of the qualities of a continuous function, as in the following example.

EXAMPLE 8-9

A careful analysis has been made of the consequences of constructing a dam in the Blue Canyon. It would be feasible to construct a dam at this site with a height anywhere from 200 to 500 feet. Using a 4% MARR and a 75-year life, the various data have been used to construct Figure 8-9. Note particularly that the dam heights are plotted on the x-axis along with the associated PW of cost. What height of dam should be constructed?

Solution: Five points have been labelled on Fig. 8-9 to aid in the discussion. Dam heights below Point A have a PW of cost $>$ PW of benefit; hence, the rate of return is less than MARR, and we would not build a dam of these heights. In the region of Point B, an increment of additional PW of cost—ΔC—produces a larger increment of PW of benefit—ΔB. These are, therefore, desirable increments of additional investment and, hence, dam height. At Point D and also at Point E, the reverse is true. An increment of additional investment —ΔC— produces a smaller increment of PW of benefit —ΔB; this is undesirable. We do not want these increments, so the dam should not be built to these heights.

At Point C, we are at the point where $\Delta B = \Delta C$. Lower dam heights have desirable increments of investment and higher dam heights have unfavorable increments of investment. The optimal dam height, therefore, is where $\Delta B = \Delta C$. On the figure, this corresponds to a height of approximately 250 feet. Another way of defining the point where $\Delta B = \Delta C$ is to describe it as the point where the slope of the curve equals the slope of the NPW = 0 line. ∎

The techniques for solving discrete alternatives or continuous function alternatives are really the same. We proceed by adding increments whenever $\Delta \text{ROR} \geq \text{MARR}$ and discarding increments when, $\Delta \text{ROR} < \text{MARR}$.

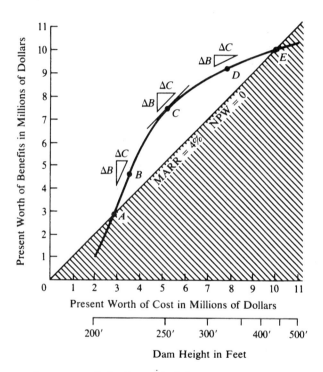

Figure 8-9 Benefit–cost graph for Example 8-9.

Present Worth Analysis With Benefit–Cost Graphs

Any of the example problems presented so far in this chapter could be solved by the present worth method. The benefit–cost graphs we introduced earlier can be used to graphically solve problems by present worth analysis.

In present worth analysis, where neither input nor output is fixed, the economic criterion is to maximize NPW. In Ex. 8-1, we had the case of two alternatives and the MARR equal to 6%:

Year	Alt. 1	Alt. 2
0	−$10	−$20
1	+15	+28

In Example 8-1 we computed,

PW of cost	$10	$20
PW of benefit	14.15	26.42

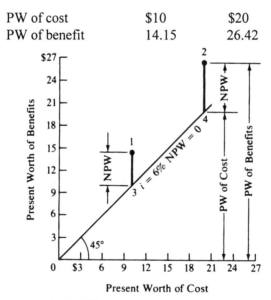

Figure 8-10 Benefit–cost graph for PW analysis.

These points are plotted in Figure 8-10. Looking at Fig. 8-10, we see that the NPW = 0 line is at 45° since identical scales were used on both axes. The point for Alt. 2 is plotted at the coordinates (PW of cost, PW of benefits). We drop a vertical line from Alt. 2 to the diagonal NPW = 0 line. The coordinates of any point on the graph are (PW of cost, PW of benefits), but along the NPW = 0 line (45° line), the *x* and *y* coordinates are equal. Thus, the coordinates of Point 4 are also (PW of cost, PW of cost). Since

$$NPW = PW \text{ of benefits} - PW \text{ of cost}$$

the vertical distance from Point 4 to Alt. 2 represents NPW. Similarly, the vertical distance between Point 3 and Alt. 1 presents NPW for Alt. 1. Since the criterion is to maximize NPW, we select Alt. 2 with larger NPW. The same technique is used in situations where there are multiple alternatives or continuous alternatives. In Ex.8-8 there were five alternatives; Fig. 8-11 shows a plot of these five alternatives. We can see that A has the greatest NPW and, therefore, is the preferred alternative.

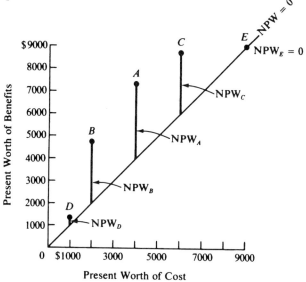

Figure 8-11 Present worth analysis of Example 8-8 using a benefit–cost plot.

Choosing An Analysis Method

At this point, we have examined in detail the three major economic analysis techniques: present worth analysis, annual cash flow analysis, and rate of return analysis. A practical question is, "Which method should be used for a particular problem?"

While the obvious answer is to use the method requiring the least computations, there are a number of factors that may affect the decision:

1. Unless the MARR—minimum attractive rate of return (or minimum required interest rate for invested money)—is known, neither present worth analysis nor annual cash flow analysis can be done.

2. Present worth analysis and annual cash flow analysis often require far less computations than rate of return analysis.

3. In some situations, a rate of return analysis is easier to explain to people unfamiliar with economic analysis. At other times, an annual cash flow analysis may be easier to explain.

4. Business enterprises generally adopt one, or at most two, analysis techniques for broad categories of problems. If you work for a corporation, and the policy manual specifies rate of return analysis, you would appear to have no choice in the matter.

Since one may not always be able to choose the analysis technique computationally best-suited to the problem, this book illustrates how to use each of the three methods in all feasible situations. Ironically, the most difficult method—rate of return analysis—is the one most frequently used by engineers in industry!

Spreadsheets and Incremental Analysis

An incremental analysis of two alternatives is easily done with the RATE or IRR functions when the lives of the alternatives are the same. However, when the lives are different the problem is more difficult. As discussed in Chapters 5 and 6, when comparing alternatives with different length lives, the usual approach is to assume that the alternatives are repeated until the least common multiple of their lives. This can be done with a spreadsheet, but Excel supports an easier approach.

Excel has a tool called GOAL SEEK that identifies a formula cell, a target value, and a variable cell. Using this tool the variable cell is changed automatically by the computer until the formula cell equals the target value. To find an IRR for an incremental analysis, the formula cell can be the difference between two equivalent annual worths with a target value of 0. Then if the variable cell is the interest rate, GOAL SEEK will find the IRR.

In Excel this tool is accessed by selecting T(ools) on the main toolbar or menu and G(oal seek) on the sub-menu. As shown in Example 8-10, the variable cell (with the interest rate) must somehow affect the formula cell (difference in EAWs or EACs), although it need not appear directly in the formula cell. In Figures 8-12a & b the interest rate (cell A1) appears in the EAC formulas (cells D3 and D4), but not in the formula cell (cell D5).

EXAMPLE 8-10

Two different asphalt mixes can be used on a highway. The good mix will last 6 years, and it will cost $600,000 to buy and lay down. The better mix will last 10 years, and it will cost

$800,000 to buy and lay down. Find the incremental IRR for using the more expensive mix.

Solution

This example would be difficult to solve without the GOAL SEEK tool. The least common multiple of 6 and 10 is 30 years, which is the comparison period. With the GOAL SEEK tool a very simple spreadsheet does the job. In Figure 8-12a the spreadsheet is shown before GOAL SEEK. Figure 8-12b shows the result after the goal (D5) is set = 0 and A1 is selected as the variable cell to change.

	A	B	C	D	E	F
1	8.00% interest rate					
2	Alternative	Cost	Life	EAC		
3	Good	600000	6	129,789	= -PMT($A$1,C3,B3)	
4	Better	800000	10	119,224	= -PMT($A$1,C4,B4)	
5			difference=	10,566	= D3-D4	

Figure 8-12a Spreadsheet before GOAL SEEK.

	A	B	C	D	E	F
1	14.52% interest rate					
2	Alternative	Cost	Life	EAC		
3	Good	600000	6	156,503	= -PMT($A$1,C3,B3)	
4	Better	800000	10	156,503	= -PMT($A$1,C4,B4)	
5			difference=	0	= D3-D4	

Figure 8-12b Spreadsheet after GOAL SEEK.

The incremental IRR found by GOAL SEEK is 14.52%. ■

Summary

A graph of the PW of benefits *vs.* PW of cost (called a benefit–cost graph) can be an effective way to examine two alternatives by incremental analysis. And since multiple-alterative incremental analysis is done by the successive analysis of two alternatives, benefit–cost graphs can be used to solve multiple alternative and continuous-function alternatives as easily as two alternative problems.

The important steps in incremental rate of return analysis are:

1. Check to see that all the alternatives in the problem are identified.

2. (Optional) Compute the rate of return for each alternative. If one or more alternatives has ROR ≥ MARR, reject any alternatives with ROR < MARR.

3. Arrange the remaining alternatives in ascending order of investment.

4. Make a two-alterative analysis of the first two alternatives.

5. Take the preferred alternative from Step 4, and the next alternative from the list in Step 3. Proceed with another two-alterative comparison.

6. Continue until all alternatives have been examined and the best of the multiple alternatives has been identified.

Decision Criteria for Increments of *Investment:*

■ If ΔROR ≥ MARR, retain the higher-cost alternative.

■ If ΔROR < MARR, retain the lower-cost alternative.

■ Reject the other alternative used in the analysis.

Decision Criteria for Increments of *Borrowing:*

■ If, ΔROR ≤ MARR, the increment is acceptable.

■ If, ΔROR > MARR, the increment is not acceptable.

Benefit–cost graphs, being a plot of the PW of benefits *vs.* the PW of cost, can also be used in present worth analysis to graphically show the NPW for each alternative.

Problems

Unless otherwise noted, all Ch. 8 problems should be solved by rate of return analysis.

8-1 A firm is considering moving its manufacturing plant from Chicago to a new location. The Industrial Engineering Department was asked to identify the various alternatives together with the costs to relocate the plant, and the benefits. They examined six likely sites, together with the do-nothing alternative of keeping the plant at its present location. Their findings are summarized below:

Plant location	First cost	Uniform annual benefit
Denver	$300 thousand	$ 52 thousand
Dallas	550	137
San Antonio	450	117
Los Angeles	750	167
Cleveland	150	18
Atlanta	200	49
Chicago	0	0

The annual benefits are expected to be constant over the eight-year analysis period. If the firm uses 10% annual interest in its economic analysis, where should the manufacturing plant be located? (*Answer:* Dallas)

8-2 In a particular situation, four mutually exclusive alternatives are being considered. Each of the alternatives costs $1300 and has no end-of-useful-life salvage value.

Alternative	Annual benefit	Useful life, in years	Calculated rate of return
A	$100 at end of first year; increasing $30 per year thereafter	10	10.0%
B	$10 at end of first year; increasing $50 per year thereafter	10	8.8%
C	Annual end of year benefit = $260	10	15.0%
D	$450 at end of first year; declining $50 per year thereafter	10	18.1%

If the MARR is 8%, which alternative should be selected? (*Answer:* Alt. C)

8-3 A more detailed examination of the situation in Problem 8-2 reveals that there are two additional mutually exclusive alternatives to be considered. Both cost more than the $1300 for the four original alternatives.

Alternative	Cost	Annual end-of-years benefit	Useful life, in years	Calculated rate of return
E	$3000	$488	10	10.0%
F	5850	1000	10	11.2%

If the MARR remains at 8%, which one of the six alternatives should be selected? Neither Alt. E nor F has any end-of-useful-life salvage value. (*Answer:* Alt. F)

8-4 The owner of a downtown parking lot has employed a civil engineering consulting firm to advise him whether or not it is economically feasible to construct an office building on the site. Bill Samuels, a newly hired civil engineer, has been assigned to make the analysis. He has assembled the following data:

Alternative	Total investment*	Total net annual revenue from property
Sell parking lot	$ 0	$ 0
Keep parking lot	200,000	22,000
Build 1-story building	400,000	60,000
Build 2-story building	555,000	72,000
Build 3-story building	750,000	100,000
Build 4-story building	875,000	105,000
Build 5-story building	1,000,000	120,000

*Includes the value of the land.

The analysis period is to be 15 years. For all alternatives, the property has an estimated resale (salvage) value at the end of 15 years equal to the present total investment. If the MARR is 10%, what recommendation should Bill make?

8-5 An oil company plans to purchase a piece of vacant land on the corner of two busy streets for $70,000. The company has four different types of businesses that it installs on properties of this type.

Plan	Cost of improvements*	
A	$ 75,000	Conventional gas station with service facilities for lubrication, oil changes, etc.
B	230,000	Automatic carwash facility with gasoline pump island in front
C	30,000	Discount gas station (no service bays)
D	130,000	Gas station with low-cost quick-carwash facility

*Cost of improvements does not include the $70,000 cost of land.

In each case, the estimated useful life of the improvements is 15 years. The salvage value for each is estimated to be the $70,000 cost of the land. The net annual income, after paying all operating expenses, is projected as follows:

Plan	Net annual income
A	$23,300
B	44,300
C	10,000
D	27,500

If the oil company expects a 10% rate of return on its investments, which plan (if any) should be selected?

8-6 A firm is considering three mutually exclusive alternatives as part of a production improvement program. The alternatives are:

	A	B	C
Installed cost	$10,000	$15,000	$20,000
Uniform annual benefit	1,625	1,625	1,890
Useful life, in years	10	20	20

For each alternative, the salvage value at the end-of-useful-life is zero. At the end of ten years, Alt. *A* could be replaced by another *A* with identical cost and benefits. The MARR is 6%. If the analysis period is twenty years, which alternative should be selected?

8-7 Given the following four mutually exclusive alternatives:

	A	B	C	D
First cost	$75	$50	$50	$85
Uniform annual benefit	16	12	10	17
Useful life, in years	10	10	10	10
End-of-useful-life salvage value	0	0	0	0
Computed rate of return	16.8%	20.2%	15.1%	15.1%

If the MARR is 8%, which alternative should be selected? (*Answer: A*)

8-8 Consider the following three mutually exclusive alternatives:

	A	B	C
First cost	$200	$300	$600
Uniform annual benefit	59.7	77.1	165.2
Useful life, in years	5	5	5
End-of-useful-life salvage value	0	0	0
Computed rate of return	15%	9%	11.7%

For what range of values of MARR is Alt. *C* the preferred alternative? Put your answer in the following form: "Alt. *C* is preferred when _____% ≤ MARR ≤ _____%."

8-9 Consider four mutually exclusive alternatives that each have an 8-year useful life:

	A	B	C	D
First cost	$1000	$800	$600	$500
Uniform annual benefit	122	120	97	122
Salvage value	750	500	500	0

If the minimum attractive rate of return is 8%, which alternative should be selected?

8-10 Three mutually exclusive projects are being considered:

	A	B	C
First cost	$1000	$2000	$3000
Uniform annual benefit	150	150	0
Salvage value	1000	2700	5600
Useful life, in years	5	6	7

When each project reaches the end of its useful life, it would be sold for its salvage value and there would be no replacement. If 8% is the desired rate of return, which project should be selected?

8-11 Consider three mutually exclusive alternatives:

Year	Buy X	Buy Y	Do nothing
0	–$100.0	–$50.0	0
1	+31.5	+16.5	0
2	+31.5	+16.5	0
3	+31.5	+16.5	0
4	+31.5	+16.5	0

Which alternative should be selected:

    a. if the minimum attractive rate of return equals 6%?

    b. if MARR = 9%?

    c. if MARR = 10%?

    d. if MARR = 14%?

    (*Answers:* *a.* X; *b.* Y; *c.* Y; *d.* Do nothing)

8-12 Consider the three alternatives:

Year	A	B	Do nothing
0	–$100	–$150	0
1	+30	+43	0
2	+30	+43	0
3	+30	+43	0
4	+30	+43	0
5	+30	+43	0

Which alternative should be selected:

    a. if MARR = 6%?

    b. if MARR = 8%?

    c. if MARR = 10%?

8-13 A firm is considering two alternatives:

	A	B
Initial cost	$10,700	$5,500
Uniform annual benefits	2,100	1,800
Salvage value at end of useful life	0	0
Useful life, in years	8	4

At the end of four years, another B may be purchased with the same cost, benefits, and so forth. If the MARR is 10%, which alternative should be selected?

8-14 Consider the following alternatives:

	A	B	C
Initial cost	$300	$600	$200
Uniform annual benefits	41	98	35

Each alternative has a ten-year useful life and no salvage value. If the MARR is 8%, which alternative should be selected?

8-15 Given the following:

Year	X	Y
0	−$10	−$20
1	+15	+28

Over what range of values of MARR is Y the preferred alternative?

8-16 Consider four mutually exclusive alternatives:

	A	B	C	D
Initial cost	$770.00	$1406.30	$2563.30	0
Uniform annual benefit	420.00	420.00	420.00	0
Useful life, in years	2	4	8	0
Computed rate of return	6.0%	7.5%	6.4%	0

The analysis period is eight years. At the end of two years, four years, and six years, Alt. *A* will have an identical replacement. Alternative *B* will have a single identical replacement at the end of four years. Over what range of values of MARR is Alt. *B* the preferred alternative?

8-17 Consider the three alternatives:

	A	B	C
Initial cost	$1500	$1000	$2035
Annual benefit in each of first 5 years	250	250	650
Annual benefit in each of subsequent 5 years	450	250	145

Each alternative has a ten-year useful life and no salvage value. Based on a MARR of 15%, which alternative should be selected? Where appropriate, use an external interest rate of 10% to transform a cash flow to one sign change before proceeding with rate of return analysis.

8-18   A new 10,000 sq. meter warehouse next door to the Tyre Corporation is for sale for $450,000. The terms offered are a $100,000 down payment with the balance being paid in sixty equal monthly payments based on 15% interest. It is estimated that the warehouse would have a resale value of $600,000 at the end of five years.

   Tyre has the needed cash available and could buy the warehouse, but does not need all the warehouse space at this time. The Johnson Company has offered to lease half the new warehouse for $2500 a month.

   Tyre presently rents and utilizes 7000 sq. meters of warehouse space for $2700 a month. It has the option of reducing the rented space to 2000 sq. meters, in which case the monthly rent would be $1000 a month. Further, Tyre could cease renting warehouse space entirely. Tom Clay, the Tyre Corp. plant engineer, is considering three alternatives:

   1. Buy the new warehouse and lease the Johnson Company half the space. In turn, the Tyre-rented space would be reduced to 2000 sq. meters.

   2. Buy the new warehouse and cease renting any warehouse space.

   3. Continue as is, with 7000 sq. meters of rented warehouse space.

Based on a 20% minimum attractive rate of return, which alternative should be selected?

8-19   Consider the alternatives below:

	A	B	C
Initial cost	$100.00	$150.00	$200.00
Uniform annual benefit	10.00	17.62	55.48
Useful life, in years	infinite	20	5

Use present worth analysis, an 8% interest rate, and an infinite analysis period. Which alternative should be selected in each of the two following situations?

   1. Alternatives B and C are replaced at the end of their useful lives with identical replacements.

   2. Alternatives B and C are replaced at the end of their useful lives with alternatives that provide an 8% rate of return.

8-20 A problem often discussed in the engineering economy literature is the "oil-well pump problem":[2] Pump 1 is a small pump; Pump 2 is a larger pump that costs more, will produce slightly more oil, and will produce it more rapidly. If the MARR is 20%, which pump should be selected? Assume any temporary external investment of money earns 10% per year.

Year	Pump 1 ($000s)	Pump 2 ($000s)
0	–$100	–$110
1	+70	+115
2	+70	+30

8-21 Three mutually exclusive alternatives are being studied. If the MARR is 12%, which alternative should be selected?

Year	A	B	C
0	–$20,000	–$20,000	–$20,000
1	+10,000	+10,000	+5,000
2	+5,000	+10,000	+5,000
3	+10,000	+10,000	+5,000
4	+6,000	0	+15,000

8-22 The South End bookstore has an annual profit of $170,000. The owner is considering opening a second bookstore on the north side of the campus. He can lease an existing building for five years with an option to continue the lease for a second five year period. If he opens the second bookstore he expects the existing store will lose some business that will he gained by the new "The North End" bookstore. It will take $500,000 of store fixtures and inventory to open The North End. He believes that the two stores will have a combined profit of $260,000 a year after paying all the expenses of both stores.

The owner's economic analysis is based on a five year period. He will be able to recover this $500,000 investment at the end of five years by selling the store fixtures and inventory. The owner will not open The North End unless he can expect a 15% rate of return. What should he do? Show computations to justify your decision.

8-23 A paper mill is considering two types of pollution control equipment.

	Neutralization	Precipitation
Initial cost	$700,000	$500,000
Annual chemical cost	40,000	110,000
Salvage value	175,000	125,000
Useful life	5 years	5 years

---

[2] One of the more interesting exchanges of opinion about this problem is in Prof. Martin Wohl's "Common Misunderstandings About the Internal Rate of Return and Net Present Value Economic Analysis Methods;" and the associated discussion by Professors Winfrey, Leavenworth, Steiner, and Bergmann, published in *Evaluating Transportation Proposals,* Transportation Research Record 731, Transportation Research Board, Washington, D.C.

The firm wants a 12% rate of return on any avoidable increments of investment. Which equipment should be purchased?

8-24 A stockbroker has proposed two investments in low-rated corporate bonds paying high interest rates and selling below their stated value (in other words, junk bonds). Both bonds are rated as equally risky. Which, if any, of the bonds should you buy if your MARR is 25%?

Bond	Stated value	Annual interest payment	Current market price, including buying commission	Bond maturity[3]
Gen Dev	$1000	$ 94	$480	15 years
RJR	1000	140	630	15

8-25 Three mutually exclusive alternatives are being considered.

	A	B	C
Initial investment	$50,000	$22,000	$15,000
Annual net income	5,093	2,077	1,643
Computed rate of return	8%	7%	9%

Each alternative has a 20-year useful life with no salvage value. If the minimum attractive rate of return is 7%, which alternative should be selected?

8-26 A firm is considering five alternatives.

	1	2	3	4	5
Initial cost	$100.00	$130.00	$200.00	$330.00	Do nothing
Uniform annual net income	26.38	38.78	47.48	91.55	
Computed rate of return	10%	15%	6%	12%	

Each alternative has a five-year useful life. The firm's minimum attractive rate of return is 8%. Which alternative should be selected?

---

[3] At maturity the bondholder receives the last interest payment plus the bond stated value.

8-27   Alternatives

	A	*B*	*C*	*D*
Initial Cost	$2000	5000	4000	3000
Annual Benefit	800	500	400	1,300
Salvage Value	2000	1500	1,400	3,000
Life in Years	5	6	7	4
MARR Required	6%	6%	6%	6%

Find the best alternative using incremental ROR analysis.

8-28 Our cat Fred's summer kitty-cottage needs a new roof. He's considering the two proposals below and feels a 15 year analysis period is in line with his remaining lives. Which roof should he choose if his MARR = 12%? What is the actual value of the ROR on the incremental cost? (There is no salvage value for old roofs.)

	Thatch	*Slate*
First Cost	$20.00	$40.00
Annual Upkeep	5.00	2.00
Service Life	3 years	5 years

8-29 Don Garlits is a landscaper. He is considering the purchase of a new commercial lawn mower. Two machines are being considered, the Atlas and the Zippy. The table shown below provides all the necessary information for the two machines. The minimum attractive rate of return is 8%.

   a. Determine the rate of return on the Atlas mower (to the nearest 1%).

   b. Does the rate of return on the Zippy mower exceed the MARR?

   c. Use incremental rate of return analysis to decide which machine to purchase.

	Atlas	*Zippy*
Initial cost	$6,700	$16,900
Annual operation and maintenance cost	$1,500	$ 1,200
Annual benefit	$4,000	$ 4,500
Salvage value	$1,000	$ 3,500
Useful life	3	6

8-30 QZY, Inc. is evaluating new widget machines offered by three companies. The machines have the following characteristics:

	Company A	Company B	Company C
First cost	$15,000	$25,000	$20,000
Maintenance & operating	1,600	400	900
Annual benefit	8,000	13,000	9,000
Salvage value	3,000	6,000	4,500
Useful life, in years	4	4	4

MARR = 15 %. Using rate of return analysis, from which company, if any, should you purchase the widget machine?

8-31 The Croc Co. is considering a new milling machine. They have narrowed the choices down to three alternatives:

	Alternative		
	Deluxe	*Regular*	*Economy*
First cost	$220,000	$125,000	$75,000
Annual benefit	79,000	43,000	28,000
Maintenance & operating costs	38,000	13,000	8,000
Salvage value	16,000	6,900	3,000

All machines have a life of ten years. Using incremental rate of return analysis, which alternative, if any, should the company choose? MARR = 15%.

8-32 Wayward Airfreight, Inc. has asked you to recommend a new automatic parcel sorter. You have obtained the following bids:

	SHIP-R	*SORT-Of*	*U-SORT-M*
First cost	$184,000	$235,000	$180,000
Salvage value	38,300	44,000	14,400
Annual benefit	75,300	89,000	68,000
Yearly maintenance and operating cost	21,000	21,000	12,000
Useful life, in years	7	7	7

Using an MARR of 15% and a rate of return analysis, which alternative, if any, should be selected?

8-33 Build a spreadsheet to find the EAC of each roof in 8-28. Use the GOAL SEEK tool of Excel to find the IRR of the incremental investment.

8-34 Build a spreadsheet to find the EAW of each lawnmower in 8-29. Use the GOAL SEEK tool of Excel to find the IRR of the incremental investment.

Other Analysis Techniques

Chapter 9 examines four topics. They are:

- Future Worth Analysis;

- Benefit–Cost Ratio Analysis;

- Payback Period;

- Sensitivity and Breakeven Analysis.

Future worth analysis is very much like present worth analysis, dealing with then (future worth) rather than with *now* (present worth) situations.

Previously, we have written economic analysis relationships based on either:

PW of cost = PW of benefit   or   EUAC = EUAB

Instead of writing it in this form, we could define these relationships as

$$\frac{\text{PW of benefit}}{\text{PW of cost}} = 1 \qquad \text{or} \qquad \frac{\text{EUAB}}{\text{EUAC}} = 1$$

When economic analysis is based on these ratios, the calculations are called benefit–cost ratio analysis.

Payback period is an approximate analysis technique, generally defined as the time required for cumulative benefits to equal cumulative costs. Sensitivity describes the relative magnitude of a particular variation in one or more elements of a problem that is sufficient to change a particular decision. Closely related is *breakeven analysis,* which determines the conditions where two alternatives are equivalent. Thus, breakeven analysis is a form of *sensitivity analysis.*

Future Worth Analysis

We have seen how economic analysis techniques resolved alternatives into comparable units. In present worth analysis, the comparison was made in terms of the present consequences of taking the feasible courses of action. In annual cash flow analysis, the comparison was in terms of equivalent uniform annual costs (or benefits). We saw that we could easily convert from present worth to annual cash flow, and *vice versa.* But the concept of resolving alternatives into comparable units is not restricted to a present or annual comparison. The comparison may be made at any point in time. There are many situations where we would like to know what the *future* situation will be, if we take some particular course of action *now.* This is called *future worth analysis.*

EXAMPLE 9-1

Ron Jamison, a twenty-year-old college student, considers himself an average cigarette smoker, for he consumes about a carton a week. He wonders how much money he could accumulate by the time he reaches 65 if he quit smoking now and put his cigarette money into a savings account. Cigarettes cost $15 per carton. Ron expects that a savings account would earn 5% interest, compounded semi-annually. Compute Ron's future worth at age 65.

Solution:

Semi-annual saving $15/carton × 26 weeks = $390

Future worth (FW) = $A(F/A,2\frac{1}{2}\%,90)$ = 390(329.2) = $128,388 ■

EXAMPLE 9-2

An East Coast firm has decided to establish a second plant in Kansas City. There is a factory for sale for $850,000 that, with extensive remodeling, could be used. As an alternative, the company can buy vacant land for $85,000 and have a new plant constructed on the property. Either way, it will be three years before the company will be able to get the plant into production. The timing and cost of the various components for the factory are given in the cash flow table below.

Year	Construct new plant		Remodel available factory	
0	Buy land	$ 85,000	Purchase factory	$850,000
1	Design and initial construction costs	200,000	Design and remodeling costs	250,000
2	Balance of construction costs	1,200,000	Additional remodeling costs	250,000
3	Setup production equipment	200,000	Setup production equipment	250,000

If interest is 8%, which alternative results in the lower equivalent cost when the firm begins production at the end of the third year?

Solution:

New plant:

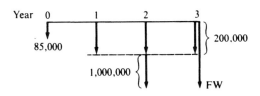

Future worth of cost (FW) $= 85,000(F/P,8\%,3) + 200,000(F/A,8\%,3)$

$$+ 1,000,000(F/P,8\%,1)$$

$$= \$1,836,000$$

Remodel available factory:

Future worth of cost (FW) $= 850,000(F/P,8\%,3) + 250,000(F/A,8\%,3)$

$$= \$1,882,000$$

The total cost of remodeling the available factory ($1,600,000) is smaller than the total cost of a new plant ($1,685,000). The timing of the expenditures, however, is less favorable than building the new plant. The new plant is projected to have the smaller future worth of cost, and thus is the preferred alternative. ∎

Benefit–Cost Ratio Analysis

At a given minimum attractive rate of return (MARR), we would consider an alternative acceptable, provided:

PW of benefits – PW of costs ≥ 0 or   EUAB – EUAC ≥ 0

These could also be stated as a ratio of benefits to costs, or

$$\text{Benefit–cost ratio } \frac{B}{C} = \frac{\text{PW of benefit}}{\text{PW of costs}} = \frac{\text{EUAB}}{\text{EUAC}} \geq 1$$

Rather than solving problems using present worth or annual cash flow analysis, we can base the calculations on the benefit–cost ratio, B/C. The criteria are presented in Table 9-1. We will illustrate "B/C analysis" by solving the same example problems worked by other economic analysis methods.

Table 9-1 BENEFIT–COST RATIO ANALYSIS

	Situation	*Criterion*
Fixed input	**Amount of money or other input resources are fixed**	**Maximize B/C**
Fixed output	**Fixed task, benefit, or other output to be accomplished**	**Maximize B/C**
Neither input nor output fixed	**Neither amount of money or other inputs nor amount of benefits or other outputs are fixed**	***Two alternatives:* Compute incremental benefit–cost ratio ($\Delta B/\Delta C$) on the increment of investment between the alternatives. If $\Delta B/\Delta C \geq 1$, choose higher-cost alternative; otherwise, choose lower-cost alternative.** ***Three or more alternatives:* Solve by benefit–cost ratio incremental analysis**

EXAMPLE 9-3

A firm is trying to decide which of two devices to install to reduce costs in a particular situation. Both devices cost $1000 and have useful lives of five years and no salvage value. Device A can be expected to result in $300 savings annually. Device B will provide cost savings of $400 the first year, but will decline $50 annually, making the second-year savings $350, the third-year savings $300, and so forth. With interest at 7%, which device should the firm purchase?

Solution: This problem was previously solved by present worth (Ex. 5-1), annual cash flow (Ex. 6-5), and rate of return (Ex. 7-9) analyses.

Device A:

$$\text{PW of cost} = \$1000$$

$$\text{PW of benefits} = 300(P/A,7\%,5) = 300(4.100) = \$1230$$

$$\frac{B}{C} = \frac{\text{PW of benefit}}{\text{PW of costs}} = \frac{1230}{1000} = 1.23$$

Device B:

$$\text{PW of cost} = \$1000$$

$$\text{PW of benefit} = 400(P/A,7\%,5) - 50(P/G,7\%,5)$$

$$= 400(4.100) - 50(7.647) = 1640 - 382 = 1258$$

$$\frac{B}{C} = \frac{\text{PW of benefit}}{\text{PW of costs}} = \frac{1258}{1000} = 1.26$$

To maximize the benefit–cost ratio, select Device B.   ■

EXAMPLE 9-4

Two machines are being considered for purchase. Assuming 10% interest, which machine should be bought?

	Machine X	*Machine Y*
Initial cost	$200	$700
Uniform, annual benefit	95	120
End-of-useful-life salvage value	50	150
Useful life, in years	6	12

Solution: Assuming a twelve-year analysis period, the cash flow table is:

Year	Machine X	Machine Y
0	−$200	−$700
1–5	+95	+120
6	$\left\{\begin{array}{l} +95 \\ -200 \\ \\ +50 \end{array}\right.$	+120
7–11	+95	+120
12	$\left\{\begin{array}{l} +95 \\ +50 \end{array}\right.$	$\begin{array}{l} +120 \\ +150 \end{array}$

We will solve the problem using

$$\frac{B}{C} = \frac{EUAB}{EUAC}$$

and considering the salvage value of the machines to be reductions in cost, rather than increases in benefits.

Machine X:

$\qquad$ EUAC = 200(*A/P*,10%,6) - 50(*A/F*, 10%,6)

$\qquad\qquad$ = 200(0.2296) - 50(0.1296) = 46 - 6 = $40

$\quad$ EUAB = $95

Note that this assumes the replacement for the last six years has identical costs. Under these circumstances, the EUAC for the first six years equals the EUAC for all twelve years.

Machine Y:

$\qquad$ EUAC = 700(*A/P*,10%,12) - 150(*A/F*,10%,12)

$\qquad\qquad$ = 700(0.1468) - 150(0.0468) = 103 - 7 = $96

$\quad$ EUAB = $120

Machine Y – Machine X:

$$\frac{\Delta B}{\Delta C} = \frac{120 - 95}{96 - 40} = \frac{25}{56} = 0.45$$

Since the incremental benefit–cost ratio is less than 1, it represents an undesirable increment of investment. We therefore choose the lower cost alternative—Machine X. If we had computed benefit–cost ratios for each machine, they would have been:

<table>
<tr><td>Machine X</td><td>Machine Y</td></tr>
<tr><td>$\dfrac{B}{C} = \dfrac{95}{40} = 2.38$</td><td>$\dfrac{B}{C} = \dfrac{120}{96} = 1.25$</td></tr>
</table>

The fact that B/C = 1.25 for Machine Y (the higher-cost alternative) must not be used as the basis for suggesting that the more expensive alternative should be selected. The incremental benefit–cost ratio, $\Delta B/\Delta C$, clearly shows that Y is a less desirable alternative than X. Also, we must not jump to the conclusion that the best alternative is always the one with the largest B/C ratio. This, too, may lead to incorrect decisions—as we shall see when we examine problems with three or more alternatives. ■

EXAMPLE 9-5

Consider the five mutually exclusive alternatives from Ex. 8-8 plus an additional alternative, F. They have twenty-year useful lives and no salvage value. If the minimum attractive rate of return is 6%, which alternative should be selected?

	A	B	C	D	E	F
Cost	$4000	$2000	$6000	$1000	$9000	$10,000
PW of benefit	7330	4700	8730	1340	9000	9,500
$\dfrac{B}{C} = \dfrac{\text{PW of benefits}}{\text{PW of cost}}$	1.83	2.35	1.46	1.34	1.00	0.95

Solution: Incremental analysis is needed to solve the problem. The steps in the solution are the same as the ones presented for incremental rate of return, except here the criterion is $\Delta B/\Delta C$, and the cutoff is 1, rather than ΔROR with a cutoff of MARR.

1. Be sure all the alternatives are identified.

2. (Optional) Compute the B/C ratio for each alternative. Since there are alternatives whose B/C ≥ 1, we will discard any with a B/C < 1. Discard Alt. F.

3. Arrange the remaining alternatives in ascending order of investment.

	D	B	A	C	E
Cost (= PW of cost)	$1000	$2000	$4000	$6000	$9000
PW of benefits	1340	4700	7330	8730	9000
B/C	1.34	2.35	1.83	1.46	1.00

	Increment B–D	Increment A–B	Increment C–A
ΔCost	$1000	$2000	$2000
ΔBenefit	3360	2630	1400
ΔB/ΔC	3.36	1.32	0.70

4. Examine each separable increment of investment. If ΔB/ΔC < 1, the increment is not attractive. If ΔB/ΔC ≥ 1, the increment of investment is desirable. The increments B–D and A–B are desirable. Thus, of the first three alternatives (D, B, and A), Alt. A is the preferred alternative. Increment C–A is not attractive as ΔB/ΔC = 0.70, which indicates that of the first four alternatives (D, B, A, and C), A continues as the best of the four. Now we want to decide between A and E, which we'll do by examining the increment of investment that represents the difference between these alternatives.

	Increment E–A
ΔCost	$5000
ΔBenefit	1670
ΔB/ΔC	0.33

The increment is undesirable. We choose Alt. A as the best of the six alternatives. One should note that the best alternative in this example does not have the highest B/C ratio. ∎

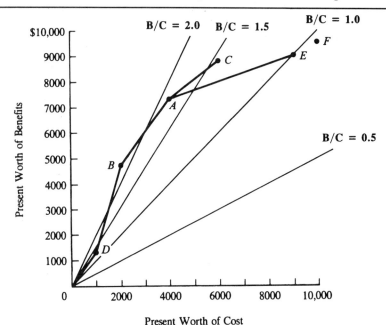

Figure 9-1 Benefit–cost ratio graph of Example 9-5.

Benefit–cost ratio analysis may be graphically represented. Figure 9-1 is a graph of Ex. 9-5. We see that *F* has a B/C ratio < 1 and can be discarded. Alt. *D* is the starting point for examining the separable increments of investment. The slope of Line *B–D* indicates a $\Delta B/\Delta C$ ratio of > 1. This is also true for Line *A–B*. Increment *C–A* has a slope much flatter than B/C = 1, indicating an undesirable increment of investment. Alt. *C* is therefore discarded and *A* retained. Increment *E–A* is similarly unattractive. Alt. *A* is therefore the best of the six alternatives.

Note particularly two additional things about Fig. 9-1: first, even if alternatives with a B/C ratio < 1 had not been initially excluded, they would have been systematically eliminated in the incremental analysis. Since this is the case, it is not essential that the B/C ratio be computed for each alternative as an initial step in incremental analysis. Nevertheless, it seems like an orderly and logical way to approach a multiple-alternative problem. Second, Alt. *B* had the highest B/C ratio (B/C = 2.35), but it is not the best of the six alternatives. We saw this same situation in rate of return analysis of three or more alternatives. The reason is the same in both analysis situations. We seek to maximize the *total* profit, not the profit rate.

Continuous Alternatives

There are times when the feasible alternatives are a continuous function. The height of a dam in Chapter 8 was an example of this situation. It was possible to build the dam anywhere from 200 to 500 feet high.

In many situations, the projected capacity of an industrial plant can be varied continuously over some feasible range. In these cases, we seek to add increments of investment where $\Delta B/\Delta C \geq 1$ and avoid increments where $\Delta B/\Delta C < 1$. The optimal size of such a project is where $\Delta B/\Delta C = 1$. Figure 9-2a shows the line of feasible alternatives with their costs and benefits. This may represent a lot of calculations to locate points through which the line passes.

Figure 9-2b shows how the incremental benefit-cost ratio ($\Delta B/\Delta C$) changes as one moves along the line of feasible alternatives. In Fig. 9-2b, the ratio of Incremental net present worth *to* Incremental cost ($\Delta NPW/\Delta C$) is also plotted. As expected, we are adding increments of NPW as long as $\Delta B/\Delta C > 1$. Finally, in Fig. 9-2c, we see the plot of (total) NPW *vs.* the size of the project.

This three-part figure demonstrates that both present worth analysis and benefit–cost ratio analysis lead to the same optimal decision. We saw in Chapter 8 that rate of return and present worth analysis led to identical decisions. Any of the exact analysis methods—present worth, annual cash flow, rate of return, or benefit–cost ratio—will lead to the same decision. Benefit–cost ratio analysis is extensively used in economic analysis at all levels of government.

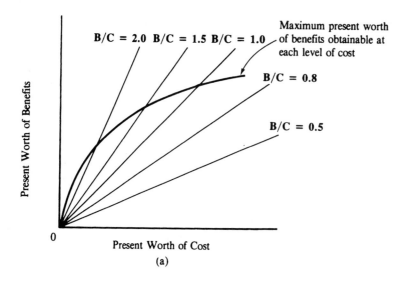

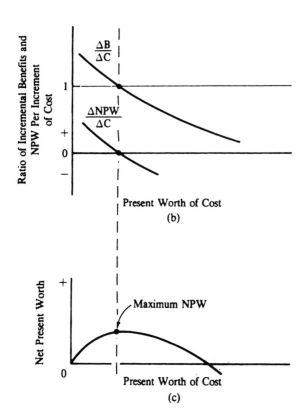

Figure 9-2a–c Selecting optimal size of project.

Payback Period

Payback period is the period of time required for the profit or other benefits from an investment to equal the cost of the investment. This is the general definition for payback period, but there are other definitions. Others consider depreciation of the investment, interest, and income taxes; they, too, are simply called "payback period." For now, we will limit our discussion to the simplest form.

> **Payback period is the period of time required for the profit or other benefits of an investment to equal the cost of the investment.**

The criterion in all situations is to minimize the payback period. The computation of payback period is illustrated in Examples 9-6 and 9-7.

EXAMPLE 9-6

The cash flows for two alternatives are as follows:

Year	A	B
0	−$1000	−$2783
1	+200	+1200
2	+200	+1200
3	+1200	+1200
4	+1200	+1200
5	+1200	+1200

You may assume the benefits occur throughout the year rather than just at the end of the year. Based on payback period, which alternative should be selected?

Solution:

Alternative A: Payback period is the period of time required for the profit or other benefits of an investment to equal the cost of the investment. In the first two years, only $400 of the $1000 cost is recovered. The remaining $600 cost is recovered in the first half of Year 3. Thus the payback period for Alt. *A* is 2.5 years.

Alternative B: Since the annual benefits are uniform, the payback period is simply

$2783/$1200 per year = 2.3 years

To minimize the payback period, choose Alt. *B*. ■

EXAMPLE 9-7

A firm is trying to decide which of two alternate weighing scales it should install to check a package-filling operation in the plant. If both scales have a six-year life, which one should be selected? Assume an 8% interest rate.

Alternative	Cost	Uniform annual benefit	End-of-useful-life salvage value
Atlas scale	$2000	$450	$100
Tom Thumb scale	3000	600	700

Solution:
Atlas scale:

$$\text{Payback period} = \frac{\text{Cost}}{\text{Uniform annual benefit}} = \frac{2000}{450} = 4.4 \text{ years}$$

Tom Thumb scale:

$$\text{Payback period} = \frac{\text{Cost}}{\text{Uniform annual benefit}} = \frac{3000}{600} = 5 \text{ years}$$

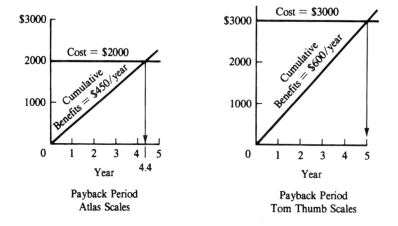

Figure 9-3 Payback period plots for Example 9-7.

Figure 9-3 illustrates the situation. To minimize payback period, select the Atlas scale. ■

There are four important points to be understood about payback period calculations:

1. **This is an approximate, rather than an exact, economic analysis calculation.**

2. **All costs and all profits, or savings of the investment, prior to payback are included *without* considering differences in their timing.**

3. **All the economic consequences beyond the payback period are completely ignored.**

4. **Being an approximate calculation, payback period may or may not select the correct alternative. That is, the payback period calculations may select a different alternative from that found by exact economic analysis techniques.**

This last point—that payback period may select the *wrong* alternative—was illustrated by Ex. 9-7. Using payback period, the Atlas scale appears to be the more attractive alternative. Yet, when this same problem was solved by the present worth method (Ex. 5-4), the Tom Thumb scale was the chosen alternative. A review of the problem reveals the reason for the different conclusions. The $700 salvage value at the end of six years for the Tom Thumb scale is a significant benefit. The salvage value occurs after the payback period, so it was ignored in the payback calculation. It *was* considered in the present worth analysis, with the result that Tom Thumb scale was more desirable.

But if payback period calculations are approximate, and are even capable of selecting the wrong alternative, why is the method used at all? There are two primary answers: first, the calculations can be readily made by people unfamiliar with economic analysis. One does not need to know how to use gradient factors, or even to have a set of Compound Interest Tables. Second, payback period is a readily understood concept. Earlier we pointed out that this was also an advantage to rate of return.

Moreover, payback period *does* give us a useful measure, telling us how long it will take for the cost of the investment to be recovered from the benefits of the investment. Businesses and industrial firms are often very interested in this time period: a rapid return of invested capital means that it can be re-used sooner for other purposes by the firm. But one must not confuse the *speed* of the return of the investment, as measured by the payback period, with economic *efficiency*. They are two distinctly separate concepts. The former emphasizes the quickness with which invested monies return to a firm; the latter considers the overall profitability of the investment.

We can create another situation to illustrate how selecting between alternatives by the payback period criterion may result in an unwise decision.

EXAMPLE 9-8

A firm is purchasing production equipment for a new plant. Two alternative machines are being considered for a particular operation.

	Tempo machine	Dura machine
Installed cost	$30,000	$35,000
Net annual benefit after deducting all annual expenses	$12,000 the first year, *declining* $3000 per year thereafter	$1000 the first year, increasing $3000 per year thereafter
Useful life, in years	4	8

Neither machine has any salvage value. Compute the payback period for each of the alternatives.

Solution:

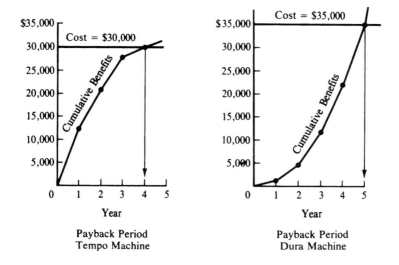

Figure 9-4 Payback period plots for Example 9-7.

The Tempo machine has a declining annual benefit, while the Dura has an increasing annual benefit. Figure 9-4 shows the Tempo has a four-year payback period and the Dura has a five-year payback period. To minimize the payback period, the Tempo is selected.

Now, as a check on the payback period analysis, compute the rate of return for each alternative. Assume the minimum attractive rate of return is 10%.

Solution based on rate of return: The cash flows for the two alternatives are as follows:

Year	Tempo machine	Dura machine
0	–$30,000	–$35,000
1	+12,000	+1,000
2	+9,000	+4,000
3	+6,000	+7,000
4	+3,000	+10,000
5	0	+13,000
6	0	+16,000
7	0	+19,000
8	0	+22,000
Σ =	0	+57,000

Tempo machine: Since the sum of the cash flows for the Tempo machine is zero, we see immediately that the $30,000 investment just equals the subsequent benefits. The resulting rate of return is 0%.

Dura machine:

$$35,000 = 1000(P/A, i, 8) + 3000(P/G, i, 8)$$

Try $i = 20\%$:

$$35,000 \stackrel{?}{=} 1000(3.837) + 3000(9.883)$$

$$\stackrel{?}{=} 3837 + 29,649 = 33,486$$

The 20% interest rate is too high. Try $i = 15\%$:

$$35,000 \stackrel{?}{=} 1000(4.487) + 3000(12.481)$$

$$\stackrel{?}{=} 4487 + 37,443 = 41,930$$

This time, the interest rate is too low. Linear interpolation would show that the rate of return is approximately 19%.

Using an exact calculation—rate of return—it is clear that Tempo is not economically very attractive. Yet it was this alternative, and not the Dura machine, that was preferred based on the payback period calculations. On the other hand, the shorter payback period for Tempo does give a measure of the speed of the return of the investment not found in the Dura. The conclusion to be drawn is that ***liquidity*** and ***profitability*** may be two quite different criteria. ■

From the discussion and the examples, we see that payback period can be helpful in providing a measure of the speed of the return of the investment. This might be quite important, for example, for a company that is short of working capital, or one where there are rapid changes in technology. This must not, however, be confused with a careful economic analysis. We have shown that a short payback period does not always mean that the investment is desirable. Thus, payback period should not be considered a suitable replacement for accurate economic analysis calculations.

Sensitivity And Breakeven Analysis

Since many data gathered in solving a problem represent *projections* of future consequences, there may be considerable uncertainty regarding the accuracy of that data. As the desired result of the analysis is decision making, an appropriate question is, "To what extent do variations in the data affect my decision?" When small variations in a particular estimate would change selection of the alternative, the decision is said to be *sensitive to the estimate.* To better evaluate the impact of any particular estimate, we compute "what variation to a particular estimate would be necessary to change a particular decision." This is called *sensitivity analysis.*

An analysis of the sensitivity of a problem's decision to its various parameters highlights the important and significant aspects of that problem. For example, one might be concerned that the estimates for annual maintenance and future salvage value in a particular problem may vary substantially. Sensitivity analysis might indicate that the decision is insensitive to the salvage-value estimate over the full range of possible values. But, at the same time, we might find that the decision is sensitive to changes in the annual-maintenance estimate. Under these circumstances, one should place greater emphasis on improving the annual maintenance estimate and less on the salvage-value estimate.

As indicated at the beginning of this chapter, breakeven analysis is a form of sensitivity analysis. To illustrate the sensitivity of a decision between alternatives to particular estimates, breakeven analysis is often presented as a *breakeven chart.*

Sensitivity and breakeven analysis frequently are useful in engineering problems called *stage construction.* Should a facility be constructed now to meet its future full-scale requirement, Or should it be constructed in stages as the need for the increased capacity arises? Three examples of this situation are:

- Should we install a cable with 400 circuits now or a 200-circuit cable now and another 200-circuit cable later?

- A 10-cm water main is needed to serve a new area of homes. Should the 10-cm main be installed now, or should a 15-cm main be installed to later provide an adequate water supply to adjoining areas when other homes are built?

■ An industrial firm needs a 10,000-m^2 warehouse now and estimates that it will need an additional 10,000 m^2 in four years. The firm could have a 10,000-m^2 warehouse built now and later enlarged, or have the 20,000-m^2 warehouse built right away.

Examples 9-9 and 9-10 illustrate sensitivity and breakeven analysis.

EXAMPLE 9-9

Consider the following situation where a project may be constructed to full capacity now or may be constructed in two stages.

Construction costs:

Two-stage construction	
Construct first stage now	$100,000
Construct second stage *n* years from now	120,000
Full-capacity construction	
Construct full capacity now	140,000

Other factors:

1. All facilities will last until forty years from now regardless of when they are installed; at that time they will have zero salvage value.

2. The annual cost of operation and maintenance is the same for both two stage construction and full-capacity construction.

3. Assume an 8% interest rate.

Plot a graph showing "age when second stage is constructed" *vs.* "costs for both alternatives." Mark the breakeven point. What is the sensitivity of the decision to second-stage construction sixteen or more years in the future?

Solution: Since we are dealing with a common analysis period, the calculations may be either annual cost or present worth. Present worth calculations appear simpler and are used here:

Construct full capacity now:

PW of cost = $140,000

Two-stage construction:

First stage constructed now and the second stage to be constructed *n* years hence. Compute the PW of cost for several values of *n* (years).

PW of cost = $100{,}000 + 120{,}000(P/F, 8\%, n)$

$n = 5$	PW $= 100{,}000 + 120{,}000(0.6806) = \$181{,}700$
$n = 10$	PW $= 100{,}000 + 120{,}000(0.4632) = \ 155{,}600$
$n = 20$	PW $= 100{,}000 + 120{,}000(0.2145) = \ 125{,}700$
$n = 30$	PW $= 100{,}000 + 120{,}000(0.0994) = \ 111{,}900$

These data are plotted in the form of a breakeven chart in Fig. 9-5.

Figure 9-5 portrays the PW of cost for the two alternatives. The *x*-axis variable is the *time* when the second stage is constructed. We see that the PW of cost for two-stage construction naturally decreases as the time for the second stage is deferred. The one-stage construction (full capacity now) is unaffected by the *x*-axis variable and, hence, is a horizontal line on the graph.

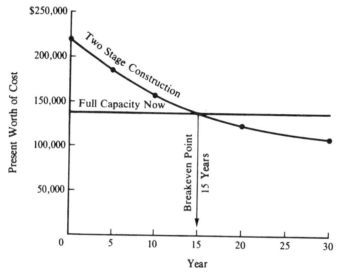

Age When Second Stage Constructed

Figure 9-5 Breakeven chart for Example 9-9.

The breakeven point on the graph is the point at which both alternatives have equivalent costs. We see that if, in two-stage construction, the second stage is deferred for 15 years, then the PW of cost of two-stage construction is equal to one-stage construction; Year 15 is the breakeven point. The graph also shows that if the second stage was needed prior to Year 15, then one-stage construction, with its smaller PW of cost, would be preferred. On the other hand, if the second stage would not be required until after fifteen years, two-stage construction is preferred.

The decision on how to construct the project is sensitive to the age at which the second stage is needed *only* if the range of estimates includes 15 years. For example, if one estimated that the second stage capacity would be needed somewhere between five and ten years hence, the decision is insensitive to that estimate. For any value within that range, the decision does not change. The more economical thing to do is to build the full capacity now. But, if the second-stage capacity were needed sometime between, say, 12 and 18 years, the decision would be sensitive to the estimate of when the full capacity would be needed.   ■

One question posed by Ex. 9-9 is *how* sensitive the decision is to the need for the second stage at 16 years or beyond. The graph shows that the decision is insensitive. In all cases for construction on or after 16 years, two-stage construction has a lower PW of cost.

EXAMPLE 9-10

Example 8-3 posed the following situation. Three mutually exclusive alternatives are given, each with a 20-year life and no salvage value. The minimum attractive rate of return is 6%.

	A	B	C
Initial cost	$2000	$4000	$5000
Uniform annual benefit	410	639	700

In Ex. 8-3 we found that Alt. *B* was the preferred alternative. Here we would like to know how sensitive the decision is to our estimate of the initial cost of *B*. If *B* is preferred at an initial cost of $4000, it will continue to be preferred at any smaller initial cost. But *how much* higher than $4000 can the initial cost be and still have *B* the preferred alternative? The computations may be done several different ways. With neither input nor output fixed, maximizing net present worth is a suitable criterion.

Alternative A:

$$\text{NPW} = \text{PW of benefit - PW of cost}$$

$$= 410(P/A,6\%,20) - 2000$$

$$= 410(11.470) - 2000 = 2703$$

Alternative B: Let x = Initial cost of *B*.

$$\text{NPW} = 639(P/A,6\%,20) - x$$

$$= 639(11.470) - x$$

$$= 7329 - x$$

Alternative C:

$$NPW = 700(P/A,6\%,20) - 5000$$

$$= 700(11.470) - 5000 = 3029$$

For the three alternatives, we see that *B* will only maximize NPW as long as its NPW is greater than 3029.

$$3029 = 7329 - x$$

$$x = 7329 - 3029 = 4300$$

Therefore, *B* is the preferred alternative if its initial cost does not exceed $4300.

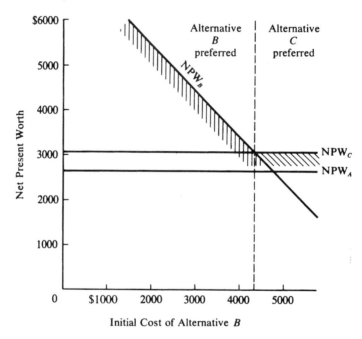

Figure 9-6 Breakeven chart for Example 9-10.

Figure 9-6 is a breakeven chart for the three alternatives. Here the criterion is to maximize NPW; as a result, the graph shows that *B* is preferred if its initial cost is less than $4300. At an initial cost above $4300, *C* is preferred. We have a breakeven point at $4300. When *B* has an initial cost of $4300, *B* and *C* are equally desirable. ■

Sensitivity analysis and breakeven point calculations can be very useful in identifying how different estimates affect the calculations. It must be recognized that these calculations assume all parameters except one are held constant, and the sensitivity of the decision to that one variable is evaluated. Later we will look further at the impact of parameter estimates on decision making.

Graphing with Spreadsheets for Sensitivity and Breakeven Analysis

Chapter 4 introduced drawing *xy* plots with spreadsheets, and Chapter 7 reviewed this for plotting present worth versus *i*. The Chapter 7 plot is an example of breakeven analysis, as it is used to determine at what interest rate does the project break even or have a present worth of 0. This section will present some of the spreadsheet tools and options that can make the *xy* plots more effective and attractive.

The spreadsheet tools and options can be used to:
- Modify the *x* or *y* axes
    - Specify the minimum or maximum value
    - Specify at what value the other axis intersects (default is 0)
- Match line types to data
    - Use line types to distinguish one curve from another
    - Use markers to show real data
    - Use lines without markers to plot curves – straight segments or smooth curves
- Match chart colors to how displayed
    - Color defaults are fine for color computer screen
    - Color defaults are ok for color printers
    - Black and white printing is better with editing (use line types not colors)
- Annotate the graph
    - Add text, arrows, and lines to graphs
    - Add data labels

In most cases the menus of Excel are self-explanatory, so the main step is deciding what you want to achieve, and then you just look for the way to do it. Left clicks are used to select the item to modify, and right clicks are used to bring up the options for that item. Example 9-11 illustrates this process.

Example 9-11

The staged construction choice described in Example 9-9 used a broad range of *x* values for the *x*-axis. Create a graph that focuses on the 10 to 20 year period, and that is designed for printing in a report. The costs are:

Year	Full capacity	Two stage
0	$140,000	$100,000
n	0	$120,000

Solution

The first step is to create a table of values that shows the present worth of the costs for different values of n – the length of time until the second stage or full capacity is needed. Notice that the full capacity is calculated at $n = 0$. The only reason to calculate the corresponding value for staged construction is to see if the formula is properly entered, since building both stages at the same time will not really cost $220,000. The values for staged construction at 5, 10, 20, and 30 years check with the values in Example 9-9.

The next step is to select cells A8:C13, which includes the x values and two series of y values. Then the ChartWizard tool is selected. In the first step, the *xy (scatter)* plot is selected with the option of smoothed lines without markers. In step two no action is required since the cells A8:C13 were selected first. In step three labels are added for the x and y axes. In step four the chart is moved around on the worksheet page, so that it does not overlap with the data. The result is shown in Figure 9-7a (except that this is the color screen version printed to a black and white printer).

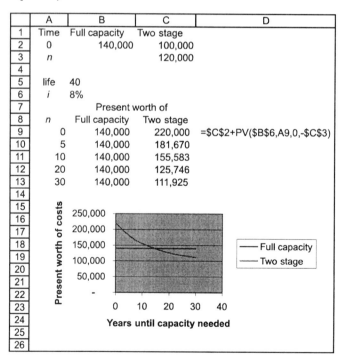

Figure 9-7a Automatic graph from spreadsheet.

Our first step in cleaning up the graph is to delete the formula in cell C9, since two stage construction will not be done at time 0. We also delete the label in the adjacent cell which explains the formula. Then we create a new label for cell C10. As shown in the appendix, the easy way to create that label is to insert an apostrophe, ', as the first entry in cell C10. This converts the formula to a label which we can copy to D10. Then we delete the apostrophe in cell C10.

The axis scales must be modified to focus on the area of concern. Select the x-axis and change the minimum from automatic to 10 and the maximum to 20. Select the y-axis and change the minimum to 125,000 and the maximum to 160,000.

Left click on the plot area to select it. Then right click to bring up the options. Select format plot area and change the area pattern to none. This will eliminate the grey fill that made Figure 9-7a difficult to read.

Left click on the two stage curve to select it. Then right click for the options. Format the data series using the "patterns" tab. Change the line style from solid to dashed, the line color from automatic to black, and increase the line weight. Similarly, increase the line weight for the full capacity line. Finally, select a grid line and change the line style to dotted. The result is far easier to read in black and white.

To further improve the graph, we can replace the legend with annotations on the graph. Left click somewhere in the white area around the graph to select "chart area." Right click and then choose the chart options on the menu. The legends tab will let us delete the legend by turning "show legend" off. Similarly, we can turn the x-axis gridlines on. The line style for these gridlines should be changed to match the y-axis gridlines. This allows us to see that the breakeven time is between 14 and 15 years.

So that the graph is less busy, change the scale on the x-axis so the interval is 5 years rather than automatic. Also eliminate the gridlines for the y-axis (by selecting the chart area, chart options, and gridlines tab). The graph size can be increased so it is easier to read as well. This may require specifying an interval of 10000 for the scale of the y-axis.

Finally to add the labels for the full capacity curve and the two stage curve, find the toolbar for graphics which is open when the chart is selected (probably along the bottom of the spreadsheet). Select the text box icon, and click on a location close to the two stage chart. Type in the label for two stage construction. Notice how including a return and a few spaces can shape the label so it fits the slanted line. Add the label for full construction. Figure 9-7b is the result.

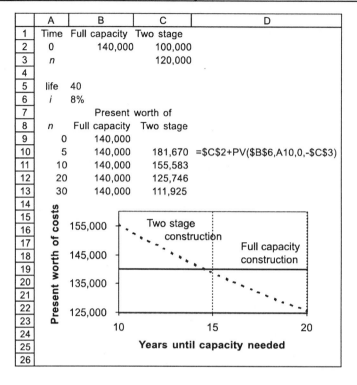

Figure 9-7b Spreadsheet with improved graph.

Summary

In this chapter, we have looked at four new analysis techniques.

Future Worth. When the point in time at which the comparison between alternatives will be made is in the future, the calculation is called future worth. This is very similar to present worth, which is based on the present, rather than a future point in time.

Benefit–Cost Ratio Analysis. This technique is based on the ratio of benefits to costs using either present worth or annual cash flow calculations. The method is graphically similar to present worth analysis. When neither input nor output is fixed, incremental benefit–cost ratios ($\Delta B/\Delta C$) are required. The method is similar in this respect to rate of return analysis. Benefit–cost ratio analysis is often used at the various levels of government.

Payback Period. Here we define payback as the period of time required for the profit or other benefits of an investment to equal the cost of the investment. Although simple to use and simple to understand, payback is a poor analysis technique for ranking alternatives. While it provides a measure of the speed of the return of the investment, it is not an accurate measure of the profitability of an investment.

Sensitivity and Breakeven Analysis. These techniques are used to see how sensitive a decision is to estimates for the various parameters. Breakeven analysis is done to locate conditions where the alternatives are equivalent. This is often presented in the form of breakeven charts. Sensitivity analysis is an examination of a range of values for some parameter to determine their effect on a particular decision.

Problems

9-1   A twenty-year-old student decided to set aside $100 on his 21st birthday for investment. Each subsequent year through his 55th birthday, he plans to increase the sum for investment on a $100 arithmetic gradient. He will not set aside additional money after his 55th birthday. If he can achieve a 12% rate of return on his investment, how much will he have accrued on his 65th birthday? (*Answer:* $1,160,700)

9-2   You have an opportunity to purchase a piece of vacant land for $30,000 cash. If you bought it, you would plan to hold the property for 15 years and then sell it at a profit. During this period, you would have to pay annual property taxes of $600. You would have no income from the property. Assuming that you would want a 10% rate of return from the investment, at what net price would you have to sell it 15 years hence?   (*Answer:* $144,373)

9-3   An individual's salary is now $32,000 per year and he anticipates retiring in thirty more years. If his salary is increased by $600 each year and he deposits 10% of his yearly salary into a fund that earns 7% interest compounded annually, what will be the amount accumulated at the time of his retirement?

9-4   A business executive is offered a management job at Generous Electric Company. They offer to give him a five-year contract which calls for a salary of $62,000 per year, plus 600 shares of their stock at the end of the five years. This executive is currently employed by Fearless Bus Company and they, too, have offered him a five-year contract. It calls for a salary of $65,000, plus 100 shares of Fearless stock each year. The stock is currently worth $60 per share and pays an annual dividend of $2 per share. Assume end-of-year payments of salary and stock. Stock dividends begin one year after the stock is received. The executive believes that the value of the stock and the dividend will remain constant. If the executive considers 9% a suitable rate of return in this situation, what must the Generous Electric stock be worth per share to make the two offers equally attractive?
    (*Answer:* $83.76)

9-5   Tom Jackson is preparing to buy a new car. He knows it represents a large expenditure of money, so he wants to do an analysis to see which of two cars is more economical. Alternative *A* is an American-built compact car. It has an initial cost of $8900 and operating costs of 9¢ per km, excluding depreciation. Tom checked automobile resale statistics. From them he estimates the American automobile can be resold at the end of three years for $1700. Alt. *B* is a foreign-built Fiasco. Its initial cost is $8000, The operating cost, also excluding depreciation, is 8¢ per km. How

low could the resale value of the Fiasco be to provide equally economical transportation? Assume Tom will drive 12,000 km per year and considers 8% as an appropriate interest rate.
 (*Answer:* $175)

9-6 A newspaper is considering purchasing locked vending machines to replace open newspaper racks for the sale of its newspapers in the downtown area. The newspaper vending machines cost $45 each. It is expected that the annual revenue from selling the same quantity of newspapers will increase $12 per vending machine. The useful life of the vending machine is unknown.

   a. To determine the sensitivity of rate of return to useful life, prepare a graph for rate of return *vs.* useful life for lives up to eight years.

   b. If the newspaper requires a 12% rate of return, what minimum useful life must it obtain from the vending machines?

   c. What would be the rate of return if the vending machines were to last indefinitely?

9-7 Able Plastics, an injection molding firm, has negotiated a contract with a national chain of department stores. A plastic pencil box is to be produced for a two-year period. Able Plastics has never produced this item before and, therefore, requires all new dies. If the firm invests $67,000 for special removal equipment to unload the completed pencil boxes from the molding machine, one machine operator can be eliminated. This would save the firm $26,000 per year. The removal equipment has no salvage value and is not expected to be used after the two-year production contract is completed. The equipment, although useless, would be serviceable for about 15 years. You have been asked to do a payback period analysis on whether or not to purchase the special removal equipment. What is the payback period? Should Able Plastics buy the removal equipment?

9-8 A cannery is considering installing an automatic case-sealing machine to replace current hand methods. If they purchase the machine for $3800 in June, at the beginning of the canning season, they will save $400 per month for the four months each year that the plant is in operation. Maintenance costs of the case-sealing machine is expected to be negligible. The case-sealing machine is expected to be useful for five annual canning seasons and will have no salvage value at the end of that time. What is the payback period? Calculate the nominal annual rate of return based on the estimates,

9-9 Consider three alternatives:

	A	B	C
First cost	$50	$150	$110
Uniform annual benefit	28.8	39.6	39.6
Useful life, in years*	2	6	4
Computed rate of return	10%	15%	16.4%

   *At the end of its useful life, an identical alternative (with the same cost, benefits, and useful life) may be installed.

All of the alternatives have no salvage value. If the MARR is 12%, which alternative should be selected?

   a. Solve the problem by future worth analysis.

   b. Solve the problem by benefit–cost ratio analysis.

   c. Solve the problem by payback period.

   d. If the answers in parts *a*, *b*, and *c* differ, explain why this is the case.

9-10   An investor is considering buying some land for $100,000 and constructing an office building on it. Three different buildings are being analyzed.

	Building height		
	2 stories	5 stories	10 stories
Cost of building (excluding cost of land)	$400,000	$800,000	$2,100,000
Resale value* of land and building at end of 20-year analysis period	200,000	300,000	400,000
Annual rental income after deducting all operating expenses	70,000	105,000	256,000

*Resale value to be considered a reduction in cost, rather than a benefit.

Using benefit–cost ratio analysis and an 8% MARR, determine which alternative, if any, should be selected.

9-11 Using benefit-cost ratio analysis, determine which one of the three mutually exclusive alternatives should be selected.

	A	B	C	do nothing
First cost	$560	$340	$120	0
Uniform annual benefit	140	100	40	0
Salvage value	40	0	0	0

Each alternative has a six-year useful life. Assume a 10% MARR.

9-12 Compute *F* for the diagram below.

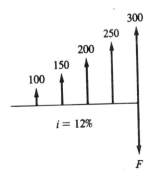

(Answer: *F* = $1199)

9-13 Sally deposited $100 a month in her savings account for 24 months. For the next five years she made no deposits. How much was in the savings account at the end of the seven years, if the account earned 6% annual interest, compounded monthly?
  (*Answer:* $3430.78)

9-14 For the diagram, compute F.

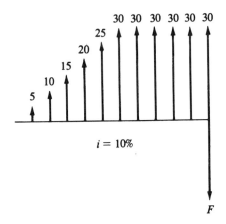

9-15 For a 12% interest rate, compute the value of F in the diagram.

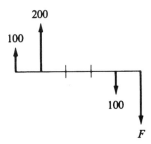

9-16 Stamp collecting has become an increasingly popular—and expensive—hobby. One favorite method is to save plate blocks (usually four stamps with the printing plate number in the margin) of each new stamp as it is issued by the post office. But with the rising postage rates and increased numbers of new stamps being issued, this collecting plan costs more each year.
  Stamp collecting, however, may have been a good place to invest money over the last ten years, as the demand for stamps previously issued has caused resale prices to increase 18% each year. Suppose a collector purchased $100 worth of stamps ten years ago, and increased his purchases by $50 per year in each subsequent year. After ten years of stamp collecting, how much would you now estimate his collection could be sold for today?

9-17 For the diagram, compute F.

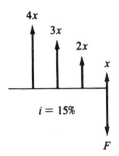

9-18 In the early 1980's, planners were examining alternate sites for a new airport to serve London. In their economic analysis, they computed the value of the structures that would need to be removed from various airport sites. At one airport site, the 12th Century Norman church of St. Michaels, in the village of Stewkley, would be demolished. The planners used the value of the fire insurance policy on the church—a few thousand pounds sterling—as the value of the structure.

An outraged antiquarian wrote to the London *Times* that an equally plausible computation would be to assume that the original cost of the church (estimated at 100 pounds sterling) be increased at the rate of 10% per year for 800 years. Based on his proposal, what would be the present value of St. Michaels? (*Note:* There was such public objection to tearing down the church, it was spared.)

9-19 Compute F for the figure.

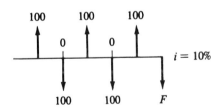

9-20 Bill made a budget and planned to deposit $150 a month in a savings account, beginning September 1st. He did this, but on the following January 1st, he reduced the monthly deposits to $100 a month. In all he made 18 deposits, four at $150 and 14 at $100. If the savings account paid 6% interest, compounded monthly, how much was in his savings account immediately after he made the last deposit? (*Answer:* $2094.42)

9-21 A company deposits $1000 in a bank at the beginning of each year for six years. The account earns 8% interest, compounded every six months. How much will be in the account at the end of six years? Make a careful, accurate computation.

9-22 Don Ball is a 55-year-old engineer. According to mortality tables, a male at 55 has an average life expectancy of 21 more years. In prior years, Don has accumulated $48,500 including interest, toward his retirement. He is now adding $5000 per year to his retirement fund. The fund earns 12% interest. Don's goal is to retire when he can obtain an annual income from his retirement fund of $20,000 per year, assuming he lives to age 76. He will make no provision for a retirement income after age 76. What is the youngest age at which Don can retire, based on his criteria?

9-23 The three alternatives shown each have a five-year useful life. If the MARR is 10%, which alternative should be selected? Solve the problem by benefit—cost ratio analysis.

	A	B	C
Cost	$600.0	$500.0	$200.0
Uniform annual benefit	158.3	138.7	58.3

(*Answer:* B)

9-24 Consider four alternatives that each have an eight-year useful life:

	A	B	C	D
Cost	$100.0	$80.0	$60.0	$50.0
Uniform annual benefit	12.2	12.0	9.7	12.2
Salvage value	75.0	50.0	50.0	0

If the MARR is 8%, which alternative should be selected? Solve the problem by benefit–cost ratio analysis.

9-25 Three mutually exclusive alternatives are being considered:

	A	B	C
Initial cost	$500	$400	$300
Benefit at end of the first year	200	200	200
Uniform benefit at end of subsequent years	100	125	100
Useful life, in years	6	5	4

At the end of its useful life, an alternative is *not* replaced. If the MARR is 10%, which alternative should be selected:

   a. based on the payback period?

   b. based on benefit-cost ratio analysis?

9-26 Consider the three alternatives that have a ten-year useful life. If the MARR is 10%, which alternative should be selected? Solve the problem by benefit–cost ratio analysis.

	A	B	C
Cost	$800	$300	$150
Uniform annual benefit	142	60	33.5

9-27 Using benefit–cost ratio analysis, a five-year useful life, and a 15% MARR, determine which of the following five alternatives should be selected.

	A	B	C	D	E
Cost	$100	$200	$300	$400	$500
Uniform annual benefit	37	69	83	126	150

9-28 The cash flows for three alternatives are as shown:

Year	A	B	C
0	–$500	–$600	–$900
1	–400	–300	0
2	200	350	200
3	250	300	200
4	300	250	200
5	350	200	200
6	400	150	200

    a. Based on payback period, which alternative should be selected?

    b. Using future worth analysis, and a 12% interest rate, determine which alternative should be selected.

9-29 A project has the following costs and benefits. What is the payback period?

Year	Costs	Benefits
0	$1400	
1	500	
2	300	$400
3-10		300 per year

9-30 Two alternatives are being considered:

	A	B
Initial cost	$500	$800
Uniform annual cost	200	150
Useful life, in years	8	8

Both alternatives provide an identical benefit.

    a. Compute the payback period if Alt. *B* is purchased rather than Alt. *A*.

    b. Using a MARR of 12% and benefit–cost ratio analysis, which alternative should be selected?

9-31 Consider three mutually exclusive alternatives. The MARR is 10%.

Year	X	Y	Z
0	-$100	-$50	-$50
1	25	16	21
2	25	16	21
3	25	16	21
4	25	16	21

a. For Alt. X, compute the benefit–cost ratio.

b. Based on the payback period, which alternative should be selected?

c. Determine the preferred alternative based on an exact economic analysis method.

9-32

Year	E	F	G	H
0	-$90	-$110	-$100	-$120
1	20	35	0	0
2	20	35	10	0
3	20	35	20	0
4	20	35	30	0
5	20	0	40	0
6	20	0	50	180

a. Using future worth analysis, which of the above four alternatives is preferred at 6% interest?

b. Using future worth analysis, which alternative is preferred at 15% interest?

c. Based on the payback period, which alternative is preferred?

d. At 7% interest, what is the benefit–cost ratio for Alt. G?

9-33 Consider four mutually exclusive alternatives:

	A	B	C	D
Cost	$75.0	$50.0	$15.0	$90.0
Uniform annual benefit	18.8	13.9	4.5	23.8

Each alternative has a five-year useful life and no salvage value. The MARR is 10%. Wh alternative should be selected, based on:

a. Future worth analysis?

b. Benefit–cost ratio analysis?

c. The payback period?

9-34 Tom Sewel has gathered data on the relative costs of a solar water heater system and on a conventional electric water heater. The data are based on a mid-American city and assume that during cloudy days an electric heating element in the solar heating system will provide the necessary heat.

The installed cost of a conventional electric water tank and heater is $200. A family of four uses an average of 300 liters of hot water a day, which takes $230 of electricity per year. The tank is glass-lined and has a twenty-year guarantee. This is probably a reasonable estimate of its actual useful life.

The installed cost of two solar panels, small electric pump, and storage tank with auxiliary electric heating element is $1400. It will cost $60 a year for electricity to run the pump and heat water on cloudy days. The solar system will require $180 of maintenance work every four years. Neither the conventional electric water heater nor the solar water heater will have any salvage value at the end of their useful lives.

  a. Using Tom's data, what is the payback period if the solar water heater system is installed, rather than the conventional electric water heater?

  b. Chris Cook studied the same situation and decided that all the data are correct, except that he believes the solar system will *not* require the $180 of maintenance every four years. Chris believes future replacements of either the conventional electric water heater, or the solar water heater system can be made at the same costs and useful lives as the initial installation. Based on a 10% interest rate, what must be the useful life of the solar system to make it no more expensive than the electric water heater system?

9-35 Data for two alternatives are as follows:

	A	B
Cost	$800	$1000
Uniform annual benefit	230	230
Useful life, in years	5	X

If the MARR is 12%, compute the value of X that makes the two alternatives equally desirable.

9-36 What is the cost of Alt. B that will make it at the breakeven point with Alt. A, assuming a 12% interest rate?

	A	B
Cost	$150	$ X
Uniform annual benefit	40	65
Salvage value	100	200
Useful life, in years	6	6

9-37 Consider two alternatives:

	A	B
Cost	$500	$300
Uniform annual benefit	75	75
Useful life, in years	infinity	X

Assume that Alt. *B* is not replaced at the end of its useful life. If the MARR is 10%, what must be the useful life of *B* to make Alternatives *A* and *B* equally desirable?

9-38 A project will cost $50,000. The benefits at the end of the first year are estimated to be $10,000, increasing at a 10% uniform rate in subsequent years. Using an eight-year analysis period and a 10% interest rate, compute the benefit–cost ratio.

9-39 Jane Chang is making plans for a summer vacation. She will take $1000 with her in the form of traveller's checks. From the newspaper, she finds that if she purchases the checks by May 31st, she will receive them without paying a service charge. That is, she will obtain $1000 worth of traveller's checks for $1000. But if she waits and buys the checks immediately before starting her summer trip, she must pay a 1% service charge. (It will cost her $1010 for $1000 of traveller's checks.)

Jane can obtain a 13% interest rate, compounded weekly, on her money. To help her with her planning, Jane decides to compute how many weeks after May 31st she can begin her trip and still justify buying the traveller's checks on May 31st. She asks you to make the computations for her. What is the answer?

9-40 A machine costs $5240 and produces benefits of $1000 at the end of each year for eight years. Assume continuous compounding and a nominal annual interest rate Of 10%.

   a. What is the payback period (in years)?

   b. What is the breakeven point (in years)?

   c. Since the answers in *a* and *b* are different, which one is "correct"?

9-41 Jean invests $100 in year one, and doubles the amount each year after that (so the investment is $100, 200, 400, 800, . . .). If she continues to do this for 10 years, and the investment pays 10% annual interest, how much will her investment be worth at the end of 10 years?

9-42 Fence posts for a particular job cost $10.50 each to install, including the labor cost. They will last 10 years. If the posts are treated with a wood preservative they can be expected to have a 15-year life. Assuming a 10% interest rate, how much could one afford to pay for the wood preservative treatment?

9-43 A motor with a 200-horsepower output is needed in the factory for intermittent use. A Graybar motor costs $7000, and has an electrical efficiency of 89%. A Blueball motor costs $6000 and has an 85% efficiency. Neither motor would have any salvage value as the cost to remove it would equal its scrap value. The annual maintenance cost for either motor is estimated at $300 per year. Electric power costs $0.072/kilowatt-hour (1 hp = 0.746 kW). If a 10% interest rate is used in the calculations, what is the minimum number of hours the higher initial cost Graybar motor must be used each year to justify its purchase?

9-44 Five mutually exclusive investment alternatives have been proposed. Based on benefit-cost ratio analysis, and a MARR of 15%, which alternative should be selected?

Year	A	B	C	D	E	F
0	–$200	–$125	–$100	–$125	–$150	–$225
1–5	+68	+40	+25	+42	+52	+68

9-45 Plan A requires a $100,000 investment now. Plan B requires an $80,000 investment now and an additional $40,000 investment at a later time. At 8% interest, compute the breakeven point for the timing of the $40,000 investment.

9-46 A piece of property is purchased for $10,000 and yields a $1000 yearly net profit. If the property is sold after five years, what is its minimum price to breakeven with interest at 10%?

9-47 If you invested $2500 in a bank 24-month certificate of deposit paying 8.65%, compounded monthly, how much would you receive when the certificate of deposit matures in two years?

9-48 Rental equipment is for sale for $110,000. A prospective buyer estimates he would keep the equipment for 12 years and spend $6000 a year on maintaining the equipment. Estimated annual net receipts from equipment rentals would be $14,400. It is estimated rental equipment could be sold for $80,000 at the end of 12 years. If the buyer wants a 7% rate of return on his investment, what is the maximum price he should pay for the equipment?

9-49 A low carbon steel machine part, operating in a corrosive atmosphere, lasts six years and costs $350 installed. If the part is treated for corrosion resistance it will cost $500 installed. How long must the treated part last to be the preferred alternative, if 10% interest is used?

9-50 A car dealer presently leases a small computer with software for $5000 per year. As an alternative he could purchase the computer for $7000 and lease the software for $3500 per year. Any time he would decide to switch to some other computer system he could cancel the software lease and sell the computer for $500. If he purchases the computer and leases the software,

    a. What is the payback period?

    b. If he kept them for six years, what would be the benefit-cost ratio, based on a

        10% interest rate?

9-51 Given the following data for two machines:

	A	*B*
Original cost	$55,000	$75,000
Annual expenses Operation	9,500	7,200
Annual expenses Maintenance	5,000	3,000
Annual expenses Taxes and insurance	1,700	2,250

The machines have no net salvage value. At what useful life are the machines equivalent if

    a. 10% interest is used in the computations?

    b. 0% interest is used in the computations?

9-52 After receiving an inheritance of $25,000 on her 21st birthday, Ayn Rand deposits the inheritance in a savings account with an effective annual interest rate of 6%. She decides that she will make regular deposits on each future birthday, beginning with a deposit $1,000 on her 22nd birthday and then increasing the amount by $200 in each following year (i.e., $1,200 on her 23rd birthday, $1,400 on her 24th birthday, etc.) How much money will she have on her 56th birthday?

9-53 An Association of General Contractors (AGC) wished to establish an endowment fund of $1,000,000 in 10 years for the Construction Engineering Technology Program at Grambling State University in Grambling, Louisiana. In doing so, the AGC established an escrow account in which 10 equal end-of-year deposits that earn 7% compound interest were made. After 7 deposits, the Louisiana Legislature revised laws relating to the licensing fees AGC charges its members, and as such, there was no deposit at the end of the 8th year. What must the amount of the remaining equal end-of-year deposits be to insure that the $1,000,000 is available to Grambling State for its Construction Engineering Technology Program?

9-54 On her birthday, a 25 year old engineer is considering investing in an Individual Retirement Account (IRA). After some research, she finds a mutual fund with an average return of 10% per year. How much will she have at age 65 if she makes annual investments of $2000 into the fund beginning on her 25th bithday? Assume the fund continues to earn an annual return of 10%?

9-55 A large project requires an investment of M$200 (M$ = Millions of Dollars). The construction will take three years. M$30 will be spent during the first year, M$100 during the second year and M$ 70 during the third year of construction. Two project operation periods are being considered: *A*) 10 years with the expected net profit of M$40 per year and *B*) 20 years with the expected net profit of M$ 32.5 per year. For simplicity of calculations it is assumed that all cash flows occur at the end of year. The company minimum required return on investment is 10%.

Calculate for each alternative:

    a. The Payback periods.

    b. The total equivalent investment cost at the end of the construction period.

    c. The equivalent uniform annual worth of the project (use the operation period of each alternative).

Make your recommendations based on the above economic parameters.

9-56 IPS Corp. will upgrade its package labeling machinery. It costs $150,000 to buy the machinery and have it installed. Operation and maintenance costs are $1500 per year for the first three years and then increase by $500 per year for the remaining years of the machine's 10 year life. The machinery has a salvage value of 5% of its initial cost. Interest is 10%. What is the Future Worth of cost of the machinery?

9-57 If the machinery in problem **9-56**, generated a costs savings of $0.05/package and the company sold 322,000 packages per year, what is the rate of return on the investment? If IPS's MARR is 15%, should they make the upgrade?

9-58 A company is considering buying a new bottle capping machine. The initial cost of the machine is $325,000 and it has a 10 year life. Monthly maintenance costs are expected to be $1200 per month for the first 7 years and $2000 per month for the remaining years. The machine requires a major overhaul costing $55,000 at the end of the fifth year of service. Assume all these costs occur at the end of the appropriate period. What is the future value of all the costs associated with owning and operating this machine if the interest rate is 7.2%?

9-59 A family starts an education fund for their son Patrick when he is 8 years old. They invest $150 on his eighth birthday and increase the yearly investment by $150 per year until Patrick is 18 years old. The fund pays 9% annual interest. How much will the fund be worth when Patrick is 18?

9-60 The interest rate is 16% per year and there are 48 compounding periods per year. The principal is $50,000. What is the future worth in 5 years?

9-61 Calculate the Present Worth and the Future Worth of a series of ten annual cash flows with the first cash flow equal to $15,000 and each successive cash flow increasing by $1,200. The interest rate is 12%. The total cash flow series is a combination of systems1 and 2 below.

System 1:   System 2:

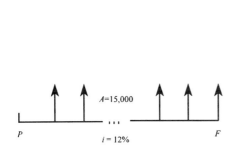

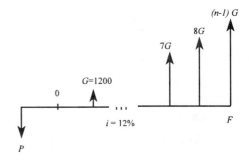

9-62 A bank account pays 19.2% interest with monthly compounding. A series of deposits started with a deposit of $5000 on January 1, 1997. Deposits in the series will occur each six months. Each deposit in the series will be for $150 less than the one before it. The last deposit in the series will occur on January 1, 2012. What balance will the account have on July 1, 2014, if the balance was zero before the first deposit and no withdrawals are made from it?

9-63 Let's assume you graduate and get a good job. You want to begin a savings account. You get paid monthly and authorize the bank to automatically withdraw $75 each month. The bank will make the first withdrawal on July 1, 1997, and you instruct them to make the last withdrawal on January 1, 2015. The bank pays a nominal interest rate of 4.5 % and compounds twice a month. What will be the balance in the account on January 1, 2015?

9-64 Bob, an engineer, decided to start a college fund for his son. Bob will deposit a series of equal, semiannual cash flows with each deposit equal to $1500. Bob will make the first deposit on July 1, 1998, and he will make the last deposit on July 1, 2018. Joe, a friend of Bob's, will receive an inheritance on April 1, 2003, and has decided to begin a college fund for his daughter. Joe wants to send his daughter to the same college as Bob's son. Therefore, Joe needs to accumulate the same amount of money on July 1, 2018 as Bob will have accumulated from his semiannual deposits. Joe had no idea how to determine the amount that should be deposited since he never took Engineering Economics. He has decided to deposit $40,000 on July 1, 2003. Will Joe's deposit be sufficient? If not, what amount should Joe deposit? Use a nominal interest of 7% with semiannual compounding on all accounts.

9-65 A firm must decide which of three alternatives to adopt to expand its capacity. The firm wishes a minimum annual profit of 20% of the initial cost of each separable increment of investment. Any money not invested in capacity expansion can be invested elsewhere for an annual yield of 20% of initial cost.

Alternative	Initial cost	Annual profit	Profit rate
A	$100,000	$30,000	30%
B	300,000	66,000	22
C	500,000	80,000	16

Which alternative should be selected?

9-66 The New England Soap Company is considering adding some processing equipment to the plant to allow it to remove impurities from some raw materials. By adding the processing equipment, the firm can purchase lower grade raw material at reduced cost and upgrade it for use in its products.

Four different pieces of processing equipment are being considered:

	A	B	C	D
Initial investment	$10,000	$18,000	$25,000	$30,000
Annual saving in materials costs	4,000	6,000	7,500	9,000
Annual operating cost	2,000	3,000	3,000	4,000

The company can obtain a 15% annual return on its investment in other projects and is willing to invest money on the processing equipment only so long as it can obtain 15% annual return on each increment of money invested. Which one, if any, of the alternatives should be selected?

CHAPTER 10

Depreciation

We have so far dealt with a variety of economic analysis problems and many techniques for their solution. In the process we have avoided income taxes, which are an important element of most economic analyses. Now, we can move to more realistic—and, unfortunately, more complex—situations.

Our government taxes individuals and businesses to support its processes—lawmaking, domestic and foreign economic policy-making, even the making and issuing of money itself. The omnipresence of taxes requires that they be included in economic analyses, which means we must understand something about the *way* taxes are imposed. For capital equipment, depreciation is required to compute income taxes. Chapter 10 examines depreciation, and Chapter 11 illustrates how depreciation is used in income tax computations. The goal is to support decision making on engineering projects, not to support final tax calculations.

Basic Aspects Of Depreciation

The word *depreciation* is defined as a "decrease in value." This is not an entirely satisfactory definition, for *value* has several meanings. In the context of economic analysis, value may refer either to *market value*—that is, the monetary value others place on property—or *value to the owner*. Thus, we now have two definitions of depreciation: a decrease in value to the market or to the owner.

Deterioration and Obsolescence

A machine may depreciate because it is ***deteriorating*** or wearing out and no longer performing its function as well as when it was new. Many kinds of machinery require increased maintenance as they age, reflecting a slow but continuing failure of individual parts. In other types of equipment, the quality of output may decline due to wear on components and resulting poorer mating of parts. Anyone who has worked to maintain an auto has observed deterioration due to failure of individual parts (such as: fan belts, mufflers, and batteries) and the wear on components (such as: bearings, piston rings, and alternator brushes).

Depreciation is also caused by ***obsolescence.*** A machine is described as obsolete when it is no longer needed or useful. A machine may be in excellent working condition, yet may still be obsolete. In the 1970s, mechanical business calculators with hundreds of gears and levers became obsolete. The advance of integrated circuits resulted in a completely different and far superior approach to calculator design. Thus, mechanical calculators rapidly declined or depreciated in value.

If your auto depreciated in the last year, that means it has declined in market value. It has less value to potential buyers. On the other hand, a manager who indicates a piece of machinery has depreciated may be describing a machine that has deteriorated because of use and/or because it has become obsolete compared to newer machinery. Both situations indicate the machine has declined in value to the owner.

The accounting profession defines depreciation in yet another way. Although in everyday conversation we are likely to use depreciation to mean a decline in market value or value to the owner, accountants define depreciation as allocating an asset's cost over its **useful** or **depreciable life**. Thus, we now have *three distinct definitions of depreciation:*

1. Decline in market value of an asset.

2. Decline in value of an asset to its owner.

3. Systematic allocation of the cost of an asset over its depreciable life.

Depreciation and Expenses

It is this third (accountant's) definition that is used to compute depreciation for business assets. Business costs are generally either ***expensed*** or ***depreciated***. **Expensed** items, such as labor, utilities, materials, and insurance, are part of regular business operations and are "consumed" over short periods of time (sometimes recurring). These costs do not lose value gradually over time. For tax purposes they are subtracted from business revenues when they occur. Expensed costs reduce income taxes because businesses are able to *write off* their full amount when they occur.

In contrast, business costs due to **depreciated** assets are not fully written off when they occur. A depreciated asset does lose value gradually, and must be written off over an extended period. For instance, consider a plastic injection machine used to produce the beverage cups found at sporting events. The plastic pellets melted into the cup shape lose their value as raw material directly after manufacturing. The raw material cost for production material (plastic pellets) is written off, or expensed, immediately. On the other hand, the plastic mold injection machine itself will lose value over time, and thus its cost (purchase price and installation expenses) are written off (or depreciated) over a period of years. The number of years over which the machine is depreciated is called its **depreciable life** or **recovery period**, which is often different than the asset's useful or most economic life. Depreciable life is determined by the depreciation method used to spread out the cost — many types of depreciated assets operate well beyond their depreciation life.

Depreciation is a **non-cash** cost that requires no exchange of dollars from one hand to another. Companies do not write a check to someone to **pay** their depreciation expenses. Rather, it is a business expense that is allowed by the government to offset the loss in value of business assets. Remember, the company has already paid for the asset up front, depreciation is simply a way to claim this "business expense" over time. Depreciation is important to the engineering economist because, even though it is a non-cash cost, it represents a cash flow on an *after-tax basis*. Depreciation deductions reduce the taxable income of businesses and thus reduce the amount of taxes paid.

In general business assets can only be depreciated if they meet the following basic requirements:

The property must be used for business purposes to produce income,

The property must have a useful life that can be determined, and this life must be longer than one year, and

The property must be an asset that decays, gets used up, wears out, becomes obsolete, or loses value to the owner from natural causes.

EXAMPLE 10-1

Consider the costs that are incurred by a local pizza business. Identify each cost as either *expensed* or *depreciated* and describe why.

- Cost for pizza dough and toppings
- Cost of new delivery van
- Cost to pay wages for janitor
- Cost of furnishings in dining room
- Cost of a new baking oven
- Utility costs for soda refrigerator

Solution

Cost Item	Type of Cost	Why
Pizza dough and toppings	expensed	Life < 1 year, loses value immediately
New delivery van	depreciated	Meets 3 requirements for depreciation
Wages for janitor	expensed	Life < 1 year, loses value immediately
Furnishings in dining room	depreciated	Meets 3 requirements for depreciation
New baking oven	depreciated	Meets 3 requirements for depreciation
Utilities for soda refrigerator	expensed	Life < 1 year, loses value immediately

---

Types of Property

The rules for depreciation are linked to the classification of business property as either *tangible* or *intangible*. Tangible property is further classified as either *real* or *personal*.

Tangible Property can be seen, touched, and felt.
> **Real Property** includes land, buildings, and all things growing on, built upon, constructed on, or attached to the land.
> **Personal Property** includes equipment, furnishings, vehicles, office machinery, and anything that is tangible excluding those assets defined as *real property*.

Intangible Property is all property that has value to the owned but can not be directly seen or touched. Examples include patents, copyrights, trademarks, trade names, and franchises.

Many different types of properties that wear out, decay, or lose value can be depreciated as business assets. This wide range includes: copy machines, helicopters, buildings, interior furnishings, production equipment, and computer networks. Almost all tangible property can be depreciated.

One important and notable exception is land, which is *never* depreciated. Land does not wear out, lose value, or have a determinable useful life and thus does not qualify as a depreciable property. Consider the aspect of loss in value. Rather than decreasing in value most land becomes more valuable as time passes. In addition to the land itself, expenses for clearing, grading, preparing, planting, and landscaping are not generally depreciated because they have no fixed useful life. Other tangible property that *can not* be depreciated includes factory inventory, containers considered as inventory, equipment used to build capital improvements, and leased property. The leased property exception highlights the fact that only the owner of property may claim depreciation expenses.

Tangible properties used in *both* business and personal activities, such as a vehicle used in a consulting engineering firm that is also used to take one's kids to school, can be depreciated. However, in such cases one can only take depreciation deductions in proportion to the use for business purposes. Accurate records indicating the portion of use for business and personal activities are required.

Depreciation Calculation Fundamentals

To understand the complexities of depreciation, the first step is to examine the fundamentals of depreciation calculations. Figure 10-1 illustrates the general depreciation problem of allocating the total depreciation charges over the asset's life. The vertical axis is labeled Money, but when we plot the curve of cost minus depreciation charges made, the vertical axis is more appropriately called ***book value***, where
> ***Book value = Cost – Depreciation charges made to date.***

Looked at another way, book value is the asset's remaining unallocated cost.

Figure 10-1 General depreciation.

In looking at Figure 10-1, *book value* goes from a value of B at time zero in the recovery period to a value of S at time 5. Thus, book value is a *dynamic* variable that changes over an asset's recovery period. The equation used to calculate an asset's book value over time is:

$$BV_t = \text{Cost Basis} - \sum_{i=1}^{t} d_i \qquad (10\text{-}1)$$

where: BV_t = Book Value of the depreciated asset at the end of time t

Cost Basis = B = The dollar amount that is being depreciated. This includes the asset's purchase price as well as any other costs necessary to make the asset "ready for use."

$\displaystyle\sum_{i=1}^{t} d_i$ = The sum of depreciation deduction taken from time 0 to time t, where: d_i is the depreciation deduction in year i.

Equation 10-1 shows that year-to-year depreciation charges reduce an asset's book value over its life. The following section describes methods that are or have been allowed under federal tax law for quantifying these yearly depreciation deductions.

Historical Depreciation Methods

Allowing businesses to deduct the cost for capital expenditures over time (depreciate business assets) has long been part of the tax code. However, over time several versions of various depreciation methods have been used to calculate these deductions. In general, accounting depreciation methods can be categorized as follows:

Pre-1981 Historical Methods These methods include the *straight line, sum-of-the-years digits*, and *declining balance* methods. Each method required estimates of an asset's useful life and salvage value. Firms could elect which method to use for assets, and thus there was little uniformity in how depreciation expenses were reported.

1981 – 1986 Method With the Economic Recovery Tax Act (ERTA) of 1981 Congress created the Accelerated Cost Recovery System (ACRS). The ACRS method had three key features: (1) property class lives were created and all depreciated assets assigned to one particular category, (2) the need to estimate salvage values was eliminated because all assets were *fully* depreciated over their recovery period, and (3) the recovery periods used to calculate annual depreciation *accelerated* the write off of capital costs more quickly than did the historical methods — thus the name.

1986 to Present The Modified Accelerated Cost Recovery System (MACRS) has been in effect since the Tax Reform Act of 1986 (TRA-86). The MACRS method is similar to the ACRS system except: (1) the number of property classes was expanded and (2) the annual depreciation percentages were modified to include a half-year convention for the first and final years.

In this chapter, our primary focus is to describe the MACRS depreciation method. However, it is useful to first describe three historical depreciation methods.

Straight Line Depreciation

The simplest and best known depreciation method is *straight line depreciation*. To calculate the constant *annual depreciation charge*, the total amount to be depreciated, $B - S$, is divided by the depreciable life, in years, N:

$$\textbf{\textit{Annual depreciation charge}} = \textbf{\textit{d}}_t = \frac{(\textbf{\textit{B}} - \textbf{\textit{S}})}{\textbf{\textit{N}}} \qquad (10\text{-}2)$$

EXAMPLE 10-2

Consider the following:

Cost of the asset, B	$900
Depreciable life, in years, N	5
Salvage value, S	$70

Compute the straight line depreciation schedule.

Solution:

$$\textbf{Annual depreciation charge} = d_t = \frac{(B - S)}{N} = \frac{900 - 70}{5} = \$166$$

Year Depreciation for year t	Sum of depreciation charges up to year t	Book value at the end of year t
(t) (d $_t$)	$\sum_{i=1}^{t} d_i$	$BV_t = B - \sum_{i=1}^{t} d_i$
1 $166	$166	900 - 166 = 734
2 166	332	900 - 332 = 568
3 166	498	900 - 498 = 402
4 166	664	900 - 664 = 236
5 166	830	900 - 830 = 70 = S

This situation is illustrated in Figure 10-2. Notice the constant $166 d_t each year for five years, and that the asset has been depreciated down to a book value of $70 which was the estimated salvage value. ∎

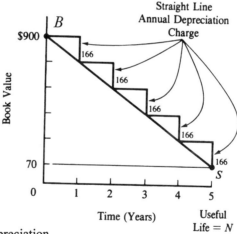

Figure 10-2 Straight line depreciation

The straight line method is often used for intangible property. Veronica's firm bought a patent in April that was not acquired as part of acquiring a business. She paid $6800 for this patent and must depreciate it using the straight line method over 17 years with no salvage value. The annual depreciation is $400 (= $6800/17). Since the patent was purchased in April the deduction must be prorated over the 9 months of ownership. This year the deduction is $300 (=$400 x 9/12), and then next year she can begin taking the full $400 per year.

Sum-Of-Years Digits Depreciation

Another method for allocating an asset's cost *minus* salvage value *over* its depreciable life is called ***Sum-Of-Years Digits—SOYD—depreciation***. This method results in larger-than-straight-line depreciation charges during an asset's early years and smaller charges as the asset nears the end of its depreciable life. Each year, the depreciation charge equals a fraction of the total amount to be depreciated $(B-S)$. The denominator of the fraction is the sum-of-years digits. For example if the depreciable life is 5 years, $1 + 2 + 3 + 4 + 5 = 15 = $ SOYD. Then 5/15, 4/15, 3/15, 2/15, and 1/15 are the fractions from year 1 to year 5. Each year the depreciation charge shrinks by 1/15th of $(B - S)$. Because this change is the same every year, SOYD depreciation can be modeled as an arithmetic gradient, G. The equations can also be written as:

$$\begin{pmatrix} \text{Sum - of - years digits} \\ \text{depreciation charge for} \\ \text{any year} \end{pmatrix} = \frac{\begin{pmatrix} \text{Remaining depreciable life} \\ \text{at beginning of year} \end{pmatrix}}{\begin{pmatrix} \text{Sum - of - years digits} \\ \text{for total depreciable life} \end{pmatrix}}(\text{Total amount depreciated})$$

$$d_t = \frac{N - t + 1}{SOYD}(B - S) \tag{10-3}$$

where: d_t = depreciation charge in any year t
                    N = number of years in depreciable life
                    SOYD = sum-of-years-digits, calculated as:

$$SOYD = \frac{N}{2}(N + 1)$$

                    B = cost of the asset made ready for use
                    S = estimated salvage value after depreciable life

---

EXAMPLE 10-3

Compute the SOYD depreciation schedule for the situation in Ex. 10-2:
        Cost of the asset, B $900
        Useful life, in years, N 5
        Salvage value, S $70

Solution:

$$SOYD = \frac{5}{2}(5 + 1) = \frac{5}{2}(6) = 15$$

$$\text{Thus, } d_1 = \frac{5-1+1}{15}(900-70) = 277$$

$$d_2 = \frac{5-2+1}{15}(900-70) = 221$$

$$d_3 = \frac{5-3+1}{15}(900-70) = 166$$

$$d_4 = \frac{5-4+1}{15}(900-70) = 166$$

$$d_5 = \frac{5-5+1}{15}(900-70) = 55$$

Year (t)	Depreciation for year t (d_t)	Sum of depreciation charges up to year t $\sum_{i=1}^{t} d_i$	Book value at the end of year t $BV_t = B - \sum_{i=1}^{t} d_i$
1	$277	$277	900 - 277 = 623
2	221	498	900 - 498 = 402
3	166	664	900 - 664 = 236
4	111	775	900 - 775 = 125
5	55	830	900 - 830 = 70 = S

These data are plotted in Figure 10-3. ■

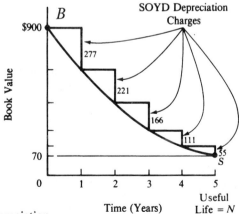

Figure 10-3 Sum-of-years digits depreciation.

Declining Balance Depreciation

Declining balance depreciation applies a *constant depreciation rate* to the property's declining book value. Two rates were commonly used before the 1981 and 1986 tax revisions, and they are used to compute MACRS depreciation percentages. These are 150% and 200% of the straight-line rate. Since 200% is twice the straight-line rate, it is called *double declining balance*, or DDB; the general equation is:

$$\text{Double declining balance } \boldsymbol{d_t} = \frac{2}{N}(\text{Book value}_{t-1}) \qquad (10\text{-}4a)$$

Since book value equals cost *minus* depreciation charges to date,

$$\text{DDB } \boldsymbol{d_t} = \frac{2}{N}(\text{Cost} - \text{Depreciation charges to date})$$

or:

$$d_t = \frac{2}{N}\left(B - \sum_{i=1}^{t} d_i\right) \qquad (10\text{-}4b)$$

EXAMPLE 10-4

Compute the DDB depreciation schedule for the situations in Examples 10-2 and 10-3.

Cost of the asset, B	$900	
Depreciable life, in years, N	5	
Salvage value, S	$70	

Solution:

Year (t)	Depreciation for year t using Eq. (10-4a) (d_t)	Sum of depreciation charges up to year t $\sum_{i=1}^{t} d_i$	Book value at the end of year t $BV_t = B - \sum_{i=1}^{t} d_i$
1	$(2/5) \cdot 900 = 360$	$360	900 - 360 = 540
2	$(2/5) \cdot 540 = 216$	576	900 - 566 = 334
3	$(2/5) \cdot 334 = 130$	706	900 - 706 = 194
4	$(2/5) \cdot 194 = 78$	784	900 - 784 = 116
5	$(2/5) \cdot 116 = 46$	830	900 - 830 = 70 $= S$

Figure 10-4 illustrates the situation. ■

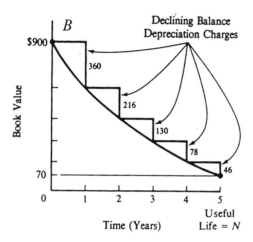

Figure 10-4 Declining balance depreciation.

If the final salvage value of Example 10-4 had not been $70, then the double declining balance method would have needed to be modified. One modification stops further depreciation once the book value equals the salvage value – this prevents taking too much depreciation. The other modification would switch from declining balance depreciation to straight line – this ensures taking enough depreciation.

These modifications are not detailed here, because (1) MACRS has been the legally appropriate system since 1986 and (2) as will be shown, MACRS incorporates the shift from declining balance to straight line depreciation.

Modified Accelerated Cost Recovery System (MACRS)

The modified accelerated cost recovery system (MACRS) depreciation method was first introduced by the Tax Reform Act of 1986 and has been continued with the Taxpayers Relief Act of 1997. Three major advantages of MACRS are that: 1) the "property class lives" are less than the "actual useful lives," 2) salvage values are assumed to be zero, and 3) tables of annual percentages simplify computations.

The definition of the MACRS classes of depreciable property is based on work by the U.S. Treasury Department. In 1971 they published guidelines for about 100 broad asset classes. For each class there was a lower limit, midpoint, and upper limit of useful life, called the *Asset Depreciation Range (ADR)*. The ADR midpoint lives were somewhat shorter than the actual average useful lives. These guidelines have been incorporated into MACRS so that the property class lives are again shorter than the ADR midpoint lives.

Use of MACRS focuses on the general depreciation system (GDS), which is based on declining balance with switch to straight line depreciation. The alternative depreciation system (ADS) provides for a longer period of recovery and uses straight line depreciation. Thus it is much less economically attractive. Under law, ADS must be used for: (1) any tangible property used primarily outside of the U.S., (2) any property that is tax-exempt or financed by tax-exempt bonds, and (3) farming property placed in service when uniform capitalization rules are not applied. The ADS may also be *elected* for property that can be depreciated using the GDS system. However, once ADS is elected for an asset, it is not possible to switch back to the GDS system. Because the ADS makes the depreciation deductions less valuable, unless ADS is specifically mentioned, subsequent discussion assumes the GDS system when reference is made to MACRS.

Once a property is determined to be eligible for depreciation, the next step is to calculate its depreciation deductions over its life. The following information is required in order to calculate these deductions:

(1) the cost basis of the property;
(2) the *property class* and *recovery period* of the asset;
(3) the asset's placed-in-service date.

Cost Basis and Placed in Service Date

The cost basis, B, is the cost to obtain and place the asset in service fit for use. However, for real property the basis may also include certain fees and charges that the buyer pays as part of the purchase. Examples of such fees include: legal and recording fees, abstract fees, survey charges, transfer taxes, title insurance, and amounts that the seller owes that you pay (such as back taxes, interest, sales commissions, etc.).

Depreciation for a business asset begins when the asset is *placed in service*. If an asset is purchased and used in a personal context, depreciation may not be taken. If that asset is later used in business for income producing activity, then depreciation may begin with the change in usage.

Property Class and Recovery Period

Each depreciated asset is placed in a *MACRS Property Class*, which defines the **recovery period** and the depreciation percentage for each year. Historically the IRS assigned each type of depreciable asset a *class life* or an *asset depreciation range*. With MACRS, asset class lives have been pooled together in the *property classes*. Table 10-1 lists the class lives and GDS and ADS property classes for several example depreciable assets. Table 10-2 lists the MACRS GDS property classes.

Table 10-1   EXAMPLE CLASS LIVES AND MACRS PROPERTY CLASSES

IRS asset class	Asset description	Class life (years) ADR	MACRS property class (years) GDS	MACRS property class (years) ADS
00.11	Office furniture, fixtures and equipment	10	7	10
00.12	Information systems: computers/peripheral	6	5	6
00.22	Automobiles, taxis	3	5	6
00.241	Light general purpose trucks	4	5	6
00.25	Railroad cars and locomotives	15	7	15
00.40	Industrial steam & electric distribution	22	15	22
01.11	Cotton gin assets	10	7	10
01.21	Cattle, breeding or dairy	7	5	7
13.00	Offshore drilling assets	7.5	5	7.5
13.30	Petroleum refining assets	16	10	16
15.00	Construction assets	6	5	6
20.10	Manufacture of grain and grain mill products	17	10	17
20.20	Manufacture of yarn, thread and woven fabric	11	7	11
24.10	Cutting of timber	6	5	6
32.20	Manufacture of cement	20	15	20
20.1	Manufacture of motor vehicles	12	7	12
48.10	Telephone distribution plant	24	15	24
48.2	Radio and television broadcasting equipment	6	5	6
49.12	Electric utility nuclear production plant	20	15	20
49.13	Electric utility steam production plant	28	20	28
49.23	Natural gas production plant	14	7	14
50.00	Municipal wastewater treatment plant	24	15	24
80.00	Theme and amusement park assets	12.5	7	12.5

The MACRS GDS property classes are described in more detail in Table 10-2. Department of the Treasury, Internal Revenue Service, *How To Depreciate Property*, Publication 946.

Table 10-2 MACRS GDS PROPERTY CLASSES

Property class	*Personal property (all property except real estate)*
Three-Year Property	Special handling devices for food and beverage manufacture;
	Special tools for the manufacture of finished plastic products, fabricated metal products, and motor vehicles;
	Property with class life of 4 years or less.
Five-Year Property	Automobiles* and trucks;
	Aircraft (of non-air-transport companies);
	Equipment used in research and experimentation;
	Computers;
	Petroleum drilling equipment;
	Property with class life of more than 4 years and less than 10 years.
Seven-Year Property	All other property not assigned to another class;
	Office furniture, fixtures, and equipment;
	Property with class life of 10 years or more and less than 16 years.
Ten-Year Property	Assets used in petroleum refining and certain food products;
	Vessels and water transportation equipment;
	Property with class life of 16 years or more and less than 20 years.
Fifteen-Year Property	Telephone distribution plants;
	Municipal sewage treatment plants;
	Property with class life of 20 years or more and less than 25 years.
Twenty-Year Property	Municipal sewers;
	Property with class life of 25 years and more.
Property class	*Real property (real estate)*
27.5 Years	Residential rental property (does not include hotels and motels)
39 Years	Nonresidential real property

*The depreciation deduction for automobiles is limited to $3160 the first tax year, $5000 the second year, $3050 the third year, and $1775 per year in subsequent years.

The proper MACRS property class can be found several different ways. In the list below, the first approach that works should be used.

1. Property class given in problem.
2. Asset is named in Table 10-2.
3. IRS tables or Table 10-1.
4. Class life.
5. 7 year property for "all other property not assigned to another class."

Once the MACRS property class is known, as well as the placed-in-service date and cost basis, the year-to-year depreciation deductions can be calculated for GDS assets over their depreciable life using:

$$d_t = B \times r_t \tag{10-5}$$

where: d_t = depreciation deduction in year t
B = cost basis being depreciated
r_t = appropriate MACRS percentage rate

Percentage Tables

The IRS has prepared tables to assist in calculating depreciation charges when MACRS GDS depreciation is used. Table 10-3 gives the yearly depreciation percentages (r_t) that are used for the six personal property classes (3, 5, 7, 10, 15, and 20 year property classes), and Table 10-4 gives the percentages for nonresidential real property. Notice the values in the table are given in *percentages* — thus, as an example, the value of 33.33% (given for year 1 for a 3-year MACRS GDS property) is used as 0.3333.

Table 10-3 MACRS DEPRECIATION
FOR PERSONAL PROPERTY—HALF-YEAR CONVENTION

The applicable percentage for the property class is:

If the recovery year is:	3-year property	5-year property	7-year property	10-year property	15-year property	20-year property
1	33.33	20.00	14.29	10.00	5.00	3.750
2	44.45	32.00	24.49	18.00	9.50	7.219
3	14.81*	19.20	17.49	14.40	8.55	6.677
4	7.41	11.52*	12.49	11.52	7.70	6.177
5		11.52	8.93*	9.22	6.93	5.713
6		5.76	8.92	7.37	6.23	5.285
7			8.93	6.55*	5.90*	4.888
8			4.46	6.55	5.90	4.522
9				6.56	5.91	4.462*
10				6.55	5.90	4.461
11				3.28	5.91	4.462
12					5.90	4.461
13					5.91	4.462
14					5.90	4.461
15					5.91	4.462
16					2.95	4.461
17						4.462
18						4.461
19						4.462
20						4.461
21						2.231

Computation Method:

The 3-,5-,7-, and 10-year classes use 200% and the 15- and 20-year classes use 150% declining balance depreciation.

All classes convert to straight line depreciation in the optimal year – shown with asterisk (*).

A half-year of depreciation is allowed in the first and last recovery years.

If more than 40 % of the year's MACRS property is placed in service in the last three months, then a mid-quarter convention must be used with depreciation tables that are not shown here.

Table 10-4 MACRS DEPRECIATION FOR REAL PROPERTY (REAL ESTATE)

Recovery Percentages for Nonresidential Real Property
Month placed in service

Recovery year	1	2	3	4	5	6	7	8	9	10	11	12
1	2.461	2.247	2.033	1.819	1.605	1.391	1.177	0.963	0.749	0.535	0.321	0.107
2-39	2.564	2.564	2.564	2.564	2.564	2.564	2.564	2.564	2.564	2.564	2.564	2.564
40	0.107	0.321	0.535	0.749	0.963	1.177	1.391	1.605	1.819	2.033	2.247	2.461

last year 40

The useful life is 39 years for nonresidential real property. Depreciation is straight line using the mid-month convention. Thus a property placed in service in January would be allowed 11½ months depreciation for recovery year 1.

Notice in Table 10-3 that the depreciation percentages continue for *one year beyond* the property class life. For example, a MACRS 10-year property has an r_t value of 3.28% in **year 11**. This is due to the *half year convention* that also halves the percentage for the first year. The half year convention assumes that all assets are placed in service at the mid-point of the first year.

Another characteristic of the MACRS percentage tables is that the sum of the r_t values in any column equal 100%. This means that assets depreciated using MACRS are *fully depreciated* at the end of the recovery period where the assumed salvage value is zero. This is a departure from the pre-1981 historical methods where an estimated salvage value was considered.

Where MACRS Percentage Rates (r_t) Come From

This section describes the connection between historical depreciation methods and the MACRS percentages that are shown in Table 10-3. Before ACRS and MACRS, the most common depreciation method was declining balance with a switch to straight line. That combined method is used for MACRS with three further assumptions.

1. Salvage values are assumed to be zero for all assets.
2. The first and last years of the recovery period are each assumed to be ½ *years*.
3. The declining balance rate is 200% for 3, 5, 7 and 10 year property and 150% for 15 and 20 year property.

As shown in Example 10-5, the MACRS percentage rates can be derived from these rules and the declining balance and straight line methods. However, it is obviously much easier to simply use the r_t values from Tables 10-3 and 10-4.

EXAMPLE 10-5

Consider a 5-Year MACRS property asset with an installed and "made ready for use" cost basis of $100. Develop the MACRS percentage rates (r_t) for the asset based on the underlying depreciation methods.

Solution

The 5-Year MACRS property percentage rates are developed using the 200% declining balance method switching over to straight line at the optimal point. Since the assumed salvage value is zero, the entire cost basis of $100 is depreciated. Also the $100 basis mimics the 100% that is used in Table 10-3.

Let's explain the table below year by year. In year one the basis is $100 − 0, and the D_t values are halved for the initial ½ year assumption. Double declining balance has the larger value, so it is chosen. The rest of the declining balance computations are simply 200%/N times the (basis minus the cumulative depreciation).

In year 2 there are 4.5 years remaining for straight line, so that is the denominator for dividing the remaining $80 in book value. Similarly in year 3 there are 3.5 years remaining. In year 4 the two calculations happen to be identical, so the switch from DDB to SL can be done in either year 4 or year 5. Once we know that the SL depreciation is 11.52 at the switch point, then the only further calculation is to halve that for the last year.

Notice that the DDB calculations get smaller every year, so that at some point the straight line calculations lead to faster depreciation. This point is the optimal switch point, and it is built into Table 10-3 for MACRS.

Year	DDB Calculation	SL Calculation	MACRS (r_t)% Rates	Cumul. Depr.
1	½(2/5)(100-0) = **20.00**	½(100-0)/5 = 10.00	20.00 (DDB)	20.00
2	(2/5)(100-20.00) = **32.00**	(100-20)/4.5 = 17.78	32.00 (DDB)	52.00
3	(2/5)(100-52.00) = **19.20**	(100-52)/3.5 = 13.71	19.20 (DDB)	71.20
4	(2/5)(100-71.20) = **11.52**	(100-71.20)/2.5 = **11.52**	11.52 (either)	82.72
5		**11.52**	11.52 (SL)	94.24
6		(½)(11.52) = **5.76**	5.76 (SL)	100.00

The values given in this example match the r_t percentage rates given in Table 10-3 for a 5-year MACRS property. ■

MACRS Method Examples

Remember the key questions in using MACRS are: (1) What type of asset and does it qualify as depreciable property? (2) What amount are you depreciating [cost basis]? and (3) When are you placing the asset in service? Let's look at several example of calculating both depreciation deductions and book values using MACRS.

EXAMPLE 10-6

Use the MACRS GDS method to calculate the yearly depreciation allowances and book values for a firm that has purchased $150,000 worth of office equipment that qualifies as depreciable property. This office furniture is estimated to have a salvage (market) value of $30,000 (20% of the original cost) after the end of its depreciable life.

Solution:
1. The assets qualify as depreciable property
2. The cost basis is given as $150,000
3. The assets are being placed in service in year 1 of our analysis
4. MACRS GDS applies
5. The salvage value is not used with MACRS to calculate depreciation or book value

Office equipment is listed in Table 10-2 as 7-year property. We now use the MACRS GDS 7-year property percentages from Table 10-3 and Eq. 10-5 to calculate the year-to-year depreciation allowances. We calculate the book value of the asset using Eq. 10-1.

Year (t)	MACRS (r_t)	Cost Basis	d_t	Cumulative d_t	$BV_t =$ B $-$cum. d_t
1	14.29% x	$150,000	$21,435	$21,435	$128,565
2	24.49	150,000	36,735	58,170	91,830
3	17.49	150,000	26,235	84,405	65,595
4	12.49	150,000	18,735	103,140	46,860
5	8.93	150,000	13,395	116,535	33,465
6	8.92	150,000	13,380	129,915	20,085
7	8.93	150,000	13,395	143,310	6,690
8	4.46	150,000	6,690	150,000	0
	100.00%		$150,000		

Notice in this example several aspects of the MACRS depreciation method: (1) the sum of the r_t values is equal to 100.00%, (2) this 7-year MACRS GDS property is depreciated over 8 years (= property class life +1), and (3) the book value after eight years is equal to $0. ■

EXAMPLE 10-7

Investors in the JMJ Group purchased a hotel resort in April. The group paid $ 2.0 million for the hotel resort and $500,000 for the grounds surrounding the resort. The Group sold the resort five years later in August. Calculate the depreciation deductions for years 1 through 6. What was the book value at the time it was sold.

Solution

Hotels are non-residential real property and depreciated over a 39-year life. Table 10-4 lists the percentages for each year. In this case the cost basis is $2.0 million, and the $500,000 paid for the land is not depreciated. JMJ's depreciation is calculated as:

Year 1 (obtained in April)	$d_1 = 2,000,000 \ (1.819\%) = \$ 36,380$
Year 2	$d_2 = 2,000,000 \ (2.564\%) = \$ 51,280$
Year 3	$d_3 = 2,000,000 \ (2.564\%) = \$ 51,280$
Year 4	$d_4 = 2,000,000 \ (2.564\%) = \$ 51,280$
Year 5	$d_5 = 2,000,000 \ (2.564\%) = \$ 51,280$
Year 6 (disposed of in August)	$d_6 = 2,000,000 \ (1.605\%) = \$ 32,100$

Thus the hotel's book value when it was sold was:

$$BV_6 = B - (d_1 + d_2 + d_3 + d_4 + d_5 + d_6) = 2,000,000 - (273,600) = \$1,726,400$$

The value of the land has not changed in terms of book value. ■

Comparing MACRS and Historical Methods

In examples 10-2 through 10-4 we illustrated how the book value of an asset that cost $900 and had a salvage value of $70 changed over its 5-year depreciation life using the *straight line*, *sum-of-the-years digits*, and *declining balance* depreciation methods. Figures 10-2 through 10-4 provided a graphical view of book value over the 5-year depreciation period using these methods. Example 10-8 below illustrates the use of the MACRS GDS depreciation method when compared directly to the historical methods.

EXAMPLE 10-8

Consider the furniture that was purchased in Example 10-6. Calculate the asset's depreciation deductions and book values over its depreciable life for MACRS and the historical methods.

Solution

Table 10-5 and Figure 10-6 compare MACRS and historical depreciation methods. MACRS depreciation is the most *accelerated* or fastest depreciation method — remember its name is the Modified *Accelerated* Cost Recovery System. The book value drops fastest and furthest with MACRS, thus the present worth is the largest for the MACRS depreciation deductions.

Depreciation deductions *benefit* a firm after taxes because they reduce taxable income and taxes. The time value of money ensures that it is better to take these deductions as soon as possible. In general, MACRS, which allocates larger deductions earlier in the depreciation life, provides more economic benefits than historical methods.

Table 10-5 COMPARISON OF MACRS AND HISTORICAL METHODS FOR EXAMPLE 10-6 ASSET

Year	MACRS		Straight Line		Double Declining		Sum of Years	
(t)	d_t	BV_t	d_t	BV_t	d_t	BV_t	d_t	BV_t
1	21,435	128,565	12,000	138,000	30,000	120,000	21,818	128,182
2	36,735	91,830	12,000	126,000	24,000	96,000	19,636	108,545
3	26,235	65,595	12,000	114,000	19,200	76,800	17,455	91,091
4	18,735	46,860	12,000	102,000	15,360	61,440	15,273	75,818
5	13,395	33,465	12,000	90,000	12,288	49,152	13,091	62,727
6	13,380	20,085	12,000	78,000	9,830	39,322	10,909	51,818
7	13,395	6,690	12,000	66,000	7,864	31,457	8,727	43,091
8	6,690	0	12,000	54,000	1,457	30,000	6,545	36,545
9	0	0	12,000	42,000	0	30,000	4,364	32,182
10	0	0	12,000	30,000	0	30,000	2,182	30,000
PW (10%)	$108,217		$73,734		$89,918		$84,118	

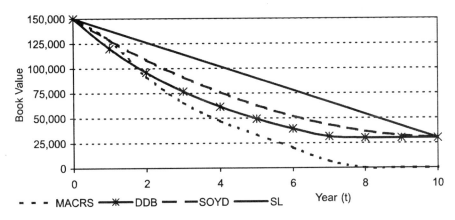

Figure 10-5 Comparing MACRS and historical depreciation methods. ■

Depreciation And Asset Disposal

When an asset is disposed of, the key question is which is larger (1) the asset's *book value, BV* (what we show in our accounting records after applying the rules set by the government) or (2) the asset's *market value , MV* (what a willing buyer pays)? If the book value is lower than the market value, then too much depreciation has been deducted from taxable income. On the other hand, if the book value is higher than the market value, then there is a *loss* on the disposal. In either case, the current level of taxes owed changes.

Depreciation Recapture (Ordinary Gains): Depreciation recapture, also called ordinary gains, is necessary when an asset is sold for more than an asset's current book value. If more than the original cost basis is received, only the amount up to the original cost basis is recaptured depreciation. Ordinary gains are also called *depreciation recapture* because the amount of gain represents the over-expense in depreciation that has been claimed. In other words we've taken too much expense for the asset's "loss in value."

Losses: A *loss* occurs when less than book value is received for a depreciated asset. In the accounting records we've exchanged an asset worth its book value for something less – which is a loss. In this case a company has not claimed enough depreciation expense.

Capital Gains: Capital gains occur when more than the asset's original cost basis is received for it. The excess over the original cost basis is the *capital gain*. As described in Chapter 11, the tax rate on such gains is sometimes lower than for ordinary income, but there are complicated rules on how long the investment has been held (short is ≤ 18 months or long is ≥ 18 months). In most engineering economic analyses capital gains are very uncommon, because business and production equipment and facilities almost always *lose* value over time. Capital gains are much more likely to occur in non-depreciated assets like: stocks, bonds, real estate, jewelry, art, and collectibles.

The relationship between these values is illustrated in Figure 10-6.

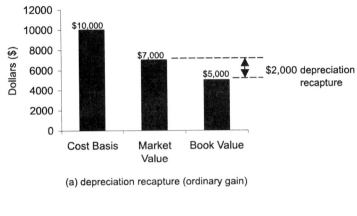

(a) depreciation recapture (ordinary gain)

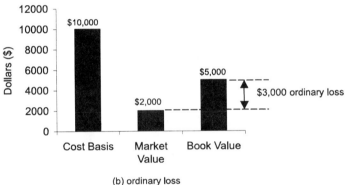

(b) ordinary loss

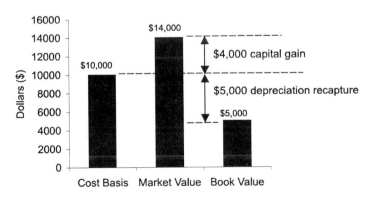

(c) capital gain and depreciation recapture

Figure 10-6 Recaptured depreciation, loss on sale, and capital gain.

Example 10-9

Consider an asset with a cost basis of $10,000 that has been depreciated using the MACRS method. This asset is a 3-Year MACRS property. What is the gain or loss if the asset is disposed of after five years of operation for: (a) $7000, (b) $0, (c) a cost of $2000.

Solution

To find *gain* or *loss* at disposal we compare *market* and *book value*. Since MACRS depreciates to a salvage value of 0, and 5 years is greater than the recovery period, the book value equals $0.

(a) Recaptured depreciation = $7000 since the book value is $7000 higher.
(b) Since market value equals salvage value, there is no recaptured depreciation or loss.
(c) Since the money is paid for disposal, there is a loss as this is less than the book value.
This general method for calculating recaptured depreciation or loss applies to all of the depreciation methods described in this chapter. ∎

If the asset is in the middle of its depreciable life, then recaptured depreciation and losses are calculated in a similar manner — compare the *market* and *book values* at the time of disposal. However, in computing the book value with MACRS depreciation a special rule must be applied for assets disposed of before the end of the recovery period. The rule is to *take one half of the allowable depreciation deduction for that year.* This basically assumes that disposals take place on average ½ way through the year. Thus for a 5-year asset disposed of in year 4, the rate allowed for MACRS depreciation is half of 11.52% or 5.76%. If the asset is disposed of in year 6, it is already past the recovery period, and a ½ year assumption has already been built into the MACRS schedule. Thus, the full r_6 is taken. Example 10-10 below illustrates several cases of disposal before the asset is fully depreciated.

EXAMPLE 10-10

Consider the asset in Example 10-9 above. Do the following:
(1) Calculate the effect of disposal if this asset is sold during year 2 for $2500.
(2) Calculate the effect of disposal if the asset is sold during year 3 for $2500.
(3) Calculate the effect of disposal if the asset is sold after year 3 for $4000 if straight line depreciation is used over the asset's 5 year life, and a salvage value of $5000 is assumed.

Solution

(1) Market Value$_2$ = $2500
    Book Value$_2$ = 10,000 - 10,000 $[r_1 + r_2/2]$ = 10,000 - 10,000[0.3333 + 0.4445/2]
    = $4444.5
    Loss = $1944.5 (= 4444.5 - 2500)   (Loss since *BV* greater than *MV*)

Note: if the full rather than ½ deduction were taken in year 2, then the book value would be $2222.50 less and the loss would become a gain of $278. Since depreciation would increase by that $2222.50, the total of (depreciation) + (loss or gain) would be $4167 in either calculation.

(2) Market Value$_3$ = $2500

Book Value$_3$ = 10,000 - 10,000 $[r_1 + r_2 + r_3/2]$

= 10,000 - 10,000[0.3333 + 0.4445 + 0.1481/2] = $1481.50

Recaptured depreciation = 2500 - 1481.50 = $1018.50   (Recaptured since $MV > BV$)

(3) Straight line rate   = $(B - S) / N$ = (10,000 - 5000)/5 = $1000

Market Value$_4$ = $4000

Book Value$_4$ = 10,000 - (4)(1000) = $6000

Loss   = 6000 - 4000 = $2000

If the asset were disposed of *during the year rather at year's end* then the straight line depreciation deduction would have to be prorated for the number of months during the year that the asset was in service. There is no half year convention with the historical depreciation methods. Example: if disposal occurred on September 30th of the 4th year, d_4 = (9/12)(1000) = $750

Note: in case (1) it was shown that the required ½ year convention did not affect the total deductions from taxable income. This is true for the other cases as well, since recaptured depreciation and losses are both treated as ordinary income.   ■

Unit Of Production Depreciation

At times, there may be situations where the recovery of depreciation on a particular asset is more closely related to use than to time. In these few situations (and they are rare), the *Unit Of Production—UOP—depreciation* in any year is:

$$\text{UOP depreciation in any year } = \frac{\text{Production for year}}{\text{Total lifetime production for asset}}(B - S) \qquad (10\text{-}6)$$

This method might be useful for machinery that processes natural resources if the resources will be exhausted before the machinery wears out. It is not considered an acceptable method for general use in depreciating industrial equipment.

EXAMPLE 10-11

For similarity with previous examples, assume that equipment costing $900 has been purchased for use in a sand and gravel pit. The pit will operate for five years, while a nearby airport is being reconstructed and paved. Then the pit will be shut down, and the equipment removed and sold for $70. Compute the unit-of-production (UOP) depreciation schedule, if the airport reconstruction schedule calls for 40,000 cubic meters of sand and gravel as follows:

Year	Required sand and gravel, in m^3
1	4,000
2	8,000
3	16,000
4	8,000
5	4,000

Solution:
The cost basis, B, is $900. The salvage value, S, is $70. The total lifetime production for the asset is 40,000 cubic meters of sand and gravel. From the airport reconstruction schedule, the first-year UOP depreciation would be:

$$\text{First-year UOP depreciation} = \frac{4000 \text{ m}^3}{40,000 \text{ m}^3}(\$900 - \$70) = \$83$$

Similar calculations for the subsequent four years gives the complete depreciation schedule:

Year	UOP depreciation
1	$ 83
2	166
3	332
4	166
5	83
	$830

It should be noted that the actual unit of production depreciation charge in any year is based on the actual production for the year rather than the scheduled production. ■

Depletion

Depletion is the exhaustion of natural resources as a result of their removal. Since depletion covers such things as mineral properties, oil and gas wells, and standing timber, removal may take the form of digging up metallic or nonmetallic minerals, producing petroleum or natural gas from wells, or cutting down trees.

Depletion is recognized for income taxes for the same reason depreciation is–capital investment is being consumed or used up. Thus a portion of the gross income should be considered a return of the capital investment. The calculation of the depletion allowance is different from depreciation as there are two distinct methods of calculating depletion: *cost depletion* and *percentage depletion*. Except for standing timber and most oil and gas wells, depletion is calculated by both methods and the larger value is taken as depletion for the year. For standing timber and most oil and gas wells, only cost depletion is permissible.

Cost Depletion

Depreciation relied on an asset's cost, depreciable life, and salvage value to apportion the cost *minus* salvage value *over* the depreciable life. In some cases where the asset is used at fluctuating rates, we might use the unit-of-production (UOP) method of depreciation. For mines, oil wells, and standing timber, fluctuating production rates are the usual situation. Thus, *cost depletion* is computed like unit-of-production depreciation using:

1. Property cost;
2. Estimated number of recoverable units (tons of ore, cubic meters of gravel, barrels of oil, million cubic feet of natural gas, thousand board-feet of timber, etc.);
3. Salvage value, if any, of the property.

EXAMPLE 10-12

A small lumber company bought a tract of timber for $35,000, of which $5000 was the land's value and $30,000 was the value of the estimated 1½ million board-feet of standing timber. The first year, the company cut 100,000 board-feet of standing timber. What was the year's depletion allowance?

Solution:

$$\text{Depletion allowance per 1000 board-ft} = \frac{\$35,000 - \$5000}{1,500 \text{ board-ft}}$$

$$= \$20 \text{ per 1000 board-ft}$$

The depletion allowance for the year would be
   100,000 board-ft × $20 per 1000 board-ft = $2000   ■

Percentage Depletion

Percentage depletion is an alternate method for mineral property and some oil or gas wells. The allowance is a certain percentage of the property's gross income during the year. This is an entirely different concept than depreciation. Unlike depreciation, which allocates cost *over useful life*, the *percentage depletion* allowance is based on the property's gross income.

Since percentage depletion is computed on the *income* rather than the property's cost, the total depletion *may exceed the cost of the property*. In computing the *allowable percentage depletion* on a property in any year, the *percentage depletion allowance* cannot exceed 50% of the property's taxable income computed without the depletion deduction. The percentage depletion calculations are illustrated by Ex. 10-13.

Table 10-6 PERCENTAGE DEPLETION ALLOWANCE FOR SELECTED ITEMS

Type of deposit	*Percent*
Lead, zinc, nickel, sulphur, uranium	22
Oil and gas (small producers only)	15
Gold, silver, copper, iron ore	15
Coal and sodium chloride	10
Sand, gravel, stone, clam & oyster shells, brick, and tile clay	5
Most other minerals and metallic ores	14

EXAMPLE 10-13

A coal mine has a gross income of $250,000 for the year. Mining expenses equal $210,000. Compute the allowable percentage depletion deduction.

Solution:
From Table 10-6, coal has a 10% depletion allowance. The percentage depletion deduction is computed from gross mining income. Then the taxable income must be computed. The allowable percentage depletion deduction is limited to the computed percentage depletion or 50% of taxable income, whichever is smaller.

Computed percentage depletion:

Gross income from mine	$250,000
Depletion percentage	× 10%
Computed percentage depletion	$ 25,000

Taxable income limitation:

Gross income from mine	$250,000
Less: expenses other than depletion	−210,000
Taxable income from mine	40,000
Deduction limitation	× 50%
Taxable income limitation	$ 20,000

Since the taxable income limitation ($20,000) is less than the computed percentage depletion ($25,000), the allowable percentage depletion deduction is $20,000. ∎

As previously stated, on mineral property and some oil and gas wells, the depletion deduction can be based on either cost or percentage depletion. Each year, depletion is computed by both methods, and the allowable depletion deduction is the larger of the two amounts.

Spreadsheets and Depreciation

The spreadsheet functions for straight-line, declining-balance, and sum-of-years digits depreciation are listed in Table 10-7. Because these techniques are simple and they were replaced by MACRS in 1986, they are not covered in detail here. All three functions include parameters for *cost* (initial book value), *salvage* (final salvage value), and *life* (depreciation period). Both DDB and SYD change depreciation amounts every year, so they include a parameter to pick the *period* (year). Finally, DDB includes a *factor*. The default value is 2 for 200% or double-declining balance, but another commonly used value is 1.5 for 150%.

Table 10-7 SPREADSHEET FUNCTIONS FOR DEPRECIATION

Depreciation technique	*EXCEL*
Straight-line	SLN(cost, salvage, life)
Declining Balance	DDB(cost, salvage, life, period, factor)
Sum-of-years digits	SYD(cost, salvage, life, period)
MACRS	VDB(cost, salvage, life, start_period, end_period, factor, no_switch)

Using VDB for MACRS

The Excel function VDB is a flexible or variable declining balance method. It includes the ability to specify the starting and ending periods, rather than simply a year. It also includes an optional no-switch for problems where a switch from declining balance to straight line depreciation is NOT desired.

To use VDB to calculate MACRS depreciation the following is true:

1. Salvage = 0, since MACRS assumes no salvage value
2. Life = recovery period of 3, 5, 7, 10, 15, or 20 years
3. First period runs from 0 to 0.5, 2nd period from 0.5 to 1.5, 3rd from 1.5 to 2.5, t^{th} from $t - 1.5$ to $t - 0.5$, and last from $life - 0.5$ to $life$
4. Factor = 2 for recovery periods of 3, 5, 7, or 10 years and = 1.5 for recovery periods of 15 or 20 years
5. No_switch can be omitted, since MACRS includes a switch to straight line.

5. No_switch can be omitted, since MACRS includes a switch to straight line.

The start_period and end_period arguments are from $t - 1.5$ to $t - 0.5$, because MACRS uses a ½ year convention for the first year. Thus the first year has 0 to 0.5 year's of depreciation, and the second year starts where the first year stops. When writing the Excel function either the first and last periods must be edited individually, or start_period must be defined with a minimum of 0 and end_period with a maximum of life. This prevents the calculation of depreciation from -0.5 to 0 and from *life* to *life* $+ 0.5$.

The results of using the VDB function match Table 10-4, except that the VDB function has more significant digits rather than being rounded to 2 decimal points. Example 10-14 illustrates the use of the VDB function.

EXAMPLE 10-14

Return to the data of Example 10-5 which had $150,000 of office equipment which is 7-year MACRS property. Use VDB to compute the depreciation amounts.

Solution
The following spreadsheet defines the start_period with a minimum of 0 and the end_period with a maximum of life. Thus this formula could be used for any year of any recovery schedule. Notice that the VDB formula uses the value 0 for the salvage value, rather than referring to the data cell for the salvage value. MACRS assumes a salvage value of zero – no matter what the value truly is.

	A	B	C	D	E	F	G
1	150,000	First cost					
2	0	Salvage					
3	7	Life					
4	200%	Factor					
5							
6	Period	Depreciation					
7	1	$21,428.57	=VDB($A$1,0,$A$3,MAX(0,A7-1.5),MIN($A$3,A7-0.5),$A$4)				
8	2	$36,734.69	or (cost, salvage, life, max(0, t-1.5), min (life, t-.5), factor)				
9	3	$26,239.07					
10	4	$18,742.19					
11	5	$13,387.28					
12	6	$13,387.28					
13	7	$13,387.28					
14	8	$6,693.64					
15		$150,000	= Sum				

Figure 10-7 Using VDB to calculate MACRS depreciation. ■

Summary

Depreciation is the foundation for including income taxes in economic analysis. There are three distinct definitions of depreciation:

1. Decline in asset's market value.
2. Decline in asset's value to its owner.
3. Allocating the asset's cost *less* its salvage value *over* its recovery period or depreciable life.

While the first two definitions are used in everyday discussions, it is the third, or accountant's, definition that is used in tax computations and this chapter. Book value is the remaining unallocated cost of an asset, or:

Book value = Cost − Depreciation charges made to date

This chapter describes how depreciable assets are *written off* (or claimed as a business expense) over a period of years instead of *expensed* in a single period (like wages, material costs, etc.). The depreciation methods described include the historical pre-1981 methods: *straight line, sum-of-the-years digits,* and *declining balance.* These methods required estimating the asset's salvage value and depreciable life.

The current tax law specifies use of the Modified Accelerated Capital Recovery System (MACRS) method. This chapter has focused on the general depreciation system (GDS) with limited discussion of the less attractive alternative depreciation system (ADS). MACRS (GDS) specifies faster *recovery periods* and a salvage value of zero, so it is generally economically more attractive than the historical methods.

The MACRS system is the current tax law, and it assumes a salvage value of zero. This is in contrast with historical methods that ensured the final book value would equal the predicted salvage value. Thus, when using MACRS it is often necessary to consider recaptured depreciation. This is the excess of salvage value over book value, and it is taxed as ordinary income. Similarly, losses on sale or disposal are taxed as ordinary income.

Integrating depreciation schedules with cash flows often involves a lot of arithmetic. Thus, the tool of spreadsheets can be quite helpful. The functions for the historical methods: straight line, sum-of-the-years digits, and declining balance are straightforward. Rather than individually entering MACRS percentages into the spreadsheet, the function VDB can be used to calculate MACRS depreciation percentages.

Unit of production (UOP) depreciation relies on usage to quantify the loss in value. UOP is appropriate for assets which lose value based on the number of units produced, the tons of gravel moved, etc. (versus the number of years in service). However, this method is not considered acceptable for most business assets.

Depletion is the exhaustion of natural resources like minerals, oil and gas wells, and standing timber. The owner of the natural resources is consuming his investment as the natural resources are removed and sold. Cost depletion is computed based on the fraction of the resource that is removed or sold. For minerals and some oil and gas wells, an alternate calculation called percentage depletion is allowed. Percentage depletion is based on income, so the total allowable depletion deductions may *exceed* the invested cost.

Problems

10-1   Some special handling devices can be obtained for $12,000. At the end of four years, they can be sold for $600. Compute the depreciation schedule for the devices using the following methods:

    a.   Straight line depreciation.
    b.   Sum-of-years digits depreciation.
    c.   Double declining balance depreciation.
    d.   MACRS depreciation.

10-2   The company treasurer is uncertain which of four depreciation methods is more desirable for the firm to use for office furniture. Its cost is $50,000, with a zero salvage value at the end of a ten-year depreciable life. Compute the depreciation schedule for the office furniture using the methods listed:

    a.   Straight line.
    b.   Double declining balance.
    c.   Sum-of-years digits.
    d.   Modified accelerated cost recovery system.

10-3   The RX Drug Company has just purchased a capsulating machine for $76,000. The plant engineer estimates the machine has a useful life of five years and little or no salvage value. He will use zero salvage value in the computations.

    a.   Compute the depreciation schedule for the machine using:
        1.   Straight line depreciation.
        2.   Sum-of-years digits depreciation.
        3.   Double declining balance depreciation.
    b.   The controller for RX Drug Co. believes that DDB with conversion to straight line depreciation may be a more desirable depreciation method. Compute the depreciation schedule for DDB depreciation with conversion to straight line depreciation at the desirable point. (Hint: see Example 10-5.)

10-4   A new machine tool is being purchased for $16,000 and is expected to have a zero salvage value at the end of its five-year useful life. Compute the DDB depreciation schedule for this capital asset. It may be desirable for this profitable firm to use double declining balance depreciation with conversion to straight line depreciation. If it is desirable, compute the depreciation schedule, making the conversion from DDB depreciation to straight line depreciation at the optimum time. Tabulate the resulting depreciation schedule. (Hint: see Example 10-5.)

10-5   A large profitable corporation purchased a small jet plane for use by the firm's executives in January. The plane cost $1,500,000 and, for depreciation purposes, is assumed to have a zero salvage value at the end of five years. Compute the MACRS depreciation schedule.

10-6 When a major highway was to be constructed nearby, a farmer realized that a dry stream bed running through his property might a valuable source of sand and gravel. He shipped samples to a testing laboratory and learned that the material met the requirements for certain low-grade fill material for the highway. The farmer contacted the highway construction contractor, who offered 65¢ per cubic meter for 45,000 cubic meters of sand and gravel. The contractor would build a haul road and would use his own equipment. All activity would take place during a single summer.

The farmer hired an engineering student for $2500 to count the truckloads of material hauled away. The farmer estimated that two acres of stream bed had been stripped of the sand and gravel. The 640-acre farm had cost him $300 per acre and the farmer felt the property had not changed in value. He knew that there had been no use for the sand and gravel prior to the construction of the highway, and he could foresee no future use for any of the remaining 50,000 cubic meters of sand and gravel. Determine the farmer's depletion allowance. (Answer: $1462.50)

10-7 Mr. H. Salt purchased a ⅛ interest in a producing oil well for $45,000. Recoverable oil reserves for the well were estimated at that time at 15,000 barrels, ⅛ of which represented Mr. Salt's share of the reserves. During the subsequent year, Mr. Salt received $12,000 as his ⅛ share of the gross income from the sale of 1000 barrels of oil. From this amount, he had to pay $3000 as his share of the expense of producing the oil. Compute Mr. Salt's depletion allowance for the year. (Answer: $3000)

10-8 A heavy construction firm has been awarded a contract to build a large concrete dam. It is expected that a total of eight years will be required to complete the work. The firm will buy $600,000 worth of special equipment for the job. During the preparation of the job cost estimate, the following utilization schedule was computed for the special equipment:

Year	Utilization (hours/year)	Year	Utilization (hours/year)
1	6000	5	800
2	4000	6	800
3	4000	7	2200
4	1600	8	2200

At the end of the job, it is estimated that the equipment can be sold at auction for $60,000.

a. Compute the sum-of-years digits depreciation schedule.

b. Compute the unit of production depreciation schedule.

10-9 Consider five depreciation schedules:

Year	A	B	C	D	E
1	$45.00	$35.00	$29.00	$58.00	$43.50
2	36.00	20.00	46.40	34.80	30.45
3	27.00	30.00	27.84	20.88	21.32
4	18.00	30.00	16.70	12.53	14.92
5	9.00	20.00	16.70	7.52	10.44
6			8.36		

They are based on the same initial cost, useful life, and salvage value. Identify each schedule as one of the following

- Straight line depreciation
- Sum-of-years digits depreciation;
- 150% declining balance depreciation
- Double declining balance depreciation
- Unit of production depreciation
- Modified accelerated cost recovery system

10-10 Using MACRS GDS depreciation for each of the (a-c) assets below, calculate items (1-3) as given below:

(a) A light general purpose truck used by a delivery business, cost = $17,000.

(b) Production equipment used by a Detroit automaker to produce vehicles, cost = $30,000.

(c) Cement production facilities used by a construction firm, cost $130,000.

(1) The MACRS GDS Property Class

(2) The depreciation deduction for year 3

(3) The book value of the asset after 6 years

10-11 Consider a $6500 piece of machinery, with a five-year depreciable life and an estimated $1200 salvage value. The projected utilization of the machinery when it was purchased, and its actual production to date, are shown below.

Year	Projected production, in tons	Actual production, in tons
1	3500	3000
2	4000	5000
3	4500	[Not
4	5000	yet
5	5500	known]

Compute the machinery depreciation schedule by each of the following methods:

Straight line.

- *a.* Sum-of-years digits.
- *b.* Double declining balance.
- *c.* Unit of production (for first two years only).
- *d.* Modified accelerated cost recovery system

10-12 A depreciable asset costs $10,000 and has an estimated salvage value of $1600 at the end of its six-year depreciable life. Compute the depreciation schedule for this asset by both SOYD depreciation and DDB depreciation.

10-13 The MACRS depreciation percentages for seven-year personal property are given in Table 10-3. Make the necessary computations to determine if the percentages shown are correct.

10-14 The Acme Chemical Company purchased $45,000 of research equipment which it believes will have zero salvage value at the end of its five-year life. Compute the depreciation schedule for the equipment by each of the following methods:
- *a.*   Straight line.
- *b.*   Sum-of-years digits.
- *c.*   Double declining balance.
- *d.*   Modified accelerated cost recovery system.

10-15 The MACRS depreciation percentages for ten-year personal property are given in Table 10-3. Make the necessary computations to determine if the percentages shown are correct.

10-16 A $1,000,000 oil drilling rig has a six-year depreciable life and a $75,000 salvage value at the end of that time. Determine which one of the following methods: DDB, DDB with conversion to straight line, or SOYD provides the preferred depreciation schedule. Show the depreciation schedule for the preferred method.

10-17 Some equipment costs $1000, has a five-year depreciable life, and an estimated $50 salvage value at the end of that time. Ann Landers has been assigned the problem to determine whether to use:
- *a.*   Straight line, or
- *b.*   SOYD depreciation.

If a 10% interest rate is appropriate, which is the preferred depreciation method for this profitable corporation? Show your computations of the difference in present worths.

10-18 The depreciation schedule for an asset, with a salvage value of $90 at the end of the recovery period, has been computed by several methods. Identify the depreciation method used for each schedule.

Depreciation Schedule

Year	A	B	C	D	E
1	$323.3	$212.0	$424.0	$194.0	$107.0
2	258.7	339.2	254.4	194.0	216.0
3	194.0	203.5	152.6	194.0	324.0
4	129.3	122.1	91.6	194.0	216.0
5	64.7	122.1	47.4	194.0	107.0
6	____	61.1	____	____	____
	970.0	1060.0	970.0	970.0	970

10-19 TELCO Corp has leased some industrial land near its plant. It is building a small warehouse on the site at a cost of $250,000. The building will be ready for use January 1st. The lease will expire fifteen years after the building is occupied. The warehouse will belong at that time to the landowner, with the result that there will be no salvage value to TELCO. The warehouse is to be depreciated either by MACRS or SOYD depreciation. If 10% interest is appropriate, which depreciation method should be selected?

10-20 The FOURX Corp. has purchased $12,000 of experimental equipment. The anticipated salvage value is $400 at the end of its five-year depreciable life. This profitable corporation is considering two methods of depreciation:

1. Sum-of-years digits;
2. Double declining balance.

If it uses 7% interest in its comparison, which method do you recommend? Show computations to support your recommendation.

10-21 The Able Corp. is buying $10,000 of special tools for its fabricated metal products that have a four-year useful life and no salvage value. Compute the depreciation charge for the *second* year by each of the following methods:

a. DDB.
b. Sum-of-years digits.
c. Modified accelerated cost recovery system..

10-22 On July lst, Nancy Regan paid $600,000 for a commercial building and an additional $150,000 for the land on which it stands. Four years later, also on July 1st, she sold the property for $850,000. Compute the modified accelerated cost recovery system depreciation for each of the *five* calendar years during which she had the property.

10-23 The White Swan Talc Company purchased $120,000 of mining equipment for a small talc mine. The mining engineer's report indicates the mine contains 40,000 cubic meters of commercial quality talc. The company plans to mine all the talc in the next five years as follows:

Year	Talc production, in cubic meters
1	15,000
2	11,000
3	4,000
4	6,000
5	4,000

At the end of five years, the mine will be exhausted and the mining equipment will be worthless. The company accountant must now decide whether to use sum-of-years digits depreciation or unit of production depreciation. The company considers 8% to be an appropriate time value of money. Compute the depreciation schedule for each of the two methods. Which method would you recommend that the company adopt? Show the computations to justify your decision.

10-24 For an asset that fits into the MACRS "All property not assigned to another class" designation, show in a table the depreciation and book value over the asset's 10 year life of use. The cost basis of the asset is $10,000.

10-25 The depreciation schedule for a microcomputer has been arrived at by several methods. The estimated salvage value of the equipment at the end of its six-year useful life is $600. Identify the resulting depreciation schedules.

Depreciation Schedule

Year	A	B	C	D
1	$2114	$2000	$1600	$1233
2	1762	1500	2560	1233
3	1410	1125	1536	1233
4	1057	844	922	1233
5	705	633	922	1233
6	352	475	460	1233

10-26 A company that manufactures food and beverages in the vending industry has purchased some handling equipment that cost $75,000 and will be depreciated using MACRS GDS. The class life of the asset is 4 years, show in a table the yearly depreciation amount and book value of the asset over its depreciation life.

10-27 A group of investors has formed Trump Corporation to purchase a small hotel. The asking price is $150,000 for the land and $850,000 for the hotel building. If the purchase takes place in June, compute the MACRS depreciation for the first three calendar years. Then assume the hotel is sold in June of the fourth year, and compute the MACRS depreciation in that year also.

10-28 Equipment costing $20,000 that is a MACRS 3-Year Property is disposed of during the second year for $14,000. Calculate any depreciation recapture, ordinary losses or capital gains associated with disposal of the equipment.

10-29 Mr. Donald Spade purchased a computer in January to keep records on all the property he owns. The computer cost $70,000 and is to be depreciated using MACRS. Donald's accountant pointed out that under a special tax rule (the rule applies when the value of property placed in service in the last three months of the tax year exceeds 40% of all the property placed in service during the tax year), the computer and all property that year would be subject to the mid-quarter convention. The mid-quarter convention assumes that all property placed in service in any quarter year is placed in service at the midpoint of the quarter. Compute Donald's MACRS depreciation for the first year, using the mid-quarter convention.

10-30 A company is considering buying a new piece of machinery. A 10% interest rate will be used in the computations. Two models of the machine are available.

	Machine I	Machine II
Initial cost	$80,000	$100,000
End of useful life salvage value, S	20,000	25,000
Annual operating cost	18,000	15,000 first 10 years 20,000 thereafter
Useful life	20 years	25 years

a. Determine which machine should be purchased, based on equivalent uniform annual cost.

b. What is the capitalized cost of Machine I?

c. If Machine I is purchased and a fund is set up to replace Machine I at the end of 20 years, compute the required uniform annual deposit.

d. Machine I will produce an annual saving of material of $28,000. What is the rate of return if Machine I is installed?

e. What will be the book value of Machine I after two years, based on sum-of-years digits depreciation?

f. What will be the book value of Machine II after three years, based on double declining balance depreciation?

g. Assuming Machine II is in the seven-year property class, what would be the MACRS depreciation in the third year?

10-31 A profitable company making earth moving equipment is considering an investment of $100,000 on equipment which will have 5 year useful life and a $20,000 salvage value. If money is worth 10%, which one of the following three methods of depreciation would be preferable?
  a. Straight Line method
  b. Double Declining Balance method
  c. MACRS method
  d. Use a spreadsheet function to compute the MACRS method

10-32 Office equipment whose initial cost is $100,000 has an estimated actual life of 6 years, with an estimated salvage value of $10,000. Prepare tables listing the annual costs of depreciation and the book value at the end of each year of the six years, based on the straight-line, sum-of-years digits, and MACRS depreciation. Use spreadsheet functions for the depreciation methods.

10-33 You are equipping an office. The total office equipment will have a first cost of $1,750,000 and a salvage value of $200,000. You expect the equipment will last 10 years. Use a spreadsheet function to compute the MACRS depreciation schedule.

10-34 An asset with a 8 year Class Life costs $50,000 and is purchased on January 1, 2001. Calculate any depreciation recapture, ordinary losses or capital gains associated with selling the equipment on Dec. 31, 2003 for (a) $15,000 , (b) $25,000, and (c) $60,000. Consider two cases of depreciation for the above problem: if (1) MACRS GDS is used, and if (2) Straightline depreciation over the Class life is used with a $10,000 salvage value.

CHAPTER 11

Income Taxes

As Benjamin Franklin said, two things are inevitable: death and taxes. In this chapter we will examine the structure of taxes in the United States. There is, of course, a wide variety of taxes ranging from sales taxes, gasoline taxes, property taxes, state and federal income taxes, and so forth. Here we will concentrate our attention on federal income taxes. Income taxes are part of most real problems and often have a substantial impact that must be considered.

First, we must understand the way in which taxes are imposed. The previous chapter concerning depreciation is an integral part of this analysis, so it is essential that the principles covered there are well understood. Then, having understood the mechanism, we will see how federal income taxes affect our economic analysis. The various analysis techniques will be used in examples of after-tax calculations.

A Partner In The Business

Probably the most straightforward way to understand the role of federal income taxes is to consider the U.S. Government as a partner in every business activity. As a partner, the Government shares in the profits from every successful venture. And in a somewhat more complex way, the Government shares in the losses of unprofitable ventures. The tax laws are complex and it is not our purpose to fully explain them.[1] Instead, we will examine the fundamental concepts of the federal income tax laws-and we must recognize that there are exceptions and variations to almost every statement we shall make!

---

[1] Both Commerce Clearing House and Research Institute of America have loose-leaf income tax reporting services that fill many binders each year with detailed tax information. One or both will be found in the reference section of most libraries.

Calculation Of Taxable Income

At the mention of income taxes, one can visualize dozens of elaborate and complex calculations. And there is some truth to that vision, for there can be all sorts of complexities in the computation of income taxes. Yet some of the difficulty is removed when one defines incomes taxes as just another type of disbursement. Our economic analysis calculations in prior chapters have dealt with all sorts of disbursements: operating costs, maintenance, labor and materials, and so forth. Now we simply add one more prospective disbursement to the list—income taxes.

Taxable Income of Individuals

The amount of federal income taxes to be paid depends on taxable income and the income tax rates. Therefore, our first concern is the definition of *taxable income.* To begin, one must compute his or her *gross income:*

> Gross income = Wages, salary, etc. + Interest income
> + Dividends
> + Capital gains
> + Unemployment compensation
> + Other income

From gross income, we subtract any allowable retirement plan contributions and other *adjustments.* The result is *adjusted gross income.* From adjusted gross income, individuals may deduct the following items:[2]

> **1.** *Personal Exemptions.* One exemption ($2750 for 1999) is provided for each person who depends on the gross income for his or her living.
>
> **2a.** *Itemized Deductions.* Some of these are:
>
> > **a.** Excessive medical and dental expenses (exceeding 7½% of adjusted gross income);
> >
> > **b.** State and local income, property and personal property tax;
> >
> > **c.** Home mortgage interest;
> >
> > **d.** Charitable contributions;
> >
> > **e.** Casualty and theft losses;

---

[2] The 1999 itemized deduction are limited if adjusted gross income is more than $126,600 ($63,300 if married filing separately.)

     f. Miscellaneous deductions (some categories must exceed 2% of adjusted gross income);

     g. Car and other business expenses;

     h. Tax benefits for Work Related Education.

2b. *Standard Deduction.* Each taxpayer may either itemize his or her deductions, or instead take a standard deduction as follows:

- Single taxpayers, $4300 (for 1999)

- Married taxpayers filing a joint return, $7200 (for 1999)

The result is *taxable income.*

For individuals:

Adjusted gross income = Gross income – Adjustments

$$Taxable\ income = \textbf{Adjusted gross income} \tag{11-1}$$
$$\textbf{– Personal exemption(s)}$$
$$\textbf{– Itemized deductions or Standard deduction}$$

Classification of Business Expenditures

When an individual or a firm operates a business, there are three distinct types of business expenditures:

     1. Expenditures for depreciable assets;

     2. Expenditures for nondepreciable assets;

     3. All other business expenditures.

Expenditures for depreciable assets. When facilities or productive equipment with useful lives in excess of one year are acquired, the taxpayer will recover his investment through depreciation charges.[3] Chapter 10 examined in great detail the several ways in which the cost of the asset could be allocated over its useful life.

Expenditures for nondepreciable assets. Land is considered a nondepreciable asset, for there is no finite life associated with it. Other nondepreciable assets are properties *not* used

---

[3] There is an exception. In 1999 businesses may immediately deduct *(expense)* up to $19,000 of business equipment (§179 property) in a year, provided their total equipment expenditure for the year does not exceed $200,000.

either in a trade, in a business, or for the production of income. An individual's home and automobile are generally nondepreciable assets. The final category of nondepreciable assets are those subject to *depletion,* rather than *depreciation.* Since business firms generally acquire assets for use in the business, their only nondepreciable assets normally are land and assets subject to depletion.

All other business expenditures. This category is probably the largest of all for it includes all the ordinary and necessary expenditures of operating a business. Labor costs, materials, all

direct and indirect costs, facilities and productive equipment with a useful life of one year or less, and so forth, are part of the routine expenditures. They are charged as a business expense—*expensed*—when they occur.

Business expenditures in the first two categories—that is, for either depreciable or nondepreciable assets—are called **capital expenditures.** In the accounting records of the firm, they are **capitalized**; all ordinary and necessary expenditures in the third category are **expensed.**

Taxable Income of Business Firms

The starting point in computing a firm's taxable income is *gross income.* All ordinary and necessary expenses to conduct the business—*except* capital expenditures—are deducted from gross income. Capital expenditures may *not* be deducted from gross income. Except for land, business capital expenditures are charged to expense period by period through depreciation or depletion charges.

For business firms:

> **Taxable income = Gross income**
>
> **– All expenditures except capital expenditures**
>
> **– Depreciation and depletion charges** (11-2)

Because of the treatment of capital expenditures for tax purposes, the taxable income of a firm may be quite different from the actual cash results.

EXAMPLE 11-1   During a three-year period a firm had the following results (in millions of dollars):

	Year 1	Year 2	Year 3
Gross income from sales	$200	$200	$200
Purchase of special tooling (useful life: 3 years)	–60	0	0
All other expenditures	–140	–140	–140
Cash results for the year	$ 0	$ 60	$ 60

Compute the taxable income for each of the three years.

Solution: The cash results for each year would suggest that Year I was a poor one, while Years 2 and 3 were very profitable. A closer look reveals that the firm's cash results were adversely affected in Year 1 by the purchase of special tooling. Since the special tooling has a three-year useful life, it is a capital expenditure with its cost allocated over the useful life. For straight line depreciation and no salvage value, the annual charge is:

$$\text{Annual depreciation charge} = \frac{P - S}{N} = \frac{60 - 0}{3} = \$20 \text{ million}$$

Applying Eq. 11-2,

Taxable income = 200 – 140 – 20 = \$40 million

In each of the three years, the taxable income is \$40 million.   ■

An examination of the cash results and the taxable income in Ex. 11-1 indicates that taxable income is a better indicator of the annual performance of the firm.

Income Tax Rates

Figure 11-1 illustrates that income tax rates for individuals have changed many times since 1960.

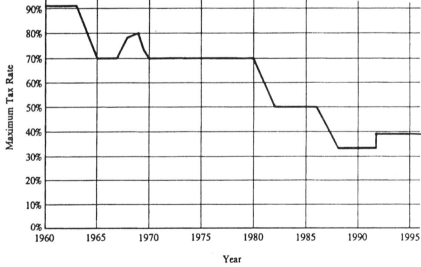

Since 1997 the individual tax rate ceiling is 39.6%

Figure 11-1 Maximum federal income tax rates for individuals.

Individual Tax Rates

There are four schedules of federal income tax rates for individuals. Single taxpayers use the Table 11-1 schedule. Married taxpayers filing a joint return use the Table 11-2 schedule. Two other schedules (not shown here) are applicable to unmarried individuals with dependent relatives ("head of household"), and married taxpayers filing separately.

Table 11-1 1999 TAX RATES–Single Filing Status

| \multicolumn{2}{}{*If your taxable income is*} | | \multicolumn{3}{}{*Your tax is*} | | |
Over	*But not over*	*This*	*Plus following percentage*	*Over this*
$ 0	$ 25,750		15%	$ 0
25,750	62,450	$3,862.50	28%	25,750
62,450	130,250	14,138.50	31%	62,450
130,250	283,150	35,156.50	36%	130,250
\multicolumn{2}{}{Over 283,150}		90,200.50	39.6%	283,150

Table 11-2 1999 TAX RATES–Married Filing Jointly

| \multicolumn{2}{}{*If your taxable income is*} | | \multicolumn{3}{}{*Your tax is*} | | |
Over	*But not over*	*This*	*Plus following percentage*	*Over this*
$ 0	41,200		15%	$ 0
43,050	104,050	$6,457.50	28%	43,050
104,050	158,550	23,537.50	31%	104,050
158,550	283,150	40,432.50	36%	158,550
\multicolumn{2}{}{Over 283,150}		85,288.50	39.6%	283,150

EXAMPLE 11-2

An unmarried student earned $8000 in the summer plus another $2000 during the rest of the year. When he files an income tax return, he will be allowed one exemption (for himself). He estimates he spent $1000 on allowable itemized deductions. How much income tax will he pay?

Solution to Ex. 11-2:

Adjusted gross income = $8000 + 2000 = $10,000

Taxable income = Adjusted gross income

– Deduction for one exemption ($2750)

− Standard deduction ($4300)

$$= 10,000 - 2750 - 4300 = \$2950$$

Federal income tax $= 15\%(2950) = \$422.50$ ■

Corporate Tax Rates

Income tax for corporations is computed in a manner similar to that for individuals. The schedule is given below and is shown graphically in Fig. 11-2.

Taxable Income	Tax Rate	Corporate Income Tax
Not over $50,000	15%	15% over $0
$50,000-75,000	25%	$7,500 + 25% over $50,000
$75,000-100,000	34%	$13,750 + 34% over $75,000
$100,000-335,000	39%*	$22,250 + 39% over $100,000
$335,000-10 million	34%	$113,900 + 34% over $335,000
$10 million-15 million	35%	$3,400,000 + 35% over $10 million
$15 million -18,333,333	38%	5,159,000 + 38% over $15 million

*The extra 5% from $100,000 to $335,000 was chaosen so that firms in the $335,000 to $10 million bracket pay a flat 34% tax rate. [(.39-.34)(335,000-100,000)=(.34-.15)(50,000)+(.34-.25)(75,000-50,000)] so tax = .34(tax income) in $335,000 to $10 million bracket.

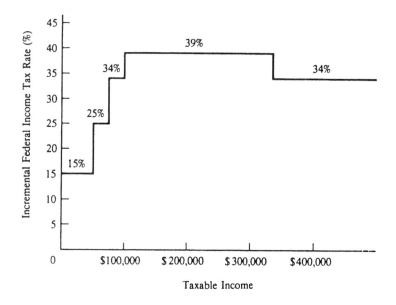

Figure 11-2 Corporation federal income tax rates (1999rates).

EXAMPLE 11-3

The French Chemical Corporation was formed to produce household bleach. The firm bought land for $220,000, had a $900,000 factory building erected, and installed $650,000 worth of chemical and packaging equipment. The plant was completed and operations begun on April 1st. The gross income for the calendar year was $450,000. Supplies and all operating expenses, excluding the capital expenditures, were $100,000. The firm will use modified accelerated cost recovery system (MACRS) depreciation.

    a.   What is the first year depreciation charge?

    b. What is the first year taxable income?

    c.   How much will the corporation pay in federal income taxes for the year?

Solution:

    a.   MACRS depreciation: Chemical equipment is personal property. From Table 10-2 it is probably in the "Seven-year, all other property" class.

        First-year depreciation = $650,000 × 14.29% = $92,885

        The building is in the 39-year real property class. Being placed in service April

1st, the appropriate

First-year depreciation = $900,000 × 1.819% = $16,371(see Table 10-4)

The land is a nondepreciable asset.

Total first year MACRS depreciation = $92,885 + $16,371 = $109,256

b. Taxable income = Gross income

– All expenditures except capital expenditures

– Depreciation and depletion charges

= $450,000 – 100,000 – 109,256 = $240,744

c. Federal income tax = $22,250 + 39%(240,744 – 100,000)

= $77,140 ∎

Combined Federal and State Income Taxes

In addition to federal income taxes, most individuals and corporations also pay state income taxes. It would be convenient if we could derive a single tax rate to represent both the state and federal incremental tax rates. In the computation of taxable income for federal taxes, the amount of state taxes paid is one of the allowable itemized deductions. Federal income taxes are not, however, generally deductible in the computation of state taxable income. Therefore, the state income tax is applied to a *larger* taxable income than is the federal income tax rate. As a result, the combined incremental tax rate will not be the sum of two tax rates.

For an increment of income (ΔIncome),and tax rate on incremental income (Δ tax rate):

State income taxes = (ΔState tax rate)(ΔIncome)

Federal taxable income = (ΔIncome)(1 – ΔState tax rate)

Federal income taxes = (ΔFederal tax rate)(ΔIncome) × (1 – ΔState tax rate)

The total of state and federal income taxes is

[ΔState tax rate + (ΔFederal tax rate)(1 – ΔState tax rate)](ΔIncome).

The terms within the left set of brackets equals the combined incremental tax rate.

Combined incremental tax rate

$$= \Delta \textbf{State tax rate} + (\Delta \textbf{Federal tax rate})(1 - \Delta \textbf{State tax rate}) \qquad (11\text{-}3)$$

EXAMPLE 11-4

An engineer has an income that puts him in the 28% federal income tax bracket and at the 10% state incremental tax rate. He has an opportunity to earn an extra $500 by doing a small consulting job. What will be his combined state and federal income tax rate on the additional income?

Solution: Using Equation 11-3, Combined incremental tax rate = 0.10 + 0.28(1 − 0.10) = 35.2% ■

Selecting an Income Tax Rate for Economy Studies

Since income tax rates vary with the level of taxable income for both individuals and corporations, one must decide which tax rate to use in a particular situation. The simple answer is the tax rate to use is the incremental tax rate that applies to the change in taxable income projected in the economic analysis. If a married couple filing jointly has a taxable income of $40,000 and can increase their income by $2000, what tax rate should be used for the $2000 of incremental income? From Table 11-2, we see the $2000 falls within the 15% tax bracket.

Now suppose this couple could increase their $40,000 income by $6000. In this situation, Table 11-2 shows that the 15% incremental tax rate should be applied to the first $3050 and a 28% incremental tax rate to the last $2950 of extra income. The appropriate incremental tax rate for corporations is equally easy to determine. For larger corporations, the federal incremental tax rate is 35%. In addition, there may be up to a 12% state tax. For computational convenience, a 50% corporate tax rate is sometimes used.

Economic Analysis Taking Income Taxes Into Account

An important step in economic analysis has been to resolve the consequences of alternatives into a cash flow. Because income taxes have been ignored, the result has been a *before-tax cash flow*. This same before-tax cash flow is an essential component in economic analysis that also considers the consequences of income tax. The principal elements in an *after-tax analysis* are:

- Before-tax cash flow;
- Depreciation;
- Taxable income (Before-tax cash flow − Depreciation);
- Income taxes (Taxable income × Incremental tax rate);
- After-tax cash flow (Before-tax cash flow − Income taxes).

These elements are usually arranged to form a *cash-flow table.* This is illustrated by Ex. 11-5.

EXAMPLE 11-5

A medium-sized profitable corporation is considering the purchase of a $3000 used pickup truck for use by the shipping and receiving department. During the truck's five-year useful life, it is estimated the firm will save $800 per year after paying all the costs of owning and operating the truck. Truck salvage value is estimated at $750.

a.   What is the before-tax rate of return?

b.   What is the after-tax rate of return on this capital expenditure? Assume straight line depreciation.

Solution to Example 11-5a: For a before-tax rate of return, we must first compute the before-tax cash flow.

Year	Before-tax cash flow
0	−$3000
1	+800
2	+800
3	+800
4	+800
5	$\begin{cases} +800 \\ +750 \end{cases}$

Solve for the rate of return. $3000 = 800(P/A, i, 5) + 750(P/F, i, 5)$

Try $i = 15\%$:

$3000 \overset{?}{=} 800(3.352) + 750(0.4972) \overset{?}{=} 2682 + 373 = 3055$

i is slightly low. Try $i = 18\%$:

$3000 \overset{?}{=} 800(3.127) + 750(0.4371) \overset{?}{=} 2502 + 328 = 2830$

$$i^* = 15\% + 3\% \left(\frac{3055 - 3000}{3055 - 2830} \right) = 15\% + 3\%(0.15) = 15.7\%$$

Solution to Example 11-5b: For an after-tax rate of return, we must set up the cash flow table (Table 11-3). The starting point is the before-tax cash flow. Then we will need the depreciation schedule for the truck:

$$\text{Straight line depreciation} = \frac{B - S}{N} = \frac{3000 - 750}{5} = \$450 \text{ per year}$$

Taxable income is the before-tax cash flow *minus* depreciation. For this medium-sized profitable corporation, the incremental federal income tax rate probably is 34%. Therefore income taxes are 34% of taxable income. Finally, the after-tax cash flow equals the before-tax cash flow *minus* income taxes. These data are used to compute Table 11-3.

Table 11-3 CASH FLOW TABLE FOR EXAMPLE 11-5

Year	Before-tax cash flow (a)	Straight line depreciation (b)	ΔTaxable income (a) – (b) (c)	34% Income taxes –0.34(c)* (d)	After-tax cash flow (a) + (d)† (e)
0	–$3000				–$3000
1	800	$450	$350	–$119	681
2	800	450	350	–119	681
3	800	450	350	–119	681
4	800	450	350	–119	681
5	$\begin{cases}800\\750\end{cases}$	450	350	–119	$\begin{cases}681\\750\end{cases}$

*Sign convention for income taxes: a minus (–) represents a disbursement of money to pay income taxes; a plus (+) represents the receipt of money by a decrease in the tax liability.

†The after-tax cash flow is the before-tax cash flow minus income taxes. Based on the income tax sign convention, this is accomplished by *adding* Columns (a) and (d).

The after-tax cash flow may be solved to find the after-tax rate of return.

Try $i = 10\%$: $3000 \overset{?}{=} 681(P/A, 10\%, 5) + 750(P/F, 10\%, 5)$

$\overset{?}{=} 681(3.791) + 750(0.6209) = 3047$

i is slightly low. Try $i = 12\%$: $3000 \overset{?}{=} 681(3.605) + 750(0.5674) = 2881$

$$i^* = 10\% + 2\% \left(\frac{3047 - 3000}{3047 - 2881} \right) = 10.6\%$$

The calculations required to compute the after-tax rate of return in Ex. 11-5 were certainly more elaborate than those for the before-tax rate of return. It must be emphasized, however, that *only* the after-tax rate of return is a meaningful value since income taxes are a major disbursement that cannot be ignored.

EXAMPLE 11-6

An analysis of a firm's sales activities indicates that a number of profitable sales are lost each year because the firm cannot deliver some of its products quickly enough. By investing an additional $20,000 in inventory it is believed that the before-tax profit of the firm will be $1000 higher the first year. The second year before-tax extra profit will be $1500. Subsequent years are expected to continue to increase on a $500 per year gradient. The investment in the additional inventory may be recovered at the end of a four-year analysis period simply by selling it and not replenishing the inventory. Compute:

a. The before-tax rate of return;

b. The after-tax rate of return assuming an incremental tax rate of 39%.

Solution: Inventory is not considered a depreciable asset, therefore, the investment in additional inventory is not depreciated. The cash flow table for the problem is presented in Table 11-4.

Table 11-4 CASH FLOW TABLE FOR EXAMPLE 11-6

Year	Before-tax cash flow (a)	Depreciation (b)	ΔTaxable income (a) – (b) (c)	39% Income taxes –0.39(c) (d)	After-tax cash flow (a) + (d) (e)
0	–$20,000				–$20,000
1	1,000	—	$1000	–$390	610
2	1,500	—	1500	–585	915
3	2,000	—	2000	–780	1,220
4	$\begin{cases} 2,500 \\ 20,000 \end{cases}$	—	2500	–975	$\begin{cases} 1,525 \\ 20,000 \end{cases}$

Solution to Example 11-6a: *Before-tax rate of return:*

$$20,000 = 1000(P/A,i,4) + 500(P/G,i,4) + 20,000(P/F,i,4)$$

Try $i = 8\%$:

$$20,000 \doteq 1000(3.312) + 500(4.650) + 20,000(0.7350)$$

$$\doteq 3312 + 2325 + 14,700 = 20,337$$

i is too low. Try $i = 10\%$:

$$20,000 \doteq 1000(3.170) + 500(4.378) + 20,000(0.6830)$$

$$\doteq 3170 + 2189 + 13,660 = 19,019$$

$$\text{Before - tax rate of return} = 8\% + 2\%\left(\frac{20{,}337 - 20{,}000}{20{,}337 - 19{,}019}\right) = 8.5\%$$

Solution to Example 11-6b: *After-tax rate of return:* The before-tax cash flow gradient is \$500. The resulting after-tax cash flow gradient is $(1 - 0.39)(500) = \$305$.

$$20{,}000 = 610(P/A,i,4) + 305(P/G,i,4) + 20{,}000(P/F,i,4)$$

Try $i = 5\%$:

$$20{,}000 \overset{?}{=} 610(3.546) + 305(5.103) + 20{,}000(0.8227) \overset{?}{=} 20{,}173$$

i is too low. Try $i = 6\%$:

$$20{,}000 \overset{?}{=} 610(3.465) + 300(4.4945) + 20{,}000(0.7921) \overset{?}{=} 19{,}304$$

$$\text{After - tax rate of return} = 5\% + 1\%\left(\frac{20{,}173 - 20{,}000}{20{,}173 - 19{,}304}\right) = 5.2\% \quad ∎$$

Capital Gains And Losses For Non-depreciated Assets

When a non-depreciated capital asset is sold or exchanged, there must be entries in the firm's accounting records to reflect the change. If the selling price of the capital asset exceeds the original cost basis, the excess is called a *capital gain.* If the selling price is less than the original cost basis, the difference is a *capital loss.* Examples of non-depreciated assets include: stocks, land, art, and collectibles.

$$\text{Capital} \begin{Bmatrix} \text{gain} \\ \text{loss} \end{Bmatrix} = \text{Selling price - Original cost basis}$$

There have been quite elaborate rules in the past for the tax treatment of capital gains and losses. For example, non-depreciated capital assets held for more than six months produced *long-term* gains or losses. Assets held less than six months produced *short-term* gains and losses.

The Taxpayer's Relief Act of 1997 set the net capital gains tax at 20% for assets held more than 18 months for individuals. This is in contrast to recaptured depreciation which is taxed at the same rate as other (ordinary) income. The tax treatment of capital gains and losses for non-depreciated assets is shown in Table 11-5.

Table 11-5 TAX TREATMENT OF CAPITAL GAINS AND LOSSES

For individuals:

	Capital gain	For assets held for less than 1 year, taxed as ordinary income . For assets held for 1 year-18 months taxed at 28% tax rate. For assets held for more than 18 months, taxed at 20% tax rate.[*]
	Capital loss	Subtract capital losses from any capital gains; balance may be deducted from ordinary income, but not more than $3000 per year

For corporations:

	Capital gain	Taxed as ordinary income
	Capital loss	Corporations may deduct capital losses only to the extent of capital gains. Any capital loss in the current year that exceeds capital gains is carried back three years, and, if not completely absorbed, is then carried forward for up to five years

[*] For taxpayers in the 15% tax bracket, long term capital gains are taxed at 10%. The Taxpayer's Relief Act of 1997 (carried into 1999) also contained a provision for exclusion of gain on the sale of a principal residence. The single exclusion is for $250,000 for the residence if held for two years or more and this exclusion could be repeatedly used. For a married couple filing joint taxes, the exclusion is $500,000. This would seem to be a bonanza for taxpayers buying fixer-uppers. While it lasts, sweat equity could be turned into tax-free dollars.

Investment Tax Credit

When the economy slows down and unemployment rises, the U.S. Government frequently alters its tax laws to promote greater industrial activity. One technique used to stimulate capital investments has been the ***investment tax credit.*** Businesses were able to deduct from 4% to 8% of their new business equipment purchases as a *tax credit.* This meant the net cost of the equipment to the firm was reduced by the amount of the investment tax credit, yet at the same time the basis for computing depreciation remained the full cost of the equipment. The Tax Reform Act of 1986 eliminated the investment tax credit. It is likely, however, that it will reappear at some future time.

Estimating The After-Tax Rate Of Return

There is no shortcut method to compute the after-tax rate of return from the before-tax rate of return. One possible exception to this statement is in the situation of nondepreciable assets. In this special case:

After-tax rate of return = (1 – Incremental tax rate) × (Before-tax rate of return)

For Ex. 11-6 we could estimate the after-tax rate of return from the before-tax rate of return as follows:

After-tax rate of return = $(1 - 0.39)(8.5\%) = 5.2\%$

This value agrees with the value computed in Example 11-6b.

This relationship may be helpful for selecting a trial after-tax rate of return where the before-tax rate of return is known. It must be emphasized, however, this relationship is only a rough approximation in almost all situations.

After-Tax Cash Flows and Spreadsheets

The starting point for after-tax analysis is to calculate before-tax cash flows (BTCF) and taxable income. Before-tax cash flows may include first costs and principal payments on loans – that are not tax deductible. Taxable income deducts depreciation – which is not a cash flow. Once these principles are understood, spreadsheets can help with the arithmetic. Taxes are computed based on taxable income, and after-tax cash flows are computed by subtracting taxes from BTCF.

Realistic after-tax analysis requires spreadsheets. Even if costs and revenues are the same every year, MACRS depreciation percentages are not. The steps for calculating an after-tax internal rate of return are illustrated in Example 11-7. Because some cash flows are taxed and some are not, the spreadsheet is easier to build if these two classes are separated. It is also easier if recaptured depreciation or other gain/loss on disposal or sale is tabulated separately.

Taxes are considered even if only the costs of a project are known. The firm that does an engineering project must generate profits – or it goes out of business. Even if a firm has an unprofitable year, the tax law includes carry forward and backward provisions to transfer deductions to profitable years. The depreciation and revenues in Example 11-7 result in a negative taxable income for year 2. Thus the *positive cash flow* due to taxes that is shown in year 2 of Example 11-7 really represents tax savings for the firm.

Example 11-7

Return to the data of Example 11-6, where the used truck had a first cost of $3000, a salvage value after 5 years of $750, and savings of $800 per year. Use MACRS depreciation and calculate the after-tax rate of return.

Solution

Under MACRS, vehicles have a 5-year recovery period. Thus the MACRS depreciation can be calculated using a VDB function (see Chapter 10's spreadsheet section) or by lookup in Table 10-3.

	3000	First cost
	800	Annual benefit
	5	Recovery period
	750	salvage value
	0.34	tax rate

year	Untaxed BTCF	Taxed BTCF	MACRS	Recaptured depreciation	Tax Income	Tax	ATCF
0	-3000						-3000.0
1		800	600.0		200.0	-68.0	732.0
2		800	960.0		(160.0)	54.4	854.4
3		800	576.0		224.0	-76.2	723.8
4		800	345.6		454.4	-154.5	645.5
5	750	800	345.6	577.2	1031.6	-350.7	1199.3
		Cum. Depr.=	2827.2			IRR=	11.24%

= Taxed BTCF - MACRS + Recapt.

= Salvage - BookValue

Figure 11-3 Spreadsheet for after-tax IRR calculation.

Note that using MACRS rather than straight-line depreciation increases the after-tax IRR from 10.6% to 11.24%. This is due solely to the faster write-off that is allowed under MACRS.

SUMMARY

Since income taxes are part of most problems, no realistic economic analysis can ignore their consequences. Income taxes make the U. S. Government a partner in every business venture. Thus the Government benefits from all profitable ventures and shares in the losses of unprofitable ventures.

The first step in computing individual income taxes is to tabulate gross income. Any adjustments—for example, allowable taxpayer contributions to a retirement fund—are subtracted to yield adjusted gross income. Personal exemptions and either itemized deductions or the standard deduction are subtracted to find taxable income. This is used, together with a tax rate table, to compute the income tax liability for the year.

For corporations, taxable income equals gross income *minus* all ordinary and necessary expenditures (except capital expenditures) and depreciation and depletion charges. The income tax computation is relatively simple with rates ranging from 15% to 39.6%. The proper rate to use in an economic analysis, whether for an individual or a corporation, is the incremental tax rate applicable to the increment of taxable income being considered.

Most individuals and corporations pay state income taxes in addition to federal income taxes. Since state income taxes are an allowable deduction in computing federal taxable income, it follows that the taxable income for the federal computation is lower than the state taxable income.

Combined state and federal incremental tax rate

$$= \Delta \text{State tax rate} + (\Delta \text{Federal tax rate})(1 - \Delta \text{State tax rate})$$

To introduce the effect of income taxes into an economic analysis, the starting point is a before-tax cash flow. Then the depreciation schedule is deducted from appropriate parts of the before-tax cash flow to obtain taxable income. Income taxes are obtained by multiplying taxable income by the proper tax rate. Before-tax cash flow less income taxes equals the after-tax cash flow.

The Taxpayer's Relief Act of 1997 decreased long term capital gains on non-depreciated assets for individuals to 20% when held for more than 18 months and provided an exclusion on the gain of the principal residence held for more than two years.

When dealing with non-depreciable assets there is a nominal relationship between before-tax and after-tax rate of return. It is

$$\text{After-tax rate of return} = (1 - \Delta \text{Tax rate})(\text{Before-tax rate of return})$$

There is no simple relationship between before-tax and after-tax rate of return in the more usual case of investments involving depreciable assets.

Problems

11-1 An unmarried taxpayer with no dependents expects an adjusted gross income of $48,000 in a given year. His nonbusiness deductions are expected to be $3400.

    a. What will his federal income tax be?

    b. He is considering an additional activity expected to increase his adjusted gross income. If this increase should be $16,000 and there should be no change in nonbusiness deductions or exemptions, what will be the increase in his federal income tax?

11-2 John Adams has a $50,000 adjusted gross income from Apple Corp. and allowable itemized deductions of $5000. Mary Eve has a $45,000 adjusted gross income and $2000 of allowable itemized deductions. Compute the total tax they would pay as unmarried individuals. Then compute their tax as a married couple filing a joint return. (*Answers:*$8482.50 +$7278.50 =$15,761; $17,447.50)

11-3   Bill Jackson worked during school and the first two months of his summer vacation. After considering his personal exemption (Bill is single) and deductions, he had a total taxable income of $1800. Bill's employer wants him to work another month during the summer, but Bill had planned to spend the month hiking. If an additional month's work would increase Bill's taxable income by $1600, how much more money would he have after paying the income tax? (*Answer:* $1360)

11-4   A prosperous businessman is considering two alternative investments in bonds. In both cases the first interest payment would be received at the end of the first year. If his personal taxable income is fixed at $40,000 and he is single, which investment produces the greater after-tax rate of return? Compute the after-tax rate of return for each bond to within ¼ of 1 percent.

Ann Arbor Municipal Bonds: A bond with a face value of $1000 pays $60 per annum. At the end of 15 years, the bond becomes due ("matures"), at which time the owner of the bond will receive $1000 plus the final $60 annual payment. The bond may be purchased for $800. Since it is a municipal bond, the annual interest is *not* subject to federal income tax. The difference between what the businessman would pay for the bond ($800) and the $1000 face value he would receive at the end of 15 years must be included in taxable income when the $1000 is received.

Southern Coal Corporation Bonds: $1000 of these bonds pay $100 per year in annual interest payments. When the bonds mature at the end of twenty years, the bondholder will receive $1000 plus the final $100 interest. The bonds may be purchased now for $1000. The income from corporation bonds must be included in federal taxable income.

11-5   Albert Chan decided to buy an old duplex as an investment. After looking for several months, he found a desirable duplex that could be bought for $93,000 cash. He decided that he would rent both sides of the duplex, and determined that the total expected income would be $800 per month. The total annual expenses for property taxes, repairs, gardening, and so forth are estimated at $600 per year. For tax purposes, Al plans to depreciate the building by the sum-of-years digits method, assuming the building has a twenty-year remaining life and no salvage value. Of the total $93,000 cost of the property, $84,000 represents the value of the building and $9000 is the value of the lot. Assume that Al is in the 38% incremental income tax bracket (combined state and federal taxes) throughout the twenty years.

In this analysis Al estimates that the income and expenses will remain constant at their present levels. If he buys and holds the property for twenty years, what after-tax rate of return can he expect to receive on his investment, using the assumptions below?

- **a.**   Al believes the building and the lot can be sold at the end of twenty years for the $9000 estimated value of the lot;

- **b.**   A more optimistic estimate of the future value of the building and the lot is that the property can be sold for $100,000 at the end of twenty years.

11-6   Mr. Sam K. Jones, a successful businessman, is considering erecting a small building on a commercial lot he owns very close to the center of town. A local furniture company is willing to lease the building for $9000 per year, paid at the end of each year. It is a net lease, which means the furniture company must also pay the property taxes, fire insurance, and all other annual costs. The furniture company will require a five-year lease with an option to buy the building and land on which it stands for $125,000 at the end of the five years. Mr. Jones could have the building constructed for

$82,000. He could sell the commercial lot now for $30,000, the same price he paid for it. Mr. Jones files a joint return and has an annual taxable income from other sources of $63,900. He would depreciate the commercial building by modified accelerated cost recovery system (MACRS) depreciation. Mr. Jones believes that at the end of the five-year lease he could easily sell the property for $125,000. What is the after-tax Present Worth of this five-year venture if Mr. Jones uses a 10% after-tax MARR?

11-7 A store owner, Joe Lang, believes his business has suffered from the lack of adequate automobile parking space for his customers. Thus, when he was offered an opportunity to buy an old building and lot next to his store, he was interested. He would demolish the old building and make off-street parking for twenty customer's cars. Joe estimates that the new parking would increase his business and produce an additional before-income-tax profit of $7000 per year. It would cost $2500 to demolish the old building. Mr. Lang's accountant advised that both costs (the property and demolishing the old building) would be considered the total value of the land for tax purposes, and it would not be depreciable. Mr. Lang would spend an additional $3000 right away to put a light gravel surface on the lot. This expenditure, he believes, may be charged as an operating expense immediately and need not be capitalized. To compute the tax consequences of adding the parking lot, Joe estimates that his combined state and federal incremental income tax rate will average 40%. If Joe wants a 15% after-tax rate of return from this project, how much could he pay to purchase the adjoining land with the old building? Assume that the analysis period is ten years, and that the parking lot could always be sold to recover the costs of buying the property and demolishing the old building. (*Answer:* $23,100)

11-8 Zeon, a large, profitable corporation, is considering adding some automatic equipment to its production facilities. An investment of $120,000 will produce an initial annual benefit of $29,000 but the benefits are expected to decline $3000 per year, making second-year benefits $26,000, third-year benefits $23,000, and so forth. If the firm uses sum-of-years digits depreciation, an eight-year useful life, and $12,000 salvage value, will it obtain the desired 6% after-tax rate of return? Assume that the equipment can be sold for its $12,000 salvage value at the end of the eight years. Also assume a 46% state-plus-federal income tax rate.

11-9 A group of businessmen formed a corporation to lease a piece of land for five years at the intersection of two busy streets. The corporation has invested $50,000 in car-washing equipment. They will depreciate the equipment by sum-of-years digits depreciation, assuming a $5000 salvage value at the end of the five-year useful life. The corporation is expected to have a before-tax cash flow, after meeting all expenses of operation (except depreciation), of $20,000 the first year, and declining $3000 per year in future years (second year = $17,000; third year = $14,000; and so forth). The corporation has other income, so it is taxed at a combined corporate tax rate of 20%. If the projected income is correct, and the equipment can be sold for $5000 at the end of five years, what after-tax rate of return would the corporation receive from this venture? (*Answer:* 14%)

11-10 The effective combined tax rate in an owner-managed corporation is 40%. An outlay of $20,000 for certain new assets is under consideration. It is estimated that for the next eight years, these assets will be responsible for annual receipts of $9000 and annual disbursements (other than for income taxes) of $4000. After this time, they will be used only for stand-by purposes and no future excess of receipts over disbursements is estimated.

a. What is the prospective rate of return before income taxes?

b. What is the prospective rate of return after taxes if these assets can be written off for tax purposes in eight years using straight line depreciation?

c. What is the prospective rate of return after taxes if it is assumed that these assets must be written off for tax purposes over the next twenty years using straight line depreciation?

11-11 In January Gerald Adair bought a small house and lot for $99,700. He estimated that $9700 of this amount represented the value of the land. He rented the house for $6500 a year during the four years he owned the house. Expenses for property taxes, maintenance, and so forth were $500 per year. For tax purposes the house was depreciated by MACRS depreciation (27.5 year straight line depreciation with a mid-month convention is used for rental property). At the end of four years the property was sold for $105,000. Gerald is married and works as an engineer. He estimates that his incremental state and federal combined tax rate is 24%. What after-tax rate of return did Gerald obtain on his investment in the property?

11-12 The management of a private hospital is considering the installation of an automatic telephone switchboard, which would replace a manual switchboard and eliminate the attendant operator's position. The class of service provided by the new equipment is estimated to be at least equal to the present method of operation. Five operators are needed to provide telephone service three shifts per day, 365 days per year. Each operator earns $14,000 per year. Company-paid benefits and overhead are 25% of wages. Money costs 8% after income taxes. Combined federal and state income taxes are 40%. Annual property taxes and maintenance are 2½% and 4% of investment, respectively. Depreciation is 15-year straight line. Disregarding inflation, how large an investment in the new equipment can be economically justified by savings obtained by eliminating the present equipment and labor costs? The existing equipment has zero salvage value.

11-13 A contractor has to choose one of the following alternatives in performing earth-moving contracts:

a. Purchase a heavy-duty truck for $13,000. Salvage value is expected to be $3000 at the end of its seven-year depreciable life. Maintenance is $1100 per year. Daily operating expenses are $35.

b. Hire a similar unit for $83 per day.

Based on a 10% after-tax rate of return, how many days per year must the truck be used to justify its purchase? Base your calculations on straight line depreciation and a 50% income tax rate. (*Answer:* 91½ days)

11-14 The Able Corporation is considering the installation of a small electronic testing device for use in conjunction with a government contract the firm has just won. The testing device will cost $20,000, and have an estimated salvage value of $5000 in five years when the government contract is finished. The firm will depreciate the instrument by the sum-of-years digits method using five years as the useful life and a $5000 salvage value. Assume Able Corp. pays 50% federal and state corporate income taxes and uses 8% *after-tax* in their economic analysis. What minimum equal annual benefit must Able obtain *before taxes* in each of the five years to justify purchasing the electronic testing device? (*Answer:* $5150)

11-15 A small business corporation is considering whether or not to replace some equipment in the plant. An analysis indicates there are five alternatives in addition to the do-nothing Alt. *A*. The alternatives have a five-year useful life with no salvage value. Straight line depreciation would be used.

Alternatives	Cost in thousands	Before-tax uniform annual benefits, in thousands
A	$ 0	$0
B	25	7.5
C	10	3
D	5	1.7
E	15	5
F	30	8.7

The corporation has a combined federal and state income tax rate of 20%. If the corporation expects a 10% after-tax rate of return for any new investments, which alternative should be selected?

11-16 A firm is considering the following investment project:

Year	Before-tax cash flow
0	−$1000
1	+500
2	+340
3	+244
4	+100
5	$\begin{cases} +100 \\ +125 \text{ Salvage value} \end{cases}$

The project has a five-year useful life with a $125 salvage value as shown. Double declining balance depreciation will be used assuming a $125 salvage value. The income tax rate is 34%. If the firm requires a 10% after-tax rate of return, should the project be undertaken?

11-17 A married couple filing jointly have a combined total adjusted gross income of $75,000. They have computed that their allowable itemized deductions are $4000. Compute their federal income tax. (*Answer:* $11,847.50)

11-18 A major industrialized state has a state corporate tax rate of 9.6% of taxable income. If a corporation has a state taxable income of $150,000, what is the total state and federal income tax it must pay? Also, compute its combined incremental state and federal income tax rate.
    (*Answers:* $50,534; 43.93%)

11-19   Jane Shay operates a management consulting business. The business has been successful and now produces a taxable income of $65,000 per year after deducting all "ordinary and necessary" expenses and depreciation. At present the business is operated as a proprietorship, that is, Jane pays personal federal income tax on the entire $65,000. For tax purposes, it is as if she had a job that pays her a $65,000 salary per year.

As an alternative, Jane is considering incorporating the business. If she does, she will pay herself a salary of $22,000 a year from the corporation. The corporation will then pay taxes on the remaining $43,000 and retain the balance of the money as a corporate asset. Thus Jane's two alternatives are to operate the business as a proprietorship or as a corporation. Jane is single and has $2500 of itemized personal deductions. Which alternative will result in a smaller total payment of taxes to the government?

(*Answer:* Now you know one of the reasons why your doctor is a corporation.)

11-20   A house and lot are for sale for $155,000. It is estimated that $45,000 is the value of the land and $110,000 is the value of the house. If purchased, the house can be rented to provide a net income of $12,000 per year after taking all expenses, except depreciation, into account. The house would be depreciated by straight line depreciation using a 27.5-year depreciable life and zero salvage value.

Mary Silva, the prospective purchaser, wants a 10% after-tax rate of return on her investment after considering both annual income taxes and a capital gain when she sells the house and lot. At what price would she have to sell the house at the end of ten years to achieve her objective? You may assume that Mary has an incremental income tax rate of 28% in each of the ten years.

11-21   Bill Alexander and his wife Valerie are both employed. Bill will have an adjusted gross income this year of $36,000. Valerie has an adjusted gross income of $2000 a month. Bill and Valerie have agreed that Valerie should continue working only until the point where the federal income tax on their joint income tax return becomes $6900. On what date should Valerie quit her job?

11-22   The Lynch Bull investment company suggests that Steven Comstock, a wealthy New York City investor (his incremental income tax rate is 50%), consider the following investment.

Buy corporate bonds on the New York Exchange with a face value (par value) of $100,000 and a 5% coupon rate (the bonds pay 5% of $100,000 which equals $5000 interest per year). These bonds can be purchased at their present market value of $75,000. At the end of each year, Steve will receive the $5000 interest, and at the end of five years, when the bonds mature, he will receive $100,000 plus the last $5000 of interest.

Steve will pay for the bonds by borrowing $50,000 at 10% interest for five years. The $5000 interest paid on the loan each year will equal the $5000 of interest income from the bonds. As a result Steve will have no net taxable income during the five years due to this bond purchase and borrowing money scheme.

At the end of five years, Steve will receive $100,000 plus $5000 interest from the bonds and will repay the $50,000 loan and pay the. last $5000 interest. The net result is that he will have a $25,000 capital gain (that is, he will receive $100,000 from a $75,000 investment). *Note:* This situation represents an actual recommendation of a stock and bond brokerage firm.

   a.   Compute Steve's after-tax rate of return on this dual bond plus loan investment package.

b. What would be Steve's after-tax rate of return if he purchased the bonds for $75,000 cash and *did not* borrow the $50,000?

11-23 A corporation with a 34% income tax rate is considering the following investment in research equipment, and has projected the benefits as follows:

Year	Before-tax cash flow
0	−$50,000
1	+2,000
2	+8,000
3	+17,600
4	+13,760
5	+5,760
6	+2,880

Prepare a cash flow table to determine the year-by-year after-tax cash flow assuming MACRS depreciation.

a. What is the after-tax rate of return?

b. What is the before-tax rate of return?

11-24 A corporation with $7 million in annual taxable income is considering two alternatives:

Year	Alt. 1 Before-tax cash flow	Alt. 2 Before-tax cash flow
0	−$10,000	−$20,000
1–10	4,500	4,500
11–20	0	4,500

Both alternatives will be depreciated by straight line depreciation using a ten-year depreciable life and no salvage value. Neither alternative is to be replaced at the end of its useful life. If the corporation has a minimum attractive rate of return of 10% *after taxes,* which alternative should it choose? Solve the problem by:

a. Present worth analysis.   **d.** Future worth analysis.

b. Annual cash flow analysis.   **e.** Benefit–cost ratio analysis.

c. Rate of return analysis.   **f.** Any method you choose.

11-25   An engineer is working on the layout of a new research and experimentation facility. Two men will be required as plant operators. If, however, an additional $100,000 of instrumentation and remote controls were added, the plant could be run by a single operator. The total before-tax cost of each plant operator is projected to be $35,000 per year. The instrumentation and controls will be depreciated by accelerated cost recovery system depreciation.

If this corporation (34% corporate tax rate) invests in the additional instrumentation and controls, how long will it take for the after-tax benefits to equal the $100,000 cost? In other words, what is the after-tax payback period? (*Answer:* 3.24 years)

11-26   A special powertool for plastic products costs $400, has a four-year useful life, no salvage value, and a two-year before-tax payback period. Assume uniform annual end-of-year benefits.

    a.   Compute the before-tax rate of return.

    b.   Compute the after-tax rate of return, based on MACRS depreciation and a 34% corporate income tax rate.

11-27   A piece of petroleum drilling equipment costs $100,000. It will be depreciated in ten years by double declining balance depreciation with conversion to straight line depreciation at the optimal point. Assume no salvage value in the depreciation computation. The equipment is owned by Shellout (assume a 34% tax rate). Shellout will lease the equipment to others and each year receive $30,000 in rent. At the end of five years, they will sell the equipment for $35,000. (Note that this is different from the zero salvage value assumption used in computing the depreciation.) What is the after-tax rate of return Shellout will receive from this equipment investment?

11-28   The Ogi Corporation, a construction company, purchased a new pickup truck for $14,000. They used MACRS depreciation in their income tax return. During the time they had the truck they estimated that it saved the firm $5000 a year. At the end of four years, Ogi sold the truck for $3000. The combined federal and state income tax rate for Ogi is 45%. Compute their after-tax rate of return for the truck. (*Answer:* 12.5%)

11-29   A profitable wood products corporation is considering buying a parcel of land for $50,000, building a small factory building at a cost of $200,000, and equipping it with $150,000 of MACRS 5-year class machinery.

If the project is undertaken, MACRS depreciation will be used. Assume the plant is put in service October 1st. The before-tax net annual benefit from the project is estimated at $70,000 per year. The analysis period is to be five years and assumes the total property (land, building, and machinery) is sold at the end of five years, also on October lst, for $328,000. Compute the after-tax cash flow based on a 34% income tax rate. If the corporation's criterion is a 15% after-tax rate of return, should it proceed with the project?

11-30   A small vessel was purchased by a chemical company for $55,000 and is to be depreciated by MACRS depreciation. When its requirements changed suddenly, the chemical company leased the vessel to an oil company for six years at $10,000 per year. The lease also provided that the vessel could be purchased at the end of six years by the oil company for $35,000. At the end of the six years, the oil company exercised its option and bought the vessel. The chemical company has a 34% incremental tax rate. Compute its after-tax rate of return on the vessel. (*Answer:* 9.86%)

11-31   A corporation is considering buying a medium-sized computer that will eliminate a task that must be performed three shifts per day, seven days per week, except for one eight-hour shift per week when the operation is shut down for maintenance. At present four people are needed to perform the day and night task. Thus the computer will replace four employees. Each employee costs the company $32,000 per year ($24,000 in direct wages plus $8000 per year in other company employee costs). It will cost $18,000 per year to maintain and operate the computer. The computer will be depreciated by sum-of-years digits depreciation using a six-year depreciable life, at which time it will be assumed to have zero salvage value.

The corporation has a combined federal and state incremental tax rate of 50%. If the firm wants a 15% rate of return, after considering both state and federal income taxes, how much can it afford to pay for the computer?

11-32   A mining corporation purchased $120,000 of production machinery. They depreciated it using SOYD depreciation, a five-year depreciable life, and zero salvage value. The corporation is a profitable one that has a 34% incremental tax rate.

At the end of five years the mining company changed its method of operation and sold the production machinery for $40,000. During the five years the machinery was used, it reduced mine operating costs by $32,000 a year, before taxes. If the company MARR is 12% after taxes, was the investment in the machinery a satisfactory one?

11-33   Two mutually exclusive alternatives are being considered by a profitable corporation with an annual taxable income between $5 million and $10 million.

Year	Alt. A Before-tax cash flow	Alt. B. Before-tax cash flow
0	−$3000	−$5000
1	1000	1000
2	1000	1200
3	1000	1400
4	1000	2600
5	1000	2800

Both alternatives have a five-year useful and depreciable life and no salvage value. Alternative *A* would be depreciated by sum-of-years digits depreciation, and Alt. *B* by straight line depreciation. If the MARR is 10% after taxes, which alternative should be selected? (*Answer:* Choose *B*.)

11-34   Xon, a small oil company, purchased a new petroleum drilling rig for $1,800,000. Xon will depreciate the drilling rig using MACRS depreciation. The drilling rig has been leased to a drilling company which will pay Xon $450,000 per year for eight years. At the end of eight years the drilling rig will belong to the drilling company. If Xon has a 34% incremental tax rate and a 10% after-tax MARR, does the investment appear to be satisfactory?

11-35   An automobile manufacturer is buying some special tools for $100,000. The tools are being depreciated by double declining balance depreciation using a four-year depreciable life and a

$6250 salvage value. It is expected the tools will actually be kept in service for six years and then sold for $6250. The before-tax benefit of owning the tools is as follows:

Year	Before-tax cash flow
1	$30,000
2	30,000
3	35,000
4	40,000
5	10,000
6	10,000
	6,250 Selling price

Compute the after-tax rate of return for this investment situation, assuming a 46% incremental tax rate. (*Answer:* 11.6%)

11-36   This is the continuation of Problem 11-35. Instead of paying $100,000 cash for the tools, the corporation will pay $20,000 now and borrow the remaining $80,000. The depreciation schedule will remain unchanged. The loan, at a 10% interest rate, will be repaid by four equal end-of-year payments of $25,240.

Prepare an expanded cash flow table that takes into account both the special tools and the loan. Note that the Year 0 cash flow is –$20,000 in this situation. *Hint:* You must determine what portion of each loan payment is interest.

a. Compute the after-tax rate of return for the tools, taking into account the $80,000 loan.

b. Explain why the rate of return obtained in Part *a* is different from the rate of return obtained in Problem 11-35.

11-37   A project will require the investment of $108,000 in equipment (sum-of-years- digits depreciation with a depreciable life of eight years and zero salvage value) and $25,000 in raw materials (which is not depreciable). The annual project income after paying all expenses, except depreciation, is projected to be $24,000. At the end of eight years the project will be discontinued and the $25,000 investment in raw materials will be recovered.

Assume a 34% income tax rate for this corporation. The corporation wants a 15% after-tax rate of return on its investments. Determine by present worth analysis whether or not this project should be undertaken.

11-38 A large profitable corporation is considering two mutually exclusive capital investments:

	Alt. A	Alt. B
Initial cost	$11,000	$33,000
Uniform annual benefit	3,000	9,000
End-of-depreciable-life salvage value	2,000	3,000
Depreciation method	SL	SOYD
Depreciable life, in years	3	4
Useful life, in years	5	5
End-of-useful-life salvage value obtained	2,000	5,000

If the firm's after-tax minimum attractive rate of return is 12% and its incremental income tax rate is 34%, which project should be selected?

11-39 A profitable incorporated business is considering an investment in equipment which has the before-tax cash flow tabulated below. The equipment will be depreciated by double declining balance depreciation with conversion, if appropriate, to straight line depreciation at the preferred time. For depreciation purposes a $700 salvage value at the end of six years is assumed. But the actual value is thought to be $1000 and it is this sum that is shown in the before-tax cash flow.

Year	Before-tax cash flow
0	12000
1	1727
2	2414
3	2872
4	3177
5	3358
6	1,997
	1,000 Salvage value

If the firm wants a 9% after-tax rate of return and its incremental income tax rate is 34%, determine by annual cash flow analysis whether or not the investment is desirable.

11-40 A salad oil bottling plant can either purchase caps for the glass bottles at 5 cents each, or install $500,000 worth of plastic molding equipment and manufacture the caps at the plant. The manufacturing engineer estimates the material, labor, and other costs would be 3 cents per cap.

  a. If 12 million caps per year are needed and the molding equipment is installed, what is the payback period?

  b. The plastic molding equipment would be depreciated by straight line depreciation using a five-year useful life and no salvage value. Assuming a 40% income tax rate, what is the after-tax payback period, and what is the after-tax rate of return?

11-41 The profitable Palmer Golf Cart Corp. is considering investing $300,000 in special tools for some of the plastic golf cart components. Executives of the company believe the present golf cart model will continue to be manufacturing and sold for five years, after which a new cart design will be needed, together with a different set of special tools.

The saving in manufacturing costs, owing to the special tools, is estimated to be $150,000 per year for five years. Assume MACRS depreciation for the special tools and a 39% income tax rate.

    a. What is the after-tax payback period for this investment?

    b. If the company wants a 12% after-tax rate of return, is this a desirable investment?

11-42 An unmarried individual in California with a taxable income of about $80,000 has a federal incremental tax rate of 31% and a state incremental tax rate of 9.3%. What is his combined incremental tax rate?

11-43 ARKO oil company purchased two large compressors for $125,000 each. One compressor was installed in their Texas refinery and is being depreciated by MACRS depreciation. The other compressor was placed in the Oklahoma refinery where it is being depreciated by sum-of-years digits depreciation with zero salvage value. Assume the company pays federal income taxes each year and the tax rate is constant. The corporate accounting department noted that the two compressors are being depreciated differently and wonders whether the corporation will wind up paying more income taxes over the life of the equipment as a result of this. What do you tell them?

11-44 Refer to Problem 11-28. To help pay for the pickup truck the Ogi Corp. obtained a $10,000 loan from the truck dealer at 10% interest, payable in four end-of-year payments of $2500 plus interest.

    a. Compute the after-tax rate of return for the truck together with the loan. Note that the interest on the loan is tax deductible, but the $2500 principal payments are not.

    b. Why is the after-tax rate of return computed in part *a* so much different from the 12.5% obtained in Problem 11-28?

11-45 A firm has invested $14,000 in machinery with a seven-year useful life. The machinery will have no salvage value, as the cost to remove it will equal its scrap value. The uniform annual benefits from the machinery are $3600. For a 47% income tax rate, and sum-of-years digits depreciation, compute the after-tax rate of return.

11-46 A sales engineer has the following two alternatives to consider in touring his sales territory.

    a. Buy a new car for $14,500. Salvage value is expected to be about $5000 after 3 years. Maintenance and insurance cost is $1000 in the first year and increases at the rate of $500/year in subsequent years. Daily operating expenses are $50/day.

b. Rent a similar car for $80/day.

Based on a 12% after-tax rate of return, how many days per year must he use the car to justify its purchase? You may assume that this sales engineer is in the 50% incremental tax bracket. Use MACRS depreciation.

11-47 Uncle Elmo is contemplating a $10,000 investment in a methane gas generator. He estimates his gross income would be $2000 the first year and increase by $200 each year over the next ten years. His expenses would be $200 the first year and increase by $200 each year over the next ten years. He would depreciate the generator by MACRS depreciation assuming a 7 year property class. A 10 year old methane generator has no market value. The income tax rate is 40%. (Remember recaptured depreciation is taxed at the same 40% rate).

    a. Construct the After Tax Cash Flow for the 10 year project life.

    b. Determine the after tax rate of return on this investment. Uncle Elmo thinks it should be at least 8%.

    c. If Uncle Elmo could sell the generator for $7000 at the end of the 5th year, would his rate of return be better than if he keeps the generator for 10 years? You don't have to actually find the rate of return, just do enough calculations to see if it is higher than that of part *b*.

11-48 Granny's Butter and Egg Business is such that she pays an effective tax rate of 40%. Granny is considering the purchase of a new Turbo Churn for $25,000. This churn is a special handling device for food manufacture and has an estimated life of 4 years and a salvage value of $5000. The new churn is expected to increase net income by $8000 per year for each of the 4 years of use. If Granny works with an after tax MARR of 10% and she uses MACRS depreciation, should she buy the churn?

11-49 Eric Heiden has a house and lot for sale for $70,000. It is estimated that $10,000 is the value of the land and $60,000 is the value of the house. Bonnie Blair is purchasing the house on January 1 to rent, and plans to own the house for five years. After five years, it is expected that the house and land can be sold on December 31st for $80,000. Total annual expenses (maintenance, property taxes, insurance, etc.) are expected to be $3000 per year. The house would be depreciated by MACRS depreciation using a 27.5 year straight line rate with mid-month convention for rental property. For depreciation, a salvage value of zero was used. Bonnie wants a 15% after-tax rate of return on her investment. You may assume that Bonnie has an incremental income tax rate of 28% in each of the five years. Determine the following:

    a. The annual depreciation (enter into worksheet)

    b. The capital gain (loss) resulting from the sale of the house (enter into worksheet)

    c. The uniform after-tax cash flow necessary to produce a rate of return of 15% (enter into worksheet)

    d. The annual rent she must charge to produce an after-tax rate of return of 15% (enter into worksheet) (Hint: write an algebraic equation to solve for rent)

Year	Income (rent)	Expenses	Before tax cash flow	MACRS Deprec	Taxable income	Income tax	After tax cash flow
0	■		-$70,000	■			-$70,000
1		-$3000					
2		-3000					
3		-3000					
4		-3000					
5		-3000					

11-50 Bill Cavitt owns a data processing company. He plans to buy an additional computer for $20,000, use the computer for three years, and sell it for $10,000. He expects that use of the computer will produce a net income of $8000 per year. The combined federal and state incremental tax rate is 45%. Using MACRS depreciation, complete the following worksheet to determine the net present worth of the after tax cash flow using an interest rate of 12%.

Year	Before tax cash flow	MACRS depreciation	Taxable income	Income tax (45%)	After tax cash flow	Present worth (12%)
0	-20,000	■				
1	+8,000					
2	+8,000					
3	+8,000					
	+10,000	■				

Net present worth = [ ]

11-51 Assume that you are debating between leasing a Mitsubishi 3000 GT or buying one to use exclusively in business. The estimated mileage in both cases will be 12,000/year. Insurance costs will be the same in either case at $600 per year. You want to examine the tax impacts of the two options to see which is preferred. Use MACRS depreciation.

Assume that you will have leased or purchased the car on June 30,1995 and keep the car for 36 months. List any assumptions that you make in your analysis. Use a 12% MARR, a 40% tax rate, and assume end of year payments.

Purchase Option: A Mitsubishi 3000 GT with manual transmission lists for $29,188 and their finance plan calls for 30 monthly payments of $973 with no money down and 0% interest. Estimated value after 36 months $15,200.

Lease Option: $2250 down and $369 a month for 36 months. Remember that the tax law limits the depreciation on cars. (See Table Below)

Depreciation Limitation on Automobiles Purchased During 1995

Year	Amount
First Tax Year	$2950
Second	4700
Third	2850
Fourth and Later	1675

11-52 Sole Brother Inc. is a shoe outlet to a major shoe manufacturing industry located in Chicago, Illinois. Sole Brother uses accounts payable as one of its financing sources. Shoes are delivered to Sole Brother with a 3% discount if payment on the invoice is received within 10 days of delivery. By paying after the 10-day period, Sole is borrowing money and paying (giving up) the 3% discount. Although Sole Brother is not required to pay interest on delayed payments, the shoe manufacturers require that payments should not be delayed beyond 45 days after the invoice date. To be sure of paying within 10 days, Sole Brothers decides to pay on the fifth day. Sole has a marginal corporate income tax of 40% (combined state and federal). By paying within the 10 day period Sole is avoiding paying a fairly high price to retain the money owed shoe manufacturers. What would have been the effective annual after tax interest rate?

11-53 To increase its market share, Sole Brother Inc. decided to borrow $5000 from its banker for the purchase of newspaper advertising for its shoe retail line. The loan is to be paid in four equal annual payments with 15% interest. The loan is discounted 6 points. The 6 "points" is an additional interest charge of 6% of the loan, deducted immediately. This additional interest 6%(5000) = $300 means the actual amount received from the $5000 loan is $4700. The $300 additional interest may be deducted as four $75 additional annual interest payments. What is the after tax interest rate on this loan?

11-54 A large profitable company, in the 40% federal/state tax bracket, is considering the purchase of a new piece of equipment. The new equipment will yield benefits of $10,000 in year 1, $15,000 in year 2, $20,000 in year 3, and $20,000 in year 4. The equipment is to be depreciated using 5-year MACRS depreciation starting in the year of purchase (year 0). It is expected that the equipment will be sold at the end of 4th year at 20% of its purchase price. What is the maximum equipment purchase price that the company can pay if its after-tax MARR is 10%?

11-55 A plant can be purchased for $1,000,000 or it can be leased for $200,000 per year. The annual income is expected to be $800,000 with the annual operating cost of $200,000. The resale value of the plant is estimated to be $400,000 at the end of its 10-year life. The company's combined federal and state income tax rate is 40%. A straight line depreciation can be used over the 10 years with the full first year depreciation rate.

> **a.** If the company uses the after-tax minimum attractive rate of return of 10%, should it lease or purchase the plant?

> **b.** What is the breakeven rate of return of purchase vs. lease?

11-56 A company wants to set up a new office in a country where the corporate tax rate is as follows: 15% of first $50,000 profits; 25% of next $25,000; 34% of next $25,000; and 39% of everything over $100,000. They estimate that they will have gross revenues of $500,000, total costs of $300,000, $30,000 in allowable tax deductions and a one time business start-up credit of $8000. What is their taxable income for the first year and how much should they expect to pay in taxes?

11-57 A firm manufactures padded shipping bags. One hundred bags are packed in a cardboard carton. At present, machine operators fill the cardboard cartons by eye, that is, when the cardboard carton looks full, it is assumed to contain 100 shipping bags. Actual inspection reveals that the cardboard carton may contain anywhere from 98 to 123 bags with an average quantity of 105.5 bags.

The management has never received complaints from its customers about cartons containing less than 100 bags. Nevertheless, management realizes that they are giving away 5½% of their output by overfilling the cartons. One solution would be to count the shipping bags to ensure that 100 are packed in each carton. Another solution would be to weigh each filled shipping carton. Underweight cartons would have additional shipping bags added, and overweight cartons would have some shipping bags removed. This would not be a perfect solution as the actual weight of the shipping bags varies slightly. If the weighing is done, it is believed that the average quantity of bags per carton could be reduced to 102, with almost no cartons containing less than 100 bags.

The weighing equipment would cost $18,600. The equipment would be depreciated by straight line depreciation using a ten-year depreciable life and a $3600 salvage value at the end of ten years. The $18,600 worth of equipment qualifies for a 10% investment tax credit. One person, hired at a cost of $16,000 per year, would be required to operate the weighing equipment and to add or remove padded bags from the cardboard cartons. 200,000 cartons will be checked on the weighing equipment each year, with an average removal of 3.5 padded bags per carton with a manufacturing cost of 3 cents per bag. This large profitable corporation has a 50% combined federal-plus-state incremental tax rate. Assume a ten-year study period for the analysis and an after-tax MARR of 20%.

Compute :

> **a.** The after-tax Present Worth of this investment.

> **b.** The after-tax internal rate of return of this investment.

> **c.** The after-tax simple payback period of this investment.

CHAPTER 12

Replacement Analysis

Up to this point we have considered the evaluation and selection of *new* alternatives in our economic analysis. Which new car or washing machine should we purchase? What new material handling system or ceramic grinder should we install? However, a choice between new alternatives is not always what we must consider — economic analysis is more frequently performed in conjunction with *existing* versus *new* facilities. For most engineers the problem is less frequently one of building a new plant; rather the goal is more often keeping a present plant operating economically. We are not choosing between new ways to perform the desired task. Instead, we have equipment performing the task, and the question is, "Should the existing equipment be retained or replaced?" This adversarial situation has given rise to the terms ***defender*** and ***challenger***. The defender is the existing equipment; the challenger is the best available replacement equipment. An economic analysis of the existing defender and the challenger replacement is the domain of ***replacement analysis***.

In this chapter we will examine the following aspects of replacement analysis:
1. Using available data to determine the analysis technique.
2. Determining the basic comparison between alternatives.
3. Using analysis techniques when:
    - defender marginal cost can be computed and is increasing.
    - defender marginal cost can be computed and is not increasing.
    - defender marginal cost data are not available.
4. Considering possible future challengers.
5. After-tax replacement analysis.

The Replacement Problem

The replacement of an existing asset may be appropriate in various situations including obsolescence, depletion, and deterioration due to aging. In each of these cases, the ability of a previously implemented business asset to produce a desired output is challenged. For cases of obsolescence, depletion and aging, it may be economical to replace the existing asset. We define each of these situations below.

Obsolescence refers to a situation where the technology of an asset has been surpassed by newer and/or different technologies. Changes in technology cause subsequent changes in the market demand for older assets. As an example, today's personal computers (PCs) with more RAM, faster clockspeeds, larger hard-drives, and more powerful central processors have made older, less powerful PCs obsolete. Thus, older obsolete assets may need to be replaced with newer more technologically advanced ones.

Depletion refers to the gradual loss of market value of an asset as it is being consumed or exhausted. Oil wells and timber tracts are examples of such assets. In most cases the asset will be used until it is depleted, at which time a replacement asset will be obtained. Depletion was treated earlier in Chapter 10.

Deterioration due to aging is the general condition of loss in value of some asset due to the aging process. Production machinery and other business assets that were once new eventually become aged. To compensate for a loss in functionality due to the aging process, additional operating and maintenance expenses are usually incurred to maintain the asset at its operating efficiency.

The normal means of monitoring expenditures in industry, as in government, are by *annual budgets*. One important facet of a budget is the allocation of money for new capital expenditures, either new facilities or replacement and upgrading of existing facilities.

Replacement analysis may, therefore, produce a recommendation that certain equipment be replaced and money for the replacement be included in the capital expenditures budget. Even if there is no recommendation to replace the equipment at the current time, such a recommendation may be made the following year or subsequently. At *some* point, the existing equipment will be replaced, either when it is no longer necessary or when better equipment is available. Thus, the question is not *if* the defender will be replaced, but *when* it will be replaced. This leads us to the first aspect of the defender-challenger comparison:

> ***Shall we replace the defender now, or shall we keep it for one or more additional years?***

Replacement Analysis Decision Map

Given in Figure 12-1 is a basic decision map for conducting a replacement analysis.

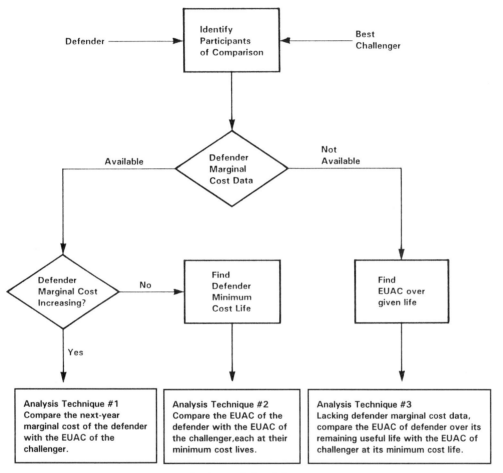

Figure 12-1 Replacement Analysis Decision Map.

Looking at the map, we can see there are three *replacement analysis techniques* that are appropriate under different circumstances. The appropriate replacement analysis technique to use in making a replacement comparison of old versus new asset is a function of the data available for the alternatives and how that data behaves over time.

What is the Basic Comparison?

By looking at the replacement analysis map we see that the first step is to identify the basic participants of the economic comparison. As was discussed previously, in replacement analysis we are interested in comparing our previously implemented asset (the *defender)* against the best current available *challenger.*

> *If the defender proves more economical, it will be retained. If the challenger proves more economical, it will be installed.*

In this comparison the challenger being evaluated against the defender has been selected from a mutually exclusive set of competing challengers. Figure 12-2 illustrates this concept in the context of a drag-race between the defender and the challenger. Notice that the challenger which is competing against the defender has emerged from an earlier competition among a set of potential challengers. Any of the methods for evaluating sets of mutually exclusive alternatives, previously discussed in this text, could be used to identify the "best" challenger to race against the defender. However, it is important to note that the comparison of these potential challenger alternatives should be made at each alternative's respective *minimum cost life.* This concept is discussed in the next section.

Minimum Cost Life of the Challenger

The *minimum cost life* of any new (or existing) asset is the number of years at which the Equivalent Uniform Annual Cost (EUAC) of ownership is minimized. This minimum cost life is often shorter than either the physical or useful life of the asset due to increasing operating and maintenance costs in the later years of asset ownership. The challenger asset selected to "race" against the defender (in Figure 12-2) is the one which has the lowest minimum cost life of all the competing mutually exclusive challengers.

To calculate the minimum cost life of an asset, determine the EUAC that results if the asset is kept for each possible life less than or equal to its useful life. As is illustrated in Example 12-1, the EUAC tends to be high if the asset is kept only a few years, then decreases to some minimum EUAC and increases again as the asset ages. By identifying the number of years at which the EUAC is a minimum and then keeping the asset for that number of years, we are minimizing the yearly cost of ownership. Example 12-1 illustrates how minimum cost life is calculated for a new asset.

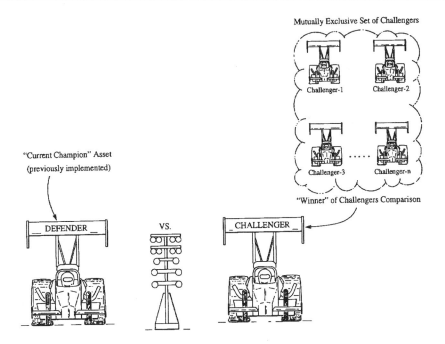

Figure 12-2 Defender versus Challenger Comparison.

EXAMPLE 12-1

A piece of machinery costs $7500 and has no salvage value after it is installed. The manufacturer's warranty will pay the first year's maintenance and repair costs. In the second year, maintenance costs will be $900, and they will increase on a $900 arithmetic gradient in subsequent years. Also, operating expenses for the machinery will be $500 the first year and will increase on a $400 arithmetic gradient in the following years. If interest is 8%, compute the useful life of the machinery that results in a minimum EUAC. That is, find its minimum cost life.

Solution:

If Retired at the end of Year n

YEAR (n)	EUAC of Capital Recovery Costs	EUAC of Maintenance and Repair Costs	EUAC of Operating Costs	EUAC Total
	$7500(A/P,8%,n)$	$900(A/G,8%,n)$	$500 + $400(A/G,8%,n)$	
1	$8100	$ 0	$500	$8600
2	4206	433	692	5331
3	2910	854	880	4644
4	2264	1264	1062	4589 ←
5	1878	1661	1238	4779
6	1622	2048	1410	5081
7	1440	2425	1578	5443
8	1305	2789	1740	5834
9	1200	3142	1896	6239
10	1117	3484	2048	6650
11	1050	3816	2196	7063
12	995	4136	2338	7470
13	948	4446	2476	7871
14	909	4746	2609	8265
15	876	5035	2738	8648

The total EUAC data are plotted in Figure 12-3. From either the tabulation or the figure, we see that the minimum cost life of the machinery is 4 years, with a minimum EUAC of $4589 for each of those 4 years. ■

Looking at Figure 12-3 a bit closer, we see the effects of each of the individual cost components on total EUAC (capital recovery, maintenance/repair, and operating expense EUACs) and how they behave over time. The total EUAC curve of most assets tends to follow this concave shape — high at the beginning due to capital recovery costs, and high at the end due to increased maintenance/repair and operating expenses. The minimum EUAC occurs somewhere between these high points.

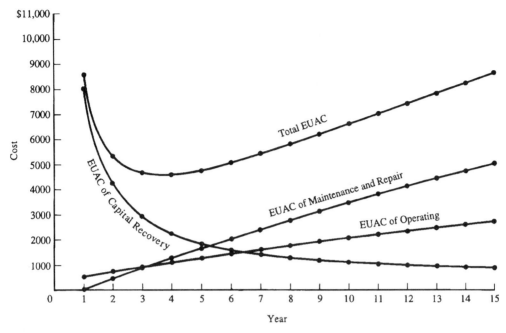

Figure 12-3 Plot of costs for Example 12-1.

Use of Marginal Cost Data

Once the basic participants in the defender-challenger comparison have been identified (see Figure 12-1), two specific questions regarding marginal costs must be answered: *"Do we have marginal cost data for the defender?"* and *"Are the defender's marginal costs increasing on a year-to-year basis?"* Let us first define marginal cost and then discuss why is it important to answer these two questions.

Marginal costs, as opposed to an EUAC, are the year-by-year costs associated with keeping an asset. Therefore, the "period" of any yearly marginal cost associated with ownership is always *1 year*. On the other hand, an EUAC can apply to any number of consecutive years. Thus, the marginal cost of ownership for any year in the life of an asset is the cost for *that year only*. In replacement problems, the total marginal cost for any year can include: the capital recovery cost (loss in market value and lost interest for the year); yearly operating and maintenance costs; yearly taxes and insurance; and any other expense that occurs during that year. To calculate the yearly marginal cost of ownership of an asset it is necessary to have estimates of an asset's market value on a year-to-year basis over its useful life, as well as ordinary yearly expenses. Example 12-2 illustrates how total marginal cost can be calculated for an asset.

EXAMPLE 12-2

A new piece of production machinery has the following costs.

Investment Cost	= $25,000
Annual Operating/Maintenance Cost	= $2000 the first year and increasing at $500 per year thereafter
Annual Insurance Cost	= $ 5000 per year for 3 years, then increasing by $1500 per year thereafter
Useful Life	= 7 years
MARR	= 15% per year

Calculate the marginal cost of keeping this asset over its useful life.

Solution:

From the problem data we can easily find the marginal O&M and insurance costs. However, to calculate the marginal capital recovery cost, we need estimates of the year-to-year market value that the production machinery could be sold for over its 7 year useful life. Market value estimates were made as given below:

Year (n)	Market Value
1	$18,000
2	13,000
3	9,000
4	6,000
5	4,000
6	3,000
7	2,500

We can now calculate the *marginal cost* (year-to-year cost of ownership) of the production machinery over its 7 year useful life.

Year (n)	Loss in Market Value in Year (n)	Foregone Interest in Year (n)	O&M Cost in Year (n)	Insurance Cost in Year (n)	Total Marginal Cost in Year (n)
1	25,000-18,000 = $7,000	25,000 (.15) = $3,750	$2,000	$5,000	$17,750
2	18,000-13,000 = 5,000	18,000 (.15) = 2,700	2,500	5,000	15,200
3	13,000 - 9,000 = 4,000	13,000 (.15) = 1,950	3,000	5,000	13,950
4	9,000 - 6,000 = 3,000	9,000 (.15) = 1,350	3,500	6,500	14,350
5	6,000 - 4,000 = 2,000	6,000 (.15) = 900	4,000	8,000	14,900
6	4,000 - 3,000 = 1,000	4,000 (.15) = 600	4,500	9,500	15,600
7	3,000 - 2,500 = 500	3,000 (.15) = 450	5,000	11,000	16,950

Notice the total marginal cost for each year is made up of: loss in market value; foregone interest; O&M cost; and insurance cost. As an example, the year 5 marginal cost of $14,900 is calculated as: 2,000 + 900 + 4,000 + 8,000. ■

Do We Have Marginal Cost Data for the Defender?

Per our decision map, it is necessary to know if marginal cost data are available for the defender asset in order to determine the appropriate replacement technique to use. Usually in engineering economic problems annual savings and expenses are given for all alternatives. However, as in the example above, it is also necessary to have year-to-year salvage value estimates to calculate total marginal costs. If the total marginal costs for the defender can be calculated, and if that data are increasing on a year-to-year basis, then it is appropriate to use *replacement analysis technique #1* for comparing the defender to the challenger.

Are These Marginal Costs Increasing?

As indicated above, it is important to know if the marginal cost for the defender is increasing on a year-to-year basis. This is determined by inspecting the total marginal cost of ownership of the defender over its remaining life. Example 12-3 illustrates the calculation of the total marginal cost for the defender asset.

EXAMPLE 12-3

An old asset purchased 5 years ago for $75,000, can be sold today for $15,000. Operating expenses in the past have been $10,000 per year, but these are estimated to increase in the future by $1500 per year each year. It is estimated that the market value of the asset will decrease by $1000 per year over the next 5 years. If the MARR used by the company is 15%, calculate the total marginal cost of ownership of this old asset for each of the next 5 years.

Solution:

We calculate the total marginal cost of maintaining the old asset for the next 5 year period as follows:

Year (n)	Loss in Market Value in Year (n)	Foregone Interest in Year (n)	Operating Cost in Year (n)	Marginal Cost in Year (n)
1	15,000-14,000 = $1000	15,000 (0.15) = $2250	$10,000	$13,250
2	14,000-13,000 = 1000	14,000 (0.15) = 2100	11,500	14,600
3	13,000-12,000 = 1000	13,000 (0.15) = 1950	13,000	15,950
4	12,000-11,000 = 1000	12,000 (0.15) = 1800	14,500	17,300
5	11,000-10,000 = 1000	11,000 (0.15) = 1650	16,000	18,650 ■

We can see that the marginal costs increase in each subsequent year of ownership. When the condition of increasing marginal costs for the defender has been met, then the defender-challenger comparison is appropriately made using *replacement analysis technique #1*.

Replacement Analysis Technique #1: Defender Marginal Costs Can Be Computed And Are Increasing

Using this method of analyzing the defender asset against the best available challenger the basic comparison involves the:

marginal cost data of the defender, and the minimum cost life data of the challenger.

In the condition where the marginal cost of the defender is increasing on a year-to-year basis, we will maintain that defender as long as the marginal cost of keeping it one more year is less than the minimum EUAC of the challenger. Thus our decision rule is:

> *Maintain the defender as long as the marginal cost of ownership for one more year is less than the minimum EUAC of the challenger. Once the marginal cost of the defender becomes greater than the minimum EUAC of the challenger, then replace the defender with the challenger.*

One can see that this technique assumes that the current best challenger, with its minimum EUAC, will be available and unchanged in the future. Example 12-4 below illustrates the use of this technique for comparing the defender and challenger assets.

EXAMPLE 12-4

Using the machinery in Example 12-2 as the *challenger* and the machinery in Example 12-3 as the *defender*, use *replacement analysis technique #1* to determine when, if at all, a replacement decision should be made.

Solution:

The *replacement analysis technique #1* is appropriate only in the condition of increasing marginal costs for the defender. Since these marginal costs are increasing for the defender (from Example 12-3), we can proceed by comparing defender marginal costs against the minimum EUAC of the challenger asset. In example 12-2 we calculated only the marginal costs of the challenger; thus it is necessary to calculate the challenger's minimum EUAC. The EUAC of keeping this asset for each year of its useful life is done below.

Year (n)	Challenger Total Marginal Cost in Year (n)		EUAC of Challenger Ownership if it is kept through Year (n)
1	$17,750	$[17,750(P/F, 15\%,1)](A/P,15\%,1)$	= $17,500
2	15,200	$[17,750(P/F,15\%,1) + 15,200(P/F,15\%,2)](A/P,15\%,2)$	= 16,560
3	13,950	$[17,750(P/F,15\%,1) + ... + 13,950(P/F,15\%,3)](A/P,15\%,3)$	= 15,810
4	14,350	$[17,750(P/F,15\%,1) + ... + 14,350(P/F,15\%,4)](A/P,15\%,4)$	= 15,520
5	14,900	$[17,750(P/F,15\%,1) + ... + 14,900(P/F,15\%,5)](A/P,15\%,5)$	= 15,430
6	15,600	$[17,750(P/F,15\%,1) + ... + 15,600(P/F,15\%,6)](A/P,15\%,6)$	= 15,450
7	16,950	$[17,750(P/F,15\%,1) + ... + 16,950(P/F,15\%,7)](A/P,15\%,7)$	= 15,580

A minimum EUAC of $15,430 is attained for the challenger at year 5, which is the challenger's *minimum cost life*. We proceed by comparing this value against the *marginal* costs of the defender from Example 12-3. This comparison is shown below.

Year (n)	Defender Total Marginal Cost in Year (n)	Challenger Minimum EUAC	Comparison Result and Recommendation
1	$13,250	$15,430	Since $13,250 is *less than* $15,430, keep defender.
2	14,600	15,430	Since $14,600 is *less than* $15,430, keep defender.
3	15,950	15,430	Since $15,950 is *greater than* $15,430, replace defender.
4	17,300		
5	18,650		

Based on the data given for the challenger and defender, we would keep the defender for 2 more years and then replace the defender with the challenger because at that point the marginal cost of 1 more year of ownership of the defender would be greater than the minimum EUAC of the challenger. ■

One may ask *"Why can't replacement analysis technique #1 be used in cases where the marginal costs of the defender do not increase?"* To answer this question we must understand that for this technique to be valid the following two basic assumptions are necessary: the best challenger will be available "with the same minimum EUAC" at any time in the future; and the period of needed service in our business is indefinitely long. In other words, we assume that once the decision is made to replace, there will be an indefinite replacement of the defender, with continuing "cycles" of the current best challenger asset. These two assumptions together are much like the repeatability assumptions that allowed us to compare competing alternatives with different useful lives using the annual cost method from earlier chapters. Taken together, we call these the *replacement repeatability assumptions.* They allow us to greatly simplify the comparison of the defender and the challenger. We state these assumptions formally below.

Replacement Repeatability Assumptions:

The two assumptions are:

1. The currently available best challenger will continue to be available in subsequent years and will be unchanged in its economic costs. When the defender is ultimately replaced, it will be replaced with this challenger. Any challengers put into service will also be replaced with this same currently available challenger.

2. The period of needed service of the asset is indefinitely long. Thus the challenger asset, once put into service, will continuously replace itself in repeating, unchanged cycles.

Given that the defender will ultimately be replaced with the current best challenger, we would not want to ever incur a defender marginal cost greater than the challenger's minimum EUAC. And because the defender's marginal costs are increasing, we can be assured that once the marginal cost of the defender is greater than the challenger's minimum EUAC, it will continue to be so in the future. However, if the marginal costs do not increase, we can not be guaranteed that *replacement analysis technique #1* will produce the alternative that is of the greatest economic advantage. One may ask, *"Are there ordinary conditions in which the marginal costs are not increasing?"* The answer to this question is yes. Consider the new asset in Example 12-2. This new asset has marginal costs that begin at a high of $17,500, then *decrease* over the next years to a low of $14,350, and then *increase* thereafter to $16,900 in year 7. If this asset were implemented and then evaluated *one year after implementation* as a defender asset, it would not have increasing marginal costs. Thus, defenders in the early stages of their respective implementations would not fit the requirements of *replacement analysis technique #1*. In considering Figure 12-3, such defender assets would be in the downward slope of a concave marginal cost curve. Example 12-5 illustrates the error that can be introduced using *replacement analysis technique #1* when defenders do not have consistently increasing marginal cost curves.

EXAMPLE 12-5

Let us look again at the defender and challenger assets in Example 12-4 . This time let us change the defender's marginal costs for its 5 year useful life. Now when, if at all, should the defender be replaced with the challenger?

Year (n)	Defender Total Marginal Cost in Year (n)
1	$16,000
2	15,750
3	13,500
4	16,750
5	18,250

Solution:

In this case the total marginal costs of the defender are *not* consistently increasing from year to year. However, if we ignore this fact and apply *replacement analysis technique #1,* the recommendation would be to replace the defender now, because the marginal cost of the defender for the first year ($16,000) is greater than the minimum EUAC of the challenger ($15,430). Let us review this decision. One can see that both the first and second year marginal costs of the defender are greater than the minimum EUAC of the challenger, but in the third year the marginal cost is less. We have a case where marginal costs decrease for three years to a minimum and then increase the following two years.

Let us calculate the EUAC of keeping the defender asset each of its remaining 5 years, at $i = 15\%$.

	EUAC of Defender	
	Ownership if it is	
Year (n)	*kept (n) years*	*Challenger minimum EUAC = $15,430*
1	$16,000	
2	15,880	
3	15,200	
4	15,510	
5	15,910	

The minimum EUAC of the defender for 3 years is $15,200, which is less than that of the challenger. But this comparison alone is not sufficient to indicate that we should keep the defender for 3 years and then replace it. In looking at a study period of 3 years we really have four options.

Option #1: Keep the defender for 3 more years (@ $15,200 /year).

Option #2: Keep the defender for 1 year (@ $16,000) and the challenger for 2 years (@ $15,430 /year).

Option #3: Keep the defender for 2 years (@ $15,880/year) and the challenger for 1 year (@ $15,430 /year).

Option #4: Implement the challenger today (@ $15,430 /year).

The EUAC of each of these options at 15% is:

Option #1   EUAC is $15,200.

Option #2   EUAC is [16,000 + 15,430(P/A,15%,2)](P/F,15%,1)(A/P,15%,3) = $ 15,650.

Option #3   EUAC is [15,880(F/A,15%,3) - 450](A/F,15%,3) = $ 15,750.

Option #4   EUAC is $15,430.

One can see that Option #1 produces the minimum EUAC. Thus we should keep the defender for 3 more years and then re-evaluate the defender-challenger decision at that time. Notice that if we replace the defender now with the challenger per *replacement decision rule #1,* we would not achieve a minimum EUAC over this 3 year period. We emphasize that this analysis is valid only because of the *replacement repeatability assumptions.* These assumptions hold that after the 3 year period, all alternatives have the exact same yearly cash flow of $15,430 (thus we can ignore them), and these identical cash flows continue indefinitely. ■

Example 12-5 demonstrates that in cases where the marginal costs of the defender are not consistently increasing from year to year, it is necessary to calculate the EUAC and minimum cost life of the defender. In the next section we describe this calculation.

Minimum Cost Life of the Defender

How long can a defender asset be kept operating? Anyone who has seen or heard old machinery in operation, whether it is a 50 year-old automobile or 20 year-old production equipment, has realized that almost any machine can be kept operating indefinitely, provided it receives proper maintenance and repair. However, even though one might be able to keep a defender going indefinitely, the cost may prove excessive. So, rather than asking what remaining *operating life* the defender may have, we really want to ask what is the *minimum cost life* of the asset. The minimum cost life of the defender is defined as the number of years of ownership in the future which result in a minimum EUAC. This calculation is much like the calculation of the minimum cost life of the challenger asset described previously, the difference being that the asset was previously installed instead of a new one.

EXAMPLE 12-6

An 11 year-old piece of equipment is being considered for replacement. It can be sold for $2000 now, and it is believed this same salvage value can also be obtained in future years. The current maintenance cost is $500 per year and is expected to increase $100 per year in future years. If the equipment is retained in service, compute the economic life that results in a minimum EUAC, based on 10% interest.

Solution: Here the salvage value is not expected to decline from its present $2000. The annual cost of this invested capital is $Si = 2000(0.10) = \$200$. The maintenance is represented by $\$500 + \$100G$. A year-by-year computation of EUAC is as follows:

Year(n)	Age of equipment, in years	EUAC of invested capital $= Si$	EUAC of maintenance $= 500$ $+ 100\,(A/G,10\%,n)$	Total EUAC
1	11	$200	$500	$700
2	12	200	548	748
3	13	200	594	794
4	14	200	638	838
5	15	200	681	881

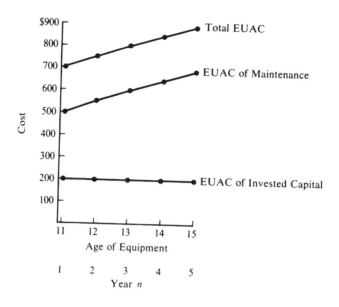

Figure 12-4 EUAC for different remaining lives.

These data are plotted in Fig. 12-4. We see that the annual cost of continuing to use the equipment is increasing. It is reasonable to assume that if the equipment is not replaced now, it will be reviewed again next year. Thus the economic life at which EUAC is a minimum is one year. ■

Example 12-6 represents a common situation. The salvage value is stable but the maintenance cost is increasing. The total EUAC will continue to increase as time passes, which means that an economic analysis which compares the defender at its most favorable remaining life will be based on retaining the defender for 1 more year. This is not always the case, as is shown in Ex. 12-7.

EXAMPLE 12-7

A 5 year-old machine, whose current market value is $5000, is being analyzed to determine its economic life in a replacement analysis. Compute its economic life using a 10% interest rate. Salvage value and maintenance estimates are given in the following table.

Years of remaining life n	Estimated salvage value (S) end of Year n	Estimated maintenance cost for year	If retired at end of Year n		Total EUAC
			EUAC *of capital recovery (P − S)* × (A/P,10%,n) + Si	EUAC *of maintenance* 100(A/G,10%,n)	
0	P=$5000				
1	4000	$ 0	$1100 + 400	$ 0	$1500
2	3500	100	864 + 350	48	1262
3	3000	200	804 + 300	94	1198
4	2500	300	789 + 250	138	1177
5	2000	400	791 + 200	181	1172
6	2000	500	689 + 200	222	1111
7	2000	600	616 + 200	262	1078
8	2000	700	562 + 200	300	1062
9	2000	800	521 + 200	337	1058←
10	2000	900	488 + 200	372	1060
11	2000	1000	462 + 200	406	1068

Solution:

A minimum EUAC of $1058 is computed at year 9 for the existing machine. ■

From Examples 12-6 and 12-7, we see that the minimum cost life remaining for an existing machine may be 1 year, or it may be longer. Looking again, we find that the two examples represent the same machine being examined at different points in its life. Example 12-7 indicates that the 5 year-old machine has a minimum cost life of 9 years, making an anticipated total of 14 years of service. This would be the point where, from age 5 onward, the total EUAC would be a minimum. It is important to recognize that this minimum EUAC is based on the projection of future costs for the 5 year-old machine, not on past or sunk costs. Therefore, the projection for the 5 year-old machine (Ex. 12-7) is different from the situation when the machine is 11 years old (Ex. 12-6): the EUAC increases when computed from age 11 onward.

For older equipment with a negligible or stable salvage value, it is likely that the operating and maintenance costs are increasing. Under these circumstances, the useful life at which EUAC is a minimum is 1 year.

It is necessary to calculate the defender's minimum cost life when marginal costs are not consistently increasing. Having made this calculation, we can now use *replacement analysis technique #2* to compare the defender against the challenger.

Replacement Analysis Technique #2:
Defender Marginal Cost Can Be Computed And Is Not Decreasing

Using this method of analyzing the defender asset against the best available challenger the basic comparison involves the:

> *EUAC of the defender asset at its minimum cost life,*
> *and the EUAC of the challenger at its minimum cost life.*

We calculate the minimum EUAC of the defender and compare this directly against the minimum EUAC of the challenger. Remember that the *replacement repeatability assumptions* allow us to do this. In this comparison we then choose the alternative with the lowest EUAC. In Example 12-5 the comparison involved the defender with an economic life at 3 years and an EUAC = $15,200 and the challenger with an economic life of 5 years and an EUAC = $15,430. Here we would recommend that the defender be retained for 3 more years, at which point its marginal costs will be increasing and *replacement analysis technique #1* would apply. Consider Example 12-8 below.

EXAMPLE 12-8

An economic analysis is to be made to determine if existing (defender) equipment in an industrial plant should be replaced. A $4000 overhaul must be done now if the equipment is to be retained in service. Maintenance is estimated at $1800 in each of the next 2 years, after which it is expected to increase annually on a $1000 arithmetic gradient. The defender has no present or future salvage value. The equipment described in Ex. 12-1 is the challenger. Make a replacement analysis to determine whether to retain the defender or replace it by the challenger if 8% interest is used.

Solution: The first step is to determine the minimum cost life of the defender. The pattern of overhaul and maintenance costs (see Fig. 12-5) suggests that if the overhaul is done, the remaining minimum cost life of the equipment will be several years. The computation is as follows: *If retired at end of Year n*

Year n	EUAC of overhaul $4000(A/P,8\%,n)$	EUAC of maintenance $1800 + $1000 gradient from Year 3 on	Total EUAC
1	$4320	$1800	$6120
2	2243	1800	4043
3	1552	1800 + 308*	3660←
4	1208	1800 + 683†	3691
5	1002	1800 + 1079	3881

*For the first 3 years, the maintenance is $1800, $1800, and $2800. Thus, EUAC = $1800 + 1000(A/F,8\%,3) = 1800 + 308.
†EUAC = 1800 + 1000(P/G,8\%,3)(P/F,8\%,1)(A/P,8\%,4) = 1800 + 683.

For minimum EUAC, the minimum cost life of the defender is 3 years. In Ex. 12-1, we determined that the minimum cost life of the challenger is 4 years and that the resulting EUAC is $4589. If we assume the equipment is needed for at least 4 years, the EUAC of the defender ($3660) is less than the EUAC of the challenger ($4589). In this situation the defender should be kept for 3 more years and be re-evaluated after that time. ■

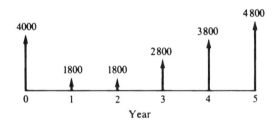

Figure 12-5 Overhaul and maintenance costs for the defender in Example 12-8.

No Defender Marginal Cost Data Available

Earlier we described replacement analysis techniques for comparing the defender and challenger if marginal cost data are available for the old asset. One may recall that yearly salvage value estimates were necessary to calculate these marginal costs. However, there may be cases in which it is difficult to obtain end-of-the-year salvage value estimates for the defender asset. A defender, based on aging technology with a shrinking market, might be such an asset. Or, from a student's problem-solving perspective, perhaps salvage value data are not given as part of the problem data. Without these data it is impossible to calculate the marginal cost of the defender and thus, impossible to compare the defender against the challenger using marginal cost data. How should one proceed?

Given the *replacement repeatability assumptions*, it is possible for us to calculate the EUAC of the defender over its remaining useful life. Then, knowing that both the defender and challenger will be replaced with indefinite cycles of the challenger, we may compare the defender and challenger using *replacement analysis technique #3.*

Replacement Analysis Technique #3:
When Defender Marginal Cost Data Are Not Available

Using this method of analyzing the defender asset against the best available challenger, the basic comparison involves the:

EUAC of the defender over its stated useful life, and the minimum EUAC of the challenger.

We will calculate the EUAC of the defender asset over its remaining useful life and compare this directly against the EUAC of the challenger at its minimum cost life, and then choose the lesser of these two values. However, in making this basic comparison an often complicating factor is deciding what first cost to assign to the challenger and the defender assets.

Defining Defender and Challenger First Costs

Because the defender is already in service, analysts often misunderstand what first cost to assign it. Example 12-9 demonstrates this problem.

EXAMPLE 12-9

A laptop word processor model SK-30 was purchased 2 years ago for $1600; it has been depreciated by straight line depreciation using a 4 year life and zero salvage value. Because of recent innovations in word processors, the current price of the SK-30 laptop has been reduced from $1600 to $995. An office equipment supply firm has offered a trade-in allowance of $350 for the SK-30 on a new $1200 model EL-40 laptop. Some discussion revealed that without a trade-in, the EL-40 can be purchased for $1050, indicating the originally quoted price of the EL-40 was overstated to allow a larger trade-in allowance. The true current market value of the SK-30 is probably only $200. In a replacement analysis, what value should be assigned to the SK-30 laptop?

Solution: In the example, five different dollar amounts relating to the SK-30 laptop have been outlined:

1. *Original cost:* The laptop cost $1600 2 years ago.

2. *Present cost:* The laptop now sells for $995.

3. *Book value:* The original cost less 2 years of depreciation is $1600 - \frac{2}{4}(1600 - 0) = \$800$.

4. *Trade-in value:* The offer was $350.

5. *Market value:* The estimate was $200.

We know that an economic analysis is based on the current situation, not on the past. We refer to past costs as *sunk* costs to emphasize that, since these costs can not be altered, they are not relevant. (The one exception is that past costs may affect present or future income taxes.) In the analysis we want to use actual cash flows for each alternative. Here the question is, "What value should be used in an economic analysis for the SK-30?" The relevant cost is the present market value for the equipment. Neither the original cost, the present cost, the book value, nor the trade-in value is relevant.   ■

At first glance, the trade-in value of an asset would appear to be a suitable present value for the equipment. Often the trade-in price is inflated *along with* the price for the new item. (This practice is so common in new-car showrooms, that the term *overtrade* is used to describe the excessive portion of the trade-in allowance. The purchaser is also quoted a higher price for the new car.) Distortion of the present value of the defender, or a distorted price for the challenger, can be serious because these distortions do not cancel out in an economic analysis.

Example 12-9 illustrates that of the several different values that can be assigned to the defender, the most appropriate is the present market value. If a trade-in value is obtained, care should be taken to ensure that it actually represents a fair market value.

Determining the value for the installed cost of the challenger asset should be less difficult. In such cases the first cost is usually made up of such items as: purchase price, sales tax, installation costs, and other items that occur initially on a one-time basis due to the selection of the challenger. These values are usually rather straightforward to obtain if a thorough analysis is conducted. One aspect to consider when assigning a first cost to the challenger is the potential disposition (or market or salvage) value of the defender. One must not arbitrarily subtract the disposition cost of the defender from the first cost of the challenger asset, as it can lead to an incorrect analysis.

As described in Example 12-9, the correct first cost to assign to the defender SK-30 laptop is its $200 current market value. This value represents the present economic benefit that we would be *foregoing* in order to keep the defender. This can be called our *opportunity first cost*. If, instead of assuming that this is an *opportunity cost* to the defender, we assume it is a *cash benefit* to the challenger, a potential error arises. Consider the following case involving the SK-30 and EL-40 laptops. Assume the following data:

	SK-30		*EL-40*
Market value	$200	First cost	$1050
Remaining life	3 years	Useful life	3 years

In this case the remaining life of the defender (SK-30) and the useful life of the challenger (EL-40) is 3 years. If we use an *opportunity cost* perspective, then the calculated capital recovery effect of first cost using an annual cost comparison is:

Annualized first cost $_{SK-30}$ = $200($A/P$,10\%,3) = $80

Annualized first cost $_{EL-40}$ = $1050($A/P$,10\%,3) = $422

The *difference* in annualized first cost between the SK-30 and EL-40 is:

$$AFC_{EL-40} - AFC_{SK-30} = \$422 - \$80 = \$342$$

Now use a *cash flow* perspective to look at the first costs of the defender and challenger. In this case we use the actual cash that changes hands when each alternative is selected. A first cost of zero (\$0) cash would be assigned to the defender and a first cost of \$850 to the challenger (-\$1050 purchase price of the challenger and +\$200 in salvage value from defender). We calculate the *difference* due to first cost between the SK-30 and EL-40 to be:

$$\text{Annualized first cost}_{SK-30} = \$0(A/P,10\%,3) = \$0$$

$$\text{Annualized first cost}_{EL-40} = (\$1050 - 200)(A/P,10\%,3) = \$342$$

$$AFC_{EL-40} - AFC_{SK-30} = \$342 - \$0 = \$342$$

When both the remaining life of the defender and the useful life of the challenger are the same, 3 years in the case above, the analysis of the first cost yields an identical (and correct) result. Both the *opportunity cost* and *cash flow* perspectives for considering first cost of the defender and challenger result in a difference of \$342 between the two alternatives on an annual cost basis.

Now look at the case where the remaining life of the defender is not equal to the useful life of the challenger. Consider the SK-30 and EL-40 word processors above, except let us assume that the lives have been changed as follows:

	SK-30		*EL-40*
Remaining life	3 years	Useful life	5 years

Looking at this second case from an *opportunity cost* perspective, the calculated capital recovery effect of the first cost using an annual cost comparison is:

$$\text{Annualized first cost}_{SK-30} = \$200(A/P,10\%,3) = \$80$$

$$\text{Annualized first cost}_{EL-40} = \$1050(A/P,10\%,5) = \$277$$

The *difference* in annualized first cost between the SK-30 and EL-40 is:

$$AFC_{EL-40} - AFC_{SK-30} = \$277 - \$80 = \$197$$

Now using a *cash flow* perspective to look at the first costs of the defender and challenger, we can calculate the *difference* due to first cost between the SK-30 and EL-40.

$$\text{Annualized first cost}_{SK-30} = \$0(A/P,10\%,3) = \$0$$

$$\text{Annualized first cost}_{EL-40} = (\$1050 - 200)(A/P,10\%,5) = \$224$$

$$AFC_{EL-40} - AFC_{SK-30} = \$224 - \$0 = \$224$$

When the remaining life of the defender (3 years) differs from that of the useful life of the challenger (5 years), an analysis of the annualized first cost yields different results. The correct difference of $197 is shown using the *opportunity cost* approach, and an inaccurate difference of $224 is obtained using the *cash flow* perspective. In the opportunity cost case the $200 is spread out over 3 years as a cost to the defender, yet in the cash flow case the opportunity cost is spread out over 5 years as a benefit to the challenger. Spreading the $200 over 3 years in one case and 5 years in the other case does not produce equivalent annualized amounts. Because of the difference in the lives of the assets, the annualized $200 opportunity cost for the defender can not be called an equivalent benefit to the challenger.

In the case of unequal lives the correct method is to assign the current market value of the defender as its time zero opportunity costs, rather than subtracting this amount from the first cost of the challenger. Because the cash flow approach yields an incorrect value when the challenger and defender have unequal lives, **the *opportunity cost* approach for assigning a first cost to the challenger and defender assets should *always* be used**.

Repeatability Assumptions not Acceptable

There are circumstances in which the repeatability assumptions described earlier may not apply for a replacement analysis. In these cases replacement analysis techniques #1, #2, and #3 may not be valid methods for comparison. For instance, a decision maker may set the study period instead of assuming that there is an indefinite need for the asset. As an example, consider the case of phasing out production after a certain number of years — perhaps a businessperson going out of business and selling all assets due to retirement. The study period could potentially be set at any number of years relative to the lives of the defender and challenger, such as: equal to the life of the defender, equal to the life of the challenger, less than the life of the defender, greater than the life of the challenger, or somewhere between the lives of the defender and challenger. The essential principle in this case is that when the decision maker is *setting the study period,* he or she must use an appropriate method as described in previous chapters. The analyst must be explicit about the economic costs and benefits of the challenger that is assumed for replacement of the defender (when replacement is made), as well as residual or salvage values of the alternatives at the end of the study period. In this case the repeatability replacement assumptions do not apply, and thus the replacement analysis techniques are not necessarily valid. Another circumstance where the analysis techniques in the decision map may not apply occurs when future challengers are not assumed to be identical to the current best challenger. This concept is discussed in the next section.

A Closer Look at Future Challengers

We defined the challenger as the best available alternative to replace the defender. But in time, the best available alternative can change. And given the trend in our technological society, it seems likely that future challengers will be better than the present challenger. If so, the prospect of improved future challengers may affect the present decision between the defender and the challenger.

Figure 12-6 illustrates two possible estimates of future challengers. In many technological areas it seems likely that the equivalent uniform annual costs associated with future challengers will decrease by a constant amount each year. There are other fields, however, where a rapidly changing technology will produce a sudden and substantially improved challenger — with decreased costs or increased benefits. The uniform decline curve of Fig. 12-6 assumes that each future challenger has a minimum EUAC that is a fixed amount less than the previous year's challenger. This assumption, of course, is only one of many possible assumptions that could be made regarding future challengers.

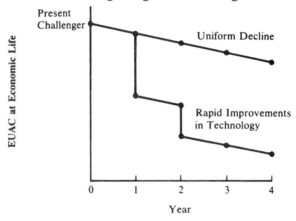

Figure 12-6 Two possible ways the EUAC of future challengers may decline.

If future challengers will be better than the present challenger, what impact will this have on an analysis now? The prospect of better future challengers may make it more desirable to retain the defender and to reject the present challenger. By keeping the defender for now, we may be able to replace it later by a better future challenger. Or, to state it another way, the present challenger may be made less desirable by the prospect of improved future challengers. As engineering economic analysts, it is important to familiarize ourselves with potential technological advances in assets targeted for replacement. This part of the decision process is much like the search for all available alternatives, from which we select the best. In finding out more about what alternatives and technologies are emerging, we will be better able to understand the repercussions of investing in the current best available challenger. Selecting the current best challenger asset can be particularly risky when: (1) the costs are very high, and/or (2) when the useful minimum cost life of that challenger is relatively long (5-10 years or more). Under circumstances where one or both of these conditions exist, it may be better to keep or even augment our defender asset until better future challengers emerge.

There are, of course, many assumptions that *could* be made regarding future challengers. However, if the replacement repeatability assumptions do not hold, the anaysis techniques described earlier may not be valid.

After-Tax Replacement Analysis

As described in Chapter 11, an after-tax analysis adds an expanded perspective to our problems because various very important effects can be included. We saw earlier that examining problems on an after-tax basis provides greater realism and insight. This advantage is also true when considering the replacement problems discussed in this chapter. Tax effects have the potential to alter recommendations made in a before-tax only analysis. After-tax effects may influence calculations in the following: remaining economic life of defender; economic life of challenger; and defender versus challenger comparisons discussed earlier. Consequently, one should always perform these analyses on an after-tax basis. In this section we illustrate the complicating circumstances that are introduced by an after-tax analysis.

Marginal Costs on an After-Tax Basis

As in the before-tax case, there are times when the defender-challenger comparison is based on the marginal costs of the defender on a year-to-year basis. Marginal costs on an after-tax basis represent the cost that would be incurred through ownership of the asset *in each year*. On an after-tax basis we must consider the effects of ordinary taxes as well as gains and losses due to asset disposal in calculating the after-tax marginal costs. Consider Example 12-10 below.

EXAMPLE 12-10

Refer to Example 12-2 where we calculated the before-tax marginal costs for a new piece of production machinery. Before-tax annual cost information was given as well as year-end salvage values for the asset over its useful life. Calculate the after-tax marginal costs of this asset considering the additional information below.

- Depreciation is by the Straight-Line Method with $S = \$0$ and $N = 5$ years.
- Ordinary Income is taxed at a rate of 40%.
- Assume all Gains and Losses are taxed at a rate of 28%.
- The after-tax MARR is 10%.

Solution:

The after-tax marginal cost of ownership will involve the following elements: foregone gain or loss, foregone interest, annual loss in after-tax value, and annual operating/maintenance and insurance. We calculate the marginal cost of each of these below.

FOREGONE CAPITAL GAINS/LOSSES

Year (n)	Market Value in Year (n)	Book Value in Year (n)	Gain or Loss if Asset Sold in Year (n)	Foregone Gain or Loss if decide to keep Year (n)		Marginal Cost in Year (n) due to Foregone Gain/Loss
1	$18,000	$20,000	$-2000		$ 0	$ 0
2	13,000	15,000	-2000	-2000(-0.28) =	560	560
3	9000	10,000	-1000	-2000(-0.28) =	560	560
4	6000	5000	1000	-1000(-0.28) =	280	280
5	4000	0	4000	1000(-0.28) =	-280	-280
6	3000	0	3000	4000(-0.28) =	-1120	-1120
7	2500	0	2500	3000(-0.28) =	-840	-840

FOREGONE INTEREST

Year (n)	Market Value Beginning of Year (n)	Foregone After-Tax Interest in Year (n)	
1	$25,000	25,000 (0.10)(0.60) =	$1500
2	18,000	18,000 (0.10)(0.60) =	1080
3	13,000	13,000 (0.10)(0.60) =	780
4	9000	9000 (0.10)(0.60) =	540
5	6000	6000 (0.10)(0.60) =	360
6	4000	4000 (0.10)(0.60) =	240
7	3000	3000 (0.10)(0.60) =	180

MARGINAL O&M, INSURANCE COSTS

Total O&M and Insurance Cost in Year (n)	Marginal Cost in Year (n) due to O&M and Insurance	
$7000	7000(0.60) =	$4200
7500	7500(0.60) =	4500
8000	8000(0.60) =	4800
10,000	10,000(0.60) =	6000
12,000	12,000(0.60) =	7200
14,000	14,000(0.60) =	8400
16,000	16,000(0.60) =	9600

LOST AFTER-TAX VALUE

Year (n)	After-Tax Value at Beginning Year (n)	After-Tax Value at End Year (n)			Loss in AT Value in Year (n)
1	$25,000	18,000 +	560 =	$18,560	$6440
2	18,560	13,000 +	560 =	13,560	5000
3	13,560	9000 +	280 =	9280	4280
4	9280	6000 -	280 =	5720	3560
5	5720	4000 -	1120 =	2880	2840
6	2880	3000 -	840 =	2160	720
7	2160	2500 -	700 =	1800	360

TOTAL MARGINAL COSTS DUE TO OWNERSHIP

Year (n)	Foregone Gain/Loss in Year (n)	Foregone After-Tax Interest in Year (n)	Marginal Cost in Year (n) due to O&M and Insurance	Loss in AT Market Value in Year (n)	Total Marginal Cost in Year (n)
1	$ 0	$ 1500	$4200	$6440	$12,140
2	560	1080	4500	5000	11,140
3	560	780	4800	4280	10,420
4	280	540	6000	3560	10,380
5	-280	360	7200	2840	10,120
6	-1120	240	8400	720	8,240
7	-840	180	9600	360	9,300

Notice that the total marginal cost for each year is much different from the marginal cost for the same asset on a before-tax basis. ∎

After-Tax Cash Flows for the Challenger

Finding the after-tax cash flows for the challenger asset are straightforward. Here we use the standard after-tax cash flow table and method developed in Chapter 11 to incorporate all of the relevant tax effects. These include: before-tax cash flows, depreciation, taxes, and gains and losses at disposal. After obtaining the challenger ATCFs it is a simple task to calculate the EUAC over the challenger's life.

After-Tax Cash Flows for the Defender

Unlike the challenger, finding the after-tax cash flows for the defender is not so straightforward. The main complicating factor is that the defender is an asset that has been previously placed in service. As such, depreciation has been taken on the cost basis (amount being depreciated) up to the present, which changes the book value of the asset. Thus, when looking at the first cost of the defender on an after-tax basis (value at time zero), we must consider the current market (salvage) value that can be obtained for the asset, as well as any depreciation recapture or losses at disposal presently associated with the defender.

As discussed previously, it is suggested that one use the *opportunity cost* perspective for assigning a first cost to both the defender and challenger. For the challenger, this first cost will be the after-tax cash flow at time zero. Assuming use of equity financing, the first cost of the challenger is the purchase price plus installation cost. However, for the defender the after-tax first cost will be made up of the *foregone market value* of the asset at the present time plus any *foregone gains or losses* associated with keeping the asset. In order to develop the ATCF of the defender at time zero it is often convenient to use the *defender sign change procedure* described below.

Defender Sign Change Procedure: The purpose of this procedure is to find the after-tax cash flow for the defender asset in after-tax defender-challenger comparisons. To find the time zero ATCF of the defender use the following steps:

1. Assume you are selling the defender now (time zero).
2. Find the ATCF for selling the defender at time zero.
3. Then, because you are actually keeping the defender, not selling it, change all of the signs (plus to minus, minus to plus) in the tax table used to develop the ATCF at time zero.
4. Thus the ATCF for selling now becomes the after-tax *opportunity cost* for keeping the defender. Assign this cost to the defender at time zero.

EXAMPLE 12-11

Find the after-tax first cost (today's cost) that should be assigned to a defender asset as described by the following data:

First Cost when Implemented 5 years ago	= $12,500
Depreciation Method	= Straight line depreciation
	(with estimated $S=$2500, $N=10$ years)
Current Market Value	= $8000
Remaining Useful Life	= 10 years
Annual Costs	= $3000
Annual Benefits	= $4500
Market Value after Useful Life	= $1500
Combined Tax Rate	= 34%
(for ordinary income and capital gains/losses)	

Solution: Using the Defender Sign Change Procedure:

Steps 1 and 2. Assume that we will sell the defender now (time zero).

Year (n)	Before Tax Cash Flow	Depreciation	Taxable Income	Income Taxes	After Tax Cash Flow
(sell) 0	+ 8000	–	+ 500	– 170	+ $7830

Where:

Yearly Depreciation = ($12,500 – $2500) / 10 = $1000 year
Before-Tax Cash Flow = + 8000
Gains/Losses = current market value - current book value
current market value = $8000
current book value = cost basis - sum of deprec. charges to date
= 12,500 – 5(1000)
= 12,500 – 5000
= $7500
= 8000 – 7500 = $500 ordinary gain (depreciation recapture)
Taxes on Gain = 500 (–0.34) = –$170
After-Tax Cash Flow = 8000 – 170 = $7830

Steps 3 and 4. Change signs. The after-tax cash flow figure is the time zero *opportunity cost* for keeping the defender.

Year (n)	Before-Tax Cash Flow	Depreciation	Taxable Income	Income Taxes	After-Tax Cash Flow
(keep) 0	– 8000	–	– 500	+ 170	–$7830

After-Tax time zero cost for the defender $= -\$7830$ ■

EXAMPLE 12-12

Determine whether the SK-30 laptop of Example 12-9 should be replaced by the EL-40 model. In addition to the data given in Example 12-9, the following estimates have been made:

> The SK-30 maintenance and service contract costs $80 a year.
> The EL-40 will require no maintenance.
> Either laptop is expected to be used for the next 5 years.
> At the end of that time, the SK-30 will have no value, but the EL-40 probably could be sold for $250.
> The EL-40 laptop is faster and easier to use than the SK-30 model. This benefit is expected to save about $120 a year by reducing the need for part-time employees.

Solve this problem with a MARR equal to 8% after taxes. Both laptop computers will be depreciated by straight line depreciation using a 4 year depreciable life. The SK-30 is already 2 years old so only 2 years of depreciation remain. The analysis period remains at 5 years. Assume a 34% corporate income tax rate.

Solution:

Alternative A: Keep the SK-30 rather than sell it. Compute the after-tax cash flows over the 5 year study period:

	Year	Before-tax cash flow	Straight line depreciation	Taxable income	34% Income taxes	After-tax cash flow
(sell)	0	$200		–$600*	$204*	$ 404
(keep)	0	–200		+600	–204*	–404**
	1	–80	$400	–480	+163	+83
	2	–80	400	–480	+163	+83
	3	–80	0	–80	+27	–53
	4	–80	0	–80	+27	–53
	5	–80	0	–80	+27	–53
	5	0 salvage		0†	0	0

*If sold for $200, there would be a $600 loss on disposal. If $600 of gains were offset by the loss during the year, there would be no gain to tax, saving 34% x $600 = $204. If the SK-30 is not sold, this loss is not realized, and the resulting income taxes will be a $204 higher than if it had been sold.
**This is the sum of the $200 selling price foregone plus the $204 income tax saving foregone.
†Gain/Loss = MV-BV = 0-[1600-(4)400] = 0.

$$\text{EUAC} = [404 - 83(P/A,8\%,2) + 53(P/A,8\%,3)(P/F\ 8\%,2)](A/P\ 8\%,5)$$
$$= \$93$$

Alternative B: Purchase an EL-40. Compute the after-tax cash flow:

Year	Before-tax cash flow	Straight line depreciation	Taxable income	34% Income taxes	After-tax cash flow
0	−$1050				−$1050
1	+120	$200	−$80	+$27	+147
2	+120	200	−80	+27	+147
3	+120	200	−80	+27	+147
4	+120	200	−80	+27	+147
5	+120	0	+120	−41	−79
5	+250		0 †	0	+250

†Gain/Loss = MV-BV = 250−[1050−(4)200] = 0

Compute the EUAC:
$$\text{EUAC} = [1050 - 147(P/A,8\%,4) - (79+250)(P/F,8\%,5)](A/P,8\%,5)$$
$$= \$85$$
Based on this after-tax analysis the EL-40 is the preferred alternative. ■

EXAMPLE 12-13
Solve Example 12-123 by computing the rate of return on the difference between alternatives. In Example 12-9, the two alternatives were "Keep the SK-30" or "Buy an EL-40." The difference between the alternatives would be:

Buy an EL-40	rather than	Keep the SK-30
Alternative *B*	*minus*	Alternative *A*

Solution: The after-tax cash flow for the difference between the alternatives may be computed as follows:

Year	A	B	B − A
0	−$404	−$1050	−$646
1	+83	+147	+64
2	+83	+147	+64
3	−53	+147	+200
4	−53	+147	+200
5	−53	+329	+382

The rate of return on the difference between the alternatives is computed as follows:

PW of cost = PW of benefit

$646 = 64(P/A,i,2) + 200(P/A,i,2)(P/F,i,2) + 382(P/F,i,5)$

Try $i = 9\%$:

$646 \stackrel{?}{=} 64(1.759) + 200(1.759)(0.8417) + 382(0.6499)$

$\qquad \stackrel{?}{=} 656.9$

Try $i = 10\%$:

$646 \stackrel{?}{=} 64(1.736) + 200(1.736)(0.8264) + 382(0.6209)$

$\qquad \stackrel{?}{=} 635.2$

The rate of return $i^* = 9\% + \left\{\dfrac{656.9 - 646.0}{656.9 - 635.2}\right\} = 9.5\%$

The rate of return is greater than the 8% after-tax MARR. The increment of investment is desirable. Buy the EL-40 model. ∎

Example 12-13 above illustrates the use of the incremental IRR method in defender-challenger comparisons on an after-tax basis. In the example this analysis method is appropriate because the life of the SK-30(defender) is equivalent to the economic life of the EL-40 (challenger). Since these lives are the same, a present or future worth analysis also could have been used. However, far more commonly the lives of the defender and the challenger are different. In such cases, if we are willing to make the assumption of a continuing requirement for the asset, then the comparison method to use is the annual cash flow method. As in our previous discussion, this method allows a direct comparison of the annual cash flow of the defender asset over its life against the annual cash flow of the challenger over its minimum cost life. The annual cash flow method is also appropriate for comparison for the case where the lives are equal (as in Example 12-12), but more importantly, it is appropriate for the case where the lives are different, under the *replacement repeatability assumptions*.

Minimum Cost Life Problems

In this section we illustrate the effect that tax considerations can have on the calculation of the minimum cost life of the defender and the challenger. The calculation of minimum EUAC on an after-tax basis can be affected by both the depreciation method used and changes in the asset's market value over time, for either the defender or challenger. Using an accelerated depreciation method (like MACRS) tends to reduce the after-tax costs early in the life of an asset. This effect alters the shape of the total EUAC curve — the concave shape can be shifted and the minimum EUAC changed. Example 12-14 illustrates the effect that taxes can have when either the straight-line or MACRS depreciation method is used.

EXAMPLE 12-14

Some new production machinery has a first cost of $100,000 and a useful life of 10 years. Its estimated operating and maintenance (O&M) costs are $10,000 the first year, which will increase annually by $4000. The asset's before-tax market value will be $50,000 at the end of the first year and then will decrease by $5000 annually. Calculate the after-tax cash flows (ATCF) using MACRS depreciation. This property is a 7-year MACRS property. The company uses a 6% after tax MARR and is subject to a combined federal/state tax rate of 40%.

Solution:

To find the minimum cost life of this new production machinery we first find the after tax cash flow effect of the O&M costs and depreciation (Table 12-1). Then, we find the ATCFs of disposal if the equipment is sold in each of the ten years (Table 12-2). Finally in the closing section on spreadsheets, we combine these two ATCFs (in Figure 12-7) and choose the minimum cost life.

In Table 12-1, the O&M expense simply starts at $10,000 and increases at $4000 per year. The depreciation entries equal the 7-year r_t MACRS depreciation values given in Table 10-3 multiplied by the $100,000 first cost. The taxable income is simply the O&M costs minus the depreciation values, which is then multiplied by minus the tax rate to determine the impact of this taxable income on taxes. The O&M expense plus taxes is the Table 10-1 portion of the total ATCF.

It should be pointed out in looking at the market value data in this problem that the initial decrease of $50,000 in year 1 is not uncommon. This is especially true for custom built equipment for a particular and unique application at a specific plant. Such equipment would not be as valuable to others in the marketplace as to the company for which it was built. Also, the $100,000 first cost (cost basis) could have included costs due to installation, facility modifications or removal of old equipment. The $50,000 is realistic for the market value of one-year old equipment.

Table 12-1 ATCF FOR O&M AND DEPRECIATION FOR EXAMPLE 12-14

Year (t)	O&M Expense	MACRS Depr. (d_t)	Taxable Income	Taxes {at 40%}	O&M & Depr. ATCF
1	$-10,000	$14,290	$-24,290	$9,716	$ -284
2	-14,000	24,490	-38,490	15,396	1,396
3	-18,000	17,490	-35,490	14,196	-3,804
4	-22,000	12,490	-34,490	13,796	-8,204
5	-26,000	8,930	-34,930	13,972	-12,028
6	-30,000	8,920	-38,920	15,568	-14,432
7	-34,000	8,930	-42,930	17,172	-16,828
8	-38,000	4,460	-42,460	16,984	-21,016
9	-42,000	0	-42,000	16,800	-25,200
10	-46,000	0	-46,000	18,400	-27,600

The next step is to determine the ATCFs that would occur in each possible year of disposal. (The ATCF for year 0 is easy, as it equals -$100,000.) For example, as shown in Table 10-2, in year 1 there is a $35,710 loss as the book value exceeds the market value. The tax savings from this loss are added to the salvage (market) value to determine the ATCF (*If the asset is disposed during year 1*).

Note, this table assumes that depreciation is taken during the year of disposal and then calculates the recaptured depreciation (gain) or loss on the book value at the end of the year.

Table 12-2 ATCF IN YEAR OF DISPOSAL FOR EXAMPLE 12-14

Year (t)	Market Value	Book Value	Gain or Loss	Gain/Loss Tax {at 40%}	ATCF if Disposed of
1	$50,000	$85,710	$-35,710	$14,284	$64,284
2	45,000	61,220	-16,220	6,488	51,488
3	40,000	43,730	-3,730	1,492	41,492
4	35,000	31,240	3,760	-1,504	33,496
5	30,000	22,310	7,690	-3,076	26,924
6	25,000	13,390	11,610	-4,644	20,356
7	20,000	4,460	15,540	-6,216	13,784
8	15,000	0	15,000	-6,000	9,000
9	10,000	0	10,000	-4,000	6,000
10	5,000	0	5,000	-2,000	3,000

Spreadsheets and After Tax Replacement Analysis

Spreadsheets are obviously useful in nearly all after-tax calculations. However, they are absolutely required for optimal life calculations in after-tax situations. Because MACRS is the tax law, the after-tax cash flows are different in every year. Thus, the NPV function and the PMT function are both needed to find the minimum EUAC after taxes. Figure 12-7 illustrates the calculation of the minimum cost life for Example 12-14.

In this figure, the NPV finds the present worth of the irregular cash flows from period 1 through period t for t = 1 to life. Then PMT can be used to find the EUAC over each potential life.

	A	B	C	D	E	F
1		Table 12-1	Table 10-2	6% interest rate		
2		O&M & Depr.	if disposed of			
3	year	ATCF	ATCF	NPV	EAC	
4	0			-$100,000		
5	1	-$284	$64,284	-39,623	$42,000	=PMT($D$1,A5,D5)
6	2	1,396	51,488	-53,201	29,018	
7	3	-3,804	41,492	-67,382	25,208	
8	4	-8,204	33,496	-82,186	23,718	
9	5	-12,028	26,924	-97,587	23,167	
10	6	-14,432	20,356	-113,530	23,088 optimal life	
11	7	-16,828	13,784	-129,904	23,270	
12	8	-21,016	9,000	-146,610	23,610	
13	9	-25,200	6,000	-163,621	24,056	
14	10	-27,600	3,000	-180,909	24,580	=PMT($D$1,A14,D14)
15						
16			=NPV($D$1,$B$5:B14)+$D$4+PV($D$1,A14,0,-C14)			
17			=NPV(*i*, B column) + year 0 + present value of a future salvage			

Figure 12-7 Spreadsheet for life with minimum after-tax cost

Summary

In selecting equipment for a new plant the question is, "Which of the machines available on the market will be more economical?" But when one has a piece of equipment that is now performing the desired task, the analysis is more complicated. The existing equipment (called the *defender*) is already in place so the question is, "Shall we replace it now, or shall we keep it for one or more years?" When a replacement is indicated, it will be by the best available replacement equipment (called the *challenger*). When we already have equipment, there may be a tendency to use past costs in the replacement analysis. But only present and future costs are relevant.

This chapter has presented three distinctly different *replacement analysis techniques* which are all relevant and appropriate depending upon the conditions of the cash flows for the defender and the challenger. In all cases of analysis the simplifying *replacement repeatability assumptions* are accepted. These state that the defender will ultimately be replaced by the current best challenger (as will any challengers implemented in the future), and that we have an indefinite need for the service of the asset in question.

In the usual case where marginal cost data are both available and increasing on a year-to-year basis, *replacement analysis technique #1* allows a comparison of

> the marginal cost data of the defender, against the minimum EUAC of the challenger.

In this case we should keep the defender as long as its marginal cost is less than the minimum EUAC of the challenger.

In the case where marginal cost data are available for the defender but are not increasing on a year-to-year basis, *replacement analysis technique #2* calls for a comparison of

> *the minimum EUAC of the defender, against the minimum EUAC of the challenger.*

If the defender's minimum EUAC is smaller, we would keep this asset for its minimum cost life and then re-evaluate. If the challenger's EUAC is less, we would select this asset in place of the defender today.

In the case where there are no marginal cost data available for the defender, *replacement analysis technique #3* prescribes a comparison of

> *the EUAC of the defender over its stated life, against the minimum EUAC of the challenger.*

As in the case of replacement analysis technique #2 we would select the alternative that has the smallest EUAC. An important concept when calculating the EUAC of both the defender and challenger is the first cost to be assigned to each alternative for calculation purposes. In the case where the lives of the two alternatives are equivalent, either an *opportunity cost* or *cash flow approach* may be used. However, in the more common case of different useful lives only the opportunity cost approach accurately assigns an investment cost to the defender and challenger assets.

It is important when performing engineering economic analyses to include the effects of taxes. In minimum economic life and marginal cost calculations, and in finding cash flows over the life of a defender or challenger, the effects of taxes can be significant. The replacement analysis techniques described on a before-tax basis are also used for the after-tax case — the difference being that after-tax cash flows are used in place of before-tax cash flows. Effects on an after-tax basis include: opportunity gains and losses at time zero; income taxes and depreciation over the assets' lives; and gains and losses at disposition time. The *sign-change procedure* can be used to determine opportunity gains/losses when assigning a first cost to the defender in after-tax problems.

Replacement analyses are vastly important, yet often ignored by companies as they invest in equipment and facilities. Investments in business and personal assets should not be forgotten once an initial economic evaluation produces a "buy" recommendation. It is important to continue to evaluate assets over their respective life cycles in order to insure that invested monies are continuing at the greatest benefit to the investor. Replacement analyses help us to insure this.

Problems

12-1 Typically there are two alternatives in a replacement analysis. One alternative is to replace the defender now. The other alternative is which one of the following?
   a. Keep the defender for its remaining useful life.
   b. Keep the defender for another year and then re-examine the situation.
   c. Keep the defender until there is an improved challenger that is better than the present challenger.
   d. The answer to this question depends on the data available for the defender and challenger as well as the assumptions made regarding the period of needed service and future challengers.

12-2 The economic life of the defender can be obtained if certain estimates about the defender can be made. Assuming those estimates prove to be exactly correct, one can accurately predict the year when the defender should be replaced, even if nothing is known about the challenger. Is the above statement true or false? Explain.

12-3 A proposal has been made to replace a large heat exchanger (initial cost was $85,000 3 years ago) with a new, more efficient unit at a cost of $120,000. The existing heat exchanger is being depreciated by the MACRS method. Its present book value is $20,400, but it has no current value as its scrap value just equals the cost to remove it from the plant. In preparing the before-tax economic analysis to see if the existing heat exchanger should be replaced, the question arises concerning the proper treatment of the $20,400 book value. Three possibilities are that the $20,400 book value of the old heat exchanger is:
   a. *Added* to the cost of the new exchanger in the economic analysis.
   b. *Subtracted* from the cost of the new exchanger in the economic analysis.
   c. *Ignored* in this before-tax economic analysis.
Which of the three possibilities is correct?

12-4 A machine tool, which has been used in a plant for 10 years, is being considered for replacement. It cost $9500 and was depreciated by MACRS depreciation using a 5 year recovery period. An equipment dealer indicates the machine has no resale value. Maintenance on the machine tool has been a problem, with an $800 cost this year. Future annual maintenance costs are expected to be higher. What is the economic life of this machine tool if it is kept in service?

12-5 A new $40,000 bottling machine has just been installed in a plant. It will have no salvage value when it is removed. The plant manager has asked you to estimate the economic service life for the machine, ignoring income taxes. He estimates that the annual maintenance cost will be constant at $2500 per year. What service life will result in the lowest equivalent uniform annual cost?

12-6 Which one of the following is the proper dollar value of defender equipment to use in replacement analysis?
    a. Original cost.
    b. Present market value.
    c. Present trade-in value.
    d. Present book value.
    e. Present replacement cost, if different from original cost.

12-7 The Ajax Corporation purchased a railroad tank car 8 years ago for $60,000. It is being depreciated by SOYD depreciation, assuming a 10 year depreciable life and a $7000 salvage value. The tank car needs to be reconditioned now at a cost of $35,000. If this is done, it is estimated the equipment will last for 10 more years and have a $10,000 salvage value at the end of the 10 years.

    On the other hand, the existing tank car could be sold now for $10,000 and a new tank car purchased for $85,000. The new tank car would be depreciated by MACRS depreciation. Its estimated actual salvage value would be $15,000. In addition, the new tank car would save $7000 per year in maintenance costs, compared to the reconditioned tank car.

    Based on a 15% before-tax rate of return, determine whether the existing tank car should be reconditioned, or a new one purchased. *Note:* The problem statement provides more data than are needed, which is typical of real situations. (*Answer:* Recondition the old tank car.)

12-8 The Clap Chemical Company needs a large insulated stainless steel tank for the expansion of its plant. Clap has located such a tank at a recently closed brewery. The brewery has offered to sell the tank for $15,000 delivered to the chemical plant. The price is so low that Clap believes it can sell the tank at any future time and recover its $15,000 investment.

    The outside of the tank is lined with heavy insulation that requires considerable maintenance with estimated costs as follows:

Year	Insulation maintenance cost
0	$2000
1	500
2	1000
3	1500
4	2000
5	2500

    a. Based on a 15% before-tax MARR, what is the economic life of the insulated tank?
    b. Is it likely that the insulated tank will be replaced by another tank at the end of its computed economic life? Explain.

12-9 The plant manager has just purchased a piece of unusual machinery for $10,000. Its resale value at the end of 1 year is estimated to be $3000 and is rising at the rate of $500 per year, because it is sought by antique collectors.

    The maintenance cost is expected to be $300 per year for each of the first 3 years, and then it is expected to double each year after that. Thus the fourth-year maintenance will be $600; the fifth-year maintenance, $1200, and so on. Based on a 15% before-tax MARR, what is the economic life of this machinery?

12-10 The Quick Manufacturing Co., a large profitable corporation, is considering the replacement of a production machine tool. A new machine would cost $3700, have a 4 year useful and depreciable life, and have no salvage value. For tax purposes, sum-of-years digits depreciation would be used. The existing machine tool was purchased 4 years ago at a cost of $4000, and has been depreciated by straight line depreciation assuming an 8 year life and no salvage value. It could be sold now to a used equipment dealer for $1000 or be kept in service for another 4 years. It would then have no salvage value. The new machine would save about $900 per year in operating costs compared to the existing machine. Assume a 40% combined state and federal tax rate.

    a.   Compute the before-tax rate of return on the replacement proposal of installing the new machine rather than keeping the existing machine.

    b.   Compute the after-tax rate of return on this replacement proposal.

(*Answer:* ***a.*** 12.6%)

12-11 The Plant Department of the local telephone company purchased four special pole hole diggers 8 years ago for $14,000 each. They have been in constant use to the present. Due to an increased workload, additional machines will soon be required. Recently it was announced that an improved model of the digger has been put on the market. The new machines have a higher production rate and lower maintenance expense than the old machines, but will cost $32,000 each. The service life of the new machines is estimated to be 8 years with a salvage estimated at $750 each. The four original diggers have an immediate salvage of $2000 each and an estimated salvage of $500 each 8 years hence. The estimated average annual maintenance expense associated with the old machines is approximately $1500 each, compared to $600 each for the new machines.

    A field study and trial indicate that the workload would require three additional new machines if the old machines are continued in service. However, if the old machines are all retired from service, the present workload plus the estimated increased load could be carried by six new machines with an annual savings of $12,000 in operation costs. A personnel training program to prepare employees to run the machines will be necessary at an estimated cost of $700 per new machine. If the MARR is 9% before taxes, what should the company do?

12-12 Fifteen years ago the Acme Manufacturing Company bought a propane powered forklift truck for $4800. The company depreciated the forklift using straight line depreciation, a 12 year life, and zero salvage value. Over the years, the forklift has been a good piece of equipment, but lately the maintenance cost has risen sharply. Estimated end-of-year maintenance costs for the next 10 years are as follows:

Year	Maintenance cost
1	$ 400
2	600
3	800
4	1000
5-10	1400 /year

The old forklift has no present or future net salvage value, as its scrap metal value just equals the cost to haul it away. A replacement is now being considered for the old forklift. A modern unit can be purchased for $6500. It has an economic life equal to its 10 year depreciable life. Straight line depreciation will be employed with zero salvage value at the end of the 10 year depreciable life. At any time the new forklift can be sold for its book value. Maintenance on the new forklift is estimated to be a constant $50 per year for the next 10 years, after which maintenance is expected to increase sharply. Should Acme Manufacturing keep its old forklift truck for the present, or replace it now with a new one? The firm expects an 8% after-tax rate of return on its investments. Assume a 40% combined state-and-federal tax rate. (*Answer:* Keep the old forklift truck.)

12-13   A firm is concerned about the condition of some of its plant machinery. Bill James, a newly hired engineer, was assigned the task of reviewing the situation and determining what alternatives are available. After a careful analysis, Bill reports that there are five feasible, mutually exclusive alternatives.

Alternative A. Spend $44,000 now repairing various items. The $44,000 can be charged as a current operating expense (rather than capitalized) and deducted from other taxable income immediately. These repairs are anticipated to keep the plant functioning for the next 7 years with operating costs remaining at present levels.

Alternative B. Purchase $49,000 of general purpose equipment. Depreciation would be straight line, with the depreciable life equal to the 7 year useful life of the equipment. The equipment will have no end-of-useful-life salvage value. The new equipment will reduce operating costs $6000 per year below the present level.

Alternative C. Purchase $56,000 of new specialized equipment. This equipment would be depreciated by sum-of-years digits depreciation over its 7 year useful life. This equipment would reduce operating costs $12,000 per year below the present level. It will have no end-of-useful-life salvage value.

Alternative D. This alternative is the same as *Alternative B,* except that this particular equipment would reduce operating costs $7000 per year below the present level.

Alternative E. This is the "do-nothing" alternative. If nothing is done, future annual operating costs are expected to be $8000 above the present level.

This profitable firm pays 40% corporate income taxes. In their economic analysis, they require a 10% after-tax rate of return. Which of the five alternatives should the firm adopt?

12-14   In a replacement analysis problem, the following facts are known:

- Initial cost: $12,000

- Annual Maintenance: None for the first 3 years;
  $2000 at the end of the fourth year;
  $2000 at the end of the fifth year;
  Increasing $2500 per year after the fifth year
  ($4500 at the end of the sixth year, $7000 at the
  end of the seventh year, and so forth).

Actual salvage value in any year is zero. Assume a 10% interest rate and ignore income taxes. Compute the economic life for this challenger. (*Answer:* 5 years)

12-15   Machine *A* has been completely overhauled for $9000 and is expected to last another 12 years. The $9000 was treated as an expense for tax purposes last year. It can be sold now for $30,000 net after selling expenses, but will have no salvage value 12 years hence. It was bought new 9 years ago for $54,000 and has been depreciated since then by straight line depreciation using a 12 year depreciable life.

Because less output is now required, Machine *A* can now be replaced with a smaller Machine *B*. Machine *B* costs $42,000, has an anticipated life of 12 years, and would reduce operating costs $2500 per year. It would be depreciated by straight line depreciation with a 12 year depreciable life and no salvage value.

Both the income tax and capital gains tax rates are 40%. Compare the after-tax annual cost of the two machines and decide whether Machine *A* should be retained or replaced by Machine *B*. Use a 10% after-tax rate of return in the calculations.

12-16   Fred's Rodent Control Corporation has been using a low frequency sonar device to locate subterranean pests. This device was purchased 5 years ago for $18,000. The device has been depreciated using SOYD depreciation with an 8 year tax life and a salvage value of $3600. Presently, it could be sold to the cat next door for $7000. If it is kept for the next 3 years, its market value is expected to drop to $1600.

A new lightweight subsurface heat sensing searcher (SHSS) is available for $10,000 that would improve the annual net income by $500 for each of the next 3 years. The SHSS would be depreciated using MACRS as a 5 year class property. At the end of 3 years, the SHSS should have a market value of $4000. Fred's Rodent Conrol is a profitable enterprise subject to a 40% tax rate.
   a.   Construct the after-tax cash flow for the old sonar unit for the next 3 years.
   b.   Construct the after-tax cash flow for the SHSS unit for the next 3 years.
   c.   Construct the after-tax cash flow for the difference between the SHSS unit and the old sonar unit for the next 3 years.
   d.   Should Fred buy the new SHSS unit if his MARR is 20%? You do not have to calculate the incremental rate of return, just show how you reach your decision.

12-17   You have the following options for a major equipment unit:
   a.   Buy new.
   b.   Trade-in and buy a similar, rebuilt equipment from the manufacturer.
   c.   Have the manufacturer rebuild your equipment with all new available options.
   d.   Have the manufacturer rebuild your equipment to the original specifications.
   e.   Buy used equipment.
State advantages and disadvantages of each option with respect to after-tax benefits.

12-18   An Engineering Economics professor owns a 1993 automobile. In the past 12 months, he has paid $2000 to replace the transmission, bought two new tires for $160, and installed a new tape deck for $110. He wants to keep the car for 2 more years because he invested money 3 years ago in a 5 year certificate of deposit, which is earmarked to pay for his dream machine, a red European sports car. Today the engine failed. The professor has two alternatives. He can have the engine overhauled at a cost of $1800, and then he expects to have to pay another $800 per year for the next two years for maintenance. The car will have no salvage value at that time. Alternatively, a colleague offered to make the professor a $5000 loan to buy another used car. He must pay the loan back in two equal installments of $2500 due at the end of year 1 and year 2, and at the end of the second year he must give the colleague the car. The "new" used car has an expected annual maintenance cost of $300. If the professor selects this alternative, he can sell his current vehicle to a junk yard for $1500. Interest is 5%. Using Present Worth analysis, which alternative should he select and why?

12-19   An injection molding machine has a first cost of $1,050,000 and a salvage value of $225,000 in any year the machine is sold. The maintenance and operating cost is $235,000 with an annual gradient of $75,000. The MARR is 10%. What is the most economic life?

12-20   VMIC Corp. has asked you to look at the following data. After considering the data, answer the questions below.

Year (n)	Defender EUAC if kept (n) years	Challenger EUAC if kept (n) years
1	$6,500	$10,600
2	5,000	9,800
3	4,800	7,500
4	4,400 ←	7,000
5	5,000	6,200 ←
6	5,600	6,800
7	7,650	7,225
8	9,800	9,010
9	—	10,022
10	—	15,000

a. What is the minimum cost life of the *Defender*?
b. What is the minimum cost life of the *Challenger*?
c. When, if at all, should we replace the *Defender* with the *Challenger*?

12-21   SHOJ Enterprises has asked you to look at the following data. After considering the data, answer the questions below.

Year (n)	Defender EUAC if kept (n) years	Challenger EUAC if kept (n) years
1	$4000	$7600
2	4400	5800
3	4800	4500
4	5400	3300
5	6000	3800
6	6600	6800

a. What is the economic life of the *Defender* ?
b. What is the economic life of the *Challenger* ?
c. When, if at all, should we replace the *Defender* with the *Challenger* ?

12-22 Thomas Martin purchased and implemented production machinery 5 years ago that had a first cost of $25,000. At the time of the initial purchase it was estimated that yearly costs would be $1250, increasing by $500 in each year that followed. It was also estimated that the market value of this machinery would be only 90% of the previous year's value. It is currently projected that this machine will be useful in operations for 5 more years. There is a new machine available now that has a first cost of $27,900 and no yearly costs over its 5 year minimum cost life. If Thomas Martin uses an 8% before-tax MARR, when, if at all, should he replace the existing machinery with the new one?

12-23 Consider the problem involving Thomas Martin above. Now change the values for the old machine to those indicated below, and suggest when, if at all, the old should be replaced with the new. The old machine retains only 70% of its value in the market from year-to-year. The yearly costs of the old machine were $3000 in year-one and increase at 10% thereafter.

12-24 As proprietor of your own business, you are considering the option of purchasing a new high efficiency machine to replace older machines currently in use. You believe that the new technology can be used to replace four of the older machines that each have a current market value of $600. The new machine will cost $5000 and will save the equivalent of 10,000 KW-hours of electricity per year over the older machines. After a period of 10 years neither option (new or old) will have any market value. If you use a before tax MARR of 25% and pay $0.075 per KW-hour of electricity, would you replace the old machines today with the new one?

12-25 BC Junction purchased some embroidering equipment for their Denver facility 3 years ago for $15,000. This equipment qualified as MACRS 5 year property. Maintenance costs are estimated to be $1000 this next year and will increase by $1000 per year thereafter. The market (salvage) value for the equipment is $10,000 at the end of this year and declines by $1000 per year in the future. If BC Junction has an after-tax MARR of 30%, a marginal tax rate of 45% on ordinary income, and depreciation recapture and losses, what is the after-tax economic life of this previously purchased equipment?

12-26 Mary O'Leary is considering some new automated sheep shearing equipment for use by her company that ships fine wool garments from County Cork, Ireland. The equipment has a first cost of $125,000 and a MACRS class life of 7 years. The annual costs for operating, maintenance, and insurance, as well as market value data for each year of the equipment's 10 year useful life, is given below. Mary O'Leary uses an after-tax MARR of 25% and a tax rate of 35% on ordinary income to evaluate this type of investment. What is the after-tax economic life of the sheep shearing equipment?

	Annual Costs in Year (n) for ...			
Year (n)	*Operating*	*Maintenance*	*Insurance*	*Market Value in Year (n)*
1	$16,000	$ 5,000	$17,000	$80,000
2	20,000	10,000	16,000	78,000
3	24,000	15,000	15,000	76,000
4	28,000	20,000	14,000	74,000
5	32,000	25,000	12,000	72,000
6	36,000	30,000	11,000	70,000
7	40,000	35,000	10,000	68,000
8	44,000	40,000	10,000	66,000
9	48,000	45,000	10,000	64,000
10	52,000	50,000	10,000	62,000

12-27 Consider the problem above regarding Mary O'Leary's sheep sheering business. Assume that Mary purchased the equipment 5 years ago, and we are now looking at the remaining 5 years. Answer the following questions (assuming the replacement repeatability assumptions are valid).

    a. Calculate the before-tax marginal costs for the remaining 5 years.

    b. Using the data from question *a* and the decision map from Figure 12-1, when, if at all, should the old sheep shearing equipment be replaced if a new challenger with a minimum EUAC of $110,000 has been identified.

    c. Rework questions *a* and *b* above on an after-tax basis. In this case the challenger equipment has an after-tax minimum EUAC of $100,000.

12-28 a. A new employee at CLL Engineering Consulting Inc., you are asked to join a team performing an economic analysis for a client. Your team seems stumped on how to assign an after-tax first cost to the defender and challenger assets under consideration. Your task is to take the data below and find the ATCF for each alternative. There is no need for a complete analysis— your colleagues can handle that responsibility — they need help only with the time zero ATCFs. CLL Inc. has a combined federal/state tax rate of 45% on ordinary income, depreciation recapture, and losses.

Defender: This asset was placed in service 7 years ago. At that time the $50,000 cost basis was set up on a straight-line depreciation schedule with an estimated salvage value of $15,000 over its 10 year ADR life. This asset has a present market value of $30,000.

Challenger: The new asset being considered has a first cost of $85,000 and will be depreciated by MACRS depreciation over its 10 year class life. This asset qualifies for a 10% Investment Tax Credit.

    b. How would your calculations change if the present market value of the *Defender* is $25,500?

    c. How would your calculations change if the present market value of the *Defender* is $18,000?

12-29 Foghorn Leghorn is considering the replacement of an old egg sorting machine used with his Foggy's Farm Fresh Eggs business. The old egg machine is not quite running eggs-actly the way it was originally designed and will require an additional investment now of $2500 (expensed at time zero) to get it back in working shape. This old machine was originally purchased 6 years ago for

$5000 and has been depreciated by the straight-line method at $500 per year. Six years ago the estimated salvage value for tax purposes was $1000. Operating expenses for the old machine are projected at $600 this next year and are increasing by $150 per year each year thereafter. Foggy projects that with refurbishing, the old egg sorting machine will last another 3 years. Foggy believes that he could sell the old egg machine as-is today for $1000 to his friend Fido to sort bones. He also believes he could sell it 3 years from now at the barnyard fleamarket for $500.

The new egg sorting machine, a deluxe model, has a purchase price of $10,000 and will last 6 years, at which time it will have a salvage value of $1000. The new machine qualifies as a MACRS 7 year property and will have operating expenses of $100 the first year, increasing by $50 per year thereafter. Foghorn uses an after tax MARR of 18% and a tax rate of 35% on original income.

a. What was the depreciation life used with the defender asset (the old egg sorter)?

b. Calculate the after-tax cash flows for both the defender and challenger assets.

c. Use the annual cash flow method to offer a recommendation to Foggy. What assumptions did you make in this analysis?

12-30 Bill's father read that, at the end of each year, an automobile is worth 25% less than it was at the beginning of the year. After a car is three years old, the rate of decline reduces to 15%. Maintenance and operating costs, on the other hand, increase as the age of the car increases. Because of the manufacturer's warranty, the first year maintenance is very low.

Age of car, in years	Maintenance expense
1	$ 50
2	150
3	180
4	200
5	300
6	390
7	500

Bill decided this is a good economic analysis problem. Bill's dad wants to keep his annual cost of automobile ownership low. The car Bill's dad prefers costs $11,200 new. Should he buy a new or a used car and, if used, when would you suggest he buy it, and how long should it be kept? Give a practical, rather than a theoretical, solution.

(Answer: Buy a three-year-old car and keep it three years.)

Inflation and Price Change

Up to this chapter we have assumed that dollars in our analyses have been unaffected by inflation or price change. However, this assumption is not always valid or realistic. If inflation occurs in the general economy, or price changes take place in economic costs and benefits, the impact can be substantial on both before and after-tax analyses. In this chapter we develop several key concepts and illustrate how inflation and price changes may be incorporated into our problems.

Meaning And Effect Of Inflation

Inflation is an important concept because the purchasing power of money used in most world economies rarely stays constant. Rather, over time the amount of goods and services that can be purchased with a fixed amount of money tends to change. Inflation causes U.S. dollars to lose **purchasing power**. That is, when prices inflate we can buy less with the same amount of money. **Inflation makes future dollars less valuable than present dollars.** Think about examples in your own life, or for an even starker comparison, ask your grandparents how much a loaf of bread or a new car cost fifty years ago. Then compare these prices with what you would pay today for these same items. This exercise will reveal the effect of inflation: as time passes, goods and services cost more, and more of the same monetary units are needed to purchase the same goods and services.

Because of inflation, dollars in one period of time are not equivalent to dollars in another. We know from our previous study that engineering economic analysis requires that comparisons be made on an **equivalent** basis. So, it is important for us to be able to incorporate the effects of inflation in our analysis of alternatives.

When the purchasing power of a monetary unit *increases* rather than decreases as time passes, the result is *deflation*. Deflation is very rare in the modern world, but nonetheless, potentially exists. Deflation has the opposite effect of inflation—one can purchase **more** with money in future years than can be purchased today. As such, deflation makes future dollars more valuable than current dollars.

How Does Inflation Happen?

Economists do not agree on all of the sources of inflation, but generally agree that the following effects influence inflation either in isolation or in combination.

Money Supply: The amount of money in our national economy is thought to have an effect on its purchasing power. If there is too much money in the system (the Federal Reserve controls the flow of money) versus goods and services to purchase with that money, it tends to decrease the value of dollars. When there are fewer dollars in the system, they become more valuable. The Federal Reserve, through its influence on the money supply, seeks to increase the volume of money in the system at the same rate that the economy is growing.

Exchange Rates: The strength of the U.S. dollar in world markets affects the profitability of international companies in those markets. Price rates may be adjusted to compensate for the relative strength or weakness of the dollar in the world market. As corporations' profits are weakened or eliminated in some markets due to exchange rates, prices may be raised in other markets to compensate.

Cost-Push Inflation: This type of inflation develops as producers of goods and services "push" their increasing operating costs along to the customer through higher prices. These operating costs include fabrication/manufacturing, marketing, and sales, among others.

Demand-Pull Inflation: This effect is realized when consumers spend money freely on goods and services. Often "free spending" is at the expense of consumer saving. As more and more people demand certain goods and services, the prices of those goods and services will rise (demand exceeding supply).

A further consideration in analyzing how inflation works is the usually different rates at which the prices of goods and services rise versus the wages of workers. Do workers benefit if their wages increase, but the price of goods and services increase, as well? In order to determine the net effect of differing rates of inflation we must be able to make comparisons and understand costs and benefits from an equivalent perspective. In the following examples we will learn how to make such comparisons.

Definitions for Considering Inflation in Engineering Economy

The definitions below are used throughout this chapter to illustrate how inflation and price change affect two quantities: interest rates and cash flows.

Inflation Rate (f): As described above, the inflation rate captures the effect of goods and services costing more — a decrease in the purchasing power of dollars. More money is required to buy a good or service whose price has inflated.

Real Interest Rate (i′): This interest rate measures the "real" growth of our money excluding the effect of inflation. Because it does not include inflation, it is sometimes called the *inflation-free interest rate.*

Market Interest Rate (i): This is the rate of interest that one obtains in the general marketplace. For instance, the interest rates on passbook savings, checking plus, and certificates of deposit quoted at the bank are all market rates. The lending interest rate for autos and boats is also a market rate. This rate is sometimes called the *combined interest rate* because it incorporates the effect of both real money growth **and** inflation. We can view *i* as follows:

Market Interest Rate	*has in it*	*"Real" Growth of Money*	*and*	*Effect of Inflation.*

The mathematical relationship between the inflation, real and market interest rates is given as:

$$i = i' + f + (i')(f) \qquad\qquad (13\text{-}1)$$

EXAMPLE 13-1

Suppose Tiger Woods wants to invest some recent golf winnings in his hometown bank for one year. Currently, the bank is paying a rate of 5.5% *compounded annually.* Assume inflation is anticipated to be: **(a)** 2% per year, and **(b)** 8% per year for the year of Tiger's investment. **In each case identify *i, f,* and *i′.***

Solution:

(a) If inflation is 2% per year: From the previous definitions the interest rate that the bank is paying is the *market rate (i).* The *inflation rate (f)* is given in the problem statment. What is left, then, is to find the *real interest rate (i′).*

$$i = 5.5\% \qquad f = 2\% \ and \qquad i' = ?$$

Solving for i' in Equation 13-1 above we have

$$i = i' + f + (i')(f)$$
$$i - f = i' + (i')(f)$$
$$i - f = i'(1 + f)$$
$$i' = (i - f) / (1 + f)$$
$$= (0.055 - 0.02) / (1 + 0.02)$$
$$= 0.034 \text{ or } \textbf{3.4\% per year}$$

This means that Tiger Woods will have 3.4% **more** purchasing power with the dollars invested in that account than he had a year previously. At the end of the year he can purchase 3.4%

more goods and services than he could have at the beginning of the year. As an example of the growth of his money, assume he was purchasing golf balls that cost $5.00 each and that he had invested $1000 in his hometown bank account.

At the *beginning* of the year he could purchase:

$$\text{Number of Balls Purchased Today} = \frac{\text{Dollars Today Available to Buy Balls}}{\text{Cost of Balls Today}}$$

$$= 1000 / 5.00 = 200 \text{ golf balls}$$

At the *end* of the year he could purchase:

$$\text{Number of Balls Purchased at End of Year} = \frac{\text{Dollars Available for Purchase at End of Year}}{\text{Cost per Ball at End of Year}}$$

In this case: Dollars Available at End of Year $= (\$1000)\,(F/P, 5.5\%, 1) = \$1055$

Ball Cost at End of Year (inflated at 2%) $= (\$5.00)\,(1+0.02)^1 = \$5.10$

Thus:

Number of Balls Purchased at End of Year $= \$1055 / 5.10 = 207 \text{ golf balls}$

Tiger Woods can, after one year, purchase 3.4% more golf balls than he could before. In this case 207 balls is about 3.4% more than 200 balls.

(b) If inflation is 8%:   As above we would solve for i'

$$i' = (i - f) / (1 + f)$$
$$= (0.055 - 0.08) / (1 + 0.08)$$
$$= -0.023 \quad \text{or} \quad -\textbf{2.3\% per year}$$

In this case we can see that the real growth in money has *decreased* by 2.3%, so that Tiger can now purchase 2.3% fewer balls with the money he had invested in the bank. Even though he has more money at the end of the year, it is worth less, so he can purchase less.

Regardless of how inflation behaves over the year, the bank will pay Tiger $1055 at the end of the year. However, as we have seen above, inflation can greatly affect the "real" growth of dollars over time. In one of his State of the Union addresses, then President George Bush called inflation "that thief" because it steals real purchasing power from our dollars. We have seen this effect in the above example. ∎

Let us continue the discussion of the effects of inflation by focusing now on cash flows in our problems. We define two "types" of dollars in our analysis:

Actual Dollars (A$): These are the dollars that we ordinarily think of when we think of money. These dollars circulate in our economy and are used for investments and payments. We can touch these dollars and often keep them in our purses and wallets — they are "actual" and exist physically. Sometimes they are called *inflated dollars* because they carry any inflation that has reduced their worth.

Real Dollars (R$): These dollars are a bit harder to define. They are always expressed in terms of some constant purchasing power "base" year. An example would be **1982-based dollars.** Real dollars are sometimes called *constant dollars* or *constant purchasing power dollars*, and because they do not carry the effects of inflation, they are also known as *inflation-free dollars*.

Having defined *market, inflation* and *real interest rates* as well as *actual* and *real dollars*, let us describe how these quantities relate. Figure 13-1 illustrates the relationship between these quantities.

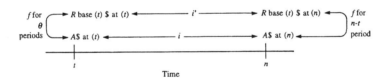

Figure 13-1 Relationship between i, f, i', A$ and R$.

Figure 13-1 illustrates the following principles:

When dealing with Actual Dollars (A$), use a Market Interest Rate (i), and when discounting A$ over time, also use i.

When dealing with Real Dollars (R$), use a Real Interest Rate (i'), and when discounting R$ over time, also use i'.

Shown in Figure 13-1 is the relationship between A$ and R$ that occur **at the same period of time**. Actual and real dollars are related by the *inflation rate*, in this case, over the period of *n-t years*. To translate between dollars of one type to dollars of the other (A$ to R$ or R$ to A$) use the inflation rate for the appropriate number of periods. The following example illustrates many of these relationships.

EXAMPLE 13-2

A university is considering replacing its stadium with a new facility. When the present building was completed in 1945, the total cost was 1.2 million dollars. At that time a wealthy alumnus gifted the university with 1.2 million dollars to be used for a future replacement. University administrators are now considering building the new facility in the year 2000. Assume the following:

- Inflation is 6.0% per year from 1945 to 2000.
- In 1945 the university invested the gift at a market interest rate of 8.0% per year.

(a) Define i, i', f, $A\$$ and $R\$$ from the problem.

Solution:

Given in the problem:

- 6.0 % is the inflation rate (f)
- 8.0 % is the market interest rate (i)

Thus $i' = (0.08 - 0.06)/(1 + 0.06) = 0.01887$ or   1.887%

- $1,200,000 was the cost of the building in 1945. These were the actual dollars ($A\$$) spent in 1945.

(b) How many actual dollars in the year 2000 will the gift be worth?

Solution:

From Figure 13-1 we are going from *actual dollars in 1945* to *actual dollars in 2000*. To do so, we would use the *market interest rate* and discount this amount forward in time 55 years, as illustrated in Figure 13-2 below.

Actual dollars in 2000 = Actual Dollars in 1945 (F/P, i, 55 years)

$$= \$1,200,000 \ (F/P, 8\%, 55)$$
$$= \$82,701,600$$

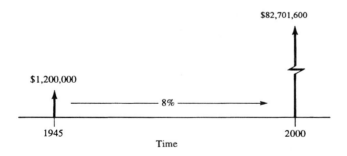

Figure 13-2 Discounting $A\$$ in 1945 to $A\$$ in 2000.

(c) How much would the actual dollars in 2000 be in terms of *1945 purchasing power?*

Solution:

Now we want to determine the amount of *real dollars— based in 1945—that occur in the year 2000* which are equivalent to the $82.7 million from above. Let us solve this problem two ways.

(1) In this approach let us directly translate the *actual dollars in the year 2000 to real 1945-based dollars in the year 2000.* From Figure 13-1 we can use the inflation rate to **strip 55 years of inflation** from the actual dollars. We do this by using the *P/F* factor for 55 years at the inflation rate. We are not physically moving the dollars in time in this case; rather we are simply removing inflation from these dollars one year at a time — the *P/F* factor does that for us. This is illustrated in Figure 13-3 below.

$$\text{Real 1945-based dollars in 2000} = (\text{Actual Dollars in 2000})(P/F, f, 55)$$
$$= (\$82,701,600)(P/F,6\%,55)$$
$$= \$3,357,000$$

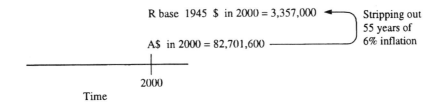

Figure 13-3 Translation of *A$* in 2000 to *R* 1945-based $ in 2000.

(2) In this method we start by recognizing that the *actual dollars in 1945* are exactly equivalent to *real 1945-based dollars that exist in 1945.* By definition, dollars that have *today* as the purchasing power base are the same as *actual dollars today.* As an example, actual dollars in 1998 are the same as real 1998-based dollars that occur in 1998. So in this example, the $1.2 million can also be said to be *real 1945-based dollars that occur in 1945.* As such, let us translate those real dollars from 1945 to the year 2000. Since they are *real dollars* we use the *real interest rate.*

$$\text{Real 1945-based dollars in 2000} = (\text{Real 1945-based Dollars in 1945})(P/F, i', 55)$$
$$= (\$1,200,000)(F/P, 1.887\%,55)$$
$$= \$3,355,000$$

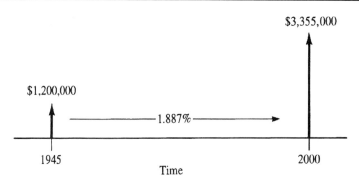

Figure 13-4 Translation of *R* 1945-based $ in 1945 to *R* 1945-based $ in 2000.

Note: The difference between the answers to parts **(1)** and **(2)** in question **(c)** above is due to rounding the market interest rate off to 1.887% versus carrying it out more significant digits. The amount of difference due to this rounding is less than 1%. If we were to carry the calculation of i' a sufficient number of digits, the answers to the two parts above would be identical. ■

From Example 13-2 we can see the relationship between dollars of different purchasing power bases, the selection of appropriate interest rates to use for moving dollars in time, and stripping out or adding in inflation. In that example the $1.2 million initially invested grew in the 55 year period to over $82.7 million, which becomes the amount available to pay for construction of the new complex. Does this mean that the new stadium will be 82.7/1.2 or about 70 times "better" than the one built in 1945? The answer to that question is "no", because in the year 2000 the purchasing power of a dollar is less than it was in the year 1945. Assuming that construction costs increased at the rate of 6% per year given in the problem, then the amount available for the project *in terms of 1945-based dollars* is almost $3.4 million. This means that the new stadium will be about 3.4/1.2 or approximately 2.8 times "better" than the original one using *real dollars* — not the 70 times ratio if *actual dollars* are used.

EXAMPLE 13-3

Mr. O'Leary buried $1000 worth of quarters in his back yard in 1924. Over the years he had always thought that the money would be a nice nest egg to give to his first grandchild. His first granddaughter, Gabrielle, arrived in 1994. At the time of her birth Mr. O'Leary was taking an economic analysis course in his spare time. He had learned the following: over the years 1924 to 1994 inflation averaged 4.5%, the stock market increased an average of 15% per year, and investments in guaranteed government obligations averaged 6.5% return per year. What was the relative purchasing power of the jar of quarters that Mr. O'Leary gave to his granddaughter Gabrielle at the time of her birth? What might have been a better choice for his "back yard investment?"

Solution:

Mr. O'Leary's $1000 dollars are *actual dollars* in both 1924 **and** in 1994.

To obtain the *real 1924-based dollar equivalent* of the $1000 that Gabrielle received in 1994 we would **strip 70 years of inflation out of those dollars**. As it turned out, Gabrielle's grandfather gave her $45.90 worth of 1924 purchasing power at the time of her birth. Because inflation has "stolen" purchasing power from his stash of quarters during the 70 year period, Mr. O'Leary gave his grandaughter much less than the amount he first spaded underground. This loss of purchasing power caused by inflation over time is calculated as follows:

$$\text{Real 1924-based dollars in 1994} = (\text{Actual dollars in 1994}) \, (P/F, f, \, 1994\text{-}1924)$$
$$= \$1000 \, (P/F, \, 4.5\%, 70) = \$45.90$$

On the other hand, if Mr. O'Leary had put his $1000 in the stock market in 1924, he would have made baby Gabrielle an instant multimillionaire by giving her $17,735,000. We calculate this as follows:

$$\text{Actual dollars in 1994} = (\text{Actual Dollars in 1924}) \, (F/P, i, \, 1994\text{-}1924)$$
$$= \$1000 \, (F/P, \, 15\%, \, 70) = \$17,735,000$$

At the time of Gabrielle's birth that $17.7 million translates to $814,069 in 1924 purchasing power. This is quite a bit different from the $45.90 in 1924 purchasing power calculated for the unearthed jar of quarters.

$$\text{Real 1924-based dollars in 1994} = \$17,735,000 \, (P/F, \, 4.5\%, 70) = \$814,069$$

Mr. O'Leary was never a risk taker, so it is doubtful he would have chosen the stock market for his future grandchild's nest egg. If he had chosen guaranteed government obligations instead of his backyard, by 1994 the investment would have grown to $59,076 (actual dollars) —the equivalent of $2712 in 1924 purchasing power.

$$\text{Actual Dollars in 1994} = (\text{Actual Dollars in 1924}) \, (F/P, i, \, 1994\text{-}1924)$$
$$= \$1000 \, (F/P, \, 6\%, \, 70) = \$59,076$$

$$\text{Real 1924-based dollars in 1994} = \$59,076 \, (P/F, \, 4.5\%, 70) = \$2712 \quad ■$$

Obviously, either option would have been better than the choice Mr. O'Leary made. This example illustrates the effects of inflation and purchasing power, as well as the power of compounded interest over time. However, in Mr. O'Leary's defense, if there had been 70 years of *deflation* instead of *inflation* in the U.S., he might have had the last laugh!

There are in general two ways to approach an economic analysis problem after recognizing the effects that inflation can have. The first is to conduct the analysis by systematically including the effects of inflation; the second is to conduct the analysis ignoring the effects of inflation. Each case requires a different approach.

Incorporating Inflation in The Analysis: Use a *market interest rate* and *actual dollars* that include inflation.

Ignoring Inflation in The Analysis: Use *real dollars* and a *real interest rate* that does not reflect inflation.

Constant Dollar Versus Then-Current Dollar Analysis

Performing an analysis requires that we distinguish cash flows as being either constant dollars (real dollars, expressed in terms of some purchasing power base) or then-current dollars (actual dollars that are then-current when they occur). As previously stated, constant (real) dollars require the use of a *real interest rate* for discounting, then-current dollars require a *market (or combined) interest rate*. It is not appropriate to mix these two types of dollars when performing an analysis. If both types of dollars are stated in the problem, one must convert either the constant dollars to then-current dollars, or the then-current dollars to constant dollars, so that a consistent comparison can be made.

EXAMPLE 13-4
The Waygate Corporation is interested in evaluating a major new video display technology (VDT). Two competing computer innovation companies have approached Waygate to develop the technology for the new VDT. Waygate believes that both companies will be able to deliver an equivalent product at the end of a five-year period. The yearly development costs of the VDT for each of the two companies is given below. Which should Waygate choose if their corporate MARR (investment market rate) is 25%, and general price inflation is assumed to be 3.5% per year over the next five years.

Company Alpha Costs: Development costs will be $150,000 the first year and increase at a rate of 5% over the five-year period.
Company Beta Costs: Development costs will be a constant $150,000 per year in terms of today's dollars over the five-year period.

Solution:
The costs for each of the two alternatives are:

Year	Then-Current Costs Stated By Alpha	Constant Dollar Costs Stated By Beta
1	$150,000 \times (1.05)^0 = \$150,000$	\$150,000
2	$150,000 \times (1.05)^1 = 157,500$	150,000
3	$150,000 \times (1.05)^2 = 165,375$	150,000
4	$150,000 \times (1.05)^3 = 173,644$	150,000
5	$150,000 \times (1.05)^4 = 182,326$	150,000

We inflate (or escalate) the stated yearly cost given by *Company Alpha* by 5% per year to obtain the then-current (actual) dollars each year. *Company Beta's* costs are given in terms of today-based constant dollars.

Using a Constant Dollar Analysis: Here we must convert the then-current costs given by *Company Alpha* to constant today-based dollars. We do this by stripping the appropriate number of years of general inflation from each year's cost using $(P/F, f, n)$ or $(1+f)^n$.

Year	Constant Dollar Costs Stated By Alpha	Constant Dollar Costs Stated By Beta
1	$\$150,000 \times (1.035)^{-1} = \$144,928$	\$150,000
2	$157,500 \times (1.035)^{-2} = 147,028$	150,000
3	$165,375 \times (1.035)^{-3} = 149,159$	150,000
4	$173,644 \times (1.035)^{-4} = 151,321$	150,000
5	$182,326 \times (1.035)^{-5} = 153,514$	150,000

We calculate the Present Worth of costs for each alternative using the *real interest rate (i')* calculated from Equation 13-1.

$$i' = (i - f) / (1+f) = (0.25 - 0.035) / (1+0.035) = 0.208$$

PW of cost (*Alpha*) = $144,928 (P/F,20.8\%,1) + \$147,028 (P/F,20.8\%,2)$
    $+\$149,159 (P/F,20.8\%,3) + \$151,321 (P/F,20.8\%,4)$
    $+\$153,514 (P/F,20.8\%,5) = \$436,000$

PW of cost (*Beta*) = $150,000 (P/A, 20.8\%, 5) = \$150,000 (2.9387) = \$441,000$

Using a Then-Current Dollar Analysis: Here we must convert the constant dollar costs of *Company Beta* to then-current dollars. We do this by "adding in" the appropriate number of years of general inflation into each year's cost using $(F/P, f, n)$ or $(1+f)^n$.

Year	*Then-Current Costs Stated By Alpha*	*Then-Current Costs Stated By Beta*
1	$150,000 \times (1.05)^0 = \$150,000$	$150,000 \times (1.035)^1 = \$155,250$
2	$150,000 \times (1.05)^1 = 157,500$	$150,000 \times (1.035)^2 = 160,684$
3	$150,000 \times (1.05)^2 = 165,375$	$150,000 \times (1.035)^3 = 166,308$
4	$150,000 \times (1.05)^3 = 173,644$	$150,000 \times (1.035)^4 = 172,128$
5	$150,000 \times (1.05)^4 = 182,326$	$150,000 \times (1.035)^5 = 178,153$

Calculate the Present Worth of costs for each alternative using the *market interest rate (i)*.

PW of cost (*Alpha*) = $150,000 (*P/F*,25%,1) + $157,500 (*P/F*,25%,2)
+$165,375 (*P/F*,25%,3) + $173,644 (*P/F*,25%,4)
+$182,326 (*P/F*,25%,5) = $436,000

PW of cost (*Beta*) = $155,250 (*P/F*,25%,1) + $160,684 (*P/F*,25%,2),
+$166,308 (*P/F*,25%,3) + $172,128 (*P/F*,25%,4),
+$178,153 (*P/F*,25%,5) = $441,000

Using either a constant dollar or then-current dollar analysis Waygate should chose *Company Alpha's* offer which has the lower present worth of costs. There may, of course, be intangible elements in the decision-making which also should be considered. ■

Price Change With Indexes

Previously, we described the effects that inflation can have on money over time. Also, several definitions and relationships regarding dollars and interest rates have been given. We have seen that it is not appropriate to compare the benefits of an investment in 1996-based dollars against costs in 1998-based dollars. This is like comparing apples to oranges. Such comparisons of benefits and costs are meaningful only if a standard purchasing power base of money is used. An often asked question is "how do I know what inflation rate to use in my studies?" This is a valid question. What **can** we use to measure price changes over time?

What is a Price Index?

Price indexes are used as a means to describe the relative price fluctuation of goods and services in our national economy. They provide a *historical* record of the behavior of these quantities over time. Price indexes are tracked for *specific commodities* as well as for *bundles (composites) of commodities*. As such, price indexes can be used to measure historical price changes for individual cost items (like labor and material costs) as well as general costs (like consumer products). In understanding the **past** price fluctuations we have more information for predicting the **future** behavior of those cash flows.

Table 13-1 below gives a list of the historic prices to send a *first class letter* in the United States via the Postal Service from 1970 to 1998. The cost is given both in terms of dollars (cents) and as measured by a fictitious price index that we could call the Letter Cost Index (LCI).

Notice two important aspects of the LCI as given below. First, as with all cost or price indexes, the numbers used to express the change in price over time are based on some **base year**. With price (or cost) indexes the base year is always assigned a value of 100. The LCI below has a base year of 1970 — thus 1970 is given an LCI=100. The letter cost index value given in subsequent years is stated in relation to 1970 as the base year. A second aspect to notice about the LCI is that it changes only when the cost of first class postage changes. In those years where this quantity does not change (in other words there was no price increase), the LCI is not affected. These general observations apply to all price indexes.

Table 13-1: HISTORIC PRICES OF FIRST CLASS MAIL, 1970-2000, and LETTER COST INDEX

Year(n)	Cost of First Class Mail	LCI	Annual Percent Increase for n	Year(n)	Cost of First Class Mail	LCI	Annual Percent Increase for n
1970	$ 0.06	100	0.00%	**1985**	$0.22	367	10.00%
1971	0.08	133	33.33	**1986**	0.22	367	0.00
1972	0.08	133	0.00	**1987**	0.22	367	0.00
1973	0.08	133	0.00	**1988**	0.25	417	13.64
1974	0.10	166	25.00	**1989**	0.25	417	0.00
1975	0.13	216	30.00	**1990**	0.25	417	0.00
1976	0.13	216	0.00	**1991**	0.29	483	16.00
1977	0.13	216	0.00	**1992**	0.29	483	0.00
1978	0.15	250	15.38	**1993**	0.29	483	0.00
1979	0.15	250	0.00	**1994**	0.29	483	0.00
1980	0.15	250	0.00	**1995**	0.32	533	10.34
1981	0.20	333	33.33	**1996**	0.32	533	0.00
1982	0.20	333	0.00	**1997**	0.32	533	0.00
1983	0.20	333	0.00	**1998**	0.33	550	3.13
1984	0.20	333	0.00	**1999**	0.33	550	0.00
				2000	0.33	550	0.00

In general, engineering economists are the "users" of cost indexes (such as the LCI above). That is to say, cost indexes are calculated or tabulated by some other party and our interest is in assessing what the index tells us about the historical prices and how these may affect our estimate of future costs. However, since we are "users" of indexes, rather than "compilers," it may be of interest to illustrate how the fictitious LCI in Table 13-1 was calculated.

In Table 13-1, the LCI is assigned a value of 100 because 1970 serves as our base year. In the following years the LCI is calculated on a year-to-year basis based on the annual percentage increase in first class mail. Equation 13-2 illustrates the equation used.

$$\text{LCI year (n)} = [((\text{cost (n)} - \text{cost } 1970) / \text{cost } 1970) \times 100\%] + 100 \qquad (13\text{-}2)$$

As an example consider the LCI for the year 1980. We calculate the LCI as follows.

$$\text{LCI year } 1980 = [((0.15 - 0.06/0.06) \times 100] + 100 = 250$$

As mentioned above, engineering economists are often in the business of using cost indexes to project future cash flows. As such, our first job is to use a cost index to **calculate** annual cost increases for the items tracked by the index. To calculate the *year-to-year* percentage increase (or *inflation*) of prices tracked by an index one can use Equation 13-3.

$$\text{Annual Percentage Increase (n)} = [\text{Index}(n) - \text{Index}(n-1)/\text{Index}(n-1)] \times 100\% \qquad (13\text{-}3)$$

To illustrate the use of this equation let us look at the percent change from 1976 to 1977 for the LCI given above.

$$\text{Annual Percentage Increase (1977)} = [250 - 217] / 217 \times 100\% \quad = \quad 15.38\%$$

For 1977 the price to mail a piece of first class mail increased by 15.38% over the previous year. This is the value tabulated in Table 13-1.

Of interest to an engineering economist is how a particular cost quantity changes over time. Often we are interested in calculating the *average* rate of price increase or inflation in some quantity, such as the cost of postage, over a period of time. For instance, one might want to know the average yearly increase in postal prices from the year 1971 to the year 1991. How do we calculate this quantity? Can we use Equation 13-2? If we were to calculate the percent change from 1971-1991 via equation 13-2, we would obtain the following:

$$\% \text{ Increase (1971 to 1991)} = [483 - 100] / 100 \times 100\% \quad = \quad 383\%$$

But how do we use this calculation to obtain the **average** rate of increase over those years? Should we divide 383% by 20 years (383/20 = 19.15%)? Of course not! As was established in earlier chapters, the concept of *compounding* precludes such a simple division. To do so would be treating the interest rate as simple interest — where compounding is not in effect. So the question remains: how do we calculate an *equivalent average rate of increase* in postage rates over a period of time? Let us start by thinking about the LCI index. We have a number (index value) of 100 in year 1971 and another number, 483, in year 1991, and we want to know the interest rate that relates these two numbers. Clearly, if we think of the index numbers as cash flows, it is easy to see that we have a simple internal rate of return problem. Given this approach, let us calculate the *average rate of increase* in postage rates for the years under consideration.

$$P = 100, \quad F = 483, \quad n = 20 \text{ years}, \quad i^* = ?$$

Using: $F = P(1+i)^n$ $483 = 100(1+i*)^{20}$ $i* = (483/100)^{1/20} - 1$ $i* = 0.082 = 8.2\%$

In this same way we can use a cost index to calculate the average rate of increase over any period of years. Understanding of how costs have behaved historically should provide insight into how they may behave in the future.

Composite Versus Commodity Indexes

Cost indexes, in general, come in two types: commodity specific indexes and composite indexes. Each type of index is useful to the engineering economist. Commodity specific indexes measure the historical change in price for specific items — such as green beans or iron ore. Common commodities that are tracked by price indexes and used in engineering economic analysis include *utility commodities*, *labor costs*, and *purchase prices*. Commodity indexes, like our Letter Cost Index, are useful when our economic analysis includes individual cost items that are tracked by such indexes. For example, if we need to estimate the direct-labor cost portion of a construction project, we could use an index that tracks the inflation, or escalation, of this particular cost over time. The U.S. Departments of Commerce and Labor track many cost quantities through the Department of Economic Analysis and Bureau of Labor Statistics. For our example construction project we would reference the appropriate labor index and investigate how this cost item had behaved in the past. This should give us valuable information about how to estimate this cost in the future.

EXAMPLE 13-5

Congratulations! Your sister just had a new baby girl named Veronica. When you hear this happy news, you fondly remember the fine pony you had as a young child, and you decide you would like to give your new niece a pony for her fifth birthday. You would like to know how much you must put into your passbook savings account (earning 4% interest per year) today to purchase a pony and saddle for Veronica five years from now. But you have no idea how much a pony and saddle might cost in five years.

Solution:

Divide the problem into the following steps:

> *Step 1.* Use commodity indexes to measure past price changes for both ponies and saddles.
> *Step 2.* Call several dealers to get prices on the current costs of ponies and saddles.
> *Step 3.* Use the *average price increase values* for each commodity (pony and saddle) to estimate what the cost of each will be in five years.
> *Step 4.* Calculate how much you would have to put into your bank account today to cover those costs in five years.

Step 1.

A trip to the library reveals that, indeed, there are price indexes for the two items that you would like to buy in five years: a pony and a saddle. From the indexes you find that in the last ten years the Little Pony Price Index (LPPI) has gone from 213 to 541, and the Leather Saddle Index (LSI) from 1046 to 1229. For each commodity you calculate the *average rate of price increase* for the past ten years as:

> *Average Price Change* (LPPI) is $541 = 213 (1 + i*)^{10}$
> solving for $i*$ for ponies = 9.8% per year

> *Average Price Change* (LSI) is $1229 = 1046 (1 + i*)^{10}$
> solving for $i*$ for saddles = 1.6% per year

Step 2.
Call a local stable and horse tackle shop to obtain the current price of a registered pony at $600 and leather saddle at $350.

Step 3.
Using the current prices for both the pony and saddle, *inflate* these costs at the respective *price change rates* calculated in Step 1 above. The assumption you are making is that the prices for ponies and saddles will change in the next five years at the same rate as the average of the last ten years. The cost for the two items, and the total, in five years will be:

> *Pony Cost in 5 Years* = $600 (1 + 0.098)^5$ = $958

> *Saddle Cost in 5 Years* = $350 (1 + 0.016)^5$ = $378

> *Total Cost in 5 Years* = $958 + $378 = $1336

Step 4.
Your final step is to calculate the amount that must be set aside today at 4.5% interest for five years to accumulate the future cost.

> *Amount Invested Now* = $1336 (1 + 0.04)^{-5}$ = $1098

So it will require **$1098** today in order to make Veronica a very happy pony rider on her fifth birthday!   ■

Composite cost indexes do not track historical prices for individual items — instead, they measure the historical prices of *groups* or *bundles* of assets. Thus, the name **composite index** is given because it measures the overall price change that is a composite of several effects. Examples of composite indexes include the *Consumer Price Index* (CPI) and the *Producer Price Index* (PPI). The CPI measures the effect of prices as experienced by consumers in the U.S. marketplace, and the PPI measures prices as felt by producers of goods in the U.S. economy.

The CPI, an index calculated by the Bureau of Labor Statistics, tracks the cost of a standard *bundle of consumer goods* from year-to-year. This "consumer bundle" or "basket of consumer goods" is made up of several common consumer expenses, including: housing, clothing, food, transportation, entertainment and others. Because of its focus on consumer goods, people often use the CPI as a substitute measure for general inflation in the economy. There are several problems with the use of the CPI in this manner, one being the assumption that all consumers purchase the same "basket of consumer goods" year after year. However, even with its deficiencies, the CPI enjoys popular identification as an "inflation" indicator. Table 13-2 gives the yearly index values and annual percent increase in the CPI for the past 20 years.

Table 13-2: CPI INDEX VALUES AND YEARLY PERCENTAGE INCREASES, 1978-2000

Year	CPI^* Value	CPI % Increase	Year	CPI^* Value	CPI % Increase
1978	65.2	7.59	1989	124.0	4.82
1979	72.6	11.35	1990	130.7	5.40
1980	82.4	13.50	1991	136.2	4.21
1981	90.9	10.32	1992	140.3	3.01
1982	96.5	6.16	1993	144.5	2.99
1983	99.6	3.21	1994	148.2	2.56
1984	103.9	4.32	1995	152.4	2.83
1985	107.6	3.56	1996	156.9	2.95
1986	109.6	1.86	1997	160.5	2.29
1987	113.6	3.65	1998	163.0	1.56
1988	118.3	4.14	1999	166.6	2.21

*Reference base: 1982-84 = 100

Composite indexes can be used in much the same way that commodity specific indexes were described earlier. That is, we can pick a single value from the table if we are interested in measuring the historic price for a single year, or we can calculate an *average inflation rate* or *average rate of price increase* as measured by the index over a several year time period.

How to Use Price Indexes in Engineering Economic Analysis

One may question the usefulness of *historical* data (such as that provided by price indexes), when engineering economic analysis deals with economic effects projected to occur in the *future*. However, both commodity specific and composite indexes indeed are useful in many analyses. Engineering economic analysis is concerned with making estimates of future events: the outcomes of yearly costs and benefits, interest rates, salvage values, and tax rates are all examples of such estimates that are made. These estimates have varying degrees of uncertainty associated with them.

The challenge for the engineering economist is to reduce this uncertainty for each estimate. Historical data provide a snapshot of how the quantities of interest have behaved in the past. Knowing this past (historical) behavior should provide insight on how to estimate their behavior in the future, and reduce the uncertainty of that estimate. This is where the data that price indexes provide come into play. Although it is very dangerous to extrapolate past data into the future on the short run, price index data can be useful in making estimates (especially when considered from a long term perspective). In this way the engineering economist can use *average historical percentage increases (or decreases)* from commodity specific and composite indexes, along with data from market analyses and other sources, to estimate how economic quantities may behave in the future.

One may wonder how both commodity specific and composite price indexes may be used in engineering economic analyses. The answer to that question is reasonably straightforward. As was established above, price indexes can be useful in making estimates of future outcomes. The following principle applies to commodity specific and composite price indexes and such estimates: **when the estimated quantities are items that are tracked by commodity specific indexes, then those indexes should be used to calculate *average historical percentage increases (or decreases)*.** If no commodity specific indexes are kept, then one should use an appropriate composite index to make this calculation. Consider the case of estimating electric usage costs for a turret lathe over a five year period. One would first want to refer to a commodity specific index that tracks this quantity. If such an index does not exist, one might use a specific index for a very closely related commodity, for example, in this case, an index of electric usage costs of screw lathes. In the absence of such substitute or related commodity indexes, one could use appropriate composite indexes. As an example, there may be a composite index that tracks electric usage costs for industrial metal cutting machinery. Or, as above, a related composite index could be used. The key point is that one should try to identify and use a price index that most closely relates to the quantity being estimated in the analysis.

Cash Flows That Inflate At Different Rates

Engineering economic analysis requires the estimation of various parameters. It is not uncommon that over time these parameters will *inflate* or *increase* (or even decrease) at different rates. For instance, one parameter might *increase* 5% per year, another 15% per year, and a third *decrease* 3.5% per year. This phenomena is important because of the various and different types of items that are sometimes included in engineering economic analyses. Since we are looking at the behavior of cash flows over time we must have a way of handling this effect.

EXAMPLE 13-6

On your first assignment as an engineer your boss asks you to develop the utility cost portion of an estimate for the cost of a new manufacturing facility. After some research you define the problem as finding the Present Worth of Utility Costs given the following data:

- Your company uses a minimum attactive rate of return (MARR) = 35%.
- The project has a Useful Life = 25 years.
- Utilities to be estimated include: electricity, water, and natural gas.
- The 35 year historical data reveal:

  electricity costs increase at 8.5% per year;

  water costs increase at 5.5% per year;

  natural gas costs increase at 6.5% per year.

- First year estimates of the utility costs (in today's dollars) are:

  electricity cost will be $55,000;

  water cost will be $18,000;

  natural gas will be $ 38,000.

Solution:

For this problem we will take each of the utilities that are used in our manufacturing facility and inflate them independently at their respective historical annual rates. Once we have these actual dollar amounts ($A\$$), we can total them and then discount each year's total at 35% for the appropriate number of periods back to the present.

Year	Electricity	Water	Natural Gas	Total
1	$55,000(1.085)^0 = 55,000$	$18,000(1.055)^0 = 18,000$	$38,000(1.065)^0 = 38,000$	111,000
2	$55,000(1.085)^1 = 59,675$	$18,000(1.055)^1 = 18,990$	$38,000(1.065)^1 = 40,470$	119,135
3	$55,000(1.085)^2 = 64,747$	$18,000(1.055)^2 = 20,034$	$38,000(1.065)^2 = 43,101$	127,882
4	$55,000(1.085)^3 = 70,251$	$18,000(1.055)^3 = 21,136$	$38,000(1.065)^3 = 45,902$	137,289
5	$55,000(1.085)^4 = 76,222$	$18,000(1.055)^4 = 22,299$	$38,000(1.065)^4 = 48,886$	147,407
6	$55,000(1.085)^5 = 82,701$	$18,000(1.055)^5 = 23,525$	$38,000(1.065)^5 = 52,063$	158,290
7	$55,000(1.085)^6 = 89,731$	$18,000(1.055)^6 = 24,819$	$38,000(1.065)^6 = 55,447$	169,997
8	$55,000(1.085)^7 = 97,358$	$18,000(1.055)^7 = 26,184$	$38,000(1.065)^7 = 59,051$	182,594
.	.	.	.	.
.	.	.	.	.
.	.	.	.	.
24	$55,000(1.085)^{23} = 359,126$	$18,000(1.055)^{23} = 61,671$	$38,000(1.065)^{23} = 161,743$	582,539
25	$55,000(1.085)^{24} = 389,652$	$18,000(1.055)^{24} = 65,063$	$38,000(1.065)^{24} = 172,256$	626,970

The Present Worth of the Total Yearly Utility Costs

$= \$111,000(P/F,35\%,1) + \$119,135(P/F,35\%,2) + \ldots + \$626,970(P/F,35\%,25)$

$= \$5,540,000$ ■

In Example 13-6 several parameters changed at different rates over the period of the economic study. By using the respective individual inflation rates, the **actual dollar** amounts for each parameter were obtained in each year. Then, these actual dollar amounts were discounted using a market interest rate. Problems of this type can be handled by: inflating the various parameters at their respective estimated inflation rates, combining these cash flows if appropriate, and then treating them as actual dollars that occur in those years.

Different Inflation Rates Per Period

In this section we address the situation where the inflation rates for the various cash flows in the analysis are changing over the life of the study period. Rather than different inflation rates for different cash flows, in the case below the *interest rate* for the same cash flow is changing over time. A method for handling this situation is much the same as in the previous section. We can simply apply the inflation rates in the years in which they are projected to occur. We would do this for each of the individual cash flows over the entirety of the study period. Once we have all of these actual dollar amounts, we can apply any of the previous measures of merit using the market interest rate and decision criteria.

EXAMPLE 13-7

While working as a clerk at Piggly Wiggly, Elvis has learned much about the cost of different vegetables. The kitchen manager at Heartbreak Hotel called recently requesting Elvis to estimate the raw material cost over the next five years to introduce succotash (lima beans and corn) to their buffet line. To develop his estimate Elvis has used his advanced knowledge regarding soil growing conditions, world demand and government subsidy programs for these two crops. He has estimated the following data:

- Costs for lima beans will inflate at 3% per year the next three years and then 4% the following two years.
- Costs for corn will inflate at 8% per year the next two years and then decrease 2% the following three years.

The kitchen manager wants to know the equivalent annual cost of providing succotash on the buffet line over the five year period. His before-tax MARR is 20%. The manager estimates that he will need an average of 50 pounds each of beans and corn each day. The hotel kitchen operates 6 days per week, 52 weeks per year. Current costs for lima beans are $0.35/lb and for corn $0.80 /lb.

Solution:
Today's cost for one year's supply of vegetables is:

- Lima Beans: 0.35 $/lb. x 50 lb./day x 6 day/wk. x 52 wk/yr. = $ 5,460 /yr.
- Corn: 0.80 $/lb. x 50 lb./day x 6 day/wk. x 52 wk/yr. = 12,480 /yr.

Year	Lima Beans	Corn	Total
0	$5,460	$12,480	
1	5,460(1.03) = 5,624	12,480(1.08) = 13,478	$19,102
2	5,624(1.03) = 5,793	13,478(1.08) = 14,556	20,349
3	5,793(1.03) = 5,967	$14,556(1.02)^{-1} = 14,271$	20,238
4	5,967(1.04) = 6,206	$14,271(1.02)^{-1} = 13,991$	20,197
5	6,206(1.04) = 6,454	$13,991(1.02)^{-1} = 13,717$	20,171

$$\text{EUAC} = [19,102(P/F,20\%,1) + 20,349(P/F,20\%,2) + 20,238(P/F,20\%,3) +$$
$$20,197(P/F,20\%,4) + 20,171(P/F,20\%,5)] \ (A/P,20\%,5)$$
$$= \$19,900 \ \text{per year} \quad \blacksquare$$

In the above example both today's cost for each vegetable and their inflation rates were used to calculate the yearly costs to purchase the desired quantities over the five year period. As in the previous example, we obtained a total marginal cost (in terms of actual dollars) by combining the two individual yearly costs. We then calculated the equivalent uniform annual cost (EUAC) using the given market interest rate.

Example 13-8 below provides another example of how the effect of changes in inflation rates over time can affect an analysis.

EXAMPLE 13-8

If general price inflation is estimated to be: 5% for the next five years, 7.5% for the three years after that, and 3% the following five years, at what market interest rate (i) would you have to invest your money to maintain a real purchasing power growth rate (i') of 10% during those years?

Solution:

In years 1-5 you must invest at: $0.10 + 0.050 + (0.10)(0.050) = 0.1150 = 11.50\%$ per year.
In years 6-8 you must invest at: $0.10 + 0.075 + (0.10)(0.075) = 0.1825 = 18.25\%$ per year.
In years 9-13 you must invest at: $0.10 + 0.030 + (0.10)(0.030) = 0.1330 = 13.30\%$ per year.

NOTE: The problem above illustrates a common problem in timing: the market interest rate necessary to maintain the real purchasing power of the investment must be calculated at a time when the inflation rate is only an estimate. Most interest bearing investments have fixed, up-front rates that the investor well understands going into the investment. Variable rate investments are the exception. On the other hand, inflation is not quantified and its effect on our real return is not measured until the end of the year. Therefore, to achieve the conditions required in the example, one would either have to anticipate inflation and adjust one's investments accordingly, or accept the fact that the real investment return (i') may not turn out to be what was originally required. ■

Inflation Effect On After-Tax Calculations

In the previous sections, we have noted the impact of inflation on before-tax calculations. We found that if the subsequent benefits brought constant quantities of dollars, then inflation will diminish the true value of the future benefits and, hence, the real rate of return. If, however, the future benefits keep up with the rate of inflation, the rate of return will not be adversely affected due to the inflation. Unfortunately, we are not so lucky when we consider a situation with income taxes, as illustrated by Example 13-9.

EXAMPLE 13-9

A $12,000 investment will return annual benefits for six years with no salvage value at the end of six years. Assume straight line depreciation and a 46% income tax rate. The problem is to be solved for both before and after-tax rates of return, the latter for two situations:

1. *No inflation:* the annual benefits are constant at $2918 per year.
2. *Inflation equal to 5%:* the benefits from the investment increase at this same rate, so that they continue to be the equivalent of $2918 in Year-0 based dollars.

The benefit schedule for the two situations is:

Year	Annual benefit for both situations, in Year-0 based dollars	No inflation, actual dollars received	5% inflation factor*	5% inflation, actual dollars received
1	$2918	$2918	$(1.05)^1$	$3064
2	2918	2918	$(1.05)^2$	3217
3	2918	2918	$(1.05)^3$	3378
4	2918	2918	$(1.05)^4$	3547
5	2918	2918	$(1.05)^5$	3724
6	2918	2918	$(1.05)^6$	3910

*May be read from the 5% Compound Interest Table as $(F/P,5\%,n)$.

Solutions: ***Before-tax rate of return:*** Since both situations (no inflation and 5% inflation) have an annual benefit, stated in Year-0 based dollars of $2918, they have the same before-tax rate of return.

PW of cost = PW of benefit

$$12,000 = 2918(P/A,i,6) \qquad (P/A,i,6) = \frac{12,000}{2918} = 4.11$$

From Interest Tables: Before-tax rate of return equals 12%.

After-tax rate of return, no inflation:

Year	Before-tax cash flow	Straight line depreciation	Taxable income	46% income taxes	Actual dollars, and Year-0 based $ after-tax cash flow
0	−$12,000				−$12,000
1–6	+2,918	$2000	$918	−$422	+2,496

PW of cost = PW of benefit

$$12,000 = 2496(P/A,i,6) \qquad (P/A,i,6) = \frac{12,000}{2496} = 4.81$$

From Interest Tables: After-tax rate of return equals 6.7%.

After-tax rate of return, 5% inflation:

Year	Before-tax cash flow	Straight line depreciation	Taxable income	46% income taxes	Actual dollars, after-tax cash flow
0	−$12000				−$12000
1	+3064	$2000	$1064	−$489	+2575
2	+3217	2000	1217	−560	+2657
3	+3378	2000	1378	−634	+2744
4	+3547	2000	1547	−712	+2835
5	+3724	2000	1724	−793	+2931
6	+3910	2000	1910	−879	+3031

Converting to Year-0 based dollars and solving for the rate of return:

Year	Actual dollars, After-tax cash flow	Conversion factor	Year-0 based dollars, after-tax cash flow	Present worth at 5%	Present worth at 4%
0	−$12000		−$12000	−$12000	−$12000
1	+2575	$\times (1.05)^{-1} =$	+2452	+2335	+2358
2	+2657	$\times (1.05)^{-2} =$	+2410	+2186	+2228
3	+2744	$\times (1.05)^{-3} =$	+2370	+2047	+2107
4	+2835	$\times (1.05)^{-4} =$	+2332	+1919	+1993
5	+2931	$\times (1.05)^{-5} =$	+2297	+1800	+1888
6	+3031	$\times (1.05)^{-6} =$	+2262	+1688	+1788
				−25	+362

Linear interpolation between 4% and 5%:

After-tax rate of return = 4% + 1%[362/(362 + 25)] = 4.9% ∎

From Example 13-9, we see that the before-tax rate of return for both situations (no inflation and 5% inflation) is the same. Equal before-tax rates of return are expected because the benefits in the inflation situation increased in proportion to the inflation. This example shows that where future benefits fluctuate with changes in inflation or deflation, the effects do not alter the Year-0 based dollar estimates. Thus, no special calculations are needed in before-tax calculations when future benefits are expected to respond to inflation or deflation rates.

The after-tax calculations illustrate a different result. The two situations, with equal before-tax rates of return, do not produce equal after-tax rates of return:

Situation	Before-tax rate of return	After-tax rate of return
No inflation	12%	6.7%
5% inflation	12%	4.9%

Thus, 5% inflation results in a smaller after-tax rate of return even though the benefits increase at the same rate as the inflation. A review of the cash flow table reveals that while benefits increase, the depreciation schedule does not. Thus, the inflation results in increased taxable income and, hence, larger income tax payments; but there are not sufficient increases in benefits to offset these additional disbursements.

The result is that while the after-tax cash flow in actual dollars increases, it is not large enough to offset *both* inflation and increased income taxes. This effect is readily apparent when the equivalent Year-0 based dollar after-tax cash flow is examined. With inflation, the Year-0 based dollar after-tax cash flow is smaller than the Year-0 based dollar after-tax cash flow without inflation. Of course, inflation might cause equipment to have a salvage value that was not forecast, or a larger one than had been projected. This effect would tend to reduce the unfavorable consequences of inflation on the after-tax rate of return.

Using Spreadsheets for Inflation Calculations

Spreadsheets are the perfect tool for analyzing economic problems where inflation is considered. For example, next year's labor costs are likely to be estimated as equal to this year's costs times $(1 + f)$, where f is the inflation rate. Thus each year's value is different, so we can't use factors for uniform flows, A. Also the formulas that link different years are easy to write. The result is problems that are very tedious to do by hand, and easy by spreadsheet.

Example 13-10 illustrates two different ways to write the equation for inflating costs. Example 13-11 illustrates that inflation reduces the after-tax rate of return, because inflation makes the depreciation deduction less valuable.

EXAMPLE 13-10

Two costs for construction of a small remote mine are for labor and transportation. Labor costs are expected to be $350,000 the first year, with inflation of 6% annually. Unit transportation costs are expected to inflate at 5% annually, but the volume of material being moved changes each year. In time-0 dollars, the transportation costs are estimated to be $40,000, $60,000, $50,000, and $30,000 in years 1 through 4. The inflation rate for the value of the dollar is 3%. If the firm uses an $i\,'$ of 7%, what is the equivalent annual cost for this four year project?

Solution

The data for labor costs is stated so that no inflation needs to be applied in year 1 – the cost is $350,000. In contrast, the transportation costs for year 1 are determined by multiplying $40,000 by 1.05 (= $1 + f$).

Also in later years the labor cost$_t$ = labor cost$_{t-1}$ · $(1 + f)$, while each transportation cost must be computed as the time 0 value times $(1 + f)^t$. The numbers in the year 0 (or real) dollar column equal the values in the actual dollars column divided by $(1.03)^t$.

	A	B	C	D	E	F	G	H
1							7% inflation free interest	
2	Inflation rate	6%		5%		3%		
3			Transportation costs		Total	Total		
4	Year	Labor costs	Year 0 $s	Actual $s	Actual $s	Real $s		
5	1	120,000	40,000	42,000	162,000	157,282	= E5/(1+$F$2)^A5	
6	2	127,200	60,000	66,150	193,350	182,251		
7	3	134,832	50,000	57,881	192,713	176,360		
8	4	142,922	30,000	36,465	179,387	159,383		
9						$571,732	= NPV(F1,F5:F8)	
10					= B8+D8	$168,791	= -PMT(F1,4,F9)	
11	=B7*(1+$B$2)		=C8*(1+$D$2)^A8					

Figure 13-1 Spreadsheet for inflation.

The equivalent annual cost equals $168,791. ■

EXAMPLE 13-11

For the data of Example 13-9, calculate the IRR with and without inflation with MACRS depreciation. How are the results affected by inflation by comparison with the earlier results.

Solution

Most of the formulas for this spreadsheet are given in rows 11 and 12 for the data in year 6. The benefits received are computed from the base value in cell B5. The depreciation is the MACRS percentage times the $120,000 spent in year 0. This value is not influenced by inflation, so the depreciation deduction is less valuable as inflation increases. The tax paid equals the tax rate times the taxable income, which equals dollars received minus the depreciation charge. Then ATCF (after-tax cash flow) equals the before-tax cash flow minus the tax paid.

Notice that in year 2 the depreciation charge is large enough that this project pays "negative" tax. For a firm this means that the deduction on this project will be used to offset income from other projects.

	A	B	C	D	E	F	G	H
1		0%	= inflation rate		46%	= tax rate		
2		Actual $s	MACRS	Actual $s	Actual $s	Actual $s	Real $s	
3	Year	received	deprec. %	deprec.	tax	ATCF	ATCF	
4	0	-12000					-12000	-12000	
5	1	2918	20.00%	2400	238	2680	2680	=F5/(1+$B$1)^A5
6	2	2918	32.00%	3840	-424	3342	3342	
7	3	2918	19.20%	2304	282	2636	2636	
8	4	2918	11.52%	1382	706	2212	2212	
9	5	2918	11.52%	1382	706	2212	2212	
10	6	2918	5.76%	691	1024	1894	1894	
11	Formulas			=-$B$4*C10		=B10-E10		
12	for Yr 6	=$B$5*(1+$B$1)^A10			=(B10-D10)*$E$1		7.29%	= IRR
13								
14		5%	= inflation rate		46%	= tax rate		
15		Actual $s	MACRS	Actual $s	Actual $s	Actual $s	Real $s	
16	Year	received	deprec. %	deprec.	tax	ATCF	ATCF	
17	0	-12000					-12000	-12000	
18	1	3064	20.00%	2400	305	2759	2627	
19	2	3217	32.00%	3840	-287	3504	3178	
20	3	3378	19.20%	2304	494	2884	2491	
21	4	3547	11.52%	1382	996	2551	2099	
22	5	3724	11.52%	1382	1077	2647	2074	
23	6	3910	5.76%	691	1481	2430	1813	
24							5.68%	= IRR

Figure 13-2 After-tax IRRs with MACRS and inflation.

The IRRs are higher in this example (7.29% without inflation versus 6.7% with straight-line depreciation in Example 13-9, and 5.68% with inflation versus 4.9%), because MACRS supports faster depreciation, so the depreciation deductions are more valuable. Also because the depreciation is faster, the results are affected somewhat less by inflation. Specifically, with MACRS 5% inflation lowers the IRR by 1.6% and with straight line 5% inflation lowers the IRR by 1.8%. ∎

Summary

Inflation is characterized by rising prices for goods and services, while deflation produces a fall in prices. An inflationary trend makes future dollars have less **purchasing power** than present dollars. Inflation benefits a long-term borrower of money because payment of debt in the future is made with dollars that have reduced purchasing power. This advantage to borrowers is at the expense of lenders.

Deflation has the opposite effect from inflation. If money is borrowed over a period of time in which deflation is occurring, then debt will be repaid with dollars that have **more** purchasing power than those originally borrowed. This condition is advantageous to lenders at the expense of borrowers. Inflation and deflation have an opposite effect on the purchasing power of a monetary unit over time.

To distinguish and account for the effect of inflation in our engineering economic analysis we define *inflation*, *real* and *market* interest rates. These interest rates are related by the following expression:

$$i = i' + f + i'f.$$

Each rate applies in a different circumstance, and it is important to apply the correct rate to the correct circumstance. Cash flows are expressed in terms of either *actual* or *real dollars*. The *market interest* rate should be used with *actual dollars* and the *real interest rate* should be used with *real dollars*.

The different cash flows in our analysis may inflate or change at different interest rates when we look over the life cycle of the investment. Also, a single cash flow may inflate or deflate at different rates over time. These two circumstances are handled easily by applying the appropriate inflation rate to each cash flow over the study period to obtain the actual dollar amounts occurring in each year. After the actual dollar quantities are calculated, the analysis proceeds, as in previous chapters, utilizing the market interest rate to calculate the measure of merit of interest.

Historical price change for single commodities and bundles of commodities are tracked with price indexes. The Consumer Price Index (CPI) is an example of a composite index formed by a bundle of consumer goods. The CPI serves as a surrogate for general inflation in our economy. Indexes can be used to calculate the *average annual increase* (or decrease) of the costs and benefits in our analysis. This historical data provides valuable information about how economic quantities may behave in the future over the long run.

The effect of inflation on the computed rate of return for an investment depends on how future benefits respond to the inflation. If benefits produce constant dollars, which are not increased by inflation, the effect of inflation is to reduce the before-tax rate of return on the investment. If, on the other hand, the dollar benefits increase to keep up with the inflation, the before-tax rate of return will not be adversely affected by the inflation. This outcome is not true when an after-tax analysis is made. Even if the future benefits increase to match the inflation rate, the allowable depreciation schedule does not increase. The result will be increased taxable income and income tax payments, which reduce the available after-tax benefits and, therefore, the after-tax rate of return. The important conclusion is that estimates of future inflation or deflation may be important in evaluating capital expenditure proposals.

Problems

13-1   One economist has predicted that there will be a 7% per year inflation of prices during the next ten years. If this proves to be correct, an item that presently sells for $10 would sell for what price ten years hence? (*Answer:* $19.67)

13-2   A man bought a 5% tax-free municipal bond. It cost $1000 and will pay $50 interest each year for twenty years. The bond will mature at the end of the twenty years and return the original $1000. If there is 2% annual inflation during this period, what real rate of return will the investor receive after considering the effect of inflation?

13-3   A firm is having a large piece of equipment overhauled. It anticipates the machine will be needed for the next twelve years. The firm has an 8% minimum attractive rate of return. The contractor has suggested three alternatives:
    a. A complete overhaul for $6000 that should permit twelve years of operation.
    b. A major overhaul for $4500 that can be expected to provide eight years of service. At the end of eight years, a minor overhaul would be needed.
    c. A minor overhaul now. At the end of four and eight years, additional minor overhauls would be needed.
If minor overhauls cost $2500, which alternative should the firm select? If minor overhauls, which now cost $2500, increase in cost at +5% per year, but other costs remain unchanged, which alternative should the firm select? (Answers: Alt. **c**; Alt. **a**)

13-4   A man wishes to set aside some money for his daughter's college education. His goal is to have a bank savings account containing an amount equivalent to $20,000 with today's purchasing power of the dollar, at the girl's 18th birthday. The estimated inflation rate is 8%. If the bank pays 5% compounded annually, what lump sum of money should he deposit in the bank savings account on his daughter's fourth birthday? (*Answer:* $29,670)

13-5   One economist has predicted that for the next five years, the United States will have an 8% annual inflation rate, followed by five years at a 6% inflation rate. This is equivalent to what average price change per year for the entire ten-year period?

13-6   A homebuilder's advertising has the caption, "Inflation to Continue for Many Years." The advertisement continues with the explanation that if one buys a home now for $97,000, and inflation continues at a 7% annual rate, the home would be worth $268,000 in 15 years. According to the advertisement, by purchasing a new home now, the buyer will realize a profit of $171,000 in fifteen years. Do you agree with the homebuilder's logic? Explain.

13-7 Sam Johnson inherited $85,000 from his father. Sam is considering investing the money in a house which he will then rent to tenants. The $85,000 cost of the property consists of $17,500 for the land, and $67,500 for the house. Sam believes he can rent the house and have $8000 a year net income left after paying the property taxes and other expenses. The house will be depreciated by straight line depreciation using a 45-year depreciable life.
    a. If the property is sold at the end of five years for its book value at that time, what after-tax rate of return will Sam receive? Assume that his incremental personal income tax rate is 34%.
    b. Now assume there is 7% per year inflation, compounded annually. Sam will increase the rent 7% per year to match the inflation rate, so that after considering increased taxes and other expenses, the annual net income will go up 7% per year. Assume Sam's incremental income tax rate remains at 34% for all ordinary taxable income related to the property. The value of the property is now projected to increase from its present $85,000 at a rate of 10% per year, compounded annually.
        If the property is sold at the end of five years, compute the rate of return on the after-tax cash flow in actual dollars. Also compute the rate of return on the after-tax cash flow in Year-0 dollars.

13-8 Tom Ward put $10,000 in a five-year certificate of deposit that pays 12% interest per year. At the end of the five years the certificate will mature and he will receive his $10,000 back. Tom has substantial income from other sources and estimates that his incremental income tax rate is 42%. If the inflation rate is 7% per year, what is his:
    a. Before-tax rate of return, ignoring inflation?
    b. After-tax rate of return, ignoring inflation?
    c. After-tax rate of return, after taking inflation into account?

13-9 Sally Seashell bought a lot at the Salty Sea for $18,000 cash. She does not plan to build on the lot, but instead will hold it as an investment for ten years. She wants a 10% after-tax real rate of return after taking the 6% annual inflation rate into account. If income taxes amount to 15% of the capital gain, at what price must she sell the lot at the end of the ten years? (*Answer:* $95,188)

13-10 A newspaper reports that in the last five years, prices have increased a total of 50%. This is equivalent to what annual inflation rate, compounded annually? (*Answer:* 8.45%)

13-11 A South American country has had a high rate of inflation. Recently, its exchange rate was 15 cruzados per dollar, that is, one dollar will buy 15 cruzados in the foreign exchange market. It is likely the country will continue to experience a 25% inflation rate, and the United States will continue at a 7% inflation rate. Assume that the exchange rate will vary the same as the inflation. In this situation, one dollar will buy how many cruzados five years from now? (*Answer:* 32.6)

13-12 A group of students decided they would lease and run a gasoline service station. The lease is for ten years. Almost immediately the students were confronted with the need to alter the gasoline pumps to read in liters. The Dayton Co. has a conversion kit available for $900 that may be expected to last ten years. As an alternative, they offer a $500 conversion kit that has a five-year useful life. The students believe that any money not invested in the conversion kits may be invested elsewhere at a 10% interest rate. Income tax consequences are to be ignored in this problem.

   a. Assuming that future replacement kits cost the same as today, which alternative should be selected?
   b. If one assumes a 7% inflation rate, which alternative should be selected?

13-13 An automobile manufacturer has an automobile that gets ten kilometers per liter of gasoline. It is estimated that gasoline prices will increase at a 12% per year rate, compounded annually, for the next eight years. This manufacturer believes that the automobile fuel consumption for its new automobiles should decline as fuel prices increase, so that the fuel cost will remain constant. To achieve this, what must be the fuel rating, in km/1, of the automobiles eight years hence?

13-14 Pollution control equipment must be purchased to remove the suspended organic material from liquid being discharged from a vegetable packing plant. Two alternative pieces of equipment are available that would accomplish the task. A Filterco unit presently costs $7000 and has a five-year useful life. A Duro unit, on the other hand, now costs $10,000 but will have a ten-year useful life.

   Equipment costs are rising at 8% per year, compounded annually, due to inflation, so when the Filterco unit would be replaced, the cost would be much more than $7000. Based on a ten-year analysis period, and a 20% minimum attractive rate of return, before taxes, which piece of pollution control equipment should be purchased?

13-15 Dick DeWolf and his wife have a total taxable income of $60,000 this year and file a joint federal income tax return. If inflation continues for the next twenty years at a 7% rate, compounded annually, Dick wonders what their taxable income must be in the future to provide them the same purchasing power, after taxes, as their present taxable income. Assuming the federal income tax rate table is unchanged, what must their taxable income be twenty years from now?

13-16 One economist has predicted that during the next six years prices in the United States will increase 55%. After that he expects a further increase of 25% in the subsequent four years, so that prices at the end of ten years will have increased to 180% of the present level. Compute the inflation rate, *f*, for the entire ten-year period.

13-17 The U.S. tax laws provide that depreciation on equipment is based on its original cost. Yet due to substantial inflation, the replacement cost of equipment is often much greater than the original cost. What effect, if any, does this have on a firm's ability to buy new equipment to replace old equipment?

13-18 The City of Columbia is trying to attract a new manufacturing business to the area. It has offered to install and operate a water pumping plant to provide service to the proposed plant site. This would cost $50,000 now, plus $5000 per year in operating costs for the next ten years, all measured in Year-0 dollars.

To reimburse the city, the new business must pay a fixed uniform annual fee, A, at the end of each year for ten years. In addition, it is to pay the city $50,000 at the end of ten years. It has been agreed that the city should receive a 3% rate of return, after taking an inflation rate, f, of 7% into account.

Determine the amount of the uniform annual fee.   (*Answer:* $12,100)

13-19   A small research device is purchased for $10,000 and depreciated by MACRS depreciation. The net benefits from the device, before deducting depreciation, are $2000 at the end of the first year, and increasing $1000 per year after that (second year equals $3000, third year equals $4000, and so on), until the device is hauled to the junkyard at the end of seven years. During the seven-year period there is an inflation rate f of 7%.

This profitable corporation has a 50% combined federal and state income tax rate. If it requires a 12% after-tax rate of return on its investment, after taking inflation into account, should the device have been purchased?

13-20   Sally Johnson loaned a friend $10,000 at 15% interest, compounded annually. She is to repay the loan in five equal end-of-year payments. Sally estimates the inflation rate during this period is 12%. After taking inflation into account, what rate of return is Sally receiving on the loan? Compute your answer to the nearest 0.1%. (*Answer:* 2.7%)

13-21   Sam purchased a home for $150,000 with some creative financing. The bank agreed to lend Sam $120,000 for six years at 15% interest. It received a first mortgage on the house. The Joneses, who sold Sam the house, agreed to lend Sam the remaining $30,000 for six years at 12% interest. They received a second mortgage on the house. Thus Sam became the owner without putting up any cash. Sam pays $1500 a month on the first mortgage and $300 a month on the second mortgage. In both cases these are "interest only" loans, and the principal is due at the end of the loan.

Sam rented the house, but after paying the taxes, insurance, and so on, he had only $800 left, so he was forced to put up $1000 a month of his own money to make the monthly payments on the mortgages. At the end of three years, Sam sold the house for $205,000. After paying off the two loans and the real estate broker, he had $40,365 left. After taking an 8% inflation rate into account, what was his before-tax rate of return?

13-22   Dale saw that the campus bookstore is having a special sale on pads of computation paper. The normal price is $3.00 a pad, but it is on sale at $2.50 a pad. This sale is unusual and Dale assumes the paper will not be put on sale again. On the other hand, Dale expects that there will be no increase in the $3.00 regular price, even though the inflation rate is 2% every three months. Dale believes that competition in the paper industry will keep wholesale and retail prices constant. He uses a pad of computation paper every three months. Dale considers 19.25% a suitable minimum attractive rate of return. Dale will buy one pad of paper for his immediate needs. How many extra pads of computation paper should he buy?   (*Answer:* 4)

13-23   When there is little or no inflation, a homeowner can expect to rent an unfurnished home for 12 percent of the market value of the property (home and land) per year. About ⅛ of the rental income is paid out for property taxes, insurance, and other operating expenses. Thus the net annual income to the owner is 10.5% of the market value of the property. Since prices are relatively stable, the future selling price of the property often equals the original price paid by the owner.

For a $150,000 property (where the land is estimated at $46,500 of the $150,000), compute the after-tax rate of return, assuming the selling price 59 months later (in December) equals the original purchase price. Use modified accelerated cost recovery system depreciation beginning January 1st. Also, assume a 35% income tax rate.   (*Answer:* 6.84%)

13-24 (This is a continuation of Problem 13-23.) As inflation has increased throughout the world, the rental income of homes has decreased and a net annual rental income of 8% of the market value of the property is common. On the other hand, the market value of homes tends to rise about 2% per year more than the inflation rate. As a result, both annual net rental income, and the resale value of the property rise faster than the inflation rate. Consider the following situation.

A $150,000 property (with the house valued at $103,500 and the land at $46,500) is purchased for cash in Year 0. Use MACRS depreciation, beginning January 1st. The market value of the property increases at a 12% annual rate. The annual rental income is 8% of the beginning-of-year market value of the property. Thus the rental income also increases each year. The general inflation rate f is 10%. The individual who purchased the property has an average income tax rate of 35%.

    a. Compute the actual dollar after-tax rate of return for the owner, assuming he sells the property 59 months later (in December).

    b. Compute the after-tax rate of return for the owner, after taking the general inflation rate into account, assuming he sells the property 59 months later.

13-25 An investor wants a real rate of return i' (rate of return without inflation) of 10% per year on any projects in which he invests. If the expected annual inflation rate for the next several years is 6%, what interest rate i should be used in project analysis calculations?

13-26 In Chapters 5 (Present Worth Analysis) and 6 (Annual Cash Flow Analysis) it is assumed that prices are stable and a machine purchased today for $5000 can be replaced for the same amount many years hence. In fact, prices have generally been rising, so the stable price assumption is incorrect. Under what circumstances is it appropriate to use the "stable price" assumption when prices actually are changing?

13-27

    a. Compute the equivalent annual inflation rate, based on the consumer price index, for the period from 1981 to 1986.

    b. Using the equivalent annual inflation rate computed in part **a**, estimate the consumer price index in 1996, working from the 1987 consumer price index.

13-28 A couple in Ruston, Louisiana is faced with a decision of whether it is more economical to buy a home or to continue to rent during an inflationary period. Presently the couple rents a one-bedroom duplex for $450 a month plus $139 a month in basic utilities (heating and cooling). These costs tend to increase with inflation and with the projected inflation rate of 5%, the couple's monthly costs per year over a 10-year planning horizon are shown below:

n =	1	2	3	4	5	6	7	8	9	10
Rent	450	473	496	521	547	574	603	633	665	698
Utilities	139	146	153	161	169	177	186	196	205	216

The couple would like to live in the northside of the town where an average home of $150 \, m^2$ of

heating area costs $75,000. A local mortgage company will provide a loan for the property provided the couple makes a down payment of 5% plus estimated closing costs of 1% cash for the home. The couple prefers a 30-year fixed-rate mortgage with an 8% interest rate. Based on the couple's gross annual income, the couple falls in the 30% marginal income tax rate (federal plus state), and as such, buying a home will provide them some tax write-off. It is also estimated that the basic utilities for the home inflating at 5% will cost $160 per month; insurance and maintenance also inflating at 5% will cost $50 per month. The home will appreciate in value about 6% per year. Assuming a nominal interest rate of 15.5%, which alternative will be more attractive to the couple on the basis of the present worth analysis? Note: Realtor's sales commission here is 5%.

13-29 How much will a $20,000 automobile cost ten years from now if inflation continues at an annual rate of 4% for the next ten years?

13-30 Consider two mutually exclusive alternatives stated in Year-0 dollars. Both alternatives have a three year life with no salvage value. Assume the annual inflation rate is 5%, an income tax rate of 25%, and straight line depreciation. The minimum attractive rate of return (MARR) is 7%. Using rate of return analysis determine which alternative is preferrable.

Year	A	B
0	-$420	-$300
1	200	150
2	200	150
3	200	150

13-31 You are considering purchasing an annuity for $15,000 which pays $2500 per year for the next ten years. You want to have a real rate of return of 5%, and you estimate inflation will average 6% per year over the next ten years. Should you buy the annuity?

13-32 Given the data below calculate the present worth of the investment.
First Cost = $60,000   Project Life = 10 years   Salvage Value = $15,000   MARR = 25%
General Price Inflation = 4% per year
Annual Cost #1 = $4500 in year 1 and inflating at 2.5% per year
Annual Cost #2 = $7000 in year 1 and inflating at 10.0% per year
Annual Cost #3 = $10,000 in year 1 and inflating at 6.5% per year
Annual Cost #4 = $8500 in year 1 and inflating at -2.5% per year

13-33 General price inflation is estimated to be: 3% for the next 5 years, 5% the five years after that, and 8% the following five years. If you invest $10,000 at 10% for those 15 years, what is the future worth of your investment: **(a)** in terms of actual dollars at that time, **(b)** in terms of real base zero dollars at that time?

13-34 Some information about a Professor Salary Index (PSI) is tabulated below.

Professor Salary Index (PSI) of Professor Salaries

Year	Professor Salary Index	Percent Change in Professor Salary Index
1991	82	3.22%
1992	89	8.50%
1993	100	*a*
1994	*b*	4.00%
1995	107	*c*
1996	116	*d*
1997	*e*	5.17%
1998	132	7.58%

 a. Calculate the unknown quantities *a, b, c, d, e* in the table above. Review Equation 13-3.
 b. What is the *base year* of the PSI? How did you determine it?
 c. Given the data for the PPI above, calculate the *average annual price increase* in salaries paid to professors for the following period of years: **(i)** 1991 - 1995, **(ii)** 1992-1998

13-35 Inflation is a reality for the general economy of the United States of America for the foreseeable future. Given this assumption, calculate the number of years it will take for the purchasing power of today's dollars to equal *one-fifth* of their present value. Assume that inflation will average 6% per year.

13-36 From the data in Table 13-1 in the text calculate the *average annual inflation rate* of first class postage as measured by the LCI for the following years:

 a. End of year 1970 to end of year 1979
 b. End of year 1980 to end of year 1989
 c. End of year 1990 to end of year 1998

13-37 From the data in Table 13-1 in the text calculate the *overall rate change* of first class postage as measured by the LCI for the following decades:

 a. The 1970's (1970-1979)
 b. The 1980's (1980-1989)
 c. The 1990's (1990-1998 best we can do for now)

13-38 From the data in Table 13-2 in the text calculate the *average annual inflation rate* as measured by the CPI for the following years:

a. End of year 1978 to end of year 1982
b. End of year 1980 to end of year 1989
c. End of year 1985 to end of year 1997

13-39 Ima Luckygirl recently found out that her grandfather passed away and has left her his Rocky Mountain Gold savings account. The account was originally opened 50 years ago when Ima's grandfather deposited $2,500. He had not added to or subtracted from the account since then. If the account has earned an average rate of 10% per year and inflation has been 4% per year, answer the following:

   a. How much money is currently in the account in *actual dollars*?
   b. Express the answer to part a. in terms of purchasing power of dollars from 50 years ago.

13-40 Auntie Frannie wants to provide tuition for her twin nephews to attend a private school. She anticipates sending a check of $2000 at the end of each of the next eight years to apply to the cost of schooling.

   a. If general price inflation, as well as tuition price inflation, is expected to average 5% per year for those eight years calculate the present worth of her gifts. Assume the real interest rate will be 3% per year.
   b.   If Auntie Frannie wants her gifts to keep pace with inflation, what would be the present worth of her gifts? Again assume inflation is 5% and the real interest rate is 3%.

13-41 As a recent graduate, you are considering employment offers from three different companies. However, in an effort to confuse you and perhaps make their offers seem better, each company has used a different *purchasing power base* for expressing your annual salary over the next five years. If you anticipate inflation to be 6% for the next five years and your personal MARR is 8%, which plan would you choose?

Company A: A constant $50,000 per year in terms of today's purchasing power.
Company B: $45,000 the first year with increases of $2500 per year thereafter.
Company C: A constant $65,000 per year in terms of year-five based purchasing power.

13-42 Calculate the future equivalent in year 15 of:
   a. Today's purchasing power dollars.
   b. Then-current purchasing power dollars, of $10,000 today. Use a market interest rate of 15% and an inflation rate of 8%.

13-43 Homeowner Henry is building a fireplace for the house he is constructing. He estimates that his fireplace will require 800 bricks. Answer the following:

   a. If the cost to purchase a chimney brick in 1978 was $2.10, calculate the material cost of Henry's project in 1998. The Chimney Brick Index (CBI) was 442 in 1970 and is

Chapter 13 Inflation and Price Change

expected to be 618 in 1998.

b. Estimate the material cost of a similar fire place to be built in the year 2008. What assumption did you make?

13-44 If a composite price index for the cost of vegetarian foods called *Eggs, Artichokes and Tofu* (EAT) was at a value of 330 ten years ago, and has averaged an increase of 12% per year after that, calculate the current value of the index.

13-45 Granny Viola has been saving money in the Bread & Butter mutual fund for 15 years. She has been a steady contributor to this fund over those years and has a pattern of putting $100 into the account every three months. If her original investment was $500 15 years ago and interest in the account has varied as below, what is the current value of her savings?

Years	Interest Earned in the Account
1-5	12% compounded quarterly
6-10	16% compounded quarterly
10-15	8% compounded quarterly

13-46 As the owner of Beanie Bob's Basement-Brewery, you are interested in a construction project to increase production to offset competition from your crosstown rival Bad Brad's Brewery and Poolhall. Construction cost percentage increases, as well as current cost estimates, for required construction costs are given in the table below over a three year period. Use a market interest rate of 25%, and assume that general price inflation is 5% over the three year period.

Cost of Item per Year	Cost if Incurred Today	Cost Percentage Increase in Year 1	Year 2	Year 3
Structural Metal/Concrete	$ 120,000	4.3%	3.2%	6.6%
Roofing Materials	14,000	2.0	2.5	3.0
Heating/Plumbing Equipment/Fixtures	35,000	1.6	2.1	3.6
Insulation Material	9,000	5.8	6.0	7.5
Labor	85,000	5.0	4.5	4.5

a. What would the costs be for labor in years one, two and three?
b. What is the *average percentage increase* of labor cost over the three year period?
c. What is the present worth of the insulation cost of this project?
d. Calculate the future worth of the labor and insulation material cost portion of the project.
e. Calculate the present worth of the total construction project for Beanie Bob.

CHAPTER 14

Estimation Of Future Events

Economic analysis, if it is to be of any value, must concern itself with present and future consequences. We know that a post audit is desirable to see how well prior estimates were actually achieved, but that is not the central thrust of economic analysis. Our task is to take the present situation and our appraisal of the future and make sound decisions based upon them. This is probably a lot easier said than done. It be relatively easy to determine the present situation; but it is not at all easy to look into the future and appraise it accurately.

One ironic aspect of an engineering education is that engineers are well trained to evaluate a situation analytically; yet there is very little attention given to learning how to predict the future. This seems strange when we recognize that engineering inevitably deals with both the present and the future. Engineering—including economic analysis—inevitably *requires* the estimation of future

In this chapter we consider the problem of evaluating the future. The easiest way to begin is to make a careful estimate. Then we examine the possibility of predicting a range of possible outcomes. Finally, we consider the situation where the probabilities of the various outcomes are known or may be estimated.

Precise Estimates

In an economic analysis, we need to evaluate the future consequences of an alternative. While that cannot be easy, it must be done. In practically every chapter of this book, there are cash flow tables where the costs and benefits for future years are precisely described. Do we really believe that we can exactly foretell a future cost? No one believes he can predict the future with certainty. Instead, the goal is to select a single value which represents the *best* estimate that can be made of the future.

Once our best estimates are made of the various future consequences, we must put them into the economic analysis. Yet the way they are used is really quite different from the way they were determined.

In estimating the future consequences, we recognize that they are not precise and that the actual values will be somewhat different from our estimates. Once these estimates are entered into the economic analysis itself, however, it is likely that we will have proceeded on the tacit assumption that these estimates *are* correct. We know the estimates will not always turn out to be correct; yet we treat them like facts once they are *in* the economic analysis. This can lead to trouble. If actual costs and benefits are different from the estimates, an undesirable alternative could be selected. This is because the variability of the future consequences is concealed by assuming that the best estimates will actually occur. The problem is illustrated by Example 14-1.

EXAMPLE 14-1

Two alternatives are being considered. The best estimates for the various consequences are as follows:

	A	*B*
Cost	$1000	$2000
Net annual benefit	150	250
Useful life, in years	10	10
End-of-useful-life salvage value	100	400

If interest is 3½%, which alternative should be selected?

Solution:

Alternative A:

$$NPW = -1000 + 150(P/A, 3\frac{1}{2}\%, 10) + 100(P/F, 3\frac{1}{2}\%, 10)$$

$$= -1000 + 150(8.317) + 100(0.7089)$$

$$= -1000 + 1248 + 71 = +\$319$$

Alternative B:

$$NPW = -2000 + 250(P/A, 3\frac{1}{2}\%, 10) + 400(P/F, 3\frac{1}{2}\%, 10)$$

$$= -2000 + 250(8.317) + 400(0.7089)$$

$$= -2000 + 2079 + 284 = +\$363$$

Alternative B, with its larger NPW, would be selected.

Alternate Formation of Ex. 14-1. Suppose that at the end of ten years, the actual salvage value turns out to be $300 instead of the $400 best estimate. If all the other estimates were correct, is B still the preferred alternative?

Solution: Corrected B:

$$\text{NPW} = -2000 + 250(P/A, 3\frac{1}{2}\%, 10) + 300(P/F, 3\frac{1}{2}\%, 10)$$

$$= -2000 + 250(8.317) + 300(0.7089)$$

$$= -2000 + 2079 + 213 = +\$292$$

Under these circumstances, A is now the preferred alternative. ∎

Example 14-1 shows that the change in the salvage value of Alternative B actually results in a change of preferred alternative.

EXAMPLE 14-2

Using Example 14-l data, compute the sensitivity of the decision to the Alt. B salvage value. For Alt. A, NPW = +319. For breakeven between the alternatives,

$$\text{NPW}_A = \text{NPW}_B$$

$$+319 = -2000 + 250(P/A, 3\frac{1}{2}\%, 10) + \text{Salvage value}_B(P/F, 3\frac{1}{2}\%, 10)$$

$$= -2000 + 250(8.317) + \text{Salvage value}_B(0.7089)$$

At the breakeven point

$$\text{Salvage value}_B = \frac{319 + 2000 - 2079}{0.7089} = \frac{240}{0.7089} = \$339$$

For Alt. B salvage value > $339, B is preferred;

salvage value < $339, A is preferred. ∎

Breakeven and sensitivity analysis provide one means of examining the impact of the variability of some estimate on the outcome. It helps by answering the question, "How much variability can a parameter have before the decision will be affected?" This approach does not, however, answer the basic problem of how to take the inherent variability of parameters into account in an economic analysis. This will be considered next.

A Range Of Estimates

Realistically, the *true* situation is that there is a range of possible values for most parameters. One could, for example, construct values for an *optimistic* estimate, the *most likely* estimate, and a *pessimistic* estimate. Then the economic analysis could be performed on each set of data to determine if the decision is sensitive to the range of projected values.

EXAMPLE 14-3
A firm is considering an investment. Three estimates for the various parameters are as follows:

	Optimistic value	Most likely value	Pessimistic value
Cost	$1000	$1000	$1052
Net annual benefit	200	198	190
Useful life, in years	12	12	9
End-of-useful-life salvage value	100	0	0

If a 10% before-tax minimum attractive rate of return is required, is the investment justified under all three estimates? Compute the rate of return for each estimate.

Solutions to Ex. 14-3.

Optimistic estimate:

   PW of cost = PW of benefit

$$\$1000 = 200(P/A,i,12) + 100(P/F,i,12)$$

Try $i = 18\%$:

$$\$1000 \doteq 200(4.793) + 100(0.1372) = 973$$

18% is too high. $i^* \approx 17\%$.

Most likely estimate:

$$\$1000 = 198(P/A,i,12)$$

$$(P/A,i,12) = \frac{1000}{198} = 5.05$$

From Compound Interest Tables, $i^* = 16.7\%$.

Pessimistic estimate:

$$\$1052 = 190(P/A,i,9)$$

$$(P/A, i, 9) = \frac{1052}{190} = 5.54$$

From Compound Interest Tables, $i^* = 11\%$.

From the calculations we conclude that the rate of return for this investment is most likely to be 16.7%, but might range from 11% to 17%. The investment meets the 10% MARR criterion for all estimates. ■

Example 14-3 required that three separate calculations be made for the investment, one each for the optimistic values, the most likely values, and the pessimistic values. This approach emphasizes the unlikely situations where all parameters prove to be very favorable or very unfavorable. Neither is likely to happen. Rather, there is likely to be a blend of results, with the parameters assuming values near to the most likely estimate, but with due consideration for the possible range of values. One way to accomplish this is to estimate an average or mean[1] value for each parameter, based on the following weighting factors:[2]

$$\text{Mean value} = \frac{\text{Optimistic value} + 4(\text{Most likely value}) + \text{Pessimistic value}}{6}$$

This approach is illustrated in Ex. 14-4.

EXAMPLE 14-4

Solve Ex. 14-3 by applying the weighting factors given above. Compute the resulting mean rate of return.

Solution:

$$\text{Mean cost} = \frac{1000 + 4(1000) + 1052}{6} = 1009$$

$$\text{Mean net annual benefit} = \frac{200 + 4(198) + 190}{6} = 197$$

$$\text{Mean useful life} = \frac{12 + 4(12) + 9}{6} = 11.5 \text{ years}$$

$$\text{End - of - useful - life salvage value} = \frac{100}{6} = 17$$

---

[1]The mean value is defined as the sum of all the values divided by the number of values.

[2]If you are interested, these weighting factors represent an approximation of the beta distribution.

Compute the mean rate of return:

PW of cost = PW of benefit

$$\$1009 = 197(P/A,i,11.5) + 17(P/F,i,11.5)$$

From Interest Tables, the mean rate of return is approximately 16%. ■

Example 14-4 gave a mean rate of return (16%) that was different from the most likely rate of return (16.7%) computed in Example 14-3. The immediate question is, "Why are these values different?" The reason for the difference can be seen from the way the two values were calculated. One rate of return was based exclusively on the most likely values. The mean rate of return, on the other hand, took into account not only the most likely values, but also the variability of the parameters.

In examining the data we see that the pessimistic values are further away from the most likely values than are the optimistic values. This causes the resulting weighted mean values to be less favorable than the most likely values. As a result, the mean rate of return, in this example, is less than the rate of return based on the most likely values.

Probability And Risk

Probability can be considered to be the long-run relative frequency of occurrence of an outcome. There are just two possible outcomes from flipping a coin (a Head or a Tail). If, for example, a fair coin is flipped over and over, we can expect in the long run that half the time Heads will appear and half the time Tails. We would say the probability of flipping a Head is 0.50 and of flipping a Tail is 0.50. Since probabilities are defined so that the sum of probabilities for all possible outcomes is 1, the situation is

Probability of flipping a Head = 0.50

Probability of flipping a Tail = 0.50

Sum of all possible outcomes = 1.00

A more complex situation is given in the following example.

EXAMPLE 14-5

If one were to roll one fair die (that is, one-half of a pair of dice), what is the probability that either a 1 or a 6 would result?

Solution: Since a die is a perfect six-sided cube, the probability of any side appearing is 1/6.

Probability of rolling a 1 = $P(1)$ = 1/6
        2 = $P(2)$ = 1/6
        3 = $P(3)$ = 1/6
        4 = $P(4)$ = 1/6
        5 = $P(5)$ = 1/6
        6 = $P(6)$ = 1/6
  Sum of all possible outcomes = 6/6 = 1

The probability of rolling either a 1 or a 6 = 1/6 + 1/6 = 1/3. ∎

In the two examples, the probability of each outcome was the same. This need not be the case.

EXAMPLE 14-6

In the game of Blackjack, a perfect hand is a Ten or Facecard plus an Ace. What is the probability of being dealt a Ten or a Facecard from a newly shuffled deck of 52 cards? What is the probability of being dealt an Ace in this same situation?

Solution: The three outcomes being examined are to be dealt a Ten or a Facecard, an Ace, or some other card. Every card in the deck represents one of these three possible outcomes. There are 4 Aces; 16 Tens, Jacks, Queens, and Kings; and 32 other cards.

The probability of being dealt a Ten or a Facecard = 16/52 = 0.31 ∎

The probability of being dealt an Ace = 4/52 = 0.08 ∎

The probability of being dealt some other card = 32/52 = 0.61

             1.00

The term *risk* has a special meaning when it is used in statistics. It is defined as a situation where *there are two or more possible outcomes and the probability associated with each outcome is known.*

In each of the two previous examples there is a risk situation. We could not know in advance what playing card would be dealt or what number would be rolled by the die. However, since the various probabilities could be computed, our definition of risk has been satisfied.

Probability and risk are not restricted to gambling games. For example:

In a particular engineering course, a student has estimated (subjectively) the probability for each of the letter grades he might receive as follows:

	Outcome	
Grade	*Grade point*	*Probability* *P(grade)*
A	4.0	0.10
B	3.0	0.30
C	2.0	0.25
D	1.0	0.20
F	0.0	<u>0.15</u>
		1.00

From the table we see that the grade with the highest probability is B. This, therefore, is the most likely grade.

   We see from the table that there is a substantial probability that some grade other than B will be received. And the probabilities indicate that if a B is not received, the grade will probably be something less than a B. But in saying the most likely grade is a B, these other outcomes are ignored. In the next section we will show that a composite statistic may be computed using all the data.

Expected Value

In the example just discussed, we saw that the most likely grade of B in an engineering class had a probability of 0.30. That is not a very high probability. In some other course, like a math class, we might estimate a probability of 0.65 of obtaining a B, making the B the most likely grade. While a B is most likely in both the classes, it is more certain in the math class.

   In the early part of this chapter, we computed a weighted mean to give a better understanding of the total situation as represented by various possible outcomes. We can do the same thing here. An obvious selection of the weighting factors is to use the probabilities for the various outcomes. Then, since the sum of the probabilities equals 1, the computation is:

$$\textbf{Weighted mean} = \frac{\textbf{Outcome}_A \times P(A) + \textbf{Outcome}_B \times P(B) + ...}{1}$$

When the probabilities are used as the weighting factors, we call the result the ***expected value*** and write the equation as:

$$\textit{Expected value} = \textbf{Outcome}_A \times P(A) + \textbf{Outcome}_B \times P(B) + \cdots$$

EXAMPLE 14-7
Compute the student's expected grade in the engineering course using the probabilities given in the text.

Solution:

Grade	Grade point	P(grade)	Grade point × P(grade)
A	4.0	0.10	0.40
B	3.0	0.30	0.90
C	2.0	0.25	0.50
D	1.0	0.20	0.20
F	0.0	0.15	0
		Expected (GP) =	2.00

The expected grade point (GP) of 2.00 indicates a grade of C. ■

From the calculations in Ex. 14-7, we find that for a given set of probabilities, the most likely grade is B and the expected grade is C. How can we resolve these conflicting results? First, the results are correct. The most likely grade *is* B and the expected grade *is* C.

The two values tell us different things about the probabilities of the outcomes.

For example, suppose 1000 students took courses in which they each believed the distribution of probabilities in Ex. 14-7 was correct. Each person could correctly state that his most likely grade would be B, with an expected grade of C. If the projected probabilities proved to be correct, suppose the 1000 students received grades as follows:

Grade	Number of students
A	100
B	300
C	250
D	200
F	150
	1000

We would note immediately that only 300 students received B grades; most students received some other grade. If the average grade were computed, what would it be?

To average A, B, C, D, and F, we would assign the numerical values 4, 3, 2, 1, and 0. The computation is:

Grade	Number of students	Grade × students
A = 4.0	100	400
B = 3.0	300	900
C = 2.0	250	500
D = 1.0	200	200
F = 0.0	150	0
	Sum =	2000

$$\text{Average grade} = \frac{\text{Sum}}{\text{Number of Students}} = \frac{2000}{1000} = 2.0$$

The average grade is C.

We recognize that the average grade is exactly the expected grade. This helps us to understand expected value. If this situation were to occur over and over again, then the accumulated results will approach the expected value. This assumes, of course, that the given probability distribution is correct. Nevertheless, B remains the most likely grade to be received.

EXAMPLE 14-8

Just before a horse race is about to begin, a spectator decides that the situation on the four-horse race is as follows:

Horse	Probability of winning	Outcome of a $10 bet if horse wins*
1	0.15	$48.00
2	0.15	58.00
3	0.50	16.50
4	0.20	42.00
	1.00	

*In horserace betting, a ticket is purchased for $10. The outcome of a winning ticket represents the refund of the $10 bet plus the amount won.

What is the expected value of the ticket if he bets $10 on Horse #3 to win?

Solution:

Expected value = Outcome if #3 wins × $P(\text{Win}_3)$

+ Outcome if #3 loses × $P(\text{Loss}_3)$

= $16.50(0.50) + $0(0.50) = $8.25

Thus a ticket purchased for $10 has an expected value of $8.25. (Is this the way to get rich?) One also notes that there is no way for the bettor to actually win the expected value. He must either win $16.50 or nothing. There are no other possibilities. Figure 14-1 illustrates the situation. In this example, we find that betting $10 on the favorite horse (the one with the highest probability of winning) could be expected to result in a net loss to the bettor.

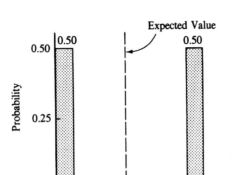

Figure 14-1 Example 14-8 situation.

Alternate Formation of Ex. 14-8: An alternate strategy would be to bet $10 on each of the horses to win. In this way one is certain to have a winning ticket. What will be the expected value of the four tickets for this betting scheme?

Solution:

Expected value = $48.00(0.15) + $58.00(0.15) + $16.50(0.50)

$$+ \$42.00(0.20)$$

$$= \$7.20 + 8.70 + 8.25 + 8.40$$

$$= \$32.55$$

Figure 14-2 is a plot of this betting plan.

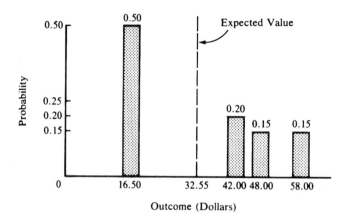

Figure 14-2 Example 14-8, alternate formation.

Regarding the two strategies offered in this example, which betting scheme is better?

Expected result from one $10 bet = $8.25 − $10.00 = $1.75 loss

Expected result from four $10 bets = $32.55 − $40.00 = $7.45 loss

On this basis, the single $10 bet is preferred. Of course, it is clear that the best decision would be to make no bet at all. (Or would it?)   ■

Expected value may be a useful analysis tool in certain kinds of situations, for it represents the long-term outcome if the particular situation occurs over and over again. If one were to put money into a slot machine and play it over and over for an hour or two, the expected value is a rather accurate estimate of the average results from playing the slot machine a couple of hundred times or more. But would it be useful in estimating the results from playing the machine once? Obviously not. The most likely result would be the loss of the coin. Much less frequently, one might receive three or more coins back.

Thus the expected value (at possibly 0.75 coin returned per play) is not very useful when we are trying to evaluate a situation that is not repeated. Example 14-8 represented a one-time event. A particular horserace will be run once. Thus the expected value cannot tell us much about the outcome from a particular race; it does say that people who continue to bet on horseraces (or play slot machines) must expect to lose money over the long term.

Distribution Of Outcomes

In the three previous sections, we have considered ways of treating situations where the outcomes vary. There was no discussion of the distributions of the outcomes as in Figures 14-1 and 14-2. In this section, we will examine two specific distributions (uniform and normal) and describe how to randomly sample from these or any other distributions. The ability to randomly sample from a distribution is prerequisite for the discussion of simulation in the following section.

Uniform Distribution

In Ex. 14-5 we saw that a die, being a perfect six-sided cube, can be rolled to provide each of six numbers with an equal probability of $\frac{1}{6}$. The distribution is discrete since only integers appear and values in between are not possible. When the outcomes have an equal probability, this is called a ***uniform distribution*** and is illustrated by Fig. 14-3.

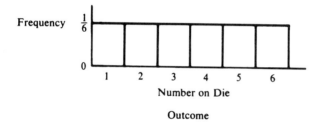

Figure 14-3 Uniform distribution—outcomes of a die.

When we would like to know what the outcome is from the roll of a die, the easiest way is simply to roll the die. On the other hand, we could simulate rolling the die in a variety of ways. One way would be to make up six slips of paper and assign the numbers I through 6 to them. Then shake them in a hat and choose one of them at random. In this manner, we would have simulated the rolling of a die. Since there are six papers in the hat and they are shaken up so each piece of paper is equally likely to be selected, we really have created another uniform distribution—Figure 14-4.

It is easy to use six pieces of paper and create a uniform distribution with outcomes of 1 through 6. A more general approach to simulating the uniform distribution can now be devised.

Suppose someone puts each of the possible two-digit numbers—00 through 99—on slips of paper and put the 100 slips in a hat. Then after careful mixing, one is chosen at random and the two-digit value written down. The slip of paper is returned to the hat and the process

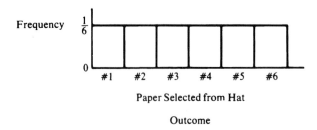

Paper Selected from Hat

Outcome

Figure 14-4 Uniform distribution—papers in a hat.

repeated. Table 14-1 is a sample of random two-digit numbers. A careful examination of Table 14-1 shows that some two-digit numbers appear several times and some do not appear at all. This is, of course, possible for the slips of paper were replaced in the hat after they were drawn.

Table 14-1   **150 TWO-DIGIT RANDOM NUMBERS**

77	41	71	94	66	14	16	77	50	17	65	61	85	23	15
06	40	75	90	34	75	45	34	96	74	34	92	24	52	99
46	19	39	60	20	50	05	80	16	33	79	03	27	22	37
02	66	67	29	92	03	24	58	08	05	56	57	47	72	02
68	40	65	20	78	69	76	30	39	21	00	22	40	61	19
43	10	23	08	48	55	14	45	52	22	90	71	15	89	04
80	87	71	81	45	19	30	72	88	08	44	24	18	41	97
35	17	82	18	84	00	77	87	38	83	42	38	55	17	31
87	21	94	49	66	74	96	10	70	09	76	34	21	06	55
15	69	32	47	88	87	14	99	19	27	41	61	40	53	03

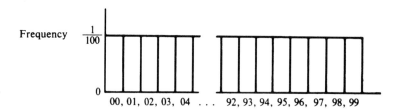

00, 01, 02, 03, 04 . . .   92, 93, 94, 95, 96, 97, 98, 99

Figure 14-5 Uniform distribution—two-digit random numbers.

The table of random numbers represents random sampling from a uniform distribution of values from 00 through 99 as depicted by Fig. 14-5. We can use the table of random numbers to simulate the outcome of a die by assigning $\frac{1}{6}$ of the random numbers to a 1 on the die, another $\frac{1}{6}$ of the random numbers to a 2 on the die, and so forth. Since $\frac{100}{6}$ equals 16.6, it will not come out even. We will let 16 random numbers represent each face of the die. The assignment of numbers is as follows:

Random numbers	*Outcome on the die*
00–15	1
16–31	2
32–47	3
48–63	4
64–79	5
80–95	6
96–99	Random numbers not used

Graphically, the situation would look like Fig. 14-6.

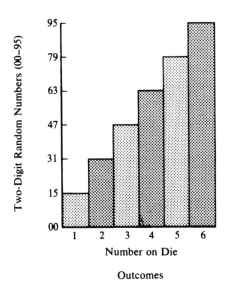

Figure 14-6 Plot of outcome on die *vs.* random numbers.

EXAMPLE 14-9

Using the table of two-digit random numbers, Table 14-1, and Figure 14-6, simulate the rolling of a die three times.

Solution: We first obtain two-digit random numbers from Table 14-1. The proper way to use a table, of random numbers is to enter the table at some random point and read in a consistent sequential manner. Following this method, we might read the numbers 50, 96, and 16 from the ninth column in the table. Enter the ordinate of Figure 14-6 graph with 50 as the random number. Read from the y-axis across to the stair-step curve and read down to 4 on the x-axis. This indicates that the first roll of the die is a 4.

The second random number, 96, is not being used in the simulation. One should discard the number and go on to the next random number selected. A random number of 16 corresponds to a 2 on the die. Another random number must be selected from Table 14-1. The selected 08 corresponds to 1 on the die. Using a table of random numbers we have simulated the rolling of a die with the numbers 4, 2, and 1 having been selected. ■

With the technique illustrated in Ex. 14-9, any uniform distribution may be simulated with a table of random numbers.

Normal Distribution

Possibly the best known of all distributions is the normal distribution. It is defined by the equation:

$$y(x) = \frac{1}{\sigma\sqrt{2\pi}} \exp{-\frac{1}{2}\left(\frac{x-\mu}{\sigma}\right)}$$

where μ = mean, σ = standard deviation

Since all values are possible, the normal distribution is a continuous distribution. We see that a particular normal distribution is defined by its mean μ and standard deviation σ. Different values of μ and σ give normal distributions that look different from one another. Two normal distributions are shown in Fig. 14-7.

Because the equation defining the normal distribution is cumbersome to use, data are usually obtained from a table of the distribution with mean μ equal to 0 and standard deviation σ equal to 1. For continuous distributions the area under the curve equals the probability. As the sum of the probabilities for all possible outcomes is 1, the total area under the normal curve is 1. We can find the area under portions of the distribution from Fig. 14-8. The figure shows that 68.3% of the area under the normal curve lies between $(\mu - 1.0\sigma)$ and $(\mu + 1.0\sigma)$.

Thus in any situation, we could expect a normally distributed variable to be within this range about 68.3% of the time.

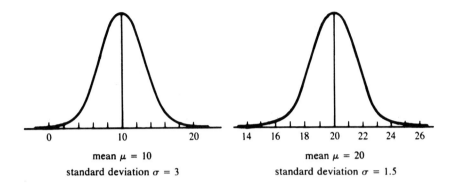

Figure 14-7 Two normal distributions with different means and standard deviations.

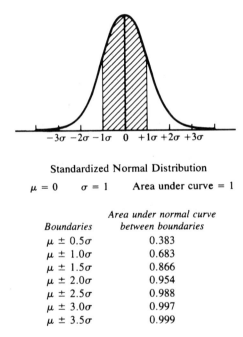

Standardized Normal Distribution

$\mu = 0$ $\sigma = 1$ Area under curve $= 1$

Boundaries	Area under normal curve between boundaries
$\mu \pm 0.5\sigma$	0.383
$\mu \pm 1.0\sigma$	0.683
$\mu \pm 1.5\sigma$	0.866
$\mu \pm 2.0\sigma$	0.954
$\mu \pm 2.5\sigma$	0.988
$\mu \pm 3.0\sigma$	0.997
$\mu \pm 3.5\sigma$	0.999

Figure 14-8 Area under the normal distribution curve.

A comment about the notation is in order. When one is referring to the normal distribution of some *population* (like the ages of *all* college students) we say the mean is μ and the standard deviationσ. But when we are referring to a *sample* of ages of college students, we say the mean is $\bar{x}$ and the standard deviation is s. Thus

	Ages of All college students	Ages of Sample of college students
Mean	μ	$\bar{x}$
Standard deviation	σ	s

EXAMPLE 14-10

The ages of twenty college students are as follows:

$$21 \quad 22 \quad 24 \quad 23 \quad 25 \quad 23 \quad 22 \quad 22 \quad 26 \quad 22$$
$$23 \quad 21 \quad 23 \quad 24 \quad 25 \quad 23 \quad 24 \quad 23 \quad 22 \quad 24$$

For this group of people, it is believed their ages are normally distributed. Compute the mean and standard deviation for the twenty students.

Solution: If all the ages are added, their sum Ex = 462. The mean of this sample of twenty students is

$$\textbf{Mean } \bar{x} = \frac{\sum x}{n} = \frac{462}{20} = \textbf{23.1 years}$$

The standard deviation was once called the *root-mean-square deviation,* for this is one of the methods of its calculation. The *standard deviation* is the square root of the sum of the square of the deviations about the mean, divided by the sample size minus 1, or

$$\textbf{Standard deviation } s = \sqrt{\frac{\sum(x - \bar{x})^2}{n - 1}}$$

The twenty ages may be grouped and this equation used to find the standard deviation.

Number in age group, n	Age group, x	$(x - \bar{x})$	$(x - \bar{x})^2$	$n(x - \bar{x})^2$
2	21	−2.1	4.4	8.8
5	22	−1.1	1.2	6.0
6	23	−0.1	0.0	0.0
4	24	+0.9	0.8	3.2
2	25	+1.9	3.6	7.2
1	26	+2.9	8.4	8.4
20			$\sum(x - \bar{x})^2$	$= 33.6$

$$s = \sqrt{\frac{\sum(x - \bar{x})^2}{n - 1}} = \sqrt{\frac{33.6}{19}} = 1.33$$

It is frequently easier to solve for the standard deviation when the equation is rewritten as follows:

$$\text{Standard deviation } s = \sqrt{\frac{\sum x^2}{n-1} - \frac{(\sum x)^2}{n(n-1)}}$$

In this problem,

$$\sum x^2 = 21^2 + 23^2 + 22^2 + 21^2 + 24^2 + \cdots = 10{,}706$$

$$\sum x = 462$$

$$s = \sqrt{\frac{10{,}706}{19} - \frac{462^2}{20(19)}} = \sqrt{563.47 - 561.69} = \sqrt{1.78}$$

$$= 1.33$$

We have computed for the twenty college students $\bar{x} = 23.1$, $s = 1.33$. The sample mean and standard deviation are our best estimates of the population. We therefore estimate that $\mu = 23.1$ and $\sigma = 1.33$. ■

To define a particular normal distribution it is necessary to specify the mean μ and standard deviation σ. Suppose, for example, you believed that the useful life of a particular type of equipment was normally distributed with a mean life of 15 years and a standard deviation of 2.4 years. How could you obtain random samples from this distribution?

A convenient method is based on the standardized normal distribution. Figure 14-9 shows both the standardized normal distribution and the particular useful life normal distribution we want. The deviation of any value of x from the mean of the distribution may be expressed in number of standard deviations.

$$z = \frac{x - m}{s} \quad \text{or} \quad x = z\,s + m$$

From this equation, we see that any Point x on a specific distribution (with μ and σ) has an equivalent Point z on the standardized normal distribution. This relationship allows us to relate the standardized normal distribution to any other normal distribution.

In our useful life example, two standard deviations above the mean would be at $x = 15 + 2(2.4) = 19.8$ years on the useful life distribution, Fig. 14-9b. The equivalent point on the standardized normal distribution is

$$z = \frac{x - \mu}{\sigma} = \frac{19.8 - 15}{2.4} = +2.0$$

This interrelationship means that if we randomly sample from the standardized normal distribution, we can relate this to an equivalent random sample for any normal distribution.

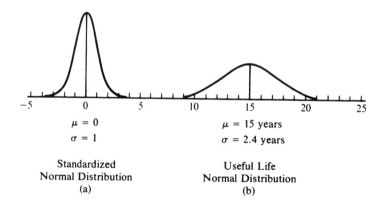

Standardized   Useful Life
Normal Distribution   Normal Distribution
(a)   (b)

Figure 14-9 Standardized normal distribution and a specific useful life normal distribution.

A point on the standardized normal distribution ($\mu = 0$, $\sigma = 1$) is fully defined by specifying the number of standard deviations the point is to the left (negative) or to the right (positive) of the mean. Thus the value + 1.02 would indicate the point is 1.02 standard deviations to the right of the mean.

Table 14-2 represents random samples from the standardized normal distribution. The use of random normal numbers is illustrated in Ex. 14-11.

Table 14-2			100 RANDOM NORMAL NUMBERS (Z)						
−0.22	−0.87	2.32	−0.94	0.63	0.81	0.74	1.08	−1.82	0.07
−0.89	−0.39	0.29	−0.27	1.06	−0.42	2.26	−0.35	1.09	−2.55
0.06	1.28	−1.74	2.47	0.58	0.69	1.41	−1.19	2.37	−0.06
−0.01	−0.40	0.64	−2.22	1.10	0.47	−0.09	−0.35	−0.72	0.30
−0.87	−1.34	0.85	0.27	−1.35	0.58	−1.72	1.88	−0.45	0.82
0.24	0.40	0.50	1.41	−1.95	−0.02	−1.00	−0.20	−1.08	−0.78
−1.05	−0.06	0.27	−0.04	0.99	−0.78	0.46	−1.18	0.37	1.07
−0.57	0.24	−1.02	0.86	0.78	−1.69	−0.11	−0.23	−0.87	−0.45
−1.24	−0.63	0.03	−0.83	0.25	−0.89	−0.77	0.90	−0.27	0.94
2.11	0.78	−1.69	−0.17	0.21	0.48	−2.82	−0.86	1.40	−1.20

EXAMPLE 14-11

On a particular portion of a highway, observations indicate that the speed of automobiles is normally distributed with mean μ = 104 kilometers per hour and standard deviation σ = 11 kilometers per hour. Obtain a random sample of five automobiles on the highway.

Solution: From Table 14-2, randomly select a value of random normal number (z). The value of z selected is − 0.94. This value of z represents a vehicle on the highway travelling at a speed of

$$x = z\sigma + \mu = -0.94(11) + 104 = 93.7 \text{ km/hr}$$

Four other random normal numbers, 1.06, 0.69, −0.09, and 1.88, are selected. The indicated automobile speeds are

$$x = 1.06(11) + 104 = 115.7 \text{ km/hr}$$

$$x = 0.69(11) + 104 = 111.6$$

$$x = -0.09(11) + 104 = 103.0$$

$$x = 1.88(11) + 104 = 124.7$$

Through the use of random normal numbers, the speed of five random automobiles has been computed. ∎

Sampling from Any Distribution Using Random Numbers

In the two previous sections we saw how to obtain a random sample from the uniform distribution and from the normal distribution. While these are two important situations, there are times when we want to obtain a random sample from some other distribution. We might, for example, wish to obtain a random sample from the grade distribution of Figure 14-10.

The procedure is to replot the data to represent a cumulative distribution of grades. This has been done in Figure 14-11.

Where the x-axis represents all possible outcomes, as in this case, the cumulative probability on the y-axis must vary from 0 to 1.00. To facilitate random sampling, random numbers may be assigned to each segment of the ordinate in proportion to its probability. Figure 14-12 shows the data of Fig. 14-11 with two-digit random numbers assigned to each segment of the y-axis.

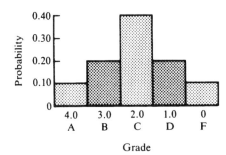

Grade	Grade point average	Probability
A	4.0	0.10
B	3.0	0.20
C	2.0	0.40
D	1.0	0.20
F	0.0	0.10
		1.00

Figure 14-10 Distribution of grades.

To obtain a random sample from the distribution of grades (Figure 14-10), the first step is to randomly select a two-digit number from the table of random numbers, Table 14-1. This number is then entered as the ordinate in Figure 14-12. Read across from the ordinate to the curve and then read the corresponding value on the x-axis. The random number 35, for example, corresponds to a grade of C on the x-axis.

The procedure described for sampling from the grade distribution can be used for any discrete or continuous distribution. In fact, looking back at the discussion of the uniform distribution reveals that we were actually using a cumulative distribution (Fig. 14-6) to randomly sample from the uniform distribution.

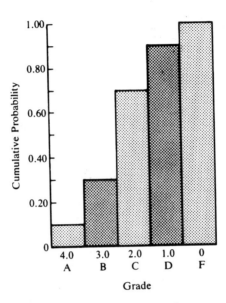

Grade	Probability	Cumulative probability
A	0.10	0.10
B	0.20	0.30
C	0.40	0.70
D	0.20	0.90
F	0.10	1.00

Figure 14-11 Cumulative distribution of grades.

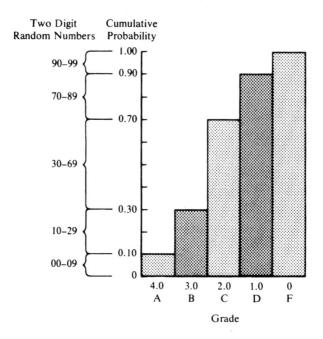

Two Digit Cumulative
Random Numbers Probability

Figure 14-12 Cumulative distribution of grades with two-digit random numbers assigned to the *y*-axis.

Simulation

Simulation may be described as the repetitive analysis of a mathematical model. When we can accurately predict all the various consequences of an investment project with precise estimates, a single computation will give us the results. Sometimes future events are stated in terms of a range of values like optimistic, most likely, and pessimistic estimates. In this circumstance, one approach is to compute the results (like rate of return) for each of the three estimates. Here, for the first time, multiple computations are used to describe the results.

When values for an economic analysis are stated in terms of one or more probability distributions, the analysis is further complicated. The number of different combinations of values quickly becomes so large—often infinite—that one can no longer consider evaluating each possibility. A practical solution is to randomly sample from each of the variables and compute the results. Is a single random sample adequate to portray the situation? The answer is, "Yes and no." Surely we know more about the situation than if we had no random sample, but not much more.

We could compare this with a random sample of a piece of chicken from a bucket full of pieces of chicken. If the one piece is a leg, does that mean the bucket consists only of chicken legs, or that all pieces are as good as the chicken leg? We simply cannot say. The random sample that produced the chicken leg proves at least that there was one in the bucket, but it says very little about the rest of the contents.

To obtain greater information about an economic analysis the procedure is to continue to take additional random samples for evaluation. This technique is simulation. In Ex. 14-9 we simulated the rolling of a die three times. We now see that this was a very simple simulation. Example 14-12 illustrates an *economic analysis simulation.*

EXAMPLE 14-12

If a more accurate scale is installed on a production line, it will reduce the error in computing postage charges and save $250 a year. The useful life of the scale is believed to be uniformly distributed and range from 12 to 16 years. The initial cost of the scale is estimated to be normally distributed with a mean of $1500 and a standard deviation of $150.

Simulate 25 random samples of the problem and compute the rate of return for each sample. Construct a graph of rate of return *vs.* frequency of occurrence.

Solution: The useful life is uniformly distributed between 12 and 16 years. This is illustrated in Fig. 14-13. The data are replotted as a cumulative distribution function in Fig. 14-14.

We can assign 20 two-digit random numbers to each of the five possible useful lives as follows:

Random numbers	Useful life, in years
00–19	12
20–39	13
40–59	14
60–79	15
80–99	16

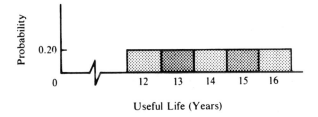

Figure 14-13 Useful life distribution for Example 14-12.

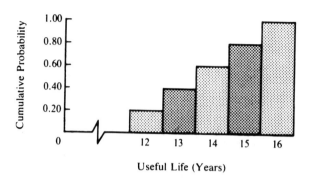

Figure 14-14 Useful life distribution for Example 14-12 replotted as a cumulative distribution function.

For the first random sample, we randomly enter the random number table (Table 14-1) and read the number 82. We previously assigned this number to represent a useful life of 16 years.

Next, to sample from the normal distribution, we read a random normal number of −1.82 from Table 14-2. To obtain the cost of the scale, multiply this value by the standard deviation and add the mean.

Cost of scale = −1.82($150) + $1500 = $1227

With these values computed, the cash flow is:

Year	Cash flow
0	−$1227
1–16	250

For the cash flow, the rate of return may be computed as follows:

PW of cost = PW of benefit

$$1227 = 250(P/A, i, 16) \qquad i^* = 19\%$$

Table 14-3 shows the results of repeating this process to obtain 25 values of rate of return. The rates of return shown in Table 14-3 are plotted in Fig. 14-15. An additional 75 random samples have been computed (but not detailed here) and the data for all 100 samples are shown in Fig. 14-16. ■

Table 14-3 RESULTS OF 25 RANDOM SAMPLES FOR EXAMPLE 14-12

Random number	Useful life, in years	Random normal number	Cost	Computed rate of return
82	16	−1.82	$1227	19%
17	12	1.08	1662	11%
35	13	0.74	1611	12%
87	16	0.81	1622	13%
21	13	−0.42	1437	14%
94	16	0.69	1604	14%
49	14	0.47	1570	13%
66	15	1.10	1665	12%
74	15	−2.22	1167	20%
96	16	0.64	1596	14%
10	12	−0.40	1440	14%
70	15	−0.01	1498	14%
09	12	0.06	1509	13%
76	15	1.28	1692	12%
34	13	−0.39	1442	14%
21	13	−0.87	1370	15%
06	12	2.32	1848	8%
55	14	0.29	1544	13%
03	12	−0.27	1460	13%
53	14	1.06	1659	12%
40	14	−0.42	1437	15%
61	15	2.26	1839	11%
41	14	1.41	1712	11%
27	13	−0.09	1486	14%
19	12	−1.72	1242	17%

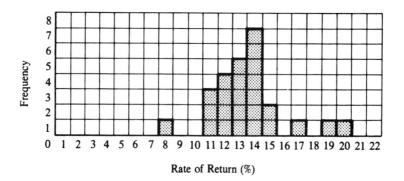

Figure 14-15 Graph of rate of return *vs.* frequency for 25 random samples in Example 14-12.

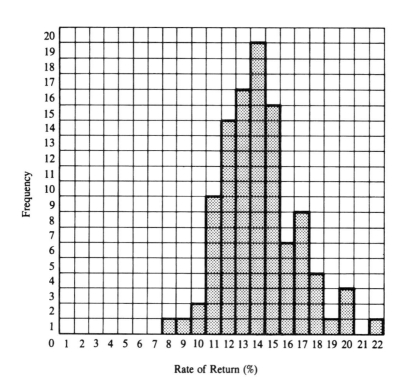

Figure 14-16 Graph of rate of return *vs.* frequency for 100 random samples in Example 14-12.

Instead of recognizing the variability of the cost of the scale in Ex. 14-12, or its useful life, we might have used the single best estimates for them. The problem would have become:

Cost of scale	$1500
Annual benefit	250
Useful life, in years	14

These values yield a 14% rate of return. Thus, all our elaborate computations in this case did not change the results very much. The computations, however, give a picture of the possible variability of the results. Without recognizing the variability of the data and doing the simulation computations we would have no way of knowing the prospective variation in the rate of return. Simulation may add a dimension to our knowledge of a problem. With it we can learn about a projected result and also the likelihood that the result or some other one will occur.

From Example 14-12, it is apparent that if one goes from a situation of best estimates to one of probability distributions, the computations are multiplied about a hundred times. Fortunately, all of these calculations may be performed on a computer.

Simulation with Spreadsheets

Example 14-12 illustrated simulation using a manual approach. There are several software packages that are "add-ins" to spreadsheets that automate this process. These packages typically have a set of probability distributions which can be used to describe the uncertainty in any cell's value. The function for the uniform, normal, exponential, lognormal, binomial, or other probability distributions is simply entered into the cell. For most packages a practically unlimited number of cells can be described with these distributions.

Another cell, such as the project's present worth or internal rate of return, is selected as the output variable. Then a simulation of 10 to 10,000 iterations can be run. For each iteration, the computer automatically selects random numbers for each cell, recalculates the spreadsheet, and records the result of the output variable(s). After all iterations are done, the software displays the results for the output variable(s).

Example 14-13 uses @Risk to do 1000 iterations for the data in Example 14-12.

EXAMPLE 14-13

A more accurate scale for a production line will save $250 annually. Its life is uniformly distributed between 12 and 16 years, and its first cost is normally distributed with a mean of $1500 and a standard deviation of $150. Generate 1000 iterations and construct a frequency distribution for the scale's rate of return.

Solution
The first IRR of 14.01% that is computed in Figure 14-17 is based on the average life and the average first cost. The second IRR of 14.01% is computed by @Risk using the average of each

distribution. This is an example of how spreadsheets with @Risk functions always display the results of using average values.

The RATE function contains two @Risk functions: RiskUniform and RiskNormal. The uniform distribution has the minimum and maximum values as parameters. The normal distribution has the average and standard deviation as parameters.

The graph in Figure 14-17 with 1000 iterations is much smoother than the graphs from Example 14-12 where 25 and 100 iterations were done.

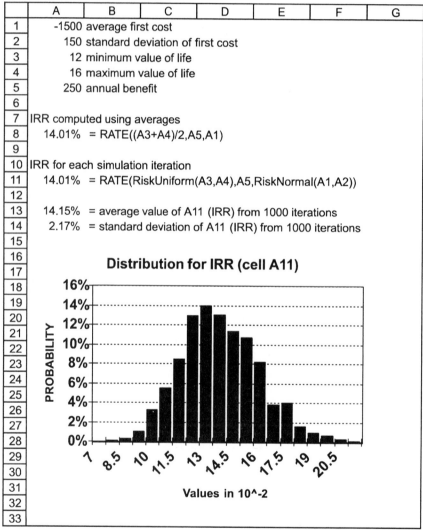

	A	B	C	D	E	F	G
1	-1500 average first cost						
2	150 standard deviation of first cost						
3	12 minimum value of life						
4	16 maximum value of life						
5	250 annual benefit						
6							
7	IRR computed using averages						
8	14.01% = RATE((A3+A4)/2,A5,A1)						
9							
10	IRR for each simulation iteration						
11	14.01% = RATE(RiskUniform(A3,A4),A5,RiskNormal(A1,A2))						
12							
13	14.15% = average value of A11 (IRR) from 1000 iterations						
14	2.17% = standard deviation of A11 (IRR) from 1000 iterations						
15							

Distribution for IRR (cell A11)

PROBABILITY

Values in 10^-2

Figure 14-17 Simulation spreadsheet for Example 14-12 and 13.

Summary

Estimation of the future is an important element of an engineer's work. In economic analysis we have several ways of describing the future. Precise estimates will not ordinarily be exactly correct, but they are considered to be the best single values to represent what we think will happen.

A simple way to represent variability is through a range of estimates for future events. Frequently this is done by making three estimates: optimistic, most likely, and pessimistic. If the problem is solved three times, using first the optimistic values, followed by the most likely and pessimistic values, the full range of prospective results may be examined. The disadvantage of this approach is that it is usually extremely unlikely that either the optimistic results (based on optimistic estimates for all the components of the problem) or the pessimistic results will occur.

A variation to the solution of the range of estimates is to assign relative weights to the various estimates and solve the problem using the weighted values. One set of weights suggested is:

Estimate	*Relative weight*
Optimistic	1
Most likely	4
Pessimistic	1

When the probability of a future event is known or may be reasonably predicted, the technique of *expected value* may be used. Here the probabilities are applied as the relative weights.

$$\text{Expected value} = \text{Outcome}_A \times \text{Probability}_B + \cdots$$
$$+ \text{Outcome}_B \times \text{Probability}_B + \cdots$$

Expected value is a useful technique in projecting the long term results when a situation occurs over and over again. For situations where the event will only occur once or a few times, expected value gives little insight into the infrequent event.

Two probability distributions are described. The uniform distribution is where each outcome has an equal probability of occurrence. The 52 different cards in a deck of cards, when shuffled, represent a uniform distribution. For computations, two-digit random numbers are a practical representation of the uniform distribution.

The normal distribution is the best known of all distributions. A normal distribution is defined by its two parameters, the mean (a measure of central tendency) and standard deviation (a measure of the variability of the distribution). A random sample from any normal distribution may be obtained through the use of random normal numbers. A random sample may also be obtained from other distributions if the cumulative distribution function can be plotted.

Where the elements of an economic analysis are stated in terms of probability distributions; a repetitive analysis of a random sample is often done. This simulation is based on the premise that a random sampling of increasing size becomes a better and better estimate of the possible outcomes. The large number of computations limit the usefulness of the technique when hand methods are used.

Problems

14-1   Two instructors announced that they "grade on the curve," that is, give a fixed percentage of each of the various letter grades to each of their classes. Their curves are:

Grade	Instructor A	Instructor B
A	10%	15%
B	15%	15%
C	45%	30%
D	15%	20%
F	15%	20%

If a random student came to you and said that his object was to enroll in the class where his expected grade point average would be greater, which instructor would you recommend?
(*Answer:* Instructor *A*)

14-2   A man wants to determine whether or not to invest $1000 in a friend's speculative venture. He will do so if he thinks he can get his money back in one year. He believes the probabilities of the various outcomes at the end of one year are:

Result	Probability
$2000 (double his money)	0.3
1500	0.1
1000	0.2
500	0.3
0 (lose everything)	0.1

What would be his expected outcome if he invests the $1000?

14-3   The M.S.U. football team has ten games scheduled for next season. The business manager wishes to estimate how much money the team can be expected to have left over after paying the season's expenses, including any post-season "bowl game" expenses. From records for the past season and estimates by informed people, the business manager has assembled the following data:

Situation	Probability	Situation	Net income
Regular season:		Regular season:	
Win 3 games	0.10	Win 5 or fewer	
Win 4 games	0.15	games	$250,000
Win 5 games	0.20		
Win 6 games.	0.15	Win 6 to 8	
Win 7 games	0.15	games	400,000
Win 8 games	0.10		
Win 9 games	0.07	Win 9 or 10	
Win 10 games	0.03	games	600,000
Post-season:		Post-season:	Additional income
Bowl game	0.10	Bowl game	of $100,000

Based on the business manager's data, what is the expected net income for the team next season? (*Answer:* $355,000)

14-4 Telephone poles are an example of items that have varying useful lives. Telephone poles, once installed in a location, remain in useful service until one of a variety of events occur.

   a. Name three reasons why a telephone pole might be removed from useful service at a particular location.

   b. You are to estimate the total useful life of telephone poles. If the pole is removed from an original location while it is still serviceable, it will be installed elsewhere. Estimate the optimistic life, most likely life, and pessimistic life for telephone poles. What percentage of all telephone poles would you expect to have a total useful life greater than your estimated optimistic life?

14-5 In the New Jersey and Nevada gaming casinos, the crap table is a popular gambling game. One of the many bets available is the "Hard-way 8." A $1 bet in this fashion will win the player $4 if in the game the pair of dice come up 4 and 4 prior to one of the other ways of totaling 8. For a $1 bet, what is the expected result? (*Answer:* 80¢)

14-6 When a pair of dice are tossed the results may be any number from 2 through 12. In the game of craps one can win by tossing either a 7 or an 11 on the first roll. What is the probability of doing this? *Hint:* There are 36 ways that a pair of six-sided dice can be tossed. What portion of them result in either a 7 or an 11? (*Answer:* $8/36$)

14-7 Your grade point average for the current school term probably cannot yet be determined with certainty.

   a. Estimate the probability of obtaining a grade point average in each of the five categories below.

Grade point average (A = 4. 00)	Probability of obtaining the GPA
0.00–0.80	
0.81–1.60	
1.61–2.40	
2.41–3.20	
3.21–4.00	

b. Plot the data from Part **a** with the grade point average as the *x*-axis and cumulative probability (of obtaining the GPA category or some lower one) as the *y*-axis.

c. Assign the values between 00 and 99 to represent the cumulative probability 0.00–1.00 on the *y*-axis. Obtain a random number from Table 14-1 and determine the matching cumulative probability. Read and record the corresponding value of the GPA on the *x*-axis. In this manner, obtain 25 values of GPA. Graph the results as a bar graph with the *x*-axis as GPA and the *y*-axis as frequency. How does this bar graph compare with the table in Part **a**?

14-8 A decision has been made to perform certain repairs on the outlet works of a small dam. For a particular 36-inch gate valve, there are three available alternatives:

1. leave the valve as it is;
2. repair the valve; or
3. replace the valve.

If the valve is left as it is, the probability of a failure of the valve seats, over the life of the project, is 60%; the probability of failure of the valve stem is 50%; and of failure of the valve body is 40%.

If the valve is repaired, the probability of a failure of the seats, over the life of the project, is 40%; of failure of the stem is 30%; and of failure of the body is 20%. If the valve is replaced, the probability of a failure of the seats, over the life of the project, is 30%; of failure of the stem is 20%; and of failure of the body is 10%.

The present worth of cost of future repairs and service disruption of a failure of the seats is $10,000; the present worth of cost of a failure of the stem is $20,000; the present worth of cost of a failure of the body is $30,000. The cost of repairing the valve now is $10,000; and of replacing it is $20,000. If the criterion is to minimize expected costs, which alternative is best?

14-9 A man went to Atlantic City with $500 and placed 100 bets of $5 each on a number on the roulette wheel. There are 38 numbers on the wheel and the gaming casino pays 35 times the amount bet if the ball drops into the bettor's numbered slot in the roulette wheel. In addition, the bettor receives back the original $5 bet. Estimate how much money the man is expected to win or lose in Atlantic City.

14-10 A heat exchanger is being installed as part of a plant modernization program. It costs $80,000, including installation, and is expected to reduce the overall plant fuel cost by $20,000 per year. Estimates of the useful life of the heat exchanger range from an optimistic 12 years to a pessimistic 4 years. The most likely value is 5 years. Using the range of estimates to compute the mean life, determine the estimated before-tax rate of return. Assume the heat exchanger has no salvage value at the end of its useful life.

14-11 A factory building is located in an area subject to occasional flooding by a nearby river. You have been brought in as a consultant to determine whether or not floodproofing of the building is economically justified. The alternatives are as follows:

    a. Do nothing. Damage in a moderate flood is $10,000 and in a severe flood, $25,000.

    b. Alter the factory building at a cost of $15,000 to withstand moderate flooding without damage and to withstand severe flooding with $10,000 damages.

    c. Alter the factory building at a cost of $20,000 to withstand a severe flood without damage.

In any year the probability of flooding is as follows: 0.70—no flooding of the river; 0.20—moderate flooding; and 0.10—severe flooding. If interest is 15% and a 15-year analysis period is used, what do you recommend?

14-12 Al took a midterm examination in physics and received a score of 65 in the exam. The mean was 60 and the standard deviation was 20. Bill received a score of 14 in mathematics where the exam mean was 12 and the standard deviation was 4. Which student ranked higher in his class? Explain.

14-13 An engineer decided to make a careful analysis of the cost of fire insurance for his $200,000 home. From a fire rating bureau he found the following risk of fire loss in any year.

Outcome	Probability
No fire loss	0.986
$ 10,000 fire loss	0.010
40,000 fire loss	0.003
200,000 fire loss	0.001

    a. Compute his expected fire loss in any year.

    b. He finds that the expected fire loss in any year is less than the $550 annual cost of fire insurance. In fact, an insurance agent explains that this is always true. Nevertheless, the engineer buys fire insurance. Explain why this is or is not a logical decision.

14-14 An industrial park is being planned for a tract of land near the river. To protect the industrial buildings that will be built on this low-lying land from flood damage, an earthen embankment can be constructed. The height of the embankment will be determined by an economic analysis of the costs and benefits. The following data have been gathered.

Embankment height above roadway (in meters)	Initial cost
2.0	$100,000
2.5	165,000
3.0	300,000
3.5	400,000
4.0	550,000

Flood level above roadway (in meters)	Average frequency that flood level will exceed height in Col. 1
2.0	Once in 3 years
2.5	Once in 8 years
3.0	Once in 25 years
3.5	Once in 50 years
4.0	Once in 100 years

The embankment can be expected to last 50 years and will require no maintenance. Anytime the flood water flows over the embankment, $300,000 of damage occurs. Should the embankment be built? If so, to which of the five heights above the roadway? A 12% rate of return is required.

14-15 Five years ago a dam was constructed to impound irrigation water and to provide flood protection for the area below the dam. Last winter a 100-year flood caused extensive damage both to the dam and to the surrounding area. This was not surprising since the dam was designed for a 50-year flood.

The cost to repair the dam not will be $250,000. Damage in the valley below amounts to $750,000. If the spillway is redesigned at a cost of $250,000 and the dam is repaired for another $250,000, the dam may be expected to withstand a 100-year flood without sustaining damage. However, the storage capacity of the dam will not be increased and the probability of damage to the surrounding area below the dam will be unchanged. A second dam can be constructed up the river from the existing dam for $1 million. The capacity of the second dam would be more that adequate to provide the desired flood protection. If the second dam is built, redesign of the existing dam spillway will not be necessary, but the $250,000 of repairs must be done.

The development in the area below the dam is expected to be complete in ten years. A new 100-year flood in the meantime will cause a $1 million loss. After ten years the loss would be $2 million. In addition there would be $250,000 of spillway damage if the spillway is not redesigned. A 50-year flood is also likely to cause about $200,000 of damage, but the spillway would be adequate. Similarly, a 25-year flood would cause about $50,000 of damage.

The three alternatives are (1) Repair the existing dam for $250,000 but make no other alterations. (2) Repair the existing dam ($250,000) and redesign the spillway to take a 100-year flood ($250,000). (3) Repair the existing dam ($250,000) and build the second dam ($1 million). Based on an expected annual cash flow analysis, and a 7% interest rate, which alternative should be selected?

14-16 The University of Detroit Mercy invested $20 million in 1998 for building and computer renovation. The University borrowed this from the National Bank of Detroit (NBD) in January, 1998 at an interest rate of 10%.

In any year (for the next 10 years) probable student population will be

Student Population	Probability
8000	0.60
9000	0.30
10,000	0.10

a. If the University wants to pay back its loan in 10 years (starting in 1999) with equal end of year annual payments. How much is the yearly payment?

b. What is the expected number of students per year for the next 10 years.

c. If borrowed money from NBD will have to be repaid in the next 10 years in 1999, then how much additional tuition must a student pay in each year, so that the University will be able to pay back its loan in 10 years with equal annual payments.

d. Generally a student spends four years at the University of Detroit Mercy. First year related costs include:

Tuition	$10,000
Room and Board	3,000
Books	500
Other	500
Total	$14,000

Second year cost will be 10 percent higher than 1st year cost. Third year cost will 20 percent higher than 1st year cost. Fourth year cost will be 30 percent higher than 1st year cost. How much should a parent deposit in the bank today, so that there will be enough money for a student entering UDM next year (for the next four years) considering interest rate 10%.

e. If in 2002, the University decides to pay the balance due on the NBD loan, what will be 2002 payment (after making 1999, 2000 and 2001 payments, as a part of equal annual payments for 10 years starting the end of 1999).

f. What parts of the 1999 payment are interest and principal payments, respectively?

g. If the students pay $10,000/yr tuition, how long will it take to pay back the NBD loan by making equal annual payments beginning the end of 1999? (Tuition is the only source of income and 10 percent is used to pay back the NBD loan. Assume the student population is 8500/yr fixed).

h. If the University wants to pay back the NBD loan by <u>one single payment</u> the end of 2001, what will be the payment amount? No other payments will be made inbetween.

i. The University tuition per student for 1998 is $7,000, when it is increased by $500 in subsequent years. What will be the equivalent uniform annual tuition payment for the five years beginning in 1999? Use 10% interest.

CHAPTER 15

Selection Of A Minimum Attractive Rate Of Return

The preceding chapters have said very little about what interest rate or minimum attractive rate of return is suitable for use in a particular situation. Since this problem is quite complex, there is no single answer that is always appropriate. A discussion of a suitable interest rate to use must inevitably begin with an examination of the sources of capital, followed by a look at the prospective investment opportunities and risk. Only in this way can an intelligent decision be made on the choice of an interest rate or minimum attractive rate of return.

Sources Of Capital

In broad terms there are four sources of capital available to a firm. They are: money generated from the operation of the firm, borrowed money, sale of mortgage bonds, and sale of capital stock.

Money Generated from the Operation of the Firm

A major source of capital investment money is through the retention of profits resulting from the operation of the firm. Since only about half of the profits of industrial firms are paid out to stockholders, the half that is retained is an important source of funds for all purposes, including capital investments. In addition to profit, there is money generated in the business equal to the annual depreciation charges on existing capital assets if the firm is profitable. In other words, a profitable firm will generate money equal to its depreciation charges *plus* its retained profits. Even a firm that earns zero profit will still generate money from operations equal to its depreciation charges. (A firm with a loss, of course, will have still less funds.)

External Sources of Money

When a firm requires money for a few weeks or months, it typically borrows the money from banks. Longer term unsecured loans (of, say, 1–4 years) may also be arranged through banks. While banks undoubtedly finance a lot of capital expenditures, regular bank loans cannot be considered a source of permanent financing.

Longer term secured loans may be obtained from banks, insurance companies, pension funds, or even the public. The security for the loan is frequently a mortgage on specific property of the firm. When sold to the public, this financing is by mortgage bonds. The sale of stock in the firm is still another source of money. While bank loans and bonds represent debt that has a maturity date, stock is considered a permanent addition to the ownership of the firm.

Choice of Source of Funds

Choosing the source of funds for capital expenditures is a decision for the board of directors; sometimes it also requires approval of the stockholders. In situations where internal operations generate adequate funds for the desired capital expenditures, external sources of money probably are seldom used. But when the internal sources are inadequate, external sources must be employed or the capital expenditures will have to be deferred or cancelled.

Cost Of Funds

Cost of Borrowed Money

A first step in deciding on a minimum attractive rate of return might be to determine the interest rate at which money can be borrowed. Longer term secured loans may be obtained from banks, insurance companies, or the variety of places where substantial amounts of money accumulates (for example, Japan or the oil-producing nations).

A large, profitable corporation might be able to borrow money at the ***prime rate,*** that is, the interest rate that banks charge their best and most sought after customers. All other firms are charged an interest rate that is higher by anywhere from one-half to several percent. In addition to the financial strength of the borrower and his ability to repay the loan, the interest rate will vary depending on the duration of the loan.

Cost of Capital

Another relevant interest rate is the ***cost of capital.*** The general assumption concerning the cost of capital is that all the money the firm uses for investments is drawn from all the components of the overall capitalization of the firm. The mechanics of the computation is given in Ex. 15-1.

EXAMPLE 15-1

For a particular firm, the purchasers of common stock require an 11% rate of return, mortgage bonds are sold at a 7% interest rate and bank loans are available at 9%. Compute the cost of capital for the following capital structure:

		Rate of return	Annual amount
$ 20 million	Bank loan	9%	$1.8 million
20 million	Mortgage bonds	7%	1.4 million
60 million	Common stock and retained earnings	11%	6.6 million
$100 million			$9.8 million

Solution: Interest payments on debt, like bank loans and mortgage bonds, are tax deductible business expenses. Thus:

After-tax interest cost = Before-tax interest cost(1 − tax rate)

If we assume the firm pays 40% income taxes, the computations become:

Bank loan:　　After-tax interest cost = 9% (1 − 0.40) = 5.4%
Mortgage bonds:　After-tax interest cost = 7%(1 − 0.40) = 4.2%

Dividends paid on the ownership in the firm (common stock + retained earnings) are not tax deductible. Combining the three components, the after-tax interest cost for the $100 million of capital is:

$20 million (5.4%) + $20 million (4.2%) + $60 million (11%) = $8.52 million

$$\text{Cost of capital} = \frac{\$8.52 \text{ million}}{\$100 \text{ million}} = 8.52\% \quad \blacksquare$$

In an actual situation, the cost of capital is quite difficult to compute. The fluctuation in the price of common stock, for example, makes it difficult to pick a cost, and the fluctuating prospects of the firm makes it even more difficult to estimate the future benefits the purchasers of the stock might expect to receive. Given the fluctuating costs and prospects of future benefits, what rate of return do stockholders require? There is no precise answer, but we can obtain an approximate answer. Similar assumptions must be made for the other components of a firm's capitalization.

Investment Opportunities

An industrial firm has many more places in which it can invest its money than does an individual. A firm has larger amounts of money and this alone makes certain kinds of investment possible that are unavailable to individual investors, with their more limited investment funds. The U.S. Government, for example, borrows money for short terms of 90 or 180 days by issuing certificates called Treasury Bills that frequently yield a greater interest rate than savings accounts. The customary minimum purchase is $25,000.

Table 15-1   A FIRM'S AVAILABLE INVESTMENT OPPORTUNITIES

Project number	Project	Cost $(\times 10^3)$	Estimated rate of return
Investment Related to Current Operations			
1	New equipment to reduce labor costs	$150	30%
2	Other new equipment to reduce labor costs	50	45%
3	Overhaul particular machine to reduced material costs	50	38%
4	New test equipment to reduce defective products produced	100	40%
New Operations:			
5	Manufacture parts that previously had been purchased	200	35%
6	Further processing of products previously sold in semi-finished form	100	28%
7	Further processing of other products	200	18%
New Production Facilities:			
8	Relocate production to new plant	250	25%
External Investments:			
9	Investment in a different industry	300	20%
10	Other investment in a different industry	300	10%
11	Overseas investment	400	15%
12	Buy Treasury Bills	Unlimited	8%

More important, however, is the fact that a firm conducts a business and this business offers many investment opportunities. While exceptions can be found, a good generalization is that the opportunities for investment of money within the firm are superior to the investment opportunities outside the firm. Consider the following situation: a tabulation of the available investment opportunities for a particular firm is outlined in Table 15-1. A plot of these projects by rate of return *vs.* investment is shown in Fig. 15-1. The cumulative investment required for all projects at or above a given rate of return is given in Fig. 15-2.

The two figures illustrate that a firm may have a broad range of investment opportunities available at varying rates of return and with varying lives and uncertainties. It may take some study and searching to identify the better investment projects available to a firm. If this is done, the available projects will almost certainly exceed the money the firm budgets for capital investment projects.

Opportunity Cost

We see that there are two aspects of investing that are basically independent. One factor is the source and quantity of money available for capital investment projects. The other aspect

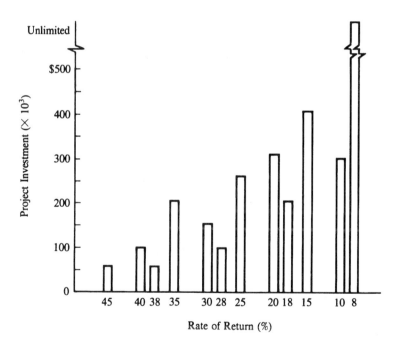

Figure 15-1 Rate of return *vs.* project investment.

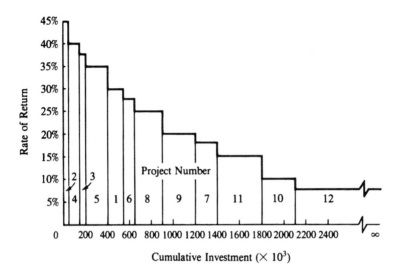

Figure 15-2 Cumulative investment required for all projects at or above a given rate of return.

is the investment opportunities themselves that are available to the firm.

These two situations are typically out of balance, with investment opportunities exceeding the available money supply. Thus some investment opportunities can be selected and many must be rejected. Obviously, we want to ensure that *all the selected projects are better than the best rejected project.* To do this, we must know something about the rate of return on the best rejected project. The best rejected project is the best opportunity foregone, and this in turn is called the ***opportunity cost.***

> ***Opportunity cost = Cost of the best opportunity foregone***
>
> ***= Rate of return on the best rejected project***

If one could predict in advance what the opportunity cost would be for some future period (like the next twelve months), this rate of return could be one way to judge whether to accept or reject any proposed capital expenditure.

___

EXAMPLE 15-2

Consider the situation represented by Figures 15-1 and 15-2. For a capital expenditure budget of $1200 ($\times$ 10^3), what is the opportunity cost?

Solution: From Fig. 15-2 we see that the eight projects with a rate of return of 20% or more require a cumulative investment of $1200 ($\times 10^3$). We would take on these projects and reject the other four (7, 11, 10, and 12) with rates of return of 18% or less. The best rejected project is #7 and it has an 18% rate of return. This indicates the opportunity cost is 18%. ∎

Selecting A Minimum Attractive Rate Of Return

Using the three concepts on the cost of money (the cost of borrowed money, the cost of capital, and opportunity cost), which, if any, of these values should be used as the minimum attractive rate of return (MARR) in economic analyses?

Fundamentally, we know that unless the benefits of a project exceed the cost of the project, we cannot add to the profitability of the firm. A lower boundary for the minimum attractive rate of return must be the cost of the money invested in the project. It would be unwise, for example, to borrow money at 8% and invest it in a project yielding a 6% rate of return.

Further, we know that no firm has an unlimited ability to borrow money. Bankers–and others who evaluate the limits of a firm's ability to borrow money–look at both the profitability of the firm and the relationship between the components in the firm's capital structure. This means that continued borrowing of money will require that additional stock must be sold to maintain an acceptable ratio between *ownership* and *debt.* In other words, borrowing for a particular investment project is only a block of money from the overall capital structure of the firm. This suggests that the MARR should not be less than the cost of capital. Finally, we know that the MARR should not be less than the rate of return on the best opportunity foregone. Stated simply,

> **Minimum attractive rate of return should be equal to the highest one of the following: cost of borrowed money; cost of capital; or opportunity cost.**

Adjusting MARR To Account For Risk And Uncertainty

We know from our previous study of estimating the future that what actually occurs is often different from the estimate. When we are fortunate enough to be able to assign probabilities to a set of possible future outcomes, we call this a *risk* situation. We saw in Chapter 14 that techniques like expected value and simulation may be used when the probabilities are known.

Uncertainty is the term used to describe the condition when the probabilities are *not* known. Thus, if the probabilities of future outcomes are known we have *risk,* and if they are unknown we have *uncertainty.*

One way to reduce the likelihood of undertaking projects that do not produce satisfactory results is to pass up marginal projects. In other words, no matter what projects are undertaken, some will turn out better than anticipated and some worse. Some undesirable results can be prevented by selecting only the best projects and avoiding those whose expected results are closer to a minimum standard: then (in theory, at least) the selected projects will provide results *above* the minimum standard even if they do considerably worse than anticipated.

In projects where there is normal business risk and uncertainty, the MARR is used without adjustment. For projects with greater risk or uncertainty, the MARR is increased. This is certainly not the best way to handle conditions of risk. A preferable way is to deal explicitly with the probabilities using one of the techniques from Chapter 14. This may be more acceptable as an adjustment for uncertainty. When the interest rate (MARR) used in economic analysis calculations is raised to adjust for risk or uncertainty, greater emphasis is placed on immediate or short term results and less emphasis on longer term results.

EXAMPLE 15-3

Consider the two following alternatives: the MARR has been raised from 10% to 15% to take into account the greater risk and uncertainty that Alt. *B*'s results may not be as favorable as indicated. What is the impact of this change of MARR on the decision?

Year	Alt. A	Alt. B
0	−80	−80
1–10	10	13.86
11–20	20	10

Solution:

Year	Alt. A	NPW at 14.05%	NPW at 10%	NPW at 15%
0	−80	−80.00	−80.00	−80.00
1–10	10	52.05	61.45	50.19
11–20	20	27.95	47.38	24.81
		0	+28.83	−5.00

Year	Alt. B	NPW at 15.48%	NPW at 10%	NPW at 15%
0	−80	−80.00	−80.00	−80.00
1–10	13.86	68.31	85.14	69.56
11–20	10	11.99	23.69	12.41
		0	+28.83	+1.97

Computations at MARR of 10% ignoring risk and uncertainty. Both alternatives have the same positive NPW (+28.83) at a MARR of 10%. Also, the differences in the benefits schedules ($A - B$) produce a 10% incremental rate of return. (The calculations are not shown here.) This must be true if NPW for the two alternatives is to remain constant at a MARR of 10%.

Considering risk and uncertainty with MARR of 10%. At 10%, both alternatives are equally desirable. Since Alt. *B* is believed to have greater risk and uncertainty, a logical conclusion is to select Alt. *A* rather than *B*.

Increase MARR to 15%. At a MARR of 15%, Alt. *A* has a negative NPW and Alt. *B* has a positive NPW. Alternative *B* is preferred under these circumstances.

Conclusion. Based on a business-risk MARR of 10%, the two alternatives are equivalent. Recognizing some greater risk of failure for Alt. *B* makes *A* the preferred alternative. If the MARR is increased to 15%, to add a margin of safety against risk and uncertainty, the computed decision is to select *B*. Since Alt. *B* has been shown to be less desirable than *A*, the decision, based on a MARR of 15%, may be an unfortunate one. The difficulty is that the same risk adjustment (increase the MARR by 5%) is applied to both alternatives even though they have different amounts of risk. ■

The conclusion to be drawn from Ex. 15-3 is that increasing the MARR to compensate for risk and uncertainty is only an approximate technique and may not always achieve the desired result. Nevertheless, adjusting the MARR upward for increased risk and uncertainty is commonly done in industry.

Inflation and the Cost of Borrowed Money

As inflation has varied, what is its effect on the cost of borrowed money? A widely held view has been that interest rates on long term borrowing, like twenty-year Treasury bonds, will be about 3% more than the inflation rate. For borrowers this is the real—that is, after-inflation—cost of money, and for lenders the real return on loans. If inflation rates increase, it would follow that borrowing rates would also increase. All this suggests a rational and orderly situation, about as we might expect.

Unfortunately, things have not worked out this way. Figure 15-3 shows that the real interest rate has not been 3% in recent times and, in fact, there have been long periods when the real interest rate was negative. Can this be possible? Would anyone invest money at an interest rate several percent less than the inflation rate? Well, consider this: when the U.S. inflation rate was 12%, savings banks were paying 5½% on regular passbook deposits. And there was a lot of money in those accounts. While there must be a relationship between interest rates and inflation, Figure 15-3 suggests that it is complex.

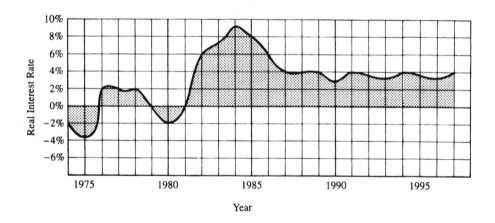

Figure 15-3 *The real interest rate.* The interest rate on twenty-year Treasury bonds *minus* the Inflation rate, *f*, as measured by changes in the Consumer Price Index.

Representative Values Of MARR Used In Industry

We argued that the minimum attractive rate of return should be established at the highest one of the following: cost of borrowed money, cost of capital, or the opportunity cost.

The cost of borrowed money will vary from enterprise to enterprise, with the lowest rate being the prime interest rate. The prime rate may change several times in a year; it is widely reported in newspapers and business publications. As we pointed out, the interest rate for firms that do not qualify for the prime interest rate may be ½% to several percent higher.

The cost of capital of a firm is an elusive value. There is no widely accepted way to compute it; we know that as a *composite value* for the capital structure of the firm, it conventionally is higher than the cost of borrowed money. The cost of capital must consider the market valuation of the shares (common stock, and so forth) of the firm, which may fluctuate widely, depending on future earnings prospects of the firm. We cannot generalize on representative costs of capital.

Somewhat related to cost of capital is the computation of the return on total capital (long term debt, capital stock, and retained earnings) actually achieved by firms. *Fortune* magazine, among others, does an annual analysis of the rate of return on total capital. The after-tax rate of return on total capital for individual firms ranges from 0% to about 40% and averages 8%. *Business Week* magazine does a periodic survey of corporate performance. It reports an after-tax rate of return on common stock and retained earnings. We would expect the values to be higher than the rate of return on total capital, and this is the case. The after-tax return on common stock and retained earnings ranges from 0% to about 65% with an average of 14%.

When discussing MARR, firms can usually be divided into two general groups. First, there are firms which are struggling along with an inadequate supply of investment capital, or are in an unstable situation or unstable industry. These firms cannot or do not invest money in anything but the most critical projects with very high rates of return and a rapid return of the capital invested. Often these firms use payback period and establish a criterion of one year or less, before income taxes. For an investment project with a five-year life, this corresponds to about a 60% after-tax rate of return. When these firms do rate of return analysis, they reduce the MARR to possibly 25% to 30% after income taxes. There is potentially a substantial difference between a one-year before tax payback period and a 30% after-tax MARR, but this apparently does not disturb firms that specify this type of dual criteria.

The second group of firms represents the bulk of all enterprises. They are in a more stable situation and take a longer range view of capital investments. Their greater money supply enables them to invest in capital investment projects that firms in the first group will reject. Like the first group, this group of firms also uses payback and rate of return analysis. When small capital investments (of about $500 or less) are considered, payback period is often the only analysis technique used. The criterion for accepting a proposal may be a before-tax payback period not exceeding one or two years. Larger investment projects are analyzed by rate of return. Where there is a normal level of business risk, an after-tax MARR of 12% to 15% appears to be widely used. The MARR is increased when there is greater risk involved.

In Chapter 9 we saw that payback period is not a proper method for the economic analysis of proposals. Thus, industrial use of payback criteria is *not* recommended. Fortunately, the trend in industry is toward greater use of accurate methods and less use of payback period.

Note that the values of MARR given above are approximations. But the values quoted appear to be opportunity costs, rather than cost of borrowed money or cost of capital. This indicates that firms cannot or do not obtain money to fund projects whose anticipated rates of return are nearer to the cost of borrowed money or cost of capital. While one could make a case that good projects are needlessly being rejected, there may be practical business reasons why firms operate as they do.

One cannot leave this section without noting that the MARR used by enterprises is so much higher than can be obtained by individuals. (Where can you get a 30% after-tax rate of return without excessive risk?) The reason appears to be that businesses are not faced with the intensively competitive situation that confronts an individual. There might be thousands of people in any region seeking a place to invest $2000 with safety; but how many people could—or would—want to invest $500,000 in a business? This diminished competition, combined with a higher risk, appears to explain at least some of the difference.

Spreadsheets, Cumulative Investments, and the Opportunity Cost of Capital

As shown in previous chapters, spreadsheets make computing rates of return dramatically easier. In addition spreadsheets can be used to sort the projects by rate of return and then calculate the cumulative first cost. This is accomplished through the following steps.

1. Enter or calculate each project's rate of return.

2. Select the data to be sorted, do *not* include headings, but do include all information on the row that goes with each project.

3. Select the SORT tool (found in the menu under DATA), identify the rate of return column as the first key, and a sort order of descending. Also ensure that row sorting is selected. Sort.

4. Add a column for the cumulative first cost. This column is compared with the capital limit to identify the opportunity cost of capital and which projects should be funded.

Example 15-4 illustrates these steps.

EXAMPLE 15-4

A firm has a budget of $800,000 for projects this year. Which of the following projects should be accepted? What is the opportunity cost of capital?

Project	First cost	Annual benefit	Salvage value	Life
A	$200,000	$25,000	$50,000	15
B	250,000	47,000	-25,000	10
C	150,000	17,500	20,000	15
D	100,000	20,000	15,000	10
E	200,000	24,000	25,000	20
F	300,000	35,000	15,000	15
G	100,000	18,000	0	10
H	200,000	22,500	15,000	20
I	350,000	50,000	0	25

Solution

The first step is to use the RATE function to find the rate of return for each project. The results of this step are shown in the top portion of Figure 15-4. Next the projects are sorted in descending order by their rates of return. Finally, the cumulative first cost is computed. Projects D, I, B, and G should be funded. The opportunity cost of capital is 12.4% if defined as the last project funded and 10.6% if defined as the first project rejected.

	A	B	C	D	E	F	G	H
1	Project	First cost	Annual benefit	Salvage value	Life	IRR	=RATE(E2,C2,-B2,D2)	
2	A	200,000	25,000	50,000	15	10.2%		
3	B	250,000	47,000	-25,000	10	12.8%		
4	C	150,000	17,500	20,000	15	8.6%		
5	D	100,000	20,000	15,000	10	16.0%		
6	E	200,000	24,000	25,000	20	10.6%		
7	F	300,000	35,000	15,000	15	8.2%		
8	G	100,000	18,000	0	10	12.4%		
9	H	200,000	22,500	15,000	20	9.6%		
10	I	350,000	50,000	0	25	13.7%		
11	Projects sorted by IRR						Cumulative first cost	
12	D	100,000	20,000	15,000	10	16.0%	100,000	
13	I	350,000	50,000	0	25	13.7%	450,000	
14	B	250,000	47,000	-25,000	10	12.8%	700,000	
15	G	100,000	18,000	0	10	12.4%	800,000	
16	E	200,000	24,000	25,000	20	10.6%	1,000,000	
17	A	200,000	25,000	50,000	15	10.2%	1,200,000	
18	H	200,000	22,500	15,000	20	9.6%	1,400,000	
19	C	150,000	17,500	20,000	15	8.6%	1,550,000	
20	F	300,000	35,000	15,000	15	8.2%	1,850,000	

Figure 15-4 Spreadsheet for finding opportunity cost of capital.

Summary

There are four general sources of capital available to an enterprise. The most important one is money generated from the operation of the firm. This has two components: there is the portion of profit that is retained in the business; in addition, a profitable firm generates funds equal to its depreciation charges that are available for reinvestment.

The three other sources of capital are from outside the operation of the enterprise:

1. Borrowed money from banks, insurance companies, and so forth.

2. Longer term borrowing from a lending institution or from the public in the form of mortgage bonds.

3. Sale of equity securities like common or preferred stock.

Retained profits and cash equal to depreciation charges are the primary sources of investment capital for most firms, and the only sources for many enterprises.

In selecting a value of MARR, three values are frequently considered:

1. Cost of borrowed money.

2. Cost of capital. This is a composite cost of the components of the overall capitalization of the enterprise.

3. Opportunity cost. This refers to the cost of the opportunity foregone; stated more simply, opportunity cost is the rate of return on the best investment project that is rejected.

The MARR should be equal to the highest one of these three values.

When there is a risk aspect to the problem (probabilities are known or reasonably estimated), this can be handled by techniques like expected value and simulation. Where there is uncertainty (probabilities of the various outcomes are not known), there are analytical techniques, but they are less satisfactory. A method commonly used to adjust for risk and uncertainty is to increase the MARR. This method has the effect of distorting the time-value-of-money relationship. The effect is to discount longer term consequences more heavily compared to short term consequences, which may or may not be desirable. Other possibilities might be to adjust the discounted cash flows or the lives of the alternatives.

Problems

15-1 Examine the financial pages of your newspaper (or *The Wall St. Journal*) and determine the current interest rate on the following securities:

U.S. Treasury bond due in five years.

a. General obligation bond of a municipal district, city, or a state due in twenty years.

b. Corporate debenture bond of a U.S. industrial firm due in twenty years.

Explain why the interest rates are different for these different bonds.

15-2 Consider four mutually exclusive alternatives:

	A	B	C	D
Initial cost	0	100	50	25
Uniform annual benefit	0	16.27	9.96	5.96
Computed rate of return	0%	10%	15%	20%

Each alternative has a ten-year useful life and no salvage value. Over what range of interest rates is *C* the preferred alternative? (*Answer:* $4.5\% < i \leq 9.6\%$)

15-3   Frequently we read in the newspaper that one should lease an automobile rather than buying it. For a typical 24-month lease on a car costing $9400, the monthly lease charge is about $267. At the end of the 24 months, the car is returned to the lease company (which owns the car). As an alternative, the same car could be bought with no down payment and 24 equal monthly payments, with interest at a 12% nominal annual percentage rate. At the end of 24 months the car is fully paid for. The car would then be worth about half of its original cost.

    a.   Over what range of nominal before-tax interest rates is leasing the preferred alternative?

    b.   What are some of the reasons that would make leasing more desirable than is indicated in *a*?

15-4   Assume you have $2000 available for investment for a five-year period. You wish to *invest* the money—not just spend it on fun things. There are obviously many alternatives available. You should be willing to assume a modest amount of risk of loss of some or all of the money if this is necessary, but not a great amount of risk (no investments in poker games or at horse races). How would you invest the money? What is your minimum attractive rate of return? Explain.

15-5   There are many venture capital syndicates that consist of a few (say, eight or ten) wealthy people who combine to make investments in small and (hopefully) growing businesses. Typically, the investors hire a young investment manager (often an engineer with an MBA) who seeks and analyzes investment opportunities for the group. Would you estimate that the MARR sought by this group is more or less than 12%? Explain.

15-6   A factory has a $100,000 capital budget. Determine which project(s) should be funded and the opportunity cost of capital.

Project	First Cost	Annual Benefits	Life	Salvage Value
A	$50,000	$13,500	5 yrs	$5000
B	50,000	9,000	10 yrs	0
C	50,000	13,250	5 yrs	1000
D	50,000	9,575	8 yrs	6000

15-7 Chips USA is considering the following projects to improve their production process. Chips have a short life, so a three year horizon is used in evaluation. Which projects should be done if the budget is $70,000? What is the opportunity cost of capital?

Project	First cost	Benefit
1	$20,000	$11,000
2	30,000	14,000
3	10,000	6,000
4	5,000	2,400
5	25,000	13,000
6	15,000	7,000
7	40,000	21,000

15-8 National Motors' Rock Creek Plant is considering the following projects to improve their production process. Which projects should be done if the budget is $500,000? What is the opportunity cost of capital?

Project	First Cost	Annual Benefit	Life (years)
1	$200,000	$50,000	15
2	300,000	70,000	10
3	100,000	40,000	5
4	50,000	12,500	10
5	250,000	75,000	5
6	150,000	32,000	20
7	400,000	125,000	5

15-9 The WhatZit Company has decided to fund six of nine project proposals for the coming budget year. Determine the next capital budget for WhatZit. What is the MARR?

Project	First cost	Annual benefits	Life
A	$15,000	$ 4,429	4 yrs
B	20,000	6,173	4 yrs
C	30,000	9,878	4 yrs
D	25,000	6,261	5 yrs
E	40,000	11,933	5 yrs
F	50,000	11,550	5 yrs
G	35,000	6,794	8 yrs
H	60,000	12,692	8 yrs
I	75,000	14,058	8 yrs

15-10 Which projects should be done if the budget is $100,000? What is the opportunity cost of capital?

Project	Life (yrs)	First cost	Annual benefit	Salvage value
1	20	20000	4000	
2	20	20000	3200	20000
3	30	20000	3300	10000
4	15	20000	4500	
5	25	20000	4500	-20000
6	10	20000	5800	
7	15	20000	4000	10000

CHAPTER 16

Economic Analysis In The Public Sector

So far we have considered economic analysis for companies in the private sector where the main objectives are to generate profits for growth and to reward current stockholders. Investment decisions for private sector companies involve evaluating the costs and benefits associated with prospective projects in terms of life-cycle cash flow streams. In previous chapters we developed several methods for calculating measures-of-merit and making decisions. Public organizations, such as federal, state and local governments, port authorities, school districts, and government agencies also make investment decisions. For these decision making bodies economic analysis is complicated by several factors which do not affect companies in the private sector. These factors include: the overall purpose of investment, project financing sources, expected project duration, effects of politics, beneficiaries of investment, and the multi-purpose nature of investments. The overall mission in the public sector is the same as that in the private sector — to make prudent investment decisions that promote the overall objectives of the organization.

The primary economic decision measure-of-merit used in the public sector is the benefit-cost (B/C) ratio. This measure is calculated as a ratio of the equivalent worth of the benefits of investment in a project to the equivalent worth of costs. If the B/C ratio is *greater than 1.0*, the project under evaluation is accepted; if not, it is rejected. The B/C ratio is used to evaluate both single investments as well as sets of mutually exclusive projects (where the incremental B/C ratio is used). The uncertainties of: quantifying cash flows, long project lives, and low interest rates, all tend to lessen the degree of reliability of the B/C ratio. There are two versions of the B/C ratio: the *conventional* and *modified*. Both provide consistent recommendations to decision makers for single investment decisions and decisions involving sets of mutually exclusive alternatives. Nevertheless, the B/C ratio is a widely used and accepted measure in government economic analysis and decision making.

Investment Objective

Organizations exist to promote the overall goals of those whom they serve. In private sector companies investment decisions are based on increasing the wealth and economic stability of the organization. Beneficiaries of investments generally are clearly identified as the owners and/or stockholders of the company.

In the public sector the purpose of investment decisions is sometimes ambiguous. For people in the United States of America the *Preamble to the U.S. Constitution* establishes the overall theme or objective for why the public body exists:

> "We the People of the United States, in order to form a more perfect Union, establish justice, insure domestic tranquillity, provide for the common defense, *promote the general welfare*, and secure the blessings of liberty to ourselves and our posterity, ordain and establish this Constitution for the United States of America."

The catch phrase *promote the general welfare* serves as a guideline for public decision making. But what does this phrase mean? At best it is a general guideline, at worst it is a vague slogan that can be used to justify any action. Projects which one segment of the citizenry deems to be useful and necessary may not be so viewed by other groups of citizens. In government economic analysis, it is not always easy to distinguish which investments promote the general welfare of the citizens and which do not.

Consider the case of a dam construction project to provide water, electricity, flood control and recreational facilities. Such a project might seem to be advantageous for the entire population of the region. But on closer inspection, decision makers must consider the fact that the dam will require the loss of land upstream due to backed up water. Farmers will lose pasture or crop land, and nature lovers will lose canyon lands. Or perhaps the land to be lost is a pivotal breeding ground for protected species, and environmentalists will oppose the project. The project may also have a negative impact on towns, cities and states downstream. How will it affect their water supply? Thus, a project initially deemed to have many benefits, on closer inspection, reveals many conflicting aspects. Such conflicting aspects are characteristic of investment and decision making in the public sector.

Public investment decisions are more difficult than those in the private sector due to the many people, organizations and political units potentially affected by investments. Opposition to a proposal is more likely in public investment decisions than in those made by private sector companies, because for every group that benefits from a particular project there is usually an opposing group. Many conflicts in opinion arise when the project involves the use of public lands, including: industrial parks, housing projects, business districts, roadways, sewage and power facilities, and landfills. Opposition may be based on the belief that development of *any* kind is bad or that the proposed development should not be near "our" homes, schools or businesses.

Consider the decision that a small town might face when deciding to establish a municipal rose garden, seemingly a beneficial public investment with no adverse consequences. However, an economic analysis of the project must consider *all* effects of the project, including potentially unforeseen outcomes. Where will visitors park their vehicles? Will increased auto travel around the park necessitate new traffic lights and signage? Will traffic and visitors to the park increase noise levels to adjacent homes? Will the garden's special varieties of roses create a disease hazard for local gardens? Will the garden require concentrated levels of fertilizers and insecticides, and how will these substances be disposed of? Clearly, many issues must be addressed. What appeared to be a simple proposal for a city rose garden, is, in fact, a decision with many aspects to consider.

Our simple rose garden illustrates how effects on *all parties involved* must be identified, even for projects that seem very useful. Public decision makers must reach a compromise between the positive effects enjoyed by some groups and the negative effects on other groups. The overall objective is to make prudent decisions that *promote the general welfare*, but in the public sector the decision process is not so straightforward as in the private sector.

The **Flood Act of 1936** specified that waterway improvements for flood control could be made so long as *the benefits to whomsoever they accrue are in excess of the estimated costs.* Perhaps, the overall general objective of investment decision analysis in government should be to promote the general welfare and assure that the value to those who can potentially benefit exceeds the overall costs to those who do not benefit.

Viewpoint For Analysis

When governmental bodies do economic analysis, an important concern is the proper viewpoint of the analysis. A look at industry will help to explain how the viewpoint, or perspective, from which an analysis is conducted influences the final recommendation. Economic analysis, both governmental and industrial, must be based on a viewpoint. In the case of industry the viewpoint is obvious — a company in the private sector pays the costs and counts *its* benefits. Thus, both the costs and benefits are measured from the perspective of the firm.

Costs and benefits that occur outside of the firm are referred to as external consequences (see Fig. 16-1). In years past, private sector companies generally ignored the external consequences of their actions. Ask anyone who has lived near a cement plant, a slaughterhouse, or a steel mill about external consequences! More recently, government has forced industry to reduce pollution and other undesirable external consequences, so that today many companies are taking a broader, or community-oriented, viewpoint in evaluating the consequences of their actions.

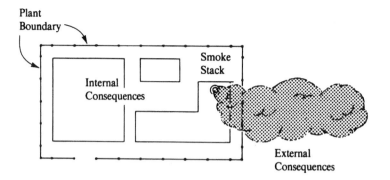

Figure 16-1 Internal and external consequences for an industrial plant.

The council members of a small town that levies taxes can be expected to take the "viewpoint of the town" in making decisions: unless it can be shown that the money from taxes can be used *effectively*, it is unlikely the town council will spend it. But what happens when the money is contributed to the town by the federal government, as in "revenue sharing" or some other federal grant? Often the federal government pays a share of project costs varying from 10% to 90%. Example 16-1 illustrates the viewpoint problem that is created.

EXAMPLE 16-1

A municipal project will cost $1 million. The federal government will pay 50% of the cost if the project is undertaken. Although the original economic analysis showed the PW of benefits was $1.5 million, a subsequent detailed analysis by the town engineer indicates a more realistic estimate of the PW of benefits is $750,000. The town council must decide whether or not to proceed with the project. What would you advise?

Solution: From the viewpoint of the town, the project is still a good one. If the town puts up half the cost ($500,000) it will receive all the benefits ($750,000). On the other hand, from an *overall* viewpoint, the revised estimate of $750,000 of benefits does not justify the $1 million expenditure. This illustrates the dilemma caused by varying viewpoints. For economic efficiency, one does not want to encourage the expenditure of money, regardless of the source, unless the benefits at least equal the costs. ∎

Possible viewpoints that may be taken include those of: an individual, a business firm or corporation, a town or city district, a city, a state, a nation, or a group of nations. To avoid suboptimizing, the proper approach is to *take a viewpoint at least as broad as those who pay the costs and those who receive the benefits*. When the costs and benefits are totally confined to a town, for example, the town's viewpoint seems an appropriate basis for the analysis. But when the costs or the benefits are spread beyond the proposed viewpoint, then the viewpoint should be enlarged to this broader view.

Other than investments in defense and social programs, most of the benefits provided by government projects are realized at a regional or local level. Projects, such as dams for electricity, flood control and recreation; and transportation facilities, such as roads, bridges and harbors all benefit most those in the region in which they are constructed. Even smaller scale projects, such as the municipal rose garden, although funded by public monies at a local or state level, provide most benefit to those near the project. As in the case of private decision making, it is important to adopt an appropriate and *consistent* viewpoint and to designate all of the costs and benefits that arise from the prospective investment based upon that perspective. It would be inappropriate to shift perspective when quantifying costs and benefits because it could greatly skew the results of the analysis and decision.

Selecting An Interest Rate

Several factors, not present for non-government firms, influence the selection of an appropriate interest rate for use in economic analysis in the government sector. Recall that for private sector companies the overall objective is wealth maximization. An appropriate interest rate for use in evaluating projects is selected consistent with this goal. Many non-government firms use *cost of capital* or *opportunity cost* concepts when setting an interest rate. The objective of public investment, on the other hand, involves the use of public resources to *promote the general welfare* of the public and invest for the benefit of *whomsoever may accrue* benefits so long as those benefits outweigh the costs. The setting of an interest rate for use in investment analysis is not so clear-cut in this case. Several alternative concepts have been suggested for how government should set this rate; these are discussed below.

No Time Value of Money Concept: In government, monies are obtained through taxation and spent about as quickly as they are obtained. Often, there is little time delay between collecting money from taxpayers and spending it. (Remember that the federal government and many states collect taxes every paycheck in the form of withholding tax.) The collection of taxes and their disbursement, although based on an annual budget, is actually a continuous process. Using this line of reasoning, some would argue that there is little or no time lag between collecting and spending these tax dollars. As such, they would advocate the use of a 0% interest rate for economic analysis of public projects.

Cost of Capital Concept: Another consideration in the determination of interest rates in public investments is that most levels of government (federal, state or local) borrow money for capital expenditures in addition to collecting taxes. Where money is borrowed for a specific project, one line of reasoning is to use an interest rate equal to the *cost of borrowed money*. This argument is less valid for state and local governments than for private firms because the federal government, through the income tax laws, subsidizes state and local bonded debt. If a state or one of its political subdivisions (like a county, city, or special assessment district) raises money through the sale of bonds, the interest paid on these bonds is exempt from federal taxes. In this way the federal government is *subsidizing* the debt, thereby encouraging investors to purchase such bonds.

These bonds, called municipal bonds, can be either *general obligation* or *revenue* bonds. General obligation municipal bonds pay interest and are retired (paid off) through taxes raised by the issuing government unit. A school district may use property taxes it receives to finance bond debt for construction of new language labs in its schools. Revenue bonds, on the other hand, are not supported by the taxing authority of the government unit; rather, they are supported by revenues earned by the project being funded. As an example, the city of Athens, Ohio would use toll revenues from a new bridge over the Hocking River to retire debt on revenue bonds sold for the bridge's construction. For those who purchase municipal bonds the tax-free status means that the expected return on this investment is somewhat less than that required of fully taxed bond investments (of similar risk). As a rough estimate, when fully taxed bonds yield an 8% interest rate, municipal bonds might make interest payments at a rate of 6%. The difference of 2% represents the effect of the preferred treatment for federal taxation, and hence a form of hidden federal subsidy on tax-free bonds. This means that the *cost of capital* approach is sometimes skewed for debt incurred through the use of long term bonds.

Opportunity Cost Concept: Opportunity cost, which relates to the interest rate on the best opportunity foregone, may take two forms in governmental economic analysis: *government opportunity cost* and *taxpayer opportunity cost.* In public decision making, if the interest rate is based on the opportunity cost to a government agency or other governing body, this interest rate is known as *government opportunity cost.* In this case the interest rate is set at that of the best prospective project for which funding is not available. One disadvantage of the *government opportunity cost* concept is that different agencies and subdivisions of government will have different opportunities. Therefore, sub-political units could potentially set different interest rates for use in economic analysis, and a project that may be rejected in one branch due to an inadequate rate of return may be accepted in another. Differing interest rates lead to inconsistent evaluation and decision making across government.

Dollars used for public investments are generally gathered through taxation of the citizenry. The concept of *taxpayer opportunity cost* suggests that a correct interest rate to use in evaluating public investments is that which the *taxpayer* could have received if the government had not collected those dollars through taxation. This philosophy holds that through taxation the government is taking away the taxpayers' opportunity to use those same dollars for investment. The interest rate that the government requires should not be less than what the taxpayer would have received. This compelling argument is supported in general by the 1972 Office of Management and Budget (OMB) directive that stipulates a 10% interest rate be used in economic analysis for a wide range of federal projects. It is not economically desirable to take money from a taxpayer with a 12% opportunity cost, for example, and invest it in a government project yielding 4%.

Recommended Concept: The general rule of thumb in setting an interest rate for government investments has been to select the *largest* of: the cost of capital; the government opportunity cost; or the taxpayer opportunity cost interest rates. However, as is the case in the private sector, there is no hard and fast rule universally applied in all decision circumstances. With the exception of projects for which the 1972 OMB directive was intended, setting an interest rate for use in economic analysis is at the discretion of the government entity performing the analysis. Consider the seven government entities given in Table 16-1. The interest rates used by these decision-making bodies for evaluating investments could all potentially be set at different levels. Setting these interest rates would involve a management dicision based on both objective (cost of capital, etc.) and subjective (risk attitudes, etc.) factors considered by each unit.

Table 16-1 EXAMPLES OF INTEREST RATES USED IN

GOVERNMENT ECONOMIC ANALYSIS

Government Entity	Interest Rate Used
United States Armed Services	4%
State Agency	6%
Federal Highway Transportation Board	3%
City Port Authority	5%
City School Board	6.5%
State Waterway Commission	5%
City of Anytown	8%

The Benefit-Cost Ratio

The benefit-cost ratio has been described briefly in Chapter 9 as one of the alternative economic analysis methods for evaluating prospective projects. This method is used almost exclusively in public investment analysis, and due to the magnitude of public dollars committed each year through such analysis, the benefit-cost ratio deserves our attention and understanding.

One of the primary reasons for the use of the benefit-cost ratio (B/C ratio) in public decision making is its simplicity. The ratio is formed by calculating the equivalent worth of the benefits accrued through investment in a project divided by the equivalent worth of the costs of the project. The benefit-cost ratio can be shown as:

$$\text{B / C ratio} = \frac{\text{Equivalent worth of net benefits}}{\text{Equivalent worth of costs}} = \frac{\text{PW benefits}}{\text{PW costs}} = \frac{\text{FW benefits}}{\text{FW costs}} = \frac{\text{AW benefits}}{\text{AW costs}}$$

Notice that *any* of the equivalent worth methods (present, future and annual) can be used to calculate this ratio. Each formulation of the ratio will produce an identical result, as illustrated in Example 16-2 below.

EXAMPLE 16-2

Consider a highway expansion project with data as given below. Demonstrate that an equivalent B/C ratio is obtained using the present, future and annual worth formulations.

Initial costs of expansion $\qquad$ = \$1,500,000

Annual costs for operating /maintenance = $\qquad$ \$65,000

Annual savings and benefits to travelers = \$225,000

Scrap value after useful life $\qquad$ = \$300,000

Useful life of investment $\qquad$ = 30 years

Interest Rate $\qquad$ = 8%

Using Present Worth

PW Benefits = 225,000(P/A,8%, 30) + 300,000(P/F,8%,30) = \$2,563,000

PW Costs = 1,500,000 + 65,000(P/A,8%,30) = \$2,232,000

Using Future Worth

FW Benefits = 225,000(F/A,8%, 30) + 300,000 = \$25,790,000

FW Costs = 1,500,000(F/P,8%,30) + 65,000(F/A,8%,30) = \$22,460,000

Using Annual Worth

AW Benefits = 225,000 + 300,000(A/F,8%,30) = \$227,600

AW Costs = 1,500,000(A/P,8%,30) + 65,000 = \$198,200

$$\text{B / C ratio} = \frac{2,563,000}{2,232,000} = \frac{25,790,000}{22,460,000} = \frac{227,600}{198,200} = 1.15$$

One can see that the ratios provided by each of these methods produces an identical ratio of 1.15. ■

An economic analysis is performed to assist in the objective of making a decision. When using the B/C ratio the decision rule is:

If the B/C ratio is > 1.0, then the decision should be to "invest".

If the B/C ratio is < 1.0, then the decision should be "do not invest".

Cases where the B/C ratio is *just equal to* 1.0 are analogous to the case where a calculated net present worth = $0 or an IRR analysis yields an $i^* = $ MARR%. In other words, the decision measure is *just at* the break-even criteria. In such cases a detailed analysis of the input variables and their estimates is necessary, and one should consider the merits of other available opportunities for the targeted funds. But, if the B/C ratio is greater than or less than 1.0, the recommendation is clear.

The B/C ratio is a numerator/denominator relationship between *benefits* and *costs*:

$$B\,/\,C \text{ ratio } = \frac{\text{EW of Net Benefits to whomsoever they may accrue}}{\text{EW of Costs to the sponsors of the project}}$$

The numerator and denominator aspects of the ratio are sometimes interpreted and used in different fashions. For instance, the *conventional B/C ratio* defines the numerator and denominator as:

$$\text{Conventional B}\,/\,C \text{ ratio } = \frac{\text{EW of Net Benefits}}{\text{EW of Initial Costs } + \text{ EW Operating and Maintenance Costs}}$$

This is the formulation of the B/C ratio used in Example 16-2. However, there is another version of the ratio called the *modified B/C ratio.* Using a *modified* version, the numerator and denominator are defined in a different manner, where the *annual operating and maintenance costs* to the users are subtracted in the numerator versus being added as a cost in the denominator. In this case the ratio becomes:

$$\text{Modified B}\,/\,C \text{ ratio } = \frac{\text{EW of Net Benefits - EW Operating and Maintenance Costs}}{\text{EW of Initial Costs}}$$

For decision making the two versions of the benefit-cost ratio will produce the same recommendation on whether to *invest* or *not invest* in the project being considered. The *numeric B/C ratio* for the two versions will not always be the same, but the recommendation will be. This fact is illustrated in Example 16-3.

EXAMPLE 16-3

Consider the highway expansion project from Example 16-2. Let us calculate the B/C ratio using the present worth formulation of *conventional* and *modified* versions.

Using the *Conventional B/C ratio*

$$B / C \text{ ratio} = \frac{225,000(P/A,8\%, 30) + 300,000(P/F,8\%,30)}{1,500,000 + 65,000(P/A,8\%,30)} = 1.15$$

Using the *Modified B/C ratio*

$$B / C \text{ ratio} = \frac{225,000(P/A,8\%, 30) + 300,000(P/F,8\%,30) - 65,000(P/A,8\%,30)}{1,500,000} = 1.22$$

Using either the conventional or modified ratio, the recommendation is to invest in the highway expansion project. The magnitude of the ratios is not identical, 1.14 versus 1.22, but the decision is the same. ■

It is important when using the conventional and modified B/C ratios that one does not directly compare the magnitude of one versus the other. Evaluating a project with one version may produce a higher ratio than produced with the other version, but this does not imply that the project is somehow better.

The *net benefits to the users* of government projects are the difference between the expected *benefits* from investment minus the expected *disbenefits*. Disbenefits are the negative effects of government projects felt by some individuals or groups. As an example, consider the National Park System in the United States. Development projects by the skiing or lumber industries might provide enormous benefits to the recreation or construction sectors while creating simultaneous disbenefits for environmental groups. Table 16-2 illustrates some of the primary benefits and disbenefits of several example public investments.

Table 16-2 EXAMPLE BENEFITS AND

DIS-BENEFITS FOR PUBLIC INVESTMENTS.

Public Project	*Primary Benefits*	*Primary Disbenefits*
New city airport on the exterior of town	More flights, new businesses	Increased travel time to airport, more traffic on outerbelt
Interstate bypass around town	Quicker commute times, reduced congestion on surface roads	Lost sales to businesses on surface roads, lost agricultural lands
New metro subway system	Faster commute times, less pollution	Lost jobs due to bus line closing, less access to service (fewer stops)
Creation of a city disposal facility versus sending waste out of state	Less costly, faster and more responsive to customers	Objectionable sight and smells, lost market value to homeowners, lost pristine forest land
Construction of a nuclear energy generation facility	Cheaper energy costs, new industry in area	Perceived environmental risks

Incremental Benefit-Cost Analysis

In chapter 9 we discussed using the incremental benefit-cost ratio in economic decision analysis. As is the case with the internal rate of return (IRR) decision method, the incremental B/C ratio should be used when comparing *sets of mutually exclusive alternatives*. This method produces a result that is consistent with the result produced by optimizing the present worth of the decision alternatives over their respective life cycles. As with the incremental IRR method, it is *not* proper to simply calculate the B/C ratio for each alternative and chose the one with the highest value — rather, an *incremental* approach is called for. The incremental approach used with the IRR method can be slightly augmented for use with the B/C ratio.

Elements of the Incremental Benefit-Cost Ratio Method

1. Identify all relevant alternatives. Decision makers identify the set of alternatives from which a choice is ultimately made. In this context it is important to identify *all* relevant and competitive alternatives. Decision rules or models can recommend a *best* course of action *only* from the set of identified alternatives. If a better alternative than those in the considered set exists, it will never be selected, and the solution will be suboptimal. For benefit-cost ratio problems, the "do-nothing" option is always the "base case" from which the incremental methodology proceeds.

2. (Optional) Calculate the B/C ratio of each competing alternative in the set. In this optional step one calculates the B/C ratio of each of the competing alternatives based on the *total cash flows for each alternative* by itself. Once the individual B/C ratios are calculated, those alternatives with a ratio *less than 1.0* are eliminated from further consideration. Those alternatives with a B/C ratio *greater than 1.0* remain in the set of feasible alternatives. This step gets the "poor performers" out of the way before the incremental procedure is initiated. However, if this step is omitted, the incremental analysis method will eliminate the sub-par alternatives in due process.

3. Rank order the projects. Using the incremental IRR method to analyze each of the mutually exclusive alternatives of the set, the alternatives are ordered in increasing first cost. However, when using the incremental B/C ratio, the alternatives must be ordered in increasing size of the *denominator of the B/C ratio.* (The rank order will be the same regardless if one uses the present worth, annual worth or future worth of costs to form the *denominator.*) To form the list the *denominator* of the B/C ratio (cost portion of the ratio) is first calculated for each of the feasible alternatives and then placed into an ascending rank order from low to high cost. The "do nothing" alternative always becomes the first on the ordered list.

4. Identify the increment under consideration. In this step the increment under consideration is identified. The first increment taken under consideration is always that of going from the "do nothing" option to the first of the feasible alternatives. As the analysis proceeds, any identified increment is always in reference to some previously justified alternative.

5. Calculate the B/C ratio on the considered incremental cash flows. Upon identifying the increment under consideration, it is necessary to calculate the *incremental benefits* as well as the *incremental costs*. This step is accomplished by finding the cash flows that represent the difference (Δ) between the two alternatives under consideration. For two alternatives X and Y, the incremental benefits (ΔB) and incremental costs (ΔC) of going from alternative X to alternative Y must be determined. The increment can be written as $(X \rightarrow Y)$ to signify *going from X to Y* or as (Y-X) to signify the *cash flows of Y **minus** cash flows of X.* Both modes identify the incremental costs and benefits of investing in alternative Y, where X is a previously justified (or base) alternative. The (ΔB) and (ΔC) values are used to calculate the overall *incremental B/C ratio* ($\Delta B/\Delta C$) of the increment.

6. Use the incremental B/C ratio to make a decision. The incremental B/C ratio ($\Delta B/\Delta C$) calculated in step 5 is evaluated as follows: if the ratio is *greater than 1.0*, then the increment is desirable or *justified*; if the ratio is *less than 1.0,* it is not desirable, or is *not justified.* If an increment is accepted, the alternative associated with that additional increment of investment becomes the base from which the next increment is formed. In the case where an increment is not justified, then the alternative associated with the additional increment is rejected and the previously justified alternative is maintained as the base for formation of the next increment.

7. Iterate to Step 4 until all increments (projects) have been considered. The incremental method requires that the entire list of ranked feasible alternatives be evaluated. All pairwise comparisons are made such that the additional increment being considered is done so from a previously justified alternative. The incremental method continues until all alternatives have been evaluated.

8. Select the appropriate alternative from the set of mutually exclusive competing projects. After all alternatives (and associated increments) have been considered, the incremental B/C ratio method calls for selection of the alternative that is *associated with the last justified increment.* In this way, it is assured that a maximum investment is made such that the ratio of equivalent worth of benefits to equivalent worth of cost is greater than 1.0. (A common error in applying the incremental B/C method is the selection of the alternative with the *largest* incremental B/C ratio, which is inconsistent with the objective of maximizing investment size above a B/C ratio of 1.0.)

Both the conventional and modified versions of the B/C ratio can be used with the incremental B/C ratio methodology described above, but the two versions should not be mixed within the same problem. Such an approach could affect the rank order and cause confusion and errors. Instead, *one* of the two versions should be *consistently* used throughout the analysis.

The following examples illustrate the use of the incremental B/C ratio and show how the conventional and modified versions can be used with this procedure to evaluate sets of mutually exclusive alternatives.

EXAMPLE 16-4

A midwestern industrial state is considering the construction and operation of facilities to provide electricity to several state owned properties. Electricity will be provided via two coal-burning power plants and a distribution network wired to the properties targeted for conversion. A group studying the proposal has indentified general cost and benefit categories for the project as follows:

Primary Costs: Construction of the power plant facilities; cost of installing the power distribution network; life-cycle maintenance and operating costs.

Primary Benefits: Elimination of payments to the current electricity provider; creation of jobs for construction, operation and maintenance of the facilities and distribution network; revenue from selling excess power to utility companies; and increased employment for coal mines within the state.

Assume that there have been four competing designs identified for the power plants. Each design affects costs and benefits in a unique way. Given the data below for the four mutually exclusive design alternatives, recommend a course of action using the *conventional* B/C ratio method.

Competing Design Alternatives   (Values in $10,000)

	I	II	III	IV
Project Costs				
Plant Construction Cost	12,500	11,000	12,500	16,800
Annual Operating & Maintenance Cost	120	480	450	145
Project Benefits				
Annual Savings from Utility Payments	580	700	950	1300
Revenue from Over-Capacity	700	550	200	250
Annual Effect of Jobs Created	400	750	150	500
Other Data				
Project Life	45 yrs.	45 yrs.	45 yrs.	45 yrs.
Discounting Rate (MARR)	8%	8%	8%	8%

Solution: Alternatives *I* through *IV* constitute a set of *mutually exclusive* choices because we will select one and only one of the design options for the power plants. Therefore, an incremental B/C ratio method is used to obtain the solution. Let us use the incremental method as described above:

Step 1 In this case the alternatives are: do nothing, and designs *I, II, III* and *IV*.

Step 2 In this optional step we calculate the *conventional* B/C ratio for each alternative based on individual cash flows. We will use the ratio of the PW of benefits to costs.

B/C ratio (*I*) $= (580 + 700 + 400)(P/A,8\%,45) / (12,500 + 120(P/A,8\%,45) = 1.46$

B/C ratio (*II*) $= (700 + 550 + 750)(P/A,8\%,45) / (11,000 + 480(P/A,8\%,45) = 1.44$

B/C ratio (*III*) $= (200 + 950 + 150)(P/A,8\%,45) / (12,500 + 325(P/A,8\%,45) = 0.96$

B/C ratio (*IV*) $= (1300 + 250 + 500)(P/A,8\%,45) / (16,800 + 145(P/A,8\%,45) = 1.34$

Alternative *I, II,* and *IV* all have B/C ratios *greater than 1.0* and thus are included in the feasible set. Alternative *III* does not meet the acceptable criteria and should be eliminated from further consideration. However, to illustrate the fact that step 2 is optional we will continue with all four design alternatives in the original feasible set.

Step 3 Here we calculate the PW of costs for each alternative in the feasible set (remember we are keeping alternative *III* along for the ride). The denominator of the *conventional* B/C ratio includes first cost and annual O&M costs. We calculate the PW of costs for each alternative as below.

PW Costs (*I*) $= 12,500 + 120(P/A,8\%,45) = \$ 13,953$

PW Costs (II) = 11,000 + 480$(P/A,8\%,45)$ = \$ 16,812

PW Costs (III) = 12,500 + 325$(P/A,8\%,45)$ = \$ 16,435

PW Costs (IV) = 16,800 + 145$(P/A,8\%,45)$ = \$ 18,556

The rank order from low to high value of the B/C ratio *denominator* would be: do nothing, I, III, II, IV

Step 4 From the ranking list, the first increment considered is that of going from "do nothing" to alternative I (do nothing → alternative I). The analysis proceeds from this point.

Steps 5 and 6 We proceed through the analysis designating the incremental cash flows and calculating the Δ B/C ratio until all alternatives on the feasible list have been considered. Each additional increment taken under consideration must be based on the last justified increment. *Increment is ....*

Incremental Effects:	(do nothing → I)	(I→ III)	(I→ II)	(II→ IV)
Δ Plant Construction Cost	12,500	0	-1500	5800
Δ Annual O & M Cost	120	205	360	-335
PW of Δ Costs	13,953	2482	2859	1744
Δ Annual Utility Payment Savings	580	370	120	600
Δ Annual Over-Capacity Revenue	700	500	-150	-300
Δ Annual Benefits of New Jobs	400	-250	350	-250
PW of Δ Benefits	20,342	-4601	3875	605
Δ B/C ratio (PW ΔB/PW ΔC)	1.46	-1.15	1.36	0.35
Is increment justified ?	Yes	No	Yes	No

As an example of the calculations in the table above, consider the third increment $(I→II)$.

Δ Plant Construction Cost	= 11,000 - 12,500 = \$ -1500
Δ Annual O & M Cost	= 480 - 120 = \$ 360
PW of Δ Costs	= -1500 + 360$(P/A,8\%,45)$ = \$ 2859
or	= 16,812 - 13,953 = \$ 2859
Δ Annual Utility Payment Savings	= 700 - 580 = \$ 120
Δ Annual Over-Capacity Revenue	= 550 - 700 = \$-150
Δ Annual Benefits of New Jobs	= 750 - 400 = \$ 350
PW of Δ Benefits	=(120-150+350)$(P/A,8\%,45)$ = \$ 3875
Δ B/C ratio (PW ΔB/PW ΔC)	= 3875 / 2850 = 1.36

Step 8 The analysis in the table above proceeded as follows: do nothing to alternative *I* was justified (Δ B/C ratio = 1.46), alternative *I* became the new base; alternative *I* to alternative *III* was not justified (Δ B/C ratio = -1.15), alternative *I* remained base; alternative *I* to alternative *II* was justified (Δ B/C ratio = 1.36), alternative *II* became the base; alternative *II* to alternative *IV* was not justified (Δ B/C ratio = 0.35), alternative *II* became the recommended power plant design alternative because it is the alternative associated with the last justified increment All alternatives in the feasible set were considered before the recommendation was produced. Notice that alternative *III*, even though included in the feasible set, did not affect the recommendation and was eliminated through the incremental method. Notice also that the first increment considered (do nothing → *I*) was not selected even though it had the *largest* Δ B/C ratio (1.45). Selection is based on the alternative associated with the *last justified increment* (in this case alternative *II*). ■

EXAMPLE 16-5

Let us consider Example 16-4 above and this time use the modified B/C ratio to analyze the set of competing design alternatives. Again we will use the present worth method.

Solution: Here we use the modified B/C ratio.

Step 1 The alternatives are still: do nothing, and designs *I, II, III* and *IV*.

Step 2 We calculate the modified B/C ratio using the PW of benefits and costs.

B/C ratio (*I*) = (580 + 700 + 400 - 120)(*P/A*,8%,45) / (12,500) = 1.51

B/C ratio (*II*) = (700 + 550 + 750 - 480)(*P/A*,8%,45) / (11,000) = 1.67

B/C ratio (*III*) = (200 + 950 + 150 - 325)(*P/A*,8%,45) / (12,500) = 0.95

B/C ratio (*IV*) = (1300 + 250 + 500 - 145)(*P/A*,8%,45) / (16,800) = 1.37

Again alternative *III* would be eliminated from further consideration because it has a B/C ratio *less than 1.0*. In this case we will eliminate it from the feasible set, which now becomes: do nothing, and alternative designs *I, II,* and *IV*.

Step 3. The PW of costs for each alternative in the feasible set are as follows.

PW Costs (*I*) = $ 12,500

PW Costs (*II*) = $ 11,000

PW Costs (*IV*) = $ 16,800

The appropriate rank order is now: do nothing, *II, I, IV*. Notice that the *modified* B/C ratio produces a different rank order than the *conventional* version did in the previous example.

Step 4 The first increment is now (do nothing → alternative *II*). The method proceeds from this point.

Steps 5 and 6 These steps are accomplished as below.

	Increment is		
Incremental Effects:	*(do nothing → II)*	*(II → I)*	*(II → IV)*
Δ Plant Construction Cost	11,000	1500	5800
PW of Δ Costs	11,000	1500	5800
Δ Annual Utility Payment Savings	700	-120	600
Δ Annual Over-Capacity Revenue	550	150	-300
Δ Annual Benefits of New Jobs	750	-350	-250
Δ Annual O & M Disbenefit	480	-360	-335
PW of Δ Benefits	18,405	484	4662
Δ B/C ratio (PW ΔB/PW ΔC)	1.67	0.32	0.80
Is increment justified ?	Yes	No	No

As an example of the calculations in the table above, consider the third increment (*II→IV*).

Δ Plant Construction Cost	= 16,800 - 11,000 = \$ 5800
PW of Δ Costs	= \$ 5800
Δ Annual Utility Payment Savings	= 1300 - 700 = \$ 600
Δ Annual Over-Capacity Revenue	= 250 - 550 = \$ -300
Δ Annual Benefits of New Jobs	= 500 - 750 = \$ -250
Δ Annual O & M Disbenefit	= 145 - 480 = \$ -335
PW of Δ Benefits	= (600 - 300 - 250 + 335)(*P/A*,8%,45) = \$ 4662
Δ B/C ratio (PW ΔB/PW ΔC)	= 5800 / 4662 = 0.80

Using the modified version of the B/C ratio, alternative *II* emerges as the recommended power plant design — just as it did using the conventional B/C ratio. ■

Other Effects of Public Projects

Three areas remain which merit discussion in describing the differences between government and non-government economic analysis: (1) financing government versus non-government projects; (2) the typical length of government versus non-government project lives; and (3) the general effects of politics on economic analysis.

Project Financing

Governmental and non-government organizations differ in the way investments in equipment, facilities and other projects are financed. In general, non-government firms rely on monies from individual investors (through stock and bond issuance), private lenders and retained earnings from operations. These sources serve as the pool from which investment dollars for projects come. Management's job in the non-government firm is to match financial resources with projects in a way that: keeps the firm growing, produces an efficient and productive environment; and continues to attract investors and lenders of future capital.

On the other hand, the government sector often uses taxation and municipal bond issuance to serve as the source of investment capital. In government, taxation and revenue from operations is adequate to finance only modest projects. However, public projects tend to be large-scale in nature (roadways, bridges, etc.), which means that for many public projects 100% of the investment costs must be borrowed — unlike those in the private sector. To prevent excessive public borrowing and assure timely debt repayment the U.S. government through constitutional and legislative channels has placed restrictions on government debt. These restrictions include:

1. Local government bodies are limited in their borrowing to a specified percentage of the assessed value of the property in their taxation district. This means government entities in areas with high property values can borrow more total monies than those in lower assessed-value areas.

2. For new construction, borrowed funds attained through the sale of bonds require the approval of local voters (sometimes by a two-thirds majority) through the election process. An example might be a $2 million levy to fund a new municipal jail for a city. If approved by voters, a $2 million increase would raise the property taxes of land owners in the city's tax district by $2 for every $1000 of assessed property value. These additional tax revenues would then be used to retire the debt on the bonds sold to finance the city jail project.

3. Repayment of public debt must be made within a preset period of time and in accordance to a specific plan. This is often the case with monies borrowed by issuing bonds, as bond interest payments and maturity dates are set at the time of issuance.

Limitations on the use and sources of borrowed monies make funding projects in the public sector much different from the private sector. Private sector firms are seldom able to borrow 100% of required funds for projects, as in the public sector, but at the same time, private entities do not face restrictions on debt retirement nor the uncertainty of voter approval.

Project Duration

Another aspect distinguishing government projects from those in the private sector is the typical duration or project life of the investment. In the private sector, projects most often have a projected or intended life ranging between 5 and 15 years. On some occasions the project life is shorter and in others longer, but a majority of projects fall in this interval. Complex advanced manufacturing technologies, like computer aided manufacturing or flexible automated manufacturing cells, typically have projects lives at the longer end of this range. In the 1980's some criticized U.S. manufacturing managers for being short-sighted in their views of capital investments in such technologies. At that time short-sightedness and lack of investments were blamed for the overall loss of competitiveness in several key U.S. industries (such as, textiles, steel, electronics, automotive, and machine tools.)

Government projects typically have lives in the range of 20 to 50 years (or longer). Typical projects include: federal highways, city water/sewer infrastructure, county dumps, and state libraries and museums. These projects, by nature, have a longer useful life than a typical project in the private sector. There are exceptions to this rule because private firms invest in facilities and other long-range projects, and government entities also invest in projects with shorter-term lives. But, in general, investment duration in the government sector is longer than in the private sector.

Government projects, because they tend to be long-range and large-scale, usually require substantial funding in the early stages. The highway, water/sewer, and library projects mentioned above could each require millions of dollars in design, surveying, and construction costs. Because of this requirement, it is in the best interest of decision makers who are advocates of such projects to spread that first cost over as many years as possible to reduce the annual cost of capital recovery. This tendency to use longer project lives to downplay the effects of a large first cost can affect the desirability of the project, as measured by the B/C ratio. Another aspect closely associated with managing the size of the capital recovery cost in a B/C ratio analysis is the interest rate used for discounting. Lower interest rates reduce the size of the capital recovery cost by reducing the penalty of having money tied up in a project. Example 16-6 below illustrates the effects that project life and interest rate can have on the analysis and acceptability of a project.

EXAMPLE 16-6

Consider a project that has been approved by local voters to build a new junior high school needed because of increased (and projected) population growth. Information for the project is as follows:

Building first costs (design, planning and construction)	= $10,000,000
Initial cost for roadway and parking facilities	= $5,500,000
First cost to equip and furnish facility	= $500,000
Annual operating and maintenance costs	= $350,000
Annual savings from rented space	= $500,000
Annual benefits to community	= $1,500,000

With this project we examine the effect that varying project lives and interest rates have on the conventional B/C ratio. Project lives at: 15, 30, and 60 years; and interest rates at: 3%, 10% and 15% are used to calculate the ratio for the investment. The ratio for each combination of project life and interest rate is given in the table below.

Conventional Benefit-Cost Ratio for various Combinations of Project Life and Interest Rate

Project Life (yrs.)	Interest at …. 3%	10%	15%
15	1.24	0.86	0.69
30	1.79	1.03	0.76
60	2.24	1.08	0.77

As an example of how the ratios are calculated, consider the case where life =30 years and interest rate =10%.

$$\text{Conventional B / C ratio} = \frac{1,500,000 + 500,000}{(10,000 + 5,500,000 + 500,000)(A / P, 10\%, 30) + 350,000} = 1.03$$

From the numbers above one can see the effect that project life and interest have on the analysis and recommendation. At the lower interest rate the project has B/C ratios above 1.0 in all cases of project life, while at the higher rate the ratios are all less than 1.0. At an interest rate of 10% the recommendation to invest changes from *no* at a life of 15 years to *yes* at 30 and 60 years. By manipulating these two parameters (project life and interest rate) it is possible to reach entirely different conclusions regarding the desirability of the project. The key point is that for those advocating "investment," using lower interest rates and longer project life is more favorable. ■

Project Politics

To some degree political influences are felt in nearly every decision made within all organizations. Predictably, some individual or group will support its own particular interests over competing views. This actuality exists in both government and non-government organizations. In government the effects of politics are continuously felt at all levels because of the large-scale and multi-purpose nature of projects and because government decision making involves the use of the citizenry's common pool of money.

To illustrate on a small scale situations faced by government, compare the decision making process a family may face when determining where to dine out. As most families can attest, this decision is not always easy — even when Dad is footing the bill! Imagine the increased level of conflict that would arise if every member of the family were to contribute to the tab. Perhaps the choices would then be: dine out; go to a ball game; have a shopping spree at the mall; loan the money to Mom; or put it into the family bank account. This scenario characterizes government investment decisions — individuals and groups with different values and views spending a common *pool* of money. As with the family decision, the parties involved often have squabbles, form alliances and maneuver politically.

The guideline for public decision making, as set forward in the *Preamble to the United States Constitution,* is to *promote the general welfare* of citizens. However, it is an impossible task to please everyone all of the time. The term *general welfare* implies that the architects of this document understood that the political process would produce opposition, but at the same time they empowered decision makers to act in a representative way.

As mentioned previously, government projects tend to be large-scale in nature. Therefore, the time required to plan, design, fund, and construct such projects is usually several years. However, the political process tends to produce government leaders who support short term decision making (because many government terms-of-office, either elected or appointed, are relatively short term.) Therein lies another difference between government and non-government decision making — short term decision making, long term projects.

Because government decision makers are in the public eye more than those in the private sector, governmental decisions are generally more affected by "politics". As such, the decisions that public officials make may not always be the best from an *overall* perspective. If a particular situation exposes a public official to ridicule, he may choose an expedient action to eliminate negative exposure (where a more careful analysis might have been better). Or, such a decision maker may placate a small, but vocal, political group over the interest of the majority of citizens by committing funds to a favored project (at the expense of other better projects). Or, a public decision maker may make no decision on an important, but politically charged, issue in order to avoid controversy (where it would be in the overall interest of the citizenry if action were taken). Indeed, compared to the private sector, the role of politics in government decision making is more complex and far ranging.

EXAMPLE 16-7

Consider again Example 16-4 where power plants designs were being evaluated. Remember that government projects are often opposed and supported by differing groups within the populace. As such, decision makers become very aware of potential political aspects when considering such projects. For the electric power plant decision, several political considerations may affect any evaluation of funding this project.

☐ The current governor has been a strong advocate of *workers rights* and has received abundant campaign support from organized labor (which is especially important in an industrialized state). By championing this project, the governor should be seen as pro-labor, thereby benefiting his bid for re-election, even if the project is not funded.

☐ The regulated electric utility providers in the state are strongly against this project, claiming it will directly compete with them and take away some of their biggest customers. The providers have a strong lobby and key contacts with the state's utilities commission. A senior state senator has already protested that this project is the first step toward "rampant socialism in this great state."

☐ Business leaders in the municipalities where the two facilities will be constructed are in favor of the project because it will create more jobs and increase the tax base. These leaders promote the project as a win-win opportunity for both government and industry, where the state can win by reducing costs, and the electric utilities can win by focusing more effectively on residential customers and their needs.

☐ The lieutenant governor is promoting this project proclaiming that it is an excellent example of "initiating proactive and creative solutions to the problems that this state faces."

☐ Federal and state regulatory agencies are closely watching this project with respect to the Clean Air Act. Speculation is that the state plans to use a high sulfur grade of local coal exclusively. Thus "stack scrubbers" would be required, or the high sulfur coal would have to be mixed with lower sulfur coal imported from other states to bring the overall air emissions in line with federal standards. The governor is using this opportunity to make the point that "the people of this state don't need regulators to tell us if we can use our own coal!"

☐ The state's coal operators and mining unions are very much in favor of this project. They see the increased demands for coal and the governor's pro-labor advocacy as very positive. They plan to lobby the legislature strongly in favor of the project.

☐ Land preservation and environmental groups are strongly opposing the proposed project. They have studied the potential negative impacts of this project on the land, water and air quality, as well as on the ecosystem and wildlife, in the areas where the two facilities will be constructed. Environmentalists have started a public awareness campaign urging the governor to act as the "chief steward" of the natural beauty and resources of the state.

Will the project be funded? We can only guess. Clearly, however, we can see the competing influences that can be, and often are, part of decision making in the public sector. ■

Summary

Economic analysis and decision making in government is notably different from that of the private sector because the basic objectives of the public and private sectors are fundamentally different. Government investments in projects seek to maximize benefits to the *greatest number of citizens*, while minimizing the *disbenefits to citizens* and *costs to the government*. Private firms, on the other hand, are focused primarily on maximizing stockholder wealth.

Several factors, not affecting private firms, enter into the decision-making process in government. The souce of capital for public projects is primarily limited to taxes and bonds. Government bonds issued for project construction are subject to legislative restrictions on debt not required for private firms. Also, raising tax and bond monies both involve sometimes long and politically charged processes not present in the private sector. In addition, government projects tend to be larger-scale than those of competitive firms and affect many more people and groups in the population. All of these factors slow down the process and make investment decision analysis more difficult for government decision makers when compared to the private sector. Another difference between the public and private sectors is how the interest rate (MARR) is set for economic studies. In the private sector, considerations for setting the rate include the cost of capital and opportunity costs. In government, establishing the interest rate is complicated by uncertainty in specifying the cost of capital and the issue of assigning opportunity costs to taxpayers or to the government.

The benefit/cost ratio is widely used to evaluate and justify government funded projects. This measure-of-merit is the ratio of the equivalent worth of benefits to the equivalent worth of costs. This ratio can be calculated using PW, AW or FW methods. A B/C ratio *greater than 1.0* indicates that a project should be invested in if funding sources are available. For considering a set of *mutually exclusive alternatives,* an incremental method should be used to evaluate the merits of additional cost. This method results in the recommendation of the project with the highest investment cost that can be incrementally justified. Two versions of the B/C ratio, the *conventional* and *modified* B/C ratios, produce identical recommendations when considering single projects or sets of competing alternatives. The difference between the two ratios is in the way that annual operating and maintenance costs are handled — as an added cost in the denominator in the former, or as a subtracted benefit in the numerator in the latter.

Problems

16-1   Consider the following investment opportunity:

Initial cost	$100,000
Additional cost at end of Year 1	150,000
Benefit at end of Year 1	0
Annual benefit per year at end of Years 2–10	20,000

With interest at 7%, what is the benefit–cost ratio for this project? (*Answer:* 0.83)

16-2   A government agency has estimated that a flood control project has costs and benefits that are parabolic, according to the equation:

$$(\text{Present worth of benefits})^2 - 22(\text{Present worth of cost}) + 44 = 0$$

where both benefits and costs are stated in millions of dollars. What is the present worth of cost for the optimal size project?

16-3   The Highridge Water District needs an additional supply of water from Steep Creek. The engineer has selected two plans for comparison:
   Gravity plan: Divert water at a point 10 miles up Steep Creek and carry it through a pipeline by gravity to the district.
   Pumping plan: Divert water at a point near the district and pump it through 2 miles of pipeline to the district. The pumping plant can be built in two stages, with half capacity installed initially and the other half 10 years later.
   Use a 40 year analysis period and 8% interest. Salvage values can be ignored. During the first 10 years, the average use of water will be less than during the remaining 30 years. Using present worth analysis, select the more economical plan.

	Gravity	Pumping
Initial investment	$2,800,000	$1,400,000
Additional investment in tenth year	0	200,000
Operation, maintenance, replacements, per year	10,000	25,000
Average Power cost per year (first 10 years)	0	50,000
Average Power cost per year (next 30 years)	0	100,000

   (*Answer:* Pumping plan)

16-4   The federal government proposes to construct a multi-purpose water project to provide water for irrigation and municipal use. In addition, flood control and recreation benefits will be realized. The estimated benefits of the project computed for 10 year periods for the next 50 years are given in the table as follows:

Purpose	1st decade	2nd decade	3rd decade	4th decade	5th decade
Municipal	$ 40,000	$ 50,000	$ 60,000	$ 70,000	$110,000
Irrigation	350,000	370,000	370,000	360,000	350,000
Flood control	150,000	150,000	150,000	150,000	150,000
Recreation	60,000	70,000	80,000	80,000	90,000
Totals:	$600,000	$640,000	$660,000	$660,000	$700,000

The annual benefits may be assumed to be one-tenth of the decade benefits. The operation and maintenance cost of the project is estimated to be $15,000 per year. Assume a 50 year analysis period with no net project salvage value.

a. If an interest rate of 5% is used, and a benefit–cost ratio of unity, what capital expenditure can be justified to build the water project now?

b. If the interest rate is changed to 8%, how does this change the justified capital expenditure?

16-5 The city engineer has prepared two plans for the construction and maintenance of roads in the city park. Both plans are designed to provide the anticipated road and road maintenance requirements for the next 40 years. The minimum attractive rate of return used by the city is 7%.

Plan *A* is a three-stage development program. $300,000 is to be spent immediately, followed by $250,000 at the end of 15 years and $300,000 at the end of 30 years. Maintenance will be $75,000 per year for the first 15 years, $125,000 per year for the next 15 years, and $250,000 per year for the final 10 years.

Plan *B* is a two-stage program. $450,000 is required immediately (including money for special equipment), followed by $50,000 at the end of 15 years. Maintenance will be $100,000 per year for the first 15 years and $125,000 for each of the subsequent years. At the end of 40 years, it is believed the equipment may be sold for $150,000.

a. Determine which plan should be chosen, using benefit–cost ratio analysis.

b. If you favored Plan *B*, what value of MARR would you use in the computations? Explain.

16-6 The state is considering eliminating a railroad grade crossing by building an overpass. The new structure, together with the needed land, would cost $1,800,000. The analysis period is assumed to be 30 years based on the projection that either the railroad or the highway above it will be relocated by then. Salvage value of the bridge (actually, the net value of the land on either side of the railroad tracks) 30 years hence is estimated to be $100,000. A 6% interest rate is to be used.

At present, about 1000 vehicles per day are delayed due to trains at the grade crossing. Trucks represent 40%, and 60% are other vehicles. Time for truck drivers is valued at $18 per hour and for other drivers at $5 per hour. Average time saving per vehicle will be two minutes if the overpass is built. No time saving occurs for the railroad.

The installation will save the railroad an annual expense of $48,000 now spent for crossing guards. During the preceding 10 year period, the railroad has paid out $600,000 in settling lawsuits and accident cases related to the grade crossing. The proposed project will entirely eliminate both these expenses. The state estimates that the new overpass will save it about $6000 per year in expenses directly due to the accidents. The overpass, if built, will belong to the state.

Should the overpass be built? If the overpass is built, how much should the railroad be asked to contribute to the state as its share of the $1,800,000 construction cost?

16-7 An existing two-lane highway between two cities is to be converted to a four-lane divided freeway. The distance between them is 10 miles. The average daily traffic (ADT) on the new freeway is forecast to average 20,000 vehicles per day over the next 20 years. Trucks represent 5% of the total traffic. Annual maintenance on the existing highway is $1500 per lane-mile. The existing accident rate is 4.58 per million vehicle miles (MVM). Three alternate plans of improvement are now under consideration.

Plan A: Improve along the existing development by adding two lanes adjacent to the existing lanes at a cost of $450,000 per mile. It is estimated that this plan will reduce auto travel time by 2 minutes, and will reduce truck travel time by 1 minute when compared to the existing highway. The Plan *A* estimated accident rate is 2.50 per MVM. Annual maintenance is estimated to be $1250 per lane-mile.

Plan B: Improve along the existing alignment with grade improvements at a cost of $650,000 per mile. Plan *B* would add two additional lanes, and it is estimated that this plan would reduce auto and truck travel time by 3 minutes each, when compared to the existing facility. The accident rate on this improved road is estimated to be 2.40 per MVM. Annual maintenance is estimated to be $1000 per lane-mile.

Plan C: Construct a new freeway on new alignment at a cost of $800,000 per mile. It is estimated that this plan would reduce auto travel time by 5 minutes and truck travel time by 4 minutes when compared to the existing highway. Plan *C* is 0.3 miles longer than *A* or *B*. The estimated accident rate for *C* is 2.30 per MVM. Annual maintenance is estimated to be $1000 per lane-mile. Plan *C* includes abandonment of the existing highway with no salvage value.

Useful data: Incremental operating cost	—autos:	6¢ per mile
	—trucks:	18¢ per mile
Time saving	—autos:	3¢ per minute
	—trucks:	15¢ per minute
Average accident cost:		$1200

If a 5% interest rate is used, which of the three proposed plans should be adopted? (*Answer:* Plan *C*)

16-8 The local highway department is preparing an economic analysis to see if reconstruction of the pavement on a mountain road is justified. The number of vehicles traveling on the road increases each year, hence the benefits to the motoring public of the pavement reconstruction also increase. Based on a traffic count, the benefits are projected as follows:

Year	End-of-year benefit
2001	$10,000
2002	12,000
2003	14,000
2004	16,000
2005	18,000
2006	20,000
	and so on, increasing $2000 per year

The reconstructed pavement will cost $275,000 when it is installed and will have a 15 year useful life. The construction period is short, hence a beginning-of-year reconstruction will result in the end-of-year benefits listed in the table. Assume a 6% interest rate. The reconstruction, if done at all, must be done not later than 2006. Should it be done, and if so, in what year?

16-9   A section of road in the state highway system needs repair at a cost of $150,000. At present, the volume of traffic on the road is low, so that few motorists would benefit from the work. However, future traffic is expected to increase with resulting increased motorist benefits. The repair work will produce benefits for 10 years after it is completed. The highway planning department is examining five mutually exclusive alternatives concerning the road repair.

Year	Do not repair	Repair now	Repair 2 years hence	Repair 4 years hence	Repair 5 years hence
0	0	−$150,000			
1	0	5,000			
2	0	10,000	−$150,000		
3	0	20,000	20,000		
4	0	30,000	30,000	−$150,000	
5	0	40,000	40,000	40,000	−$150,000
6	0	50,000	50,000	50,000	50,000
7	0	50,000	50,000	50,000	50,000
8	0	50,000	50,000	50,000	50,000
9	0	50,000	50,000	50,000	50,000
10	0	50,000	50,000	50,000	50,000
11	0	0	50,000	50,000	50,000
12	0	0	50,000	50,000	50,000
13	0	0	0	50,000	50,000
14	0	0	0	50,000	50,000
15	0	0	0	0	50,000

Should the road be repaired and, if so, when should the work be done? Use a 15% MARR.

16-10   A 50 meter tunnel must be constructed as part of a new aqueduct system for a city. Two alternatives are being considered. One is to build a full-capacity tunnel now for $500,000. The other alternative is to build a half-capacity tunnel now for $300,000 and then to build a second parallel half-capacity tunnel 20 years hence for $400,000. The cost to repair the tunnel lining every 10 years is estimated to be $20,000 for the full-capacity tunnel and $16,000 for each half-capacity tunnel.

Determine whether the full-capacity tunnel or the half-capacity tunnel should be constructed now. Solve the problem by benefit–cost ratio analysis, using a 5% interest rate, and a 50-year analysis period. There will be no tunnel lining repair at the end of the 50 years.

16-11   Think about a major government construction project currently underway in your state/city/region. Are the decision makers who originally analyzed and initiated the project currently in office? How can politicians use "political posturing" with respect to government projects?

16-12 Describe how a decision maker can use each of the following to "skew" the results of a B/C ratio analysis in favor of his or her own position on funding projects:
  a. Conventional versus modified ratios.
  b. Interest rates.
  c. Project duration.
  d. Benefits, costs and disbenefits.

16-13 List the potential costs, benefits and disbenefits that should be considered when evaluating a nuclear power plant construction.

16-14 Given the data below, calculate the: **a.** conventional, and **b.** modified benefit-cost ratio for the investment.

Required First Costs	$1,200,000
Annual Benefits to Users	$ 500,000
Annual Disbenefits to Users	$ 25,000
Annual Cost to Government	$ 125,000
Project Life	35 years
Interest Rate	10%

16-15 For the data given in problem 16-14 above, for handling benefits and costs, demonstrate that the calculated B/C ratio is the same using the each of the following methods: present worth, annual worth and future worth.

16-16 Big City Carl, a local politician, is advancing a project for the construction of a new dock and pier system at the river to attract new commerce to the city. A committee appointed by the mayor (an opponent of Carl's) has developed estimates for the effects of the project. This data is given below:

Cost to wreck and remove current facilities	$ 750,000	
Material, labor and overhead for new construction	$ 2,750,000	
Annual operating and maintenance expenses		$ 185,000
Annual benefits from new commerce	$ 550,000	
Annual disbenefits to sportsmen in area	$ 35,000	
Project Life	20 years	
Interest Rate	8%	

  a. Using the *conventional* B/C ratio, determine if the project should be funded.
  b. After studying the numbers given by the committee, Big City Carl has argued that the project life should be *at least* 25 years and is more likely closer to 30 years. How did he arrive at this estimate, and why is he making this statement?

16-17   Two different routes, which entail driving across a mountainous section, are being considered for a highway construction project. The first route (the **high road**) will require building several bridges and navigates around the highest mountain points, thus requiring more roadway. The second alternative (the **low road**) will require the construction of several tunnels, but takes a more direct approach through the mountainous area. Projected travel volume for this new section of road is 2500 cars per day. Given the data below, use the *modified* B/C ratio to determine which alternative should be recommended. Assume the project life is 45 years and i = 6%.

	The High Road	*The Low Road*
Average Construction Cost Per mile	$200,000 /mile	$450,000 /mile
Number of Miles Required	35 miles	10 miles
Annual Benefit Per Car Per Mile	$0.015	$0.045
Annual O&M Costs Per Mile	$2000	$10,000

16-18 The Fishery and Wildlife Agency of Ireland is considering four mutually exclusive design alternatives for a major salmon hatchery. This agency of the Irish government uses the following B/C ratio for decision making:

$$B \, / \, C \text{ ratio} \; = \; \frac{EW(\text{Net Benefits})}{EW \, (\text{Capital Recovery Cost}) \; - \; EW \, (O\&M \text{ Cost})}$$

Using an interest rate of 8% and a project life of 30 years, recommend which of the designs is best.

Irish Fishery Design Alternatives (in thousands of dollars)

	A	*B*	*C*	*D*
First Cost	9500	12,500	14,000	15,750
Annual Benefits	2200	1500	1000	2500
Annual O&M Costs	550	175	325	145
Annual Disbenefits	350	150	75	700
Salvage Value	1000	6000	3500	7500

16-19 Six mutually exclusive investments have been identified for evaluation with the benefit-cost ratio method. Data for each is given below. Assume a MARR of 10% and an equal project life of 25 years for all alternatives.

Mutually exclusive alternatives ...

	1	*2*	*3*	*4*	*5*	*6*
Annualized net costs to sponsor	15.5	13.7	16.8	10.2	17.0	23.3
Annualized net benefits to users	20.0	16.0	15.0	13.7	22.0	25.0

   a.   Using annual worth and the B/C ratio, which alternative is best?
   b.   If this were a set of *independent* alternatives, how would you conduct a comparison?

16-20   Mr. D. O'Gratias, a top manager in his company, has been asked to consider the three mutually exclusive investment alternatives given below.

	A	B	C
Initial Investment	9,500	18,500	22,000
Annual Savings	3,200	5,000	9,800
Annual Costs	1,000	2,750	6,400
Salvage Value	6,000	4,200	14,000
Project Life	15 yrs.	15 yrs.	15 yrs.
MARR	12%	12%	12%

Answer the following questions.

a. Use the *conventional* B/C ratio to evaluate the alternatives and make a recommendation.
b. Use the *modified* B/C ratio to evaluate the alternatives and make a recommendation.
c. Use a present worth analysis to evaluate the alternatives and make a recommendation.
d. Use an internal rate of return analysis to evaluate the alternatives and make a recommendation.
e. Use the simple payback period to evaluate the alternatives and make a recommendation.

CHAPTER 17

Rationing Capital Among Competing Projects

We have until now dealt with situations where, at some interest rate, there is an ample amount of money to make all desired capital investments. But the concept of *scarcity of resources* is fundamental to a free market economy. It is through this mechanism that more economically attractive activities are encouraged at the expense of less desirable activities. For industrial firms, there are often more ways of spending money than there is money that is available. The result is that we must select from available alternatives the more attractive projects and reject—or, at least, delay—the less attractive projects.

This problem of rationing capital among competing projects is one part of a two-part problem called *capital budgeting.* In planning its capital expenditures, an industrial firm is faced with two questions: "Where will money for capital expenditures come from?" and, "How shall we allocate available money among the various competing projects?" In Chapter 15 we discussed the sources of money for capital expenditures as one aspect in deciding on an appropriate interest rate for economic analysis calculations. Thus, the first problem has been treated.

Throughout this book, we have examined for any given project two or more feasible alternatives. We have, therefore, sought to identify in each project the most attractive alternative. For the sake of simplicity, we have looked at these projects in an isolated setting—almost as if a firm had just one project it was considering. In the business world, we know that this is rarely the case. A firm will find that there are a great many projects that are economically attractive. This situation raises two problems not previously considered:

1. How do you rank projects to show their order of economic attractiveness?

2. What do you do if there is not enough money to pay the costs of all economically attractive projects?

In this chapter we will look at the typical situation faced by a firm: multiple attractive projects, with an inadequate money supply to fund all of them. To do this, we must review our concepts of capital expenditure situations and available alternatives. Then we can summarize the various techniques that have been presented for determining if an alternative is economically attractive, first by screening all alternatives to find those that merit further consideration. Following this, we will select the best alternative from each project, assuming there is no shortage of money. The next step will be the addition of a budget constraint.

When there is not enough money to fund the best alternative from each project, we will have to do what we can with the limited amount of money available. It will become important that we have a technique for accurately ranking the various competing projects in their order of economic attractiveness. All this is designed to answer the question, "How shall we allocate available money among the various competing projects?"

Capital Expenditure Project Proposals

At the beginning of the book, we described decision making as the process of selecting the best alternative to achieve the desired objective in a given situation or problem. By carefully defining our objective, the model, and the choice of criteria, the given situation is reduced to one of selecting the best from the feasible alternatives. In this chapter we call the engineering decision-making process for a given situation or problem a *project proposal.* Associated with various project proposals are their particular available alternatives. For a firm with many project proposals, the following situation may result:

Capital Expenditure Proposals.

Project 1—Additional manufacturing facility:

Alternative   *A.* Lease an existing building.

B. Construct a new building.

C. Contract for the manufacturing to be done overseas.

Project 2—Replace old grinding machine:

Alternative   *A.* Purchase semi-automatic machine.

B. Purchase automatic machine.

Project 3—Production of parts for the assembly line:

Alternative   *A.* Make the parts in the plant.

B. Buy the parts from a subcontractor.

Our task is to apply economic analysis techniques to this more complex problem.

Mutually Exclusive Alternatives and Single Project Proposals

Until now we have dealt with mutually exclusive alternatives, that is, where selecting one alternative results in rejecting the other alternatives being considered. Even in the simplest problems encountered, the question was one of selection *between* alternatives. Should, for example, Machine *A* or Machine *B* be purchased to perform the necessary task? Clearly, the purchase of one of the machines meant that the other one would not be purchased. Since either machine would perform the task, the selection of one precludes the possibility of selecting the other one as well.

Even in the case of multiple alternatives, we have been considering mutually exclusive alternatives. A typical example was: "What size pipeline should be installed to supply water to a remote construction site?" Only one alternative is to be selected. This is different from the situation for single proposals where only one course of action is outlined. Consider Example 17-1.

EXAMPLE 17-1

The general manager of a manufacturing plant has received the following project proposals from the various operating departments:

1. The foundry wishes to purchase a new ladle to speed up their casting operation.

2. The machine shop has asked for some new inspection equipment.

3. The painting department reports they must make improvements to the spray booth to conform with new air pollution standards.

4. The office manager wants to buy a larger, more modern safe.

Each project consists of a single course of action. Note that the single project proposals are also independent, for there is no interrelationship or interdependence among them. The general manager can decide to allocate money for none, some, or all of the various project proposals.

Solution: *Do-Nothing Alternative.* The four project proposals above each have a single course of action. The general manager could, for example, buy the office manager a new safe and buy the inspection equipment for the machine shop. But he could also decide to *not* buy the office manager a safe or the equipment for the machine shop. There is, then, an alternative to buying the safe for the office manager: not to buy him the safe—to do nothing. Similarly, he could decide to do nothing about the request for the machine shop inspection equipment. Naturally, there are do-nothing alternatives for each of the four single project proposals:

1*A*. Purchase the foundry a new ladle.

1*B*. Do nothing. (Do not purchase a new ladle.)

2*A*. Obtain the inspection equipment for the machine shop.

2*B*. Do nothing. (Do not obtain the inspection equipment.)

3*A*. Make improvements to the spray booth in the painting department.

3*B*. Do nothing. (Make no improvements.)

4*A*. Buy a new safe for the office manager.

4*B*. Do nothing. (Let him use the old safe!)

One can adopt Alt. 1*A* (buy the ladle) or 1*B* (do not buy the ladle), but not both. We find that what we considered to be a single course of action is really a pair of mutually exclusive alternatives. Even Alt. 3 is in this category. The originally stated single proposal was:

> "The painting department reports that they must make improvements to the
> spray booth to conform with new air pollution standards."

Since the painting department reports they must *make* the improvements, is there actually another alternative? Although at first glance we might not think so, the company does have one or more alternatives available. It may be possible to change the paint, or the spray equipment, and thereby solve the air pollution problem without any improvements to the spray booth. In this situation, there does not seem to be a practical do-nothing alternative, for failure to comply with the air pollution standards might result in large fines or even shutting down the plant. But if there is not a practical do-nothing alternative, there might be a number of do-something-else alternatives.

We conclude that all project proposals may be considered to have mutually exclusive alternatives.   ■

Identifying and Rejecting Unattractive Alternatives

It is clear that no matter what the circumstances may be, we want to eliminate from further consideration any alternative that fails to meet the minimum level of economic attractiveness, provided one of the other alternatives does meet the criterion. Table 17-1 summarizes five techniques that may be used.

At first glance it appears that many calculations are required, but this is not the situation. *Any* of the five techniques listed in Table 17-1 may be used to determine whether or not to reject an alternative. Each will produce the same decision regarding *Reject–Don't reject*.

Selecting the Best Alternative from Each Project Proposal

The task of selecting the best alternative from among two or more mutually exclusive alternatives has been a primary subject of this book. Since a project proposal is the same form of problem, we may use any of the several methods discussed in Chapters 5 through 9. The criteria are summarized in Table 17-2.

Table 17-1 CRITERIA FOR REJECTING UNATTRACTIVE ALTERNATIVES

For each alternative compute	Reject alternative when	Do not reject alternative when
Rate of return, i	$i <$ MARR	$i \geq$ MARR
Present worth, PW	PW of benefits $<$ PW of costs	PW of benefits $\geq$ PW of costs
Annual cost, EUAC Annual benefit, EUAB	EUAC $>$ EUAB	EUAC $\leq$ EUAB
Benefit–cost ratio, B/C	B/C < 1	B/C ≥ 1
Net present worth, NPW	NPW < 0	NPW ≥ 0

Table 17-2   CRITERIA FOR CHOOSING THE BEST ALTERNATIVE FROM AMONG MUTUALLY EXCLUSIVE ALTERNATIVES

Analysis method	Situation		
	Fixed input (The cost of each alternative is the same)	Fixed output (The benefits from each alternative are the same)	Neither input nor output fixed (Neither the costs nor the benefits for each alternative are the same)
Present worth	Maximize present worth of benefits	Minimize present worth of cost	Maximize net present worth
Annual cash flow	Maximize equivalent uniform annual benefits	Minimize equivalent uniform annual cost	Maximize (EAUB – EAUC)
Benefit–cost ratio	Maximize benefit–cost ratio	Maximize benefit–cost ratio	Incremental benefit–cost ratio analysis is required
Rate of return	Incremental rate of return analysis is required		

Rationing Capital By Rate Of Return

One way of looking at the capital rationing problem is through the use of rate of return. The technique for selecting from among independent projects may be illustrated by, an example.

EXAMPLE 17-2

Nine independent projects are being considered. Figure 17-1 may be prepared from the following data.

Project	Cost	Uniform annual benefit	Useful life, in years	Salvage value	Computed rate of return
1	$100	$23.85	10	$ 0	20%
2	200	39.85	10	0	15%
3	50	34.72	2	0	25%
4	100	20.00	6	100	20%
5	100	20.00	10	100	20%
6	100	18.00	10	100	18%
7	300	94.64	4	0	10%
8	300	47.40	10	100	12%
9	50	7.00	10	50	14%

If a capital budget of $650 is available, which projects should be selected?

Solution: Looking at the nine projects, we see that some are expected to produce a larger rate

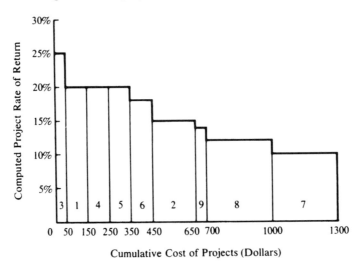

Figure 17-1 Cumulative cost of projects *vs.* rate of return.

of return than others. It is natural that if we are to select from among them, we will pick those with a higher rate of return. When the projects are arrayed by rate of return, as in Fig. 17-1, the choice of Projects 3, 1, 4, 5, 6, and 2 is readily apparent, and is a correct decision.■

In Example 17-2, the rate of return was computed for each project and then the projects were arranged in order of decreasing rate of return. For a fixed amount of money in the capital budget, the projects are selected by going down the list until the money is exhausted. Using this procedure, the point where the money runs out is where we cut off approving projects. This point is called the *cutoff rate of return.* Figure 17-2 illustrates the general situation.

For any set of ranked projects and any capital budget, the rate of return at which the budget is exhausted is the cutoff rate of return. In Fig. 17-2 the cost of each individual project is small compared to the capital budget. The cumulative cost curve is a relatively smooth curve producing a specific cutoff rate of return. Looking back at Fig. 17-1, we see the curve is actually a step function. For Example 17-2, the cutoff rate of return is between 14% and 15% for a capital budget of $650.

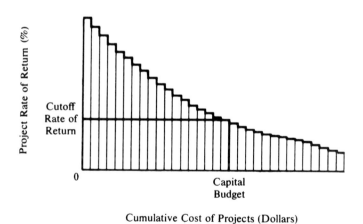

Figure 17-2 Location of the cutoff rate of return.

Significance of the Cutoff Rate of Return

Cutoff rate of return is determined by the comparison of an established capital budget and the available projects. One must examine all the projects and all the money for some period of time (like an annual budget) to compute the cutoff rate of return. It is a computation relating known projects with a known money supply. For this period of time, the cutoff rate of return is the opportunity cost (rate of return on the opportunity or project foregone) and also the minimum attractive rate of return. In other words, the minimum attractive rate of return to get a project accomplished *is* the cutoff rate of return.

MARR = Cutoff rate of return = Opportunity cost

We generally use the minimum attractive rate of return to decide whether or not to approve an individual project even though we do not know exactly what other projects will be proposed during the year. In this situation, we cannot know if the MARR is equal to the cutoff rate of return. When the MARR is different from the cutoff rate of return, incorrect decisions may occur. This will be illustrated in the next section.

Rationing Capital By Present Worth Methods

Throughout this book we have chosen from among project alternatives to maximize net present worth. If we can do the same thing for a group of projects, and we do not exceed the available money supply, then the capital budgeting problem is solved.

But more frequently, the capital budgeting problem is one where we will be unable to accept all desirable projects. We, therefore, have a task not previously encountered. We must choose the best from the larger group of acceptable projects.

Lorie and Savage[1] showed that a proper technique is to use a multiplier, *p,* to decrease the attractiveness of an alternative in proportion to its use of the scarce supply of money. The revised criterion is

NPW − *p*(PW of cost)   (17-1)

   where *p* is a multiplier

---

[1] Lorie, J. and L. Savage, "Three Problems in Rationing Capital," *Journal of Business,* October, 1955, pp. 229–239.

If a value of *p* were selected (say, 0.1), then some alternatives with a positive NPW will have a negative [NPW – *p*(PW of cost)]. This new criterion will reduce the number of favorable alternatives and thereby reduce the combined cost of the projects meeting this more severe criterion. By trial and error, the multiplier *p* is adjusted until the total cost of the projects meeting the [NPW – *p*(PW of cost)] criterion equals the available money supply–the capital budget.

EXAMPLE 17-3

Using the present worth method, determine which of the nine independent projects of Ex. 17-2 should be included in a capital budget of $650. The minimum attractive rate of return has been set at 8%.

Project	Cost	Uniform annual benefit	Useful life, in years	Salvage value	Computed NPW
1	$100	$23.85	10	$ 0	$60.04
2	200	39.85	10	0	67.40
3	50	34.72	2	0	11.91
4	100	20.00	6	100	55.48
5	100	20.00	10	100	80.52
6	100	18.00	10	100	67.10
7	300	94.64	4	0	13.46
8	300	47.40	10	100	64.38
9	50	7.00	10	50	20.13

Solution:

Locating a value of *p* in [NPW – *p*(PW of cost)] by trial and error:

			Trial *p* = 0.20		Trial *p* = 0.25	
		Computed	[NPW –		[NPW –	
Project	*Cost*	*NPW*	*p*(PW *of cost*)]	*Cost*	*p*(PW *of cost*)]	*Cost*
1	$ 100	$60.04	$40.04	$ 100	$35.04	$100
2	200	67.40	27.40	200	17.40	200
3	50	11.91	1.91	50	−0.59	
4	100	55.48	35.48	100	30.48	100
5	100	80.52	60.52	100	55.52	100
6	100	67.10	47.10	100	42.10	100
7	300	13.46	−46.54		−61.54	
8	300	64.38	4.38	300	−10.62	
9	50	20.13	10.13	50	7.63	50
	$1300			$1000		$650

For a value of *p* equal to 0.25, the best selection is computed to be Projects 1, 2, 4, 5, 6, and 9.

Alternate Formation of Example 17-3: This answer does not agree with the solution obtained in Ex. 17-2. The difficulty is that the interest rate used in the present worth calculations is not equal to the computed cutoff rate of return. In Ex. 17-2 the cutoff rate of return was between 14% and 15%, say 14.5%. We will recompute the present worth solution using MARR = 14.5%.

		Computed	*Cost of*
		NPW	*projects with*
Project	*Cost*	*at 14.5%*	*positive* NPW
1	$100	$22.01	$100
2	200	3.87	200
3	50	6.81	50
4	100	21.10	100
5	100	28.14	100
6	100	17.91	100
7	300	−27.05	
8	300	−31.69	
9	50	−1.28	
			$650

Solution:

At a MARR of 14.5% the best set of projects is the same as computed in Ex. 17-2, namely, Projects 1, 2, 3, 4, 5, and 6, and their cost equals the capital budget. One can see that only projects with a rate of return greater than MARR can have a positive NPW at this interest rate. With MARR equal to the cutoff rate of return, we *must* obtain the same solution by either the rate of return or present worth methods. ■

Figure 17-4 outlines the present worth method for the more elaborate case where there are independent projects each with mutually exclusive alternatives.

EXAMPLE 17-4

A company is preparing its capital budget for next year. The amount has been set at $250 by the Board of Directors. The MARR of 8% is believed to be close to the cutoff rate of return. The following project proposals are being considered.

Project proposals	Cost	Uniform annual benefit	Salvage value	Useful life, in years	Computed NPW
Proposal 1					
Alt. *A*	$100	$23.85	$0	10	$60.04
B	150	32.20	0	10	66.06
C	200	39.85	0	10	67.40
D	0	0			0
Proposal 2					
Alt. *A*	50	14.92	0	5	9.57
B	0	0			0
Proposal 3					
Alt. *A*	100	18.69	25	10	36.99
B	150	19.42	125	10	38.21
C	0	0			0

Which project alternatives should be selected, based on present worth methods?

Solution: The tabulation below shows that to maximize NPW, we would choose Alternatives 1*C*, 2*A*, and 3*B*. The total cost of these three projects is $400. Since the capital budget is only $250, we cannot fund these projects. To penalize all projects in proportion to their cost, we will use Equation 17-1 with its multiplier, p. As a first trial, a value of $p = 0.10$ is selected and the alternatives with the largest [NPW, $-p$(PW of cost)] selected.

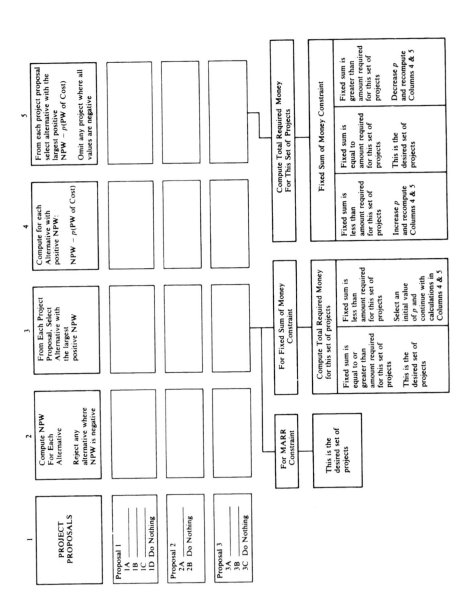

Figure 17-3 Steps in computing a capital budget.

$$p = 0.10$$

Project proposals	Cost	NPW	Alternative with largest positive NPW		[NPW − p(PW of cost)]	Alternative with largest positive [NPW − p(PW of cost)]	
			Alt.	Cost		Alt.	Cost
Proposal 1							
Alt. A	$100	$60.04			$50.04		
B	150	66.06			51.06	1B	$150
C	200	67.40	1C	$200	47.40		
D	0	0			0		
Proposal 2							
Alt. A	50	9.57	2A	50	4.57	2A	50
B	0	0			0		
Proposal 3							
Alt. A	100	36.99			26.99	3A	100
B	150	38.21	3B	150	23.21		
C	0	0			0		
				$400			$300

The first trial with $p = 0.10$ selects Alternatives 1B, 2A, and 3A with a total cost of $300. This still is greater than the $250 capital budget. Another trial is needed with a larger value of p. Select $p = 0.15$ and recompute.

$$p = 0.15$$

Project proposals	Cost	NPW	[NPW − p(PW of cost)]	Alternative with largest positive [NPW − p(PW of cost)]	
				Alt.	Cost
Proposal 1					
Alt. A	$100	$60.04	$45.04	1A	$100
B	150	66.06	43.56		
C	200	67.40	37.40		
D	0	0	0		
Proposal 2					
Alt. A	50	9.57	2.07	2A	50
B	0	0	0		
Proposal 3					
Alt. A	100	36.99	21.99	3A	100
B	150	38.21	15.71		
C	0	0	0		
					$250

The second trial, with $p = 0.15$, points to Alternatives 1*A*, 2*A*, and 3*A* for a total cost of $250. This equals the capital budget, hence is the desired set of projects. ■

EXAMPLE 17-5

Solve Ex. 17-4 by the rate of return method. For project proposals with two or more alternatives, incremental rate of return analysis is required. The data from Ex. 17-4 and the computed rate of return for each alternative and each increment of investment is shown in the tabulation below.

Solution:

	Cost	Uniform annual benefit	Salvage value	Computed rate of return	Cost	Uniform annual benefit	Salvage value	Computed rate of return
						Incremental analysis		
Proposal 1								
A	$100	$23.85	$ 0	20.0%				
B − A					$50	$8.35	$ 0	10.6%
B	150	32.20	0	17.0%				
C − B					50	7.65	0	8.6%
C − A					100	16.00	0	9.6%
C	200	39.85	0	15.0%				
D	0	0	0	0%				
Proposal 2								
A	50	14.92	0	15.0%				
B	0	0	0	0%				
Proposal 3								
A	100	18.69	25	15.0%				
B − A					50	0.73	100	8.3%
B	150	19.42	125	12.0%				
C	0	0	0	0%				

The various separable increments of investment may be ranked by rate of return. They are plotted in a cumulative cost *vs.* rate of return graph in Fig. 17-4. The ranking of projects by rate of return gives the following:

Project

1*A*
2*A*
3*A*
1*B* in place of 1*A*
1*C* in place of 1*B*
3*B* in place of 3*A*

For a budget of $250, the selected projects are 1*A*, 2*A*, and 3*A*. Note that if a budget of $300 was available, 1*B* would replace 1*A*, making the proper set of projects 1*B*, 2*A*, and 3*A*. At a budget of $400, 1*C* would replace 1*B*; and 3*B* would replace 3*A*, making the selected projects 1*C*, 2*A*, and 3*B*. These answers agree with the computations in Ex. 17-4. ■

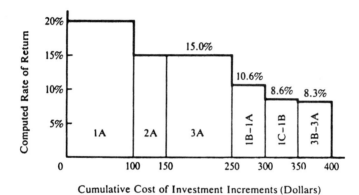

Figure 17-4 Cumulative cost *vs.* incremental rate of return.

Ranking Project Proposals

Closely related to the problem of capital budgeting is the matter of ranking project proposals. We will first examine a method of ranking by present worth methods and then show that project rate of return is not a suitable method of ranking projects.

Anyone who has ever bought firecrackers probably used the practical ranking criterion of "biggest bang for the buck" in selecting the fireworks. This same criterion—stated more eloquently—may be used to correctly rank independent projects.

> **Rank independent projects according to their value of net present worth divided by the present worth of cost. The appropriate interest rate is MARR (as a reasonable estimate of the cutoff rate of return).**

Example 17-6 illustrates the method of computation.

EXAMPLE 17-6

Rank the following nine independent projects in their order of desirability, based on a 14.5% minimum attractive rate of return. (To facilitate matters the necessary computations are included in the tabulation.)

Project	Cost	Uniform annual benefit	Useful life, in years	Salvage value	Computed rate of return	Computed NPW at 14.5%	Computed NPW/Cost
1	$100	$23.85	10	$0	20%	$22.01	0.2201
2	200	39.85	10	0	15%	3.87	0.0194
3	50	34.72	2	0	25%	6.81	0.1362
4	100	20.00	6	100	20%	21.10	0.2110
5	100	20.00	10	100	20%	28.14	0.2814
6	100	18.00	10	100	18%	17.91	0.1791
7	300	94.64	4	0	10%	−27.05	−0.0902
8	300	47.40	10	100	12%	−31.69	−0.1056
9	50	7.00	10	50	14%	−1.28	−0.0256

Solution: Ranked by NPW/PW of cost, the projects are listed:

Project	NPW / PW of cost	Rate of return, %
5	0.2814	20
1	0.2201	20
4	0.2110	20
6	0.1791	18
3	0.1362	25
2	0.0194	15
9	−0.0256	14
7	−0.0902	10
8	−0.1056	12

The rate of return tabulation illustrates that it is not a satisfactory ranking criterion and would have given a different ranking from the present worth criterion. ■

In Example 17-6, the projects are ranked according to the ratio NPW/PW of cost. In Fig. 17-3, the criterion used is [NPW $- p$(PW of cost)]. If one were to compute the value of p at which [NPW $- p$(PW of cost)] = 0, we would obtain p = (NPW/PW of cost). Thus the multiplier p is the ranking criterion at the point where [NPW $- p$(PW of cost)] = 0.

If independent projects can be ranked in their order of desirability, then the selection of projects to be included in a capital budget is a simple task. One may proceed down the list of ranked projects until the capital budget is exhausted. The only difficulty with this scheme occasionally occurs when the capital budget is more than enough for n projects, but too little for $(n + 1)$ projects.

In Example 17-6, a capital budget of $300 is just right to fund the top three projects. But a capital budget of $550 is more than enough for the top five projects (sum = $450) but not enough for the top six projects (sum = $650). When we have this lumpiness problem, one cannot always say with certainty that the best use of a capital budget of $550 is to fund the top five projects. There may be some other set of projects that makes better use of the available $550. While some trial and error computations may indicate the proper set of projects, more elaborate techniques are needed to prove optimality.

As a practical matter, a capital budget probably has some flexibility. If in Ex. 17-6 the tentative capital budget is $550, then a careful examination of Project 2 will dictate whether to expand the capital budget to $650 (to be able to include Project 2) or to drop back to $450 (and leave Project 2 out of the capital budget).

Summary

Prior to this chapter we have assumed that all worthwhile projects are approved and implemented. But industrial firms, like individuals and governments, are typically faced with more good projects than there is money available. The task is to select the best projects and reject, or at least delay, the rest.

Mutually exclusive alternatives are those where the acceptance of one alternative effectively prevents the adoption of the other alternatives. This could be because the alternatives perform the same function (like Pump *A vs.* Pump *B*) or occupy the same physical location (like a gas station *vs.* a hamburger stand). If a project has a single alternative of doing something, we know there is likely to be a mutually exclusive alternative of doing nothing—or possibly doing something else. A project proposal may be thought of as having two or more mutually exclusive alternatives. Projects are assumed in this chapter to be independent.

Capital may be rationed among competing investment opportunities by either rate of return or present worth methods. The results may not always be the same for these two methods in many practical situations.

If projects are ranked by rate of return, a proper procedure is to go down the list until the capital budget is exhausted. The rate of return at this point is the cutoff rate of return. This procedure gives the best group of projects, but does not necessarily have them in the proper priority order.

Maximizing NPW is an appropriate present worth selection criterion where the available projects do not exhaust the money supply. But if the amount of money required for the best alternative from each project exceeds the available money, a more severe criterion is imposed: adopt only those alternatives and projects that have a positive [NPW – p(PW of cost)]. The value of the multiplier p is chosen by trial and error until the alternatives and projects meeting the criterion just equal the available capital budget money.

It has been shown in earlier chapters that the usual business objective is to maximize NPW, and this is not necessarily the same as maximizing rate of return. One suitable procedure is to use the ratio (NPW/PW of cost) to rank the projects. This present worth ranking method will order the projects so that, for a limited capital budget, NPW will be maximized. We know that MARR must be adjusted from time to time to reasonably balance the cost of the projects that meet the MARR criterion and the available supply of money. This adjustment of the MARR to equal the cutoff rate of return is essential for the rate of return and present worth methods to yield compatible results.

Another way of ranking is by incremental rate of return analysis. Once a ranking has been made, we can go down the list and accept the projects until the money runs out. There is a theoretical difficulty if the capital budget contains more money than is required for n projects, but not enough for one more, or $(n + 1)$ projects. As a practical matter, capital budgets are seldom inflexible, with the result that some additional money may be allocated if the $(n + 1)$ project looks like it should be included.

Problems

17-1   The following ten independent projects each have a ten-year life and no salvage value

Project	Cost, in thousands	Uniform annual benefits, in thousands	Computed rate of return, %
1	$ 5	$1.03	16
2	15	3.22	17
3	10	1.77	12
4	30	4.88	10
5	5	1.19	20
6	20	3.83	14
7	5	1.00	15
8	20	3.69	13
9	5	1.15	19
10	10	2.23	18

The projects have been proposed by the staff of the Ace Card Company. The MARR of Ace has been 12% for several years.

    a. If there is ample money available, what projects should Ace approve?

    b. Rank order all the acceptable projects in their order of desirability.

    c. If only $55,000 is available, which projects should be approved?

17-2 At Miami Products, four project proposals (three with mutually exclusive alternatives) are being considered. All the alternatives have a ten-year useful life and no salvage value.

Project proposal	Cost, in thousands	Uniform annual benefits, in thousands	Computed rate of return, %
Project 1			
Alt. *A*	$25	$4.61	13
B	50	9.96	15
C	10	2.39	20
Project 2			
Alt. *A*	20	4.14	16
B	35	6.71	14
Project 3			
Alt. *A*	25	5.56	18
B	10	2.15	17
Project 4	10	1.70	11

    a. Using rate of return methods, determine which set of projects should be undertaken if the MARR is 10%.

    b. Using rate of return methods, which set of projects should be undertaken if the capital budget is limited to $100,000?

    c. For a budget of $100,000, what interest rate should be used in rationing capital by present worth methods? (Limit your answer to a value for which there is a Compound Interest Table available at the back of the book.)

    d. Using the interest rate determined in *c*, rank order the eight different investment opportunities by the present worth method.

    e. For a budget of $100,000 and the ranking in *d*, which of the investment opportunities should be selected?

17-3   Al Dale is planning his Christmas shopping as he must buy gifts for seven people. To quantify how much the various people would enjoy receiving a list of prospective gifts, Al has assigned appropriateness units (called "ohs") for each gift if given to each of the seven people. A rating of five ohs represents a gift that the recipient would really like. A rating of four ohs indicates the recipient would like it four-fifths as much; three ohs, three-fifths as much, and so forth. A zero rating indicates an inappropriate gift that cannot be given to that person. These data are tabulated below.

	Prospective gift	"Oh" rating of gift if given to various family members						
		Father	Mother	Sister	Brother	Aunt	Uncle	Cousin
1.	$20 box of candy	4	4	2	1	5	2	3
2.	$12 box of cigars	3	0	0	1	0	1	2
3.	$16 necktie	2	0	0	3	0	3	2
4.	$20 shirt or blouse	5	3	4	4	4	1	4
5.	$24 sweater	3	4	5	4	3	4	2
6.	$30 camera	1	5	2	5	1	2	0
7.	$ 6 calendar	0	0	1	0	1	0	1
8.	$16 magazine subscription	4	3	4	4	3	1	3
9.	$18 book	3	4	2	3	4	0	3
10.	$16 game	2	2	3	2	2	1	2

The objective is to select the most appropriate set of gifts for the seven people (that is, maximize total ohs) that can be obtained with the selected budget.

  a. How much will it cost to buy the seven gifts the people would like best, if there is ample money for Christmas shopping?

  b. If the Christmas shopping budget is set at $112, which gifts should be purchased, and what is their total appropriateness rating in ohs?

  c. If the Christmas shopping budget must be cut to $90, which gifts should be purchased, and what is their total appropriateness rating in ohs?

  (*Answer: a.* $168)

The following facts are to be used in solving Problems 17-4 through 17-7:

In assembling data for the Peabody Company annual capital budget, five independent projects are being considered. Detailed examination by the staff has resulted in the identification of from three to six mutually exclusive do-something alternatives for each project. In addition, each project has a do-nothing alternative. The projects and their alternatives are listed below.

Project proposal	Cost in thousands	Uniform annual benefit, in thousands	Useful life, in years	End-of-useful-life salvage value, in thousands	Computed rate of return, %
Project 1					
Alt. *A*	$40	$13.52	2	$20	10
B	10	1.87	16	5	18
C	55	18.11	4	0	12
D	30	6.69	8	0	15
E	15	3.75	2	15	25
Project 2					
Alt. *A*	10	1.91	16	2	18
B	5	1.30	8	0	20
C	5	0.97	8	2	15
D	15	5.58	4	0	18
Project 3					
Alt. *A*	20	2.63	16	10	12
B	5	0.84	16	0	15
C	10	1.28	16	0	10
D	15	2.52	16	0	15
E	10	3.50	4	0	15
F	15	2.25	16	15	15
Project 4					
Alt. *A*	10	2.61	8	0	20
B	5	0.97	16	0	18
C	5	0.90	16	5	18
D	15	3.34	8	0	15
Project 5					
Alt. *A*	5	0.75	8	5	15
B	10	3.50	4	0	15
C	15	2.61	8	5	12

Each project concerns operations at the St. Louis brewery. The plant was leased from another firm many years ago and the lease expires 16 years from now. For this reason, the analysis period for all projects is 16 years. Peabody considers 12% to be the minimum attractive rate of return.

In solving the Peabody Co. problems, an important assumption concerns the situation at the end of the useful life of an alternative when the alternative has a useful life less than the 16-year analysis period. Two replacement possibilities are listed.

Assumption 1: When an alternative has a useful life less than 16 years, it will be replaced by a new alternative with the same useful life as the original. This may need to occur more than once. The new alternative will have a 12% computed rate of return and, hence, a NPW = 0 at 12%.

Assumption 2: When an alternative has a useful life less than 16 years, it will be replaced at the end of its useful life by an identical alternative (one with the same cost, uniform annual benefit, useful life, and salvage value as the original alternative).

17-4 For an unlimited supply of money, and replacement Assumption 1, which project alternatives should Peabody select? Solve the problem by present worth methods.
(*Answer:* Project Alternatives 1*B*, 2*A*, 3*F*, 4*A*, and 5*A*)

17-5 For an unlimited supply of money, and replacement Assumption 2, which project alternatives should Peabody select? Solve the problem by present worth methods.

17-6 For an unlimited supply of money, and replacement Assumption 2, which project alternatives should Peabody select? Solve the problem by rate of return methods. (*Hint:* By careful inspection of the alternatives, you should be able to reject about half of them. Even then the problem requires lengthy calculations.)

17-7 For a capital budget of $55,000, and replacement Assumption 2, which project alternatives should Peabody select?
(*Answer:* Project Alternatives 1*E*, 2*A*, 3*F*, 4*A*, and 5*A*)

17-8 A financier has a staff of three people whose job it is to examine possible business ventures for him. Periodically they present him their findings concerning business opportunities. On a particular occasion, they presented the following investment opportunities:

Project A: This is a project for the utilization of the commercial land the financier already owns. Three mutually exclusive alternatives are:
 Project A1. Sell the land for $500,000.
 Project A2. Lease the property for a car-washing business. An annual income, after all costs, like property taxes, and so on, of $98,700 would be received at the end of each year for twenty years. At the end of the twenty years, it is believed the property could be sold for $750,000.
 Project A3. Construct an office building on the land. The building will cost $4,500,000 to construct and will not produce any net income for the first two years. The probabilities of various levels of rental income, after all expenses, for the subsequent 18 years are as follows:

Annual rental income	Probability
$1,000,000	0.1
1,100,000	0.3
1,200,000	0.4
1,900,000	0.2

The property (building. and land) probably can be sold for $3 million at the end of twenty years.

Project B: An insurance company is seeking to borrow money for ninety days. They offer to pay 13¾% per annum, compounded continuously.

Project C: The financier owns a manufacturing company. The firm desires additional working capital to allow it to increase its inventories of raw materials and finished products. An investment of $2,000,000 will allow the company to obtain sales that in the past the company had to forgo. The additional capital will increase company profits by $500,000 a year. The financier can recover this additional investment by ordering the company to reduce its inventories and to return the $2,000,000. For planning purposes, assume the additional investment will be returned at the end of ten years.

Project D: The owners of Sunrise magazine are seeking a loan of $500,000 for ten years at a 16% interest rate.

Project E: The Galveston Bank has indicated they are willing to accept a deposit of any sum of money over $100,000, for any desired duration, at a 14.06% interest rate, compounded monthly. It seems likely that this interest rate will be available from Galveston, or some other bank, for the next several years.

Project F: A car rental company is seeking a loan of $2,000,000 to expand their fleet of automobiles. They offer to repay the loan by paying $1,000,000 at the end of one year, and $1,604,800 at the end of two years.

a. If there is $4 million available for investment now (or $4.5 million if the Project *A* land is sold), which projects should be selected? What is the MARR in this situation?

b. If there is $9 million available for investment now (or $9.5 million if the Project *A* land is sold), which projects should be selected?

17-9 The Raleigh Soap Company has been offered a five-year contract to manufacture and package a leading brand of soap for Taker Bros. It is understood the contract will not be extended past the five years as Taker Bros. plans to build their own plant nearby. The contract calls for 10,000 metric tons (one metric ton equals 1000 kilograms) of soap a year. Raleigh normally produces 12,000 metric tons of soap a year, so production for the five-year period would be increased to 22,000 metric tons. Raleigh must decide what changes, if any, to make to accommodate this increased production. Five projects are under consideration.

Project 1: Increase liquid storage capacity.

At present, Raleigh has been forced to buy caustic soda in tank truck quantities due to inadequate storage capacity. If another liquid caustic soda tank is installed to hold 1000 cubic meters, the caustic soda may be purchased in railroad tank car quantities at a more favorable price. The result would be a saving of 0.1 cent per kilogram of soap. The tank, which would cost $83,400, has no net salvage value.

Project 2: Another sulfonation unit.

The present capacity of the plant is limited by the sulfonation unit. The additional 12,000 metric tons of soap cannot be produced without an additional sulfonation unit. Another unit can be installed for $320,000.

Project 3: Packaging department expansion.

With the new contract, the packaging department must either work two 8-hour shifts, or have another packaging line installed. If the two-shift operation is used, a 20% wage premium must be paid for the second shift. This premium would amount to $35,000 a year. The second packaging line could be installed for $150,000. It would have a $42,000 salvage value at the end of five years.

Project 4: New warehouse.

The existing warehouse will be inadequate for the greater production. It is estimated that 400 square meters of additional warehouse is needed. A new warehouse can be built on a lot beside the existing warehouse for $225,000, including the land. The annual taxes, insurance, and other ownership costs would be $5000 a year. It is believed the warehouse could be sold at the end of five years for $200,000.

Project 5: Lease a warehouse.

An alternative to building an additional warehouse would be to lease warehouse space. A suitable warehouse one mile away could be leased for $15,000 per year. The $15,000 includes taxes, insurance, and so forth. The annual cost of moving materials to this more remote warehouse would be $34,000 a year.

The contract offered by Taker Bros. is a favorable one which Raleigh Soap plans to accept. Raleigh management has set a 15% before-tax minimum attractive rate of return as the criterion for any of the projects. Which projects should be undertaken?

17-10 Ten capital spending proposals have been made to the budget committee as they prepare the annual budget for their firm. The independent projects each have a 5-year life and no salvage value.

Project	Initial cost in thousands	Uniform annual benefit in thousands	Computed rate of return
A	$10	$2.98	15%
B	15	5.58	25
C	5	1.53	16
D	20	5.55	12
E	15	4.37	14
F	30	9.81	19
G	25	7.81	17
H	10	3.49	22
I	5	1.67	20
J	10	3.20	18

a. Based on a MARR of 14%, which projects should be approved?

b. Rank order all the projects in their order of desirability.

c. If only $85,000 is available, which projects should be approved?

17-11 Mike Moore's microbrewery is considering production of a new ale called Mike's Honey Harvest Brew. To produce this new offering he is considering two independent projects. Each of these projects has two mutually exclusive alternatives and each alternative has a useful life of 10 years and no salvage value. Mike's MARR is 8%. Information regarding the projects and alternatives are given in the following table:

Project/Alternative	Cost	Annual benefit
Project 1. Purchase new fermenting tanks		
Alt. *A.* 5000 gallon tank	$ 5000	$1192
Alt. *B.* 15,000 gallon tank	10,000	1992
Project 2. Purchase bottle filler and capper		
Alt. *A* 2500 bottle/hour machine	15,000	3337
Alt. *B.* 5000 bottle/hour machine	25,000	4425

Use incremental rate-of-return analysis to complete the partially filled out worksheet given below.

Proj/Alt	Cost (P)	Annual benefit (A)	(A/P, i, 10)	ROR
1*A*	$5,000	$1192	0.2385	20%
1*B*-1*A*	5,000	800	0.1601	
2*A*	15,000	3337		
2*B*-2*A*	10,000			

Use this information to determine:

a. Which projects should be funded if only $15,000 dollars are available.

b. The cut-off rate of return if only $15,000 dollars are available.

c. Which projects should be funded if $25,000 dollars are available.

CHAPTER 18

A Further Look
At Rate Of Return

In Chapter 7, the rate of return is defined for an *investment* as follows:

> ***Internal rate of return* is the interest rate earned on the unrecovered investment such that the payment schedule makes the unrecovered investment equal to zero at the end of the life of the investment.**

In a *borrowing* situation, the definition is:

> ***Internal rate of return* is the interest rate paid on the unpaid balance of a loan such that the payment schedule makes the unpaid loan balance equal to zero when the final payment is made.**

In the actual calculation of internal rate of return, we wrote one equation relating costs and benefits (for example, Present worth of cost equals Present worth of benefits) and solved the equation for the unknown rate of return. This works fine if the situation represents either a pure investment or a pure borrowing situation and there is a single positive rate of return.

Unfortunately, there are times when these two conditions are not met. A remedy was suggested in Chapter 7A: by application of an external interest rate, the cash flow was adjusted until the number of sign changes was reduced to 1. The *cash flow rule of signs* then tells us there is either none or one positive rate of return. And with only one sign change in the cash flow, the situation must be one of either pure investment or of pure borrowing. This remedy works. But, as we will find in this chapter, we may be adjusting the cash flow when no adjustment is necessary, and even when an adjustment is required, we may be making too large an adjustment. The resulting computed rate of return is thus affected by adjustments which are either unnecessary or too much.

Cash Flow Situations

A cash flow may represent any kind of situation. It may be a pure borrowing, pure investment, or a mixture of the two. This is illustrated by Table 18-1. In Case *A*, the typical investment situation is represented by the investment of $50 at Year 0, followed by the return of the resulting benefits in Years 1 through 4. Case *B* represents a borrowing situation. Fifty dollars is received in Year 0, followed by four payments of $15 each in Years 1–4 to repay the loan. Note that *A* and *B* are mirror images of one another. That is, changing the sign of all the cash flows in *A* gives us *B*, and vice versa.

Case *C* represents a mixed situation. Initially there is a receipt of benefits (like a borrowing situation), followed by investments in Years 1 and 2. Finally, there are benefits in Years 3 and 4. The result is a mixed situation with Years 0, 1, and 2 looking like a borrowing situation and, at the same time, Years 1, 2, 3, and 4 look like an investment situation.

In computing the rate of return for a cash flow, one must carefully decide what one wishes the number to represent. Is it to be the internal rate of return earned on invested money, the external rate of return paid on borrowed money, or what? As this is a book on capital expenditure analysis, the view here is that we want to determine the rate of return earned on the money invested in the project while it is actually in the project. Money that is invested outside of the project will generally be assumed to earn some established external rate of return.

Table 18-1   EXAMPLES OF DIFFERENT CASH FLOW SITUATIONS

Year	Case A *pure investment* Cash flow	Case B *pure borrowing* Cash flow	Case C *mixed borrowing and investment* Cash flow
0	−$50	+$50	+$50
1	+15	−15	−30
2	+15	−15	−30
3	+15	−15	+15
4	+15	−15	+15

Analysis Of A Cash Flow As An Investment Situation

A good deal can be learned about the desirability of a cash flow as an investment situation. The goal is to produce a single value that accurately portrays the profitability of the investment opportunity reflected by the cash flow. We do not want multiple rates of return. (After all, how can an investment have both a 20% and a 60% rate of return at the same time?) And we do not want a rate of return that assumes that one must temporarily invest money outside of the investment project at an unrealistic interest rate. If a suitable external investment rate is 6%, for example, then the computations must not be based on some other rate. In short, we want a single, realistic value representing the rate of return (note that this could also be called the *profitability rate*) on the investment.

There are four tests that will help us to understand the investment situation represented by any cash flow. They are:

1. Cash flow rule of signs.
2. Accumulated cash flow sign test.
3. Algebraic sum of the cash flow.
4. Net investment conditions.

These will be examined one by one.

Cash Flow Rule of Signs

There may be as many positive rates of return as there are sign changes in the cash flow.

If we let a_i represent the cash flow in Year i, then the entire cash flow could be represented as follows:

Year	Cash flow
0	a_0
1	a_1
2	a_2
3	a_3
.	.
.	.
.	.
n	a_n

As described in detail in Chapter 7A, a *sign change* is where successive terms in the cash flow (ignoring zeros) have different signs.

Number of sign changes in cash flow	*Number of positive rates of return*
0	0 (or $i^* = \infty$)
1	1, or 0
2	2, 1, or 0
3	3, 2, 1, or 0
.	.
.	.
.	.

Accumulated Cash Flow Sign Test

The accumulated cash flow is the algebraic sum of the cash flow to that point in time. If we let a_i represent the cash flow in a year and A_i represent the accumulated cash

flow, the situation is:

Year	*Cash flow*	*Accumulated cash flow*
0	a_0	$A_0 = a_0$
1	a_1	$A_1 = a_0 + a_1$
2	a_2	$A_2 = a_0 + a_1 + a_2$
3	a_3	$A_3 = a_0 + a_1 + a_2 + a_3$
.	.	.
.	.	.
.	.	.
n	a_n	$A_n = a_0 + a_1 + a_2 + a_3 + \cdots + a_n$

The sequence of accumulated cash flows $(A_0, A_1, A_2, A_3, \ldots, A_n)$ is examined to determine the number of sign changes. As before, a sign change is where successive terms in the accumulated cash flow (ignoring zeros) have different signs.

Norström[1] proved that sufficient (but not necessary) conditions for a single positive rate of return are:

1.   The accumulated cash flow in Year n is greater than zero ($A_n > 0$).

---

[1] Norström, Carl J., "A Sufficient Condition for a Unique Nonnegative Internal Rate of Return," *Journal of Financial and Quantitative Analysis*, VII (June, 1972), pp. 1835–9.

2. There is exactly one sign change in the sequence of accumulated cash flows.

Algebraic Sum of the Cash Flow

$$\text{Algebraic sum} \ = \sum_{i=0}^{n} a_i$$

or looking at the accumulated cash flow,

$$\text{Algebraic sum} = A_n$$

A positive algebraic sum ($A_n > 0$) suggests a positive rate of return and an algebraic sum equal to zero ($A_n = 0$) suggests a 0% rate of return. In either case, however, there may be multiple positive rates of return.

A negative algebraic sum means the costs exceed the benefits of the project. This would seem to immediately mean that a positive rate of return would be impossible in an investment situation, which is usually true. It has been shown by Merrett and Sykes[2], however, that where there are two or more sign changes in the cash flow, and a substantial outlay near the end of the life of a project, the result may be one or more positive rates of return.

Net Investment Conditions

Two conditions have been shown[3] to be sufficient (but not necessary) to establish that a computed positive rate of return, i^*, is the only positive rate of return. The conditions are as follows:

1. The cash flow in the *n*th year must be positive ($a_n > 0$).

2. Given that a positive rate of return i^* has been computed, there is a net investment throughout the life of the project until the end of the *n*th year when the net investment becomes zero. Mathematically, this is:

$$\text{Net investment in any Year k} \ = \sum_{j=0}^{k} a_j \left(1 + i^*\right)^{k-j}$$
$$\text{for } k = 0,1,2,\ldots,n$$

---

[2] Merrett, A. J. and Allen Sykes. The Finance and Analysis of Capital Projects, 2nd ed. London: Longman, 1973,p.135

[3] Soper,C.S., "The Marginal Efficiency of Capital: A Further Note," Economic Journal LXIX,pp.174-7.

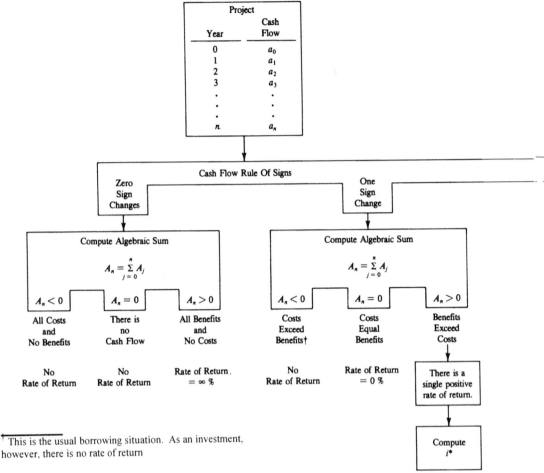

Figure 18-1 Computing a rate of return for an investment.

When either of these two conditions is not met, we know a net investment does not exist throughout the life of the project. This means there are one or more periods when the project has a net outflow of money *which will later be required to be returned to the project.* This money can be put into an external investment until such time as it is needed in the project. The

interest rate of the external investment ($e*$) will be the interest rate at which the money can in fact be invested outside the project. The external interest rate ($e*$) is unrelated to the internal rate of return ($i*$) on the project. If there is no external investment, then no value of $e*$ is required in the computations.

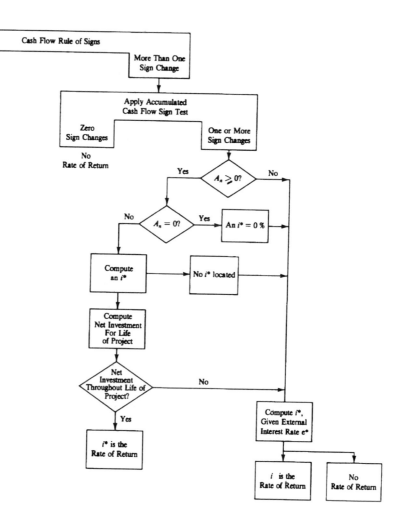

Application of the Four Tests of a Cash Flow

The four tests can be used to produce a chart for computing the rate of return for any investment project. Figure 18-1 is such a chart. The required computations may be done by hand or on a microcomputer. A microcomputer program called RORiPC has been written to perform all the computations. It is described in Chapter 19. The computation of the rate of return for any investment project may be illustrated by the following example problems.

EXAMPLE 18-1

Given the following cash flow, compute the internal rate of return i^*. If needed, use an external interest rate e^* of 6%.

Year	Cash flow
0	−$2000
1	+1200
2	+400
3	+400
4	−200
5	+400

First, we will write out the cash flow and accumulated cash flow and then determine the sign changes in both, along with the algebraic sum of the cash flow.

Solution:

Year	Cash flow a_i	Accumulated cash flow
0	−$2000	−$2000
1	+1200	−800
2	+400	−400
3	+400	0
4	−200	−200
5	+400	+200 = A_n
	+200	

The cash flow rule of signs indicates there may be as many as three positive rates of return. Since there is one sign change in the accumulated cash flow and $A_n > 0$, these are sufficient conditions for a single positive rate of return. It may be computed in the usual manner.

$$\text{PW of cost} = \text{PW of benefits}$$

$$2000 + 200(P/F,i,4) = 1200(P/F,i,1) + 400[(P/F,i,2) + (P/F,i,3) + (P/F,i,5)]$$

Try $i = 5\%$:

$$2000 + 200(0.8227) = 1200(0.9524) + 400(0.9070 + 0.8638 + 0.7835)$$

$$2165 = 2165$$

Thus, $i^* = 5\%$.

Also: Compute the net investment for the project at each time period:

Year	Cash flow		Net investment
0	−$2000		−$2000.0
1	+1200	−2000.0(1 + 0.05) + 1200.0 =	−900.0
2	+400	−900.0(1 + 0.05) + 400. 0 =	−545.0
3	+400	−545.0(1 + 0.05) + 400.0 =	−172.3
4	−200	−172.3(1 + 0.05) −200.0 =	−380.9
5	+400	−380.9(1 + 0.05) + 400.0 =	0

The fact that the net investment becomes zero at the end of the fifth year proves that 5% is a rate of return for the cash flow. We also see that there is a continuing net investment in the project for all time periods until the end of the project. This means there is no external investment of money and, hence, no need for an external interest rate. We conclude that $i^* = 5\%$ is a correct measure of the profitability of the project represented by the cash flow. ■

EXAMPLE 18-2

Given the following cash flow, compute the internal rate of return i^*. If needed, use an external interest rate e^* of 6%.

Year	Cash flow
0	−$100
1	+50
2	+50
3	+50
4	+50
5	−73.25

Solution: As in the previous example, we first determine the sign changes in the cash flow and the accumulated cash flow and compute the sum of the cash flow.

Year	Cash flow a_i	Accumulated cash flow A_i
0	−$100	−$100
1	+50	−50
2	+50	0
3	+50	+50
4	+50	+100
5	−73.25	+26.75 = A_n
	+$26.75	

Sign changes = 2   Sign changes =1

With one sign change in the accumulated cash flow and $A_n > 0$, we know there is a single positive rate of return. The problem was devised with a 20% interest rate as is shown:

$$PW \text{ of cost} = PW \text{ of benefits}$$

$$100 + 73.25(P/F,20\%,5) = 50(P/A,20\%,4)$$

$$100 + 73.25(0.4019) = 50(2.589)$$

$$129.45 = 129.45$$

Thus, $i^* = 20\%$. Check net investment:

Year	Cash flow		Unrecovered net investment
0	−$100		−$100
1	+50	−100.00(1 + 0.20) + 50.00 =	−70
2	+50	−70.00(1 + 0.20) + 50.00 =	−34
3	+50	−34.00(1 + 0.20) + 50.00 =	+9.20
4	+50	+ 9.20(1 + 0.20) + 50. 00 =	+61.04
5	−73.25	+61.04(1 + 0.20) − 73.25 =	0

Note that the fact that net investment becomes zero at the end of the last year confirms that 20% is a rate of return for the cash flow. It is also clear that there is not net investment throughout the life of the project. At the end of Year 3, there is in fact +9.20 that must be invested outside of the project represented by the cash flow. And in Year 4 the amount for external investment has increased to +61.04. The money invested externally must be returned to the project in Year 5 to provide the needed final disbursement of 73.25. In the table above, the money has been assumed to be invested at a 20% interest rate. More importantly, this same assumption of a 20% external investment is implicit in the PW of cost = PW of benefits calculation that also indicated that $i^* = 20\%$.

The problem statement specifies that external investment may be expected to earn a 6% interest rate. Since this is less than the 20% in the calculations above, this means that a benefit–the external interest earned–of the cash flow is reduced. For $e* = 6\%$, we can see that i is less than 20%. The calculation must be done by trial and error. Try $i = 15\%$:

Year	Cash flow		Net investment
0	−$100		−$100.00
1	+50	−100.00(1 + 0.15) + 50.00 =	−65.00
2	+50	−65.00(1 + 0.15) + 50.00 =	−24.75
3	+50	−24.75(1 + 0.15) + 50.00 =	+21.54
4	+50	+21.54(1 + 0.06) + 50.00 =	+72.83
5	−73.25	+72.83(1 + 0.06) − 73.25 =	+3.95

The net investment does not equal zero at the end of the project. This indicates our trial $i = 15\%$ is in error. The remaining positive net investment signals that i should be increased.

Further trials would guide us to $i = 16.5\%$. For $e* = 6\%$ and $i = 16.5\%$ we have:

Year	Cash flow		Net investment
0	−$100		−$100.00
1	+50	−100.00(1 + 0.165) + 50.00 =	−66.50
2	+50	− 66.50(1 + 0.165) + 50.00 =	−27.47
3	+50	−27.47(1 + 0.165) + 50.00 =	+17.99
4	+50	+ 17.99(1 + 0.060) + 50.00 =	+69.07
5		+69.07(1 + 0.060) −73.25 =	−0.03
	−73.25		

We see that with $e* = 6\%$, i is very close to 16.5%. ■

The results of Example 18-2 are important. The two sign changes in the cash flow warned us that there might be as many as two positive rates of return. From the accumulated cash flow sign test we proved that in reality there is one positive rate of return. We then showed that 20% is the single positive rate of return. One would be tempted to stop at this point, satisfied that $i* = 20\%$ is a proper measure of the profitability of the project represented by the cash flow.

Examination of the net investment throughout the life of the project reveals that external investments occur at the end of Years 3 and 4. Using an external interest rate $e*$, we found the true profitability of the project is $i = 16.5\%$. The surprising discovery is that where there are multiple sign changes in an investment project cash flow, there may be a unique positive rate of return, but the existence of external investment may mean that the unique positive rate of return is *not* a suitable measure of project profitability.

EXAMPLE 18-3

Given the following cash flow, compute the internal rate of return i^*. If needed, use an external interest rate e^* of 6%.

Year	Cash flow
0	−$100
1	+330
2	−362
3	+132

Solution: We will begin the solution by writing the accumulated cash flow and checking the sign changes.

Year	Cash flow	Accumulated cash flow
0	−$100	−$100
1	+330	+230
2	−362	−132
3	+132	$0 = A_n$
	0	
	Sign changes = 3	Sign changes = 2

Careful examination of these data reveal the following:

1. There may be as many as three positive rates of return.

2. One rate of return is 0%.

3. The accumulated cash flow does not meet the sufficient conditions for a single positive rate of return.

4. At $i^* = 0\%$ there is external investment at the end of Year 1. (The accumulated cash flow sequence becomes the net investment sequence at 0%.)

At this point we have two choices. We can search to find out how many positive rates of return there are and their numerical values. As an alternative we can seek i^* for the given $e^* = 6\%$.

This particular cash flow was devised by picking the desired roots of 0%, 10%, and 20% for a third order polynomial as follows: let $x = 1 + i$. For $i = 0.00, 0.10$, and 0.20,

$$(x - 1.0)(x - 1.1)(x - 1.2) = 0$$

Multiplying we obtain:

$$x^3 - 3.3x^2 + 3.62x - 1.32 = 0$$

Multiplying by -100 gives:

$$-100x^3 + 330x^2 - 362x + 132 = 0$$

$$-100(1+i)^3 + 330(1+i)^2 - 362(1+i) + 132 = 0$$

This represents the future worth of the following cash flow.

Year	Cash flow
0	−$100
1	+330
2	−362
3	+132

Thus the three positive rates of return for the cash flow are 0%, 10%, and 20%. Knowing that there are three positive rates of return, and their values, has not helped us in our search for a true measure of project profitability.

A trial and error search is made for i^*, given that $e^* = 6\%$. For a trial $i = 5.98\%$:

Year	Cash flow		Net investment
0	−$100		−$100.00
1	+330	$-100.00(1 + 0.0598) + 330 =$	+224.02
2	−362	$-224.02(1 + 0.0600) - 362 =$	−124.54
3	+132	$-124.54(1 + 0.0598) + 132 =$	+0.01

For $e^* = 6\%$, i is about 5.98%. We see that the external investment is more desirable than the internal project. ■

EXAMPLE 18-4

Given the following cash flow, compute the internal rate of return i^*. If needed, use an external interest rate e^* of 6%. Note that this cash flow pattern might occur in situations with terminal cleanup costs.

Year	Cash flow	Accumulated cash flow
0	–$200	–$200
1	+150	–50
2	+50	0
3	+50	+50
4	+50	+100
5	+50	+150
6	+50	+200
7	+50	+250
8	<u>–285</u>	$-35 = A_n$
	–$35	
	Sign changes = 2	Sign changes = 2

Solution: Although the algebraic sum of the cash flow is negative, this is one of the cash flows where there are two positive rates of return (at about 8.7% and 17.4%). The accumulated cash flow shows that an important benefit is the substantial external investment of money for relatively long periods of time. If we compute the external investment at $e^* = 6\%$ (rather than 8.7% or 17.4%), the benefits will be substantially reduced and the computed internal rate of return will decline. The internal rate of return is computed by trial and error. Try $i = 4\%$:

Year	Cash flow		Net investment
0	–$200		–$200.0
1	+150	–200.0(1.04) +150 =	–58.0
2	+50	–58.0(1.04) +50 =	–10.3
3	+50	–10.3(1.04) +50 =	+39.3
4	+50	+39.3(1.06) +50 =	+91.7
5	+50	+91.7(1.06) +50 =	+147.2
6	+50	+147.2(1.06) +50 =	+206.0
7	+50	+206.0(1.06) +50 =	+268.4
8	–285	+268.4(1.06) –285 =	–0.5

With $e^* = 6\%$, i is about 4%. ∎

EXAMPLE 18-5

Consider the following project cash flow. The external interest rate $e^* = 6\%$.

Year	Cash flow
0	−$500
1	+300
2	+300
3	+300
4	−600
5	+200
6	+135.66

a. Solve for the project internal rate of return ($i*$) using the methods described in Chapter 7A.

b. Solve for the project internal rate of return ($i*$) using the methods described in this chapter.

c. Explain the difference in the results obtained in Parts **a** and **b**.

Solution to Example 18-5a: There are three sign changes in the cash flow. We wish to reduce this to one sign change. We will eliminate the −600 at the end of Year 4 by accumulating money at the external interest rate ($e* = 6\%$) to equal the required $600. The computations are as follows:

Year	Cash flow	External investment	Transformed cash flow
0	−$500		−$500
1	+300		+300
2	+300	$x(1.06)^2\ x = 250.98$	+49.02
3	+300		0
		+300(1.06) ⎯ ↓ ↓	
4	−600	+318 +282	0
5	+200		+200
6	+135.66		+135.66

As described in the computations above, all the money at Year 3 is assumed to be invested externally at 6%. The +$300 increases to +$318 at the end of one year and can be brought back to the cash flow at the end of Year 4. This makes the internal investment cash flow have a zero for Year 3 and −600 + 318 = −$282 for Year 4.

A further alteration of the cash flow is needed to reduce the number of sign changes to 1. Part of the Year 2 cash flow must be set aside in an external investment so that its accumulated sum at the end of two years will be +282. This can be brought back to the cash flow at the end of Year 4 with the result that the internal investment cash flow will have a zero for Year 4. The amount to invest externally in Year 2 is $x = +282/(1.06)^2 = +250.98$. The result is the transformed cash flow. The transformed cash flow has one sign change and, therefore, cannot have more than one positive rate of return. Using the technique described in Chapter 7A, we will see that i is very close to 11.1%.

Year	Transformed cash flow	Present worth at 11%	Present worth at 12%	Present worth at 11.1%
0	-$500	-$500	-$500	-$500
1	+300	+270.27	+267.86	+270.03
2	+49.02	+39.79	+39.08	+39.71
3	0	0	0	0
4	0	0	0	0
5	+200	+118.69	+113.49	+118.16
6	+135.66	+72.53	+68.73	+72.14
	+$184.68	+$1.28	-$10.86	+$0.04

Solution to Example 18-5b: The procedure will be to solve the cash flow for a rate of return, check net investment, and proceed using $e*$, if necessary.

Some preliminary computations, not shown here, lead us to a 15% rate of return. This may be verified, and net investment computed for $i* = 15\%$.

Year	Cash flow	Present worth at 15%		Net investment at 15%
0	-$500	-$500		-$500.00
1	+300	260.87	-500.00(1.15) + 300.00 =	-275.00
2	+300	+226.84	-275.00(1.15) + 300.00 =	-16.25
3	+300	+197.25	-16.25(1.15) + 300.00 =	+281.31
4	-600	-343.05	+281.31(1.15) - 600.00 =	-276.49
5	+200	+99.44	-276.49(i.15) + 200.00 =	-117.96
6	+135.66	+58.65	-117.96(1.15) + 135.66 =	0
	+$135.66	$0		

The computation of net investment at 15% indicates there is not net investment throughout the life of the project. At the end of Year 3 there is an external investment of $281.31. The computation is based on $e* = i* = 15\%$. For $e* = 6\%$, we must compute i.

For $e* = 6\%$, and Trial $i = 13\%$:

Year	Cash flow		Net investment
0	−$500		−$500.00
1	+300	−500.00(1.13) +300.00 =	−265.00
2	+300	−265.00(1.13) +300.00 =	+0.55
3	+300	+0.55(1.06) +300.00 =	+300.58
4	−600	+300.58(1.06) −600-00 =	−281.38
5	+200	−281.38(1.13) +200.00 =	−117.96
6	+135.66	−117.96(1.13) +135.66 =	+2.36

The end of Year 6 net investment is not zero, indicating that i is not equal to our 13% Trial i. Select 13.2% as another Trial i and repeat the computation:

Year	Cash flow		Net investment
0	−$500		−$500.00
1	+300	−500.00(1.132) + 300.00 =	−266.00
2	+300	−266.00(1.132) + 300.00 =	−1.11
3	+300	−1.11(1.132) + 300.00 =	+298.74
4	−600	+298.74(1.06) − 600.00 =	−283.33
5	+200	−283.33(1.132) + 200.00 =	−120.73
6	+135.66	−120.73(1.132) + 135.66 =	−1.01

Results of the two trials:

Trial i	End of Year 6 net investment
13.0%	+2.36
13.2%	−1.01

We conclude that for $e^* = 6\%$, i is approximately 13.13%.
At 13.13% the net investment is:

Year	Cash flow		Net investment
0	+$500		−$500.00
1	+300	−500.00(1.1313) +300.00 =	−265.65
2	+300	−265.65(1.1313) +300.00 =	−0.53
3	+300	−0.53(1.1313) +300.00 =	+299.40
4	−600	+299.40(1.06) − 600.00 =	−282.64
5	+200	−282.64(1.1313) +200.00 =	−119.75
6	+135.66	−119.75(1.1313) +135.66 =	+0.19 ≅ 0

Solution to Example 18-5c: The two computation methods transform the project cash flow into different amounts of internal and external investment. This is illustrated by the following tabulation:

Year	Cash flow	Part a Internal investment	Part a External investment	Part b Internal investment	Part b External investment
0	−$500	−$500.00		−$500.00	
1	+300	+300		+300.00	
2	+300	+49.02	+$250.98	+300.00	
3	+300		+300.00	+0.60	+$299.40
4	−600		−600.00	−282.64	−317.36
5	+200	+200.00		+200.00	
6	+135.66	+135.66		+135.66	

The tabulation shows that the methods outlined in this chapter do not necessarily assume as much external investment as is required by the methods of Chapter 7A. We now see clearly that the criterion on how much to alter a cash flow is properly based on maintaining a net investment—not on reducing the number of sign changes in the cash flow.

Based on net investment a smaller portion of the benefits are assumed in an external investment at $e^* = 6\%$. For values of i^* greater than e^*, we will find Part **b**'s i^* to be larger than Part **a**'s i^*. ∎

From what has been said, plus the five example problems, there are some important conclusions to be noted. When there are multiple positive rates of return for a cash flow, in general, none of them are a suitable measure of project profitability. Even in the situation where there is only one positive rate of return, that value still may not be a good indicator of project profitability. The critical question is whether the rate of return is exclusively the return on funds invested in the project or is the return on the combination of funds in the project and funds temporarily invested outside the project. If the project cash flow reflects both internal and external investments, the one or more positive rates of return assume that the internal rate of return i^* equals the external rate of return e^*. This is seldom a valid assumption. Where it is not, an external rate of return e^* should be selected and i computed, given e^*.

In a situation where an initial investment is made, followed by subsequent benefits (only one sign change in the cash flow), there is no temporary external investment and no difficulty in computing a suitable i^*. There will be other situations where in the later years of a cash flow an additional net investment is required. This may or may not result in multiple positive rates of return. It will, however, mean that an external interest rate e^* is needed to compute a suitable rate of return.

Summary

We can now see that the computation of rate of return for an investment project is far more complex than was stated in Chapter 7.

The cash flow rule of signs is the simple test of the number of changes of sign in the project cash flow. Zero sign changes is the unusual case of all benefits or all disbursements. This is immediately seen as either good or a disaster. One sign change is the conventional situation. This is the case that has been discussed throughout most of this book. It generally leads to a single positive rate of return that is a valid measure of project profitability. Multiple sign changes are a warning sign that a valid measure of project profitability may be difficult to locate.

The accumulated cash flow sign test is a test of the accumulated cash flows ($A_0, A_1, A_2, \ldots, A_n$). The algebraic sum of the cash flow is simply the sum of the project cash flow and equals the accumulated cash flow, A_n. If there is one sign change in the sequence and $A_n > 0$, these are sufficient (but not necessary) conditions for a single positive rate of return. When there is more than one sign change in the cash flow, the existence of a single positive rate of return does not necessarily give us a suitable measure of project profitability. The critical test is net investment.

Net investment is the computation to determine whether or not there is a continuing net investment throughout the life of the project for a value of i^*. Thus the first step in computing net investment is to locate a value of i^*. Then net investment is computed year-by-year with the net investment earning i^* each year. A cash flow will either have net investment throughout the life of the project, or it will not have it. A cash flow that does not have net investment is one where there is a net outflow of money from the project which will later be required to be returned to the project. At this point we need to invest the outflow of money someplace else (at external interest rate e^*) until it is required back in the project.

If there is net investment throughout the life of the project, we know this is sufficient (but not necessary) for a single positive rate of return, and this i^* is a suitable measure of project profitability. When there is not net investment throughout the life of the project, another rate of return i must be computed, with an external interest rate e^* assumed for project money temporarily invested externally. This revised i is an appropriate measure of project profitability. As we have seen, the calculation of i, using an external interest rate e^*, can be lengthy. The personal computer program ROR, described in the next chapter, performs the necessary computations.

Problems

18-1 Consider the following situation:

Year	Cash flow
0	−$500
1	+2000
2	−1200
3	−300

    a. For the cash flow, what information can be learned from—
      1. Cash flow rule of signs?
      2. Accumulated cash flow sign test?
      3. Algebraic sum of the cash flow?
      4. Net investment conditions?
    b. Compute the internal rate of return for the cash flow. If there is external investment, assume it is made at the same interest rate as the internal rate of return.
    c. Compute the internal rate of return for the cash flow. This time assume a 6% external interest rate.

18-2 Repeat Problem 18-1 for the following cash flow:

Year	Cash flow
0	−$500
1	+200
2	−500
3	+1200

(*Answers:* *b.* 21.1%; *c.* 21.1%)

18-3 Repeat Problem 18-1 for the following cash flow:

Year	Cash flow
0	−$500
1	+200
2	−500
3	+200

18-4 Repeat Problem 18-1 for the following cash flow:

Year	Cash Flow
0	−$100
1	+360
2	−570
3	+360

18-5   Given the following cash flow:

Year	Cash flow
0	−$200
1	+100
2	+100
3	+100
4	−300
5	+100
6	+200
7	+200
8	−124.5

Compute the internal rate of return assuming—

a.   External interest rate equals the internal rate of return.
b.   External interest rate equals 6%.

(*Answer:* *a.* 20%; *b.* 18.9%)

18-6   In Examples 7A-1 and 7A-2 (in Chapter 7A), an analysis was made of the proposed sale of an airplane by Going Aircraft Co. to Interair. In Ex. 7A-2, the rate of return was computed to be 8.4%, given a 6% external interest rate. Using the methods described in this chapter, compute i^*, given $e^* = 6\%$. Explain why your answer is or is not different from the 8.4% computed in Ex. 7A-2.

18-7   Refer to Example. 5-9 in Chapter 5.

a.   Construct a plot of NPW *vs. i* for the cash flow.
b.   Compute the rate of return for the investment project, assuming if necessary a 10% interest rate on external investments.
c.   Would the method described in Chapter 7A produce a different answer from that obtained in *b*? Explain.

18-8   Consider the following situation:

Year	Cash flow	Present worth at 70.7%
0	−$200	−$200.00
1	+400	+234.33
2	−100	−34.32
		+$0.01

a.   What is i^* if e^* equals 70.7%?

b.   What is i^* if e^* equals 0%?

18-9   An investor is considering two mutually exclusive projects in which to invest his money. He can obtain a 6% before-tax rate of return on external investments, but he requires a minimum attractive rate of return of 7% for these projects. Use a ten-year analysis period.

	Project A: Build drive-up photo shop	Project B: Buy land in Hawaii
Initial capital investment	$58,500	$ 48,500
Net uniform annual income	6,648	0
Salvage value ten years hence	30,000	138,000
Computed rate of return	8%	11%

Compute the incremental rate of return from investing in Project *A* rather than Project *B*.

18-10   In January, 1983, an investor purchased a convertible debenture bond issued by the XLA Corporation. The bond cost $1000 and paid $60 per year interest in annual payments on December 31. Under the convertible feature of the bond, it could be converted into twenty shares of common stock by tendering the bond, together with $400 cash. The day after the investor received the December 31, 1985, interest payment, he submitted the bond together with $400 to the XLA Corporation. In return, he received the twenty shares of common stock. The common stock paid no dividends. On December 31, 1987, the investor sold the stock for $1740, terminating his five-year investment in XLA Corporation. What rate of return did he receive?

Appendix

Introduction to Spreadsheets

Computerized spreadsheets are available nearly everywhere, and they can be easily applied to economic analysis. In fact, spreadsheets were originally developed to analyze financial data, and they are often credited with initiating the explosive growth in demand for desktop computing.

A spreadsheet is a two-dimensional table, whose cells can contain numerical values, labels, or formulas. The software automatically updates the table when an entry is changed, and there are powerful tools for copying formulas, creating graphs, and formatting results.

The Elements of a Spreadsheet

A spreadsheet is a two-dimensional table that labels the columns in alphabetical order A to Z, AA to AZ, BA to BZ, up to IV (256 columns). The rows are numbered from 1 to 65,536. Thus a *cell* of the spreadsheet is specified by its column letter and row number. For example, A3 is the third row in column A and AA6 is the sixth row in the twenty-seventh column. Each cell can contain a label, a numerical value, or a formula.

A *label* is any cell where the contents should be treated as text. Arithmetic cannot be performed on labels. Labels are used for variable names, row and column headings, and explanatory notes. For Excel™ any cell which contains more than a simple number, such as 3.14159, is treated as a label, unless it begins with an =, which is the signal for a formula. Thus 2*3 and B1+B2 are labels. Meaningful labels can be wider than a normal column. One solution is to allow those cells to "wrap" text, which is one of the "alignment" options. The table heading row in Example A-1 has turned this on by selecting the row, right clicking on the row, and selecting wrap text under the alignment tab.

A *numerical value* is any number. Acceptable formats for entry or display include percentages, currency, accounting, scientific, fractions, date, and time. In addition the number of decimal digits, the display of $ symbols, and commas for "thousands" separators can be adjusted. The format for cells can be changed by selecting a cell, a block of cells, a row, a column, or the entire spreadsheet. Then right click on the selected area, and a menu that includes "format cells" will appear. Then number formats, alignment, borders, fonts, and patterns can be selected.

Formulas must begin with an =, such as =3*4^2 or =B1+B2. They can include many functions – financial, statistical, trigonometric, etc. (and others can be defined by the user). The formula for the "current" cell is displayed in the formula bar at the top of the spreadsheet. The value resulting from the formula is displayed in the cell in the spreadsheet.

Often the printed out spreadsheet will be part of a report or a homework assignment and the formulas must be explained. Here is an easy way to place a copy of the formula in an adjacent or nearby cell. (1) Convert the cell with the formula to a label by either inserting a space before the = sign or by deleting the =. (2) Copy that label to an adjacent cell by using cut and paste. Do not drag the cell to copy it, as any formula ending with a number (even an address like B4) will have the number automatically incremented. (3) Convert the original formula back into a formula by deleting the space or inserting the =.

Defining Variables in a Data Block

The cell A1, top left corner, is the HOME cell for a spreadsheet. Thus, the top left area is where the data block should be placed. This data block should have every variable in the spreadsheet with an adjacent label for each. This data block supports a basic principle of good spreadsheet modeling, which is to use variables in your models.

The data block in Example A-1 contains *entered data* – the loan amount (A1), the number of payments (A2), and the interest rate (A3), and *computed data* – the payment (A4). Then instead of using the loan amount of $5000 in a formula, the cell reference A1 is used. Even if a value is only referenced once, it is better to include it in the data block. By using one location to define each variable, you can change any value one place in the spreadsheet and have the entire spreadsheet instantly re-computed.

Even for simple homework problems you should use a data block.
1. You may be able to use it for another problem.
2. Solutions to simple problems may grow into solutions for complex problems.
3. Good habits, like using data blocks, are easy to maintain once they are established.
4. It makes the assumptions clear if you've estimated a value or for grading.

In the real world, data blocks are even more important. Most problems are solved more than once, as more and more accurate values are estimated. Often the spreadsheet is revised to add other variables, time periods, locations, etc. Without data blocks, it is hard to change a spreadsheet and the likelihood of missing a required change skyrockets.

If you want your formulas to be easier to read, you can name your variables. Note: in current versions of Excel, the cell's location or name is displayed at the left of the formula bar. Variable names can be entered here. They will then automatically be applied if cell addresses are entered by point and click. If cell addresses are entered as A2, then A2 is what is displayed. To change a displayed A2, to the name of the cell (LoanAmount), the process is to click on insert, click on name, click on apply, and then select the names to be applied.

Copy Command

The copy command and relative/absolute addressing make spreadsheet models easy to build. If the range of cells to be copied contains only labels, numbers, and functions, then the copy command is easy to use and understand. For example, the formula =EXP(1.9) would be copied unchanged to a new location. However, cell addresses are usually part of the range being copied, and their absolute and relative addresses are treated differently.

An *absolute address* is denoted by adding $ signs before the column and/or row. For example in Figure A-1a, $A$4 is the absolute address for the interest rate. When an absolute address is copied, the column and/or row that is fixed is copied unchanged. Thus $A$4 is completely fixed, $A4 fixes the column, and A$4 fixes the row. One common use for absolute addresses is any data block entry, such as the interest rate. Changing between A4, $A$4, A$4, $A4, and A4 is most easily done using the F4 key, which scrolls an address through the choices.

In contrast a *relative address* is best interpreted as directions from one cell to another. For example in Figure A-1a, the balance due in year *t* equals the balance due in year *t* − 1 minus the principal payment in year *t*. Specifically for the balance due in year 1, D10 contains =D9-C10. From cell D10, cell D9 is one row up and C10 is one column to the left, so the formula is really (contents of 1 up) minus (contents of 1 to the left). When a cell containing a relative address is copied to a new location, it is these directions that are copied to determine any new relative addresses. So if cell D10 is copied to cell F14, the formula is =F13-E14.

Thus to calculate a loan repayment schedule, as in Figure A-1, the row of formulas is created and then copied for the remaining years.

EXAMPLE A-1

Table 3-1 shows four repayment schedules for a loan of $5000 to be repaid over 5 years at an interest rate of 8%. Use a spreadsheet to calculate the amortization schedule for the constant principal payment option.

Solution

The first step is to enter the loan amount, number of periods, and interest rate into a data block in the top left part of the spreadsheet. The next step is to calculate the constant principal payment amount, which was given as $1252.28 in Table 3-1. The factor approach to finding this value is given in Chapter 3 and the spreadsheet function is explained in Chapter 4.

The next step is to identify the columns for the amortization schedule. These are the year, interest owed, principal payment, and balance due. Because some of these labels are wider than a normal column, the cells are formatted so that the text wraps (row height increases automatically). The initial balance is shown in the year 0 row.

Next, the formulas for the first year are written, as shown in Figure A-1a. The interest owed (cell B10) equals the interest rate ($A$4) times the balance due for year 0 (D9). The principal payment (cell C10) equals the annual payment ($A$6) minus the interest owed and paid (B10). Finally, the balance due (cell B10) equals the balance due for the previous year (D9) minus the principal payment (C10). The results are shown in Figure A-1a.

Now cells A10 to D10 are selected for year 1. By dragging down on the right corner of D10, the entire row can be copied for years 2 through 5. Note if cut and paste is used, then it is necessary to complete the year column separately (dragging increments the year, but cutting and pasting does not). The results are shown in Figure A-1b.

	A	B	C	D	E
1	Entered Data				
2	5000	loan amount			
3	5	number of payments			
4	8%	interest rate			
5	Computed Data				
6	$1,252.28	loan payment			
7					
8	Year	Interest owed	Principal payment	Balance due	
9	0			5000.00	
10	1	400.00	852.28	4147.72	=D9-C10
11					
12		=$A$4*D9		=$A$6-B10	

Figure A-1a Year 1 amortization schedule.

	A	B	C	D	E
8	Year	Interest owed	Principal payment	Balance due	
9	0			5000.00	
10	1	400.00	852.28	4147.72	
11	2	331.82	920.46	3227.25	
12	3	258.18	994.10	2233.15	
13	4	178.65	1073.63	1159.52	
14	5	92.76	1159.52	0.00	=D13-C14
15					
16		=$A$4*D13		=$A$6-B14	

Figure A-1b Completed amortization schedule.

This appendix has introduced the basics of spreadsheets. Chapter 2 uses spreadsheets and simple bar charts to draw cash flow diagrams. Chapters 4-15 each have spreadsheet sections. These are designed to develop spreadsheet modeling skills and to reinforce your understanding of engineering economy. As current spreadsheet packages are built around using mice to click on cells and items in charts, there is usually an intuitive connection between what you would like to do and how to do it. It seems the best way to learn the "how-to's" is to simply play around with the spreadsheet. In addition, as you look at the menu choices, you will find "what's" that you hadn't thought of but find useful.

References

AASHTO. *A Manual on User Benefit Analysis of Highway and Bus-Transit Improvements.* Washington, D.C.: American Association of State Highway and Transportation Officials, 1978.

American Telephone and Telegraph Co. *Engineering Economy,* 3rd ed. New York: McGraw-Hill, 1977.

Au, T., and Au, P. *Engineering Economics for Capital Investment Analysis.* Boston: Allyn and Bacon, 1983.

Barish, N. N., and Kaplan, S. *Economic Analysis for Engineering and Managerial Decision Making,* 2nd ed. New York: McGraw-Hill, 1978.

Benford, H. A *Naval Architect's Introduction to Engineering Economics.* Ann Arbor: Dept. of Naval Architecture and Marine Engineering Report 282, 1983.

Bernhard, R. H. "A Comprehensive Comparison and Critique of Discounting Indices Proposed for Capitol Investment Evaluation," *The Engineering Economist.* Vol. 16, No. 3, pp. 157-186.

Blank, L. T., and Tarquin, A. J. *Engineering Economy,* 3rd ed. New York: McGraw-Hill, 1989.

Bussey, L. E., and Eschenbach, T. E., The *Economic Analysis of Industrial Projects.* Englewood Cliffs, New Jersey: Prentice-Hall, 1992.

Buxton, 1. L. *Engineering Economics and Ship Design,* 3rd ed. Time and Wear, British Maritime Technology Ltd., 1987.

Cassimatis, P. *A Concise Introduction to Engineering Economics.* Boston: Unwin Hyman, 1988.

Collier, C. A., and Ledbetter, W. B. *Engineering Economic and Cost Analysis,* 2nd ed. New York: Harper and Row, 1988.

DeGarmo, E. P., Sullivan, W. G., and Bontadelli, J. A. *Engineering Economy,* 8th ed. New York: MacMillan, 1988.

Engineering Economist, 7he. A quarterly journal of the Engineering Economy Divisions of ASEE and IIE. Norcross, Georgia: Institute of Industrial Engineers.

Eschenbach, T. *Cases in Engineering Economy.* New York: John Wiley & Sons, 1989.

Fleischer, G.A. *Engineering Economy: Capital Allocation Theory.* Boston: PWS Engineering, 1984.

Grant, E. L., Ireson, W. G., and Leavenworth, R. S. *Principles of Engineering Economy,* 8th ed. New York: John Wiley & Sons, 1990.

Jones, B. W. *Inflation in Engineering Economic Analysis.* New York: John Wiley & Sons, 1982.

Kleinfeld, 1. *Engineering and Managerial Economics.* New York: Holt, Rinehart and Winston, 1986.

Lorie, J. H., and Savage, L. J. "Three Problems in Rationing Capital," *The Journal of Business.* Vol. 28, No. 4, pp. 229-239.

Mallik, A. K. *Engineering Economy with Computer Applications.* Mahomet, Illinois: Engineering Technology, 1979.

Meffett, A. J., and Sykes, A. *7he Finance and Analysis of Capital Projects,* 2nd ed. London: Longman, 1973.

Newnan, D. G., editor. *Engineering Economy Exam File.* San Jose, California: Engineering Press, 1984.

_____. "Determining Rate of Return by Means of Payback Period and Useful Life," *The Engineering Economist.* Vol. 15, No. 1, pp. 29-39.

Oglesby, C. H., and Hicks, R. G. *Highway Engineering,* 4th ed. New York: John Wiley & Sons, 1982.

Park, Chan. S. *Contemporary Engineering Economics.* New York: Addison-Wesley, 1993.

Park, W. R., and Jackson, a E. *Cost Engineering Analysis,* 2nd ed. New York: John Wiley & Sons, 1984.

Riggs, J. L., and West, T. M. *Engineering Economics,* 3rd ed. New York: McGraw-Hill, 1986.

Smith, G. W. *Engineering Economy: Analysis of Capital Expenditures,* 4th ed. Ames, Iowa: The Iowa State University Press, 1987.

Steiner, H. M. *Public and Private Investments: Socioeconomic Analysis.* New York: John Wiley & Sons, 1980.

_____. *Engineering Economic Principles.* New York: McGraw-Hill, 1992.

Stevens, G. T. *Economic and Financial Analysis of Capital Investments.* New York: John Wiley & Sons, 1979.

Swalm, R. O., and Lopez-Leautaud, J. L. *Engineering Economic Analysis: A Future Wealth Approach.* New York: John Wiley & Sons, 1984.

Terborgh, G. *Business Investment Management.* Washington, D.C.: Machinery and Allied Products Institute, 1967.

Thuesen, H. G., Fabrycky, W. J., and Thuesen, G. J. *Engineering Economy,* 7th ed. Englewood Cliffs, New Jersey: Prentice-Hall, 1988.

Wellington, A. M. *The Economic Theory of Railway Location.* New York: John Wiley & Sons, 1887.

White, J. A., Agee, M. H., and Case, K. E. *Principles of Engineering Economic Analysis,* 3rd ed. New York: John Wiley & Sons, 1989.

Compound Interest Tables

From
Engineering Economic Analysis
Donald G. Newnan

Values Of Interest Factors
When N Equals Infinity

Single Payment:

$(F/P,i,\infty) = \infty$

$(P/F,i,\infty) = 0$

Arithmetic Gradient Series:

$(A/G,i,\infty) = 1/i$

$(P/G,i,\infty) = 1/i^2$

Uniform Payment Series:

$(A/F,i,\infty) = 0$

$(A/P,i,\infty) = i$

$(F/A,i,\infty) = \infty$

$(P/A,i,\infty) = 1/i$

	Single Payment		Uniform Payment Series				Arithmetic Gradient		
	Compound Amount Factor	Present Worth Factor	Sinking Fund Factor	Capital Recovery Factor	Compound Amount Factor	Present Worth Factor	Gradient Uniform Series	Gradient Present Worth	
n	Find F Given P F/P	Find P Given F P/F	Find A Given F A/F	Find A Given P A/P	Find F Given A F/A	Find P Given A P/A	Find A Given G A/G	Find P Given G P/G	n
1	1.003	.9975	1.0000	1.0025	1.000	0.998	0	0	1
2	1.005	.9950	.4994	.5019	2.003	1.993	0.504	1.005	2
3	1.008	.9925	.3325	.3350	3.008	2.985	1.005	2.999	3
4	1.010	.9901	.2491	.2516	4.015	3.975	1.501	5.966	4
5	1.013	.9876	.1990	.2015	5.025	4.963	1.998	9.916	5
6	1.015	.9851	.1656	.1681	6.038	5.948	2.498	14.861	6
7	1.018	.9827	.1418	.1443	7.053	6.931	2.995	20.755	7
8	1.020	.9802	.1239	.1264	8.070	7.911	3.490	27.611	8
9	1.023	.9778	.1100	.1125	9.091	8.889	3.987	35.440	9
10	1.025	.9753	.0989	.1014	10.113	9.864	4.483	44.216	10
11	1.028	.9729	.0898	.0923	i1.139	10.837	4.978	53.950	11
12	1.030	.9705	.0822	.0847	12.167	11.807	5.474	64.634	12
13	1.033	.9681	.0758	.0783	13.197	12.775	5.968	76.244	13
14	1.036	.9656	.0703	.0728	14.230	13.741	6.464	88.826	14
15	1.038	.9632	.0655	.0680	15.266	14.704	6.957	102.301	15
16	1.041	.9608	.0613	.0638	16.304	15.665	7.451	116.716	16
17	1.043	.9584	.0577	.0602	17.344	16.624	7.944	132.063	17
18	1.046	.9561	.0544	.0569	18.388	17.580	8.437	148.319	18
19	1.049	.9537	.0515	.0540	19.434	18.533	8.929	165.492	19
20	1.051	.9513	.0488	.0513	20.482	19.485	9.421	183.559	20
21	1.054	.9489	.0464	.0489	21.534	20.434	9.912	202.531	21
22	1.056	.9465	.0443	.0468	22.587	21.380	10.404	222.435	22
23	1.059	.9442	.0423	.0448	23.644	22.324	10.894	243.212	23
24	1.062	.9418	.0405	.0430	24.703	23.266	11.384	264.854	24
25	1.064	.9395	.0388	.0413	25.765	24.206	11.874	287.407	25
26	1.067	.9371	.0373	.0398	26.829	25.143	12.363	310.848	26
27	1.070	.9348	.0358	.0383	27.896	26.078	12.852	335.150	27
28	1.072	.9325	.0345	.0370	28.966	27.010	13.341	360.343	28
29	1.075	.9301	.0333	.0358	30.038	27.940	13.828	386.366	29
30	1.078	.9278	.0321	.0346	31.114	28.868	14.317	413.302	30
36	1.094	.9140	.0266	.0291	37.621	34.387	17.234	592.632	36
40	1.105	.9049	.0238	.0263	42.014	38.020	19.171	728.882	40
48	1.127	.8871	.0196	.0221	50.932	45.179	23.025	1 040.22	48
50	1.133	.8826	.0188	.0213	53.189	46.947	23.984	1 125.96	50
52	1.139	.8782	.0180	.0205	55.458	48.705	24.941	1 214.76	52
60	1.162	.8609	.0155	.0180	64.647	55.653	28.755	1 600.31	60
70	1.191	.8396	.0131	.0156	76.395	64.144	33.485	2 147.87	70
72	1.197	.8355	.0127	.0152	78.780	65.817	34.426	2 265.81	72
80	1.221	.8189	.0113	.0138	88.440	72.427	38.173	2 764.74	80
84	1.233	.8108	.0107	.0132	93.343	75.682	40.037	3 030.06	84
90	1.252	.7987	.00992	.0124	100.789	80.504	42.820	3 447.19	90
96	1.271	.7869	.00923	.0117	108.349	85.255	45.588	3 886.62	96
100	1.284	.7790	.00881	.0113	113.451	88.383	47.425	4 191.60	100
104	1.297	.7713	.00843	.0109	118.605	91.480	49.256	4 505.93	104
120	1.349	.7411	.00716	.00966	139.743	103.563	56.512	5 852.52	120
240	1.821	.5492	.00305	.00555	328.306	180.312	107.590	19 399.75	240
360	2.457	.4070	.00172	.00422	582.745	237.191	152.894	36 264.96	360
480	3.315	.3016	.00108	.00358	926.074	279.343	192.673	53 821.93	480

	Single Payment		Uniform Payment Series				Arithmetic Gradient		
	Compound Amount Factor	Present Worth Factor	Sinking Fund Factor	Capital Recovery Factor	Compound Amount Factor	Present Worth Factor	Gradient Uniform Series	Gradient Present Worth	
n	Find *F* Given *P* *F/P*	Find *P* Given *F* *P/F*	Find *A* Given *F* *A/F*	Find *A* Given *P* *A/P*	Find *F* Given *A* *F/A*	Find *P* Given *A* *P/A*	Find *A* Given *G* *A/G*	Find *P* Given *G* *P/G*	*n*
1	1.005	.9950	1.0000	1.0050	1.000	0.995	0	0	1
2	1.010	.9901	.4988	.5038	2.005	1.985	0.499	0.991	2
3	1.015	.9851	.3317	.3367	3.015	2.970	0.996	2.959	3
4	1.020	.9802	.2481	.2531	4.030	3.951	1.494	5.903	4
5	1.025	.9754	.1980	.2030	5.050	4.926	1.990	9.803	5
6	1.030	.9705	.1646	.1696	6.076	5.896	2.486	14.660	6
7	1.036	.9657	.1407	.1457	7.106	6.862	2.980	20.448	7
8	1.041	.9609	.1228	.1278	8.141	7.823	3.474	27.178	8
9	1.046	.9561	.1089	.1139	9.182	8.779	3.967	34.825	9
10	1.051	.9513	.0978	.1028	10.228	9.730	4.459	43.389	10
11	1.056	.9466	.0887	.0937	11.279	10.677	4.950	52.855	11
12	1.062	.9419	.0811	.0861	12.336	11.619	5.441	63.218	12
13	1.067	.9372	.0746	.0796	13.397	12.556	5.931	74.465	13
14	1.072	.9326	.0691	.0741	14.464	13.489	6.419	86.590	14
15	1.078	.9279	.0644	.0694	15.537	14.417	6.907	99.574	15
16	1.083	.9233	.0602	.0652	16.614	15.340	7.394	113.427	16
17	1.088	.9187	.0565	.0615	17.697	16.259	7.880	128.125	17
18	1.094	.9141	.0532	.0582	18.786	17.173	8.366	143.668	18
19	1.099	.9096	.0503	.0553	19.880	18.082	8.850	160.037	19
20	1.105	.9051	.0477	.0527	20.979	18.987	9.334	177.237	20
21	1.110	.9006	.0453	.0503	22.084	19.888	9.817	195.245	21
22	1.116	.8961	.0431	.0481	23.194	20.784	10.300	214.070	22
23	1.122	.8916	.0411	.0461	24.310	21.676	10.781	233.680	23
24	1.127	.8872	.0393	.0443	25.432	22.563	11.261	254.088	24
25	1.133	.8828	.0377	.0427	26.559	23.446	11.741	275.273	25
26	1.138	.8784	.0361	.0411	27.692	24.324	12.220	297.233	26
27	1.144	.8740	.0347	.0397	28.830	25.198	12.698	319.955	27
28	1.150	.8697	.0334	.0384	29.975	26.068	13.175	343.439	28
29	1.156	.8653	.0321	.0371	31.124	26.933	13.651	367.672	29
30	1.161	.8610	.0310	.0360	32.280	27.794	14.127	392.640	30
36	1.197	.8356	.0254	.0304	39.336	32.871	16.962	557.564	36
40	1.221	.8191	.0226	.0276	44.159	36.172	18.836	681.341	40
48	1.270	.7871	.0185	.0235	54.098	42.580	22.544	959.928	48
50	1.283	.7793	.0177	.0227	56.645	44.143	23.463	1 035.70	50
52	1.296	.7716	.0169	.0219	59.218	45.690	24.378	1 113.82	52
60	1.349	.7414	.0143	.0193	69.770	51.726	28.007	1 448.65	60
70	1.418	.7053	.0120	.0170	83.566	58.939	32.468	1 913.65	70
72	1.432	.6983	.0116	.0166	86.409	60.340	33.351	2 012.35	72
80	1.490	.6710	.0102	.0152	98.068	65.802	36.848	2 424.65	80
84	1.520	.6577	.00961	.0146	104.074	68.453	38.576	2 640.67	84
90	1.567	.6383	.00883	.0138	113.311	72.331	41.145	2 976.08	90
96	1.614	.6195	.00814	.0131	122.829	76.095	43.685	3 324.19	96
100	1.647	.6073	.00773	.0127	129.334	78.543	45.361	3 562.80	100
104	1.680	.5953	.00735	.0124	135.970	80.942	47.025	3 806.29	104
120	1.819	.5496	.00610	.0111	163.880	90.074	53.551	4 823.52	120
240	3.310	.3021	.00216	.00716	462.041	139.581	96.113	13 415.56	240
360	6.023	.1660	.00100	.00600	1 004.5	166.792	128.324	21 403.32	360
480	10.957	.0913	.00050	.00550	1 991.5	181.748	151.795	27 588.37	480

3/4%   Compound Interest Factors   3/4%

| | Single Payment | | Uniform Payment Series | | | | Arithmetic Gradient | | |
| | Compound Amount Factor | Present Worth Factor | Sinking Fund Factor | Capital Recovery Factor | Compound Amount Factor | Present Worth Factor | Gradient Uniform Series | Gradient Present Worth | |
n	Find F Given P F/P	Find P Given F P/F	Find A Given F A/F	Find A Given P A/P	Find F Given A F/A	Find P Given A P/A	Find A Given G A/G	Find P Given G P/G	n
1	1.008	.9926	1.0000	1.0075	1.000	0.993	0	0	1
2	1.015	.9852	.4981	.5056	2.008	1.978	0.499	0.987	2
3	1.023	.9778	.3308	.3383	3.023	2.956	0.996	2.943	3
4	1.030	.9706	.2472	.2547	4.045	3.926	1.492	5.857	4
5	1.038	.9633	.1970	.2045	5.076	4.889	1.986	9.712	5
6	1.046	.9562	.1636	.1711	6.114	5.846	2.479	14.494	6
7	1.054	.9490	.1397	.1472	7.160	6.795	2.971	20.187	7
8	1.062	.9420	.1218	.1293	8.213	7.737	3.462	26.785	8
9	1.070	.9350	.1078	.1153	9.275	8.672	3.951	34.265	9
10	1.078	.9280	.0967	.1042	10.344	9.600	4.440	42.619	10
11	1.086	.9211	.0876	.0951	11.422	10.521	4.927	51.831	11
12	1.094	.9142	.0800	.0875	12.508	11.435	5.412	61.889	12
13	1.102	.9074	.0735	.0810	13.602	12.342	5.897	72.779	13
14	1.110	.9007	.0680	.0755	14.704	13.243	6.380	84.491	14
15	1.119	.8940	.0632	.0707	15.814	14.137	6.862	97.005	15
16	1.127	.8873	.0591	.0666	16.932	15.024	7.343	110.318	16
17	1.135	.8807	.0554	.0629	18.059	15.905	7.822	124.410	17
18	1.144	.8742	.0521	.0596	19.195	16.779	8.300	139.273	18
19	1.153	.8676	.0492	.0567	20.339	17.647	8.777	154.891	19
20	1.161	.8612	.0465	.0540	21.491	18.508	9.253	171.254	20
21	1.170	.8548	.0441	.0516	22.653	19.363	9.727	188.352	21
22	1.179	.8484	.0420	.0495	23.823	20.211	10.201	206.170	22
23	1.188	.8421	.0400	.0475	25.001	21.053	10.673	224.695	23
24	1.196	.8358	.0382	.0457	26.189	21.889	11.143	243.924	24
25	1.205	.8296	.0365	.0440	27.385	22.719	11.613	263.834	25
26	1.214	.8234	.0350	.0425	28.591	23.542	12.081	284.421	26
27	1.224	.8173	.0336	.0411	29.805	24.360	12.548	305.672	27
28	1.233	.8112	.0322	.0397	31.029	25.171	13.014	327.576	28
29	1.242	.8052	.0310	.0385	32.261	25.976	13.479	350.122	29
30	1.251	.7992	.0298	.0373	33.503	26.775	13.942	373.302	30
36	1.309	.7641	.0243	.0318	41.153	31.447	16.696	525.038	36
40	1.348	.7416	.0215	.0290	46.447	34.447	18.507	637.519	40
48	1.431	.6986	.0174	.0249	57.521	40.185	22.070	886.899	48
50	1.453	.6882	.0166	.0241	60.395	41.567	22.949	953.911	50
52	1.475	.6780	.0158	.0233	63.312	42.928	23.822	1 022.64	52
60	1.566	.6387	.0133	.0208	75.425	48.174	27.268	1 313.59	60
70	1.687	.5927	.0109	.0184	91.621	54.305	31.465	1 708.68	70
72	1.713	.5839	.0105	.0180	95.008	55.477	32.289	1 791.33	72
80	1.818	.5500	.00917	.0167	109.074	59.995	35.540	2 132.23	80
84	1.873	.5338	.00859	.0161	116.428	62.154	37.137	2 308.22	84
90	1.959	.5104	.00782	.0153	127.881	65.275	39.496	2 578.09	90
96	2.049	.4881	.00715	.0147	139.858	68.259	41.812	2 854.04	96
100	2.111	.4737	.00675	.0143	148.147	70.175	43.332	3 040.85	100
104	2.175	.4597	.00638	.0139	156.687	72.035	44.834	3 229.60	104
120	2.451	.4079	.00517	.0127	193.517	78.942	50.653	3 998.68	120
240	6.009	.1664	.00150	.00900	667.901	111.145	85.422	9 494.26	240
360	14.731	.0679	.00055	.00805	1 830.8	124.282	107.115	13 312.50	360
480	36.111	.0277	.00021	.00771	4 681.5	129.641	119.662	15 513.16	480

Compound Interest Factors

	Single Payment		Uniform Payment Series				Arithmetic Gradient		
	Compound Amount Factor	Present Worth Factor	Sinking Fund Factor	Capital Recovery Factor	Compound Amount Factor	Present Worth Factor	Gradient Uniform Series	Gradient Present Worth	
n	Find F Given P F/P	Find P Given F P/F	Find A Given F A/F	Find A Given P A/P	Find F Given A F/A	Find P Given A P/A	Find A Given G A/G	Find P Given G P/G	n
1	1.010	.9901	1.0000	1.0100	1.000	0.990	0	0	1
2	1.020	.9803	.4975	.5075	2.010	1.970	0.498	0.980	2
3	1.030	.9706	.3300	.3400	3.030	2.941	0.993	2.921	3
4	1.041	.9610	.2463	.2563	4.060	3.902	1.488	5.804	4
5	1.051	.9515	.1960	.2060	5.101	4.853	1.980	9.610	5
6	1.062	.9420	.1625	.1725	6.152	5.795	2.471	14.320	6
7	1.072	.9327	.1386	.1486	7.214	6.728	2.960	19.917	7
8	1.083	.9235	.1207	.1307	8.286	7.652	3.448	26.381	8
9	1.094	.9143	.1067	.1167	9.369	8.566	3.934	33.695	9
10	1.105	.9053	.0956	.1056	10.462	9.471	4.418	41.843	10
11	1.116	.8963	.0865	.0965	11.567	10.368	4.900	50.806	11
12	1.127	.8874	.0788	.0888	12.682	11.255	5.381	60.568	12
13	1.138	.8787	.0724	.0824	13.809	12.134	5.861	71.112	13
14	1.149	.8700	.0669	.0769	14.947	13.004	6.338	82.422	14
15	1.161	.8613	.0621	.0721	16.097	13.865	6.814	94.481	15
16	1.173	.8528	.0579	.0679	17.258	14.718	7.289	107.273	16
17	1.184	.8444	.0543	.0643	18.430	15.562	7.761	120.783	17
18	1.196	.8360	.0510	.0610	19.615	16.398	8.232	134.995	18
19	1.208	.8277	.0481	.0581	20.811	17.226	8.702	149.895	19
20	1.220	.8195	.0454	.0554	22.019	18.046	9.169	165.465	20
21	1.232	.8114	.0430	.0530	23.239	18.857	9.635	181.694	21
22	1.245	.8034	.0409	.0509	24.472	19.660	10.100	198.565	22
23	1.257	.7954	.0389	.0489	25.716	20.456	10.563	216.065	23
24	1.270	.7876	.0371	.0471	26.973	21.243	11.024	234.179	24
25	1.282	.7798	.0354	.0454	28.243	22.023	11.483	252.892	25
26	1.295	.7720	.0339	.0439	29.526	22.795	11.941	272.195	26
27	1.308	.7644	.0324	.0424	30.821	23.560	12.397	292.069	27
28	1.321	.7568	.0311	.0411	32.129	24.316	12.852	312.504	28
29	1.335	.7493	.0299	.0399	33.450	25.066	13.304	333.486	29
30	1.348	.7419	.0287	.0387	34.785	25.808	13.756	355.001	30
36	1.431	.6989	.0232	.0332	43.077	30.107	16.428	494.620	36
40	1.489	.6717	.0205	.0305	48.886	32.835	18.178	596.854	40
48	1.612	.6203	.0163	.0263	61.223	37.974	21.598	820.144	48
50	1.645	.6080	.0155	.0255	64.463	39.196	22.436	879.417	50
52	1.678	.5961	.0148	.0248	67.769	40.394	23.269	939.916	52
60	1.817	.5504	.0122	.0222	81.670	44.955	26.533	1 192.80	60
70	2.007	.4983	.00993	.0199	100.676	50.168	30.470	1 528.64	70
72	2.047	.4885	.00955	.0196	104.710	51.150	31.239	1 597.86	72
80	2.217	.4511	.00822	.0182	121.671	54.888	34.249	1 879.87	80
84	2.307	.4335	.00765	.0177	130.672	56.648	35.717	2 023.31	84
90	2.449	.4084	.00690	.0169	144.863	59.161	37.872	2 240.56	90
96	2.599	.3847	.00625	.0163	159.927	61.528	39.973	2 459.42	96
100	2.705	.3697	.00587	.0159	170.481	63.029	41.343	2 605.77	100
104	2.815	.3553	.00551	.0155	181.464	64.471	42.688	2 752.17	104
120	3.300	.3030	.00435	.0143	230.039	69.701	47.835	3 334.11	120
240	10.893	.0918	.00101	.0110	989.254	90.819	75.739	6 878.59	240
360	35.950	.0278	.00029	.0103	3 495.0	97.218	89.699	8 720.43	360
480	118.648	.00843	.00008	.0101	11 764.8	99.157	95.920	9 511.15	480

1¼%   Compound Interest Factors   1¼%

	Single Payment		Uniform Payment Series				Arithmetic Gradient		
	Compound Amount Factor	Present Worth Factor	Sinking Fund Factor	Capital Recovery Factor	Compound Amount Factor	Present Worth Factor	Gradient Uniform Series	Gradient Present Worth	
n	Find F Given P F/P	Find P Given F P/F	Find A Given F A/F	Find A Given P A/P	Find F Given A F/A	Find P Given A P/A	Find A Given G A/G	Find P Given G P/G	n
1	1.013	.9877	1.0000	1.0125	1.000	0.988	0	0	1
2	1.025	.9755	.4969	.5094	2.013	1.963	0.497	0.976	2
3	1.038	.9634	.3292	.3417	3.038	2.927	0.992	2.904	3
4	1.051	.9515	.2454	.2579	4.076	3.878	1.485	5.759	4
5	1.064	.9398	.1951	.2076	5.127	4.818	1.976	9.518	5
6	1.077	.9282	.1615	.1740	6.191	5.746	2.464	14.160	6
7	1.091	.9167	.1376	.1501	7.268	6.663	2.951	19.660	7
8	1.104	.9054	.1196	.1321	8.359	7.568	3.435	25.998	8
9	1.118	.8942	.1057	.1182	9.463	8.462	3.918	33.152	9
10	1.132	.8832	.0945	.1070	10.582	9.346	4.398	41.101	10
11	1.146	.8723	.0854	.0979	11.714	10.218	4.876	49.825	11
12	1.161	.8615	.0778	.0903	12.860	11.079	5.352	59.302	12
13	1.175	.8509	.0713	.0838	14.021	11.930	5.827	69.513	13
14	1.190	.8404	.0658	.0783	15.196	12.771	6.299	80.438	14
15	1.205	.8300	.0610	.0735	16.386	13.601	6.769	92.058	15
16	1.220	.8197	.0568	.0693	17.591	14.420	7.237	104.355	16
17	1.235	.8096	.0532	.0657	18.811	15.230	7.702	117.309	17
18	1.251	.7996	.0499	.0624	20.046	16.030	8.166	130.903	18
19	1.266	.7898	.0470	.0595	21.297	16.819	8.628	145.119	19
20	1.282	.7800	.0443	.0568	22.563	17.599	9.088	159.940	20
21	1.298	.7704	.0419	.0544	23.845	18.370	9.545	175.348	21
22	1.314	.7609	.0398	.0523	25.143	19.131	10.001	191.327	22
23	1.331	.7515	.0378	.0503	26.458	19.882	10.455	207.859	23
24	1.347	.7422	.0360	.0485	27.788	20.624	10.906	224.930	24
25	1.364	.7330	.0343	.0468	29.136	21.357	11.355	242.523	25
26	1.381	.7240	.0328	.0453	30.500	22.081	11.803	260.623	26
27	1.399	.7150	.0314	.0439	31.881	22.796	12.248	279.215	27
28	1.416	.7062	.0300	.0425	33.280	23.503	12.691	298.284	28
29	1.434	.6975	.0288	.0413	34.696	24.200	13.133	317.814	29
30	1.452	.6889	.0277	.0402	36.129	24.889	13.572	337.792	30
36	1.564	.6394	.0222	.0347	45.116	28.847	16.164	466.297	36
40	1.644	.6084	.0194	.0319	51.490	31.327	17.852	559.247	40
48	1.815	.5509	.0153	.0278	65.229	35.932	21.130	759.248	48
50	1.861	.5373	.0145	.0270	68.882	37.013	21.930	811.692	50
52	1.908	.5242	.0138	.0263	72.628	38.068	22.722	864.960	52
60	2.107	.4746	.0113	.0238	88.575	42.035	25.809	1 084.86	60
70	2.386	.4191	.00902	.0215	110.873	46.470	29.492	1 370.47	70
72	2.446	.4088	.00864	.0211	115.675	47.293	30.205	1 428.48	72
80	2.701	.3702	.00735	.0198	136.120	50.387	32.983	1 661.89	80
84	2.839	.3522	.00680	.0193	147.130	51.822	34.326	1 778.86	84
90	3.059	.3269	.00607	.0186	164.706	53.846	36.286	1 953.85	90
96	3.296	.3034	.00545	.0179	183.643	55.725	38.180	2 127.55	96
100	3.463	.2887	.00507	.0176	197.074	56.901	39.406	2 242.26	100
104	3.640	.2747	.00474	.0172	211.190	58.021	40.604	2 355.90	104
120	4.440	.2252	.00363	.0161	275.220	61.983	45.119	2 796.59	120
240	19.716	.0507	.00067	.0132	1 497.3	75.942	67.177	5 101.55	240
360	87.543	.0114	.00014	.0126	6 923.4	79.086	75.840	5 997.91	360
480	388.713	.00257	.00003	.0125	31 017.1	79.794	78.762	6 284.74	480

	Single Payment		Uniform Payment Series				Arithmetic Gradient		
	Compound Amount Factor	Present Worth Factor	Sinking Fund Factor	Capital Recovery Factor	Compound Amount Factor	Present Worth Factor	Gradient Uniform Series	Gradient Present Worth	
n	Find F Given P F/P	Find P Given F P/F	Find A Given F A/F	Find A Given P A/P	Find F Given A F/A	Find P Given A P/A	Find A Given G A/G	Find P Given G P/G	n
1	1.015	.9852	1.0000	1.0150	1.000	0.985	0	0	1
2	1.030	.9707	.4963	.5113	2.015	1.956	0.496	0.970	2
3	1.046	.9563	.3284	.3434	3.045	2.912	0.990	2.883	3
4	1.061	.9422	.2444	.2594	4.091	3.854	1.481	5.709	4
5	1.077	.928ͻ	.1941	.2091	5.152	4.783	1.970	9.422	5
6	1.093	.9145	.1605	.1755	6.230	5.697	2.456	13.994	6
7	1.110	.9010	.1366	.1516	7.323	6.598	2.940	19.400	7
8	1.126	.8877	.1186	.1336	8.433	7.486	3.422	25.614	8
9	1.143	.8746	.1046	.1196	9.559	8.360	3.901	32.610	9
10	1.161	.8617	.0934	.1084	10.703	9.222	4.377	40.365	10
11	1.178	.8489	.0843	.0993	11.863	10.071	4.851	48.855	11
12	1.196	.8364	.0767	.0917	13.041	10.907	5.322	58.054	12
13	1.214	.8240	.0702	.0852	14.237	11.731	5.791	67.943	13
14	1.232	.8118	.0647	.0797	15.450	12.543	6.258	78.496	14
15	1.250	.7999	.0599	.0749	16.682	13.343	6.722	89.694	15
16	1.269	.7880	.0558	.0708	17.932	14.131	7.184	101.514	16
17	1.288	.7764	.0521	.0671	19.201	14.908	7.643	113.937	17
18	1.307	.7649	.0488	.0638	20.489	15.673	8.100	126.940	18
19	1.327	.7536	.0459	.0609	21.797	16.426	8.554	140.505	19
20	1.347	.7425	.0432	.0582	23.124	17.169	9.005	154.611	20
21	1.367	.7315	.0409	.0559	24.470	17.900	9.455	169.241	21
22	1.388	.7207	.0387	.0537	25.837	18.621	9.902	184.375	22
23	1.408	.7100	.0367	.0517	27.225	19.331	10.346	199.996	23
24	1.430	.6995	.0349	.0499	28.633	20.030	10.788	216.085	24
25	1.451	.6892	.0333	.0483	30.063	20.720	11.227	232.626	25
26	1.473	.6790	.0317	.0467	31.514	21.399	11.664	249.601	26
27	1.495	.6690	.0303	.0453	32.987	22.068	12.099	266.995	27
28	1.517	.6591	.0290	.0440	34.481	22.727	12.531	284.790	28
29	1.540	.6494	.0278	.0428	35.999	23.376	12.961	302.972	29
30	1.563	.6398	.0266	.0416	37.539	24.016	13.388	321.525	30
36	1.709	.5851	.0212	.0362	47.276	27.661	15.901	439.823	36
40	1.814	.5513	.0184	.0334	54.268	29.916	17.528	524.349	40
48	2.043	.4894	.0144	.0294	69.565	34.042	20.666	703.537	48
50	2.105	.4750	.0136	.0286	73.682	35.000	21.428	749.955	50
52	2.169	.4611	.0128	.0278	77.925	35.929	22.179	796.868	52
60	2.443	.4093	.0104	.0254	96.214	39.380	25.093	988.157	60
70	2.835	.3527	.00817	.0232	122.363	43.155	28.529	1 231.15	70
72	2.921	.3423	.00781	.0228	128.076	43.845	29.189	1 279.78	72
80	3.291	.3039	.00655	.0215	152.710	46.407	31.742	1 473.06	80
84	3.493	.2863	.00602	.0210	166.172	47.579	32.967	1 568.50	84
90	3.819	.2619	.00532	.0203	187.929	49.210	34.740	1 709.53	90
96	4.176	.2395	.00472	.0197	211.719	50.702	36.438	1 847.46	96
100	4.432	.2256	.00437	.0194	228.802	51.625	37.529	1 937.43	100
104	4.704	.2126	.00405	.0190	246.932	52.494	38.589	2 025.69	104
120	5.969	.1675	.00302	.0180	331.286	55.498	42.518	2 359.69	120
240	35.632	.0281	.00043	.0154	2 308.8	64.796	59.737	3 870.68	240
360	212.700	.00470	.00007	.0151	14 113.3	66.353	64.966	4 310.71	360
480	1 269.7	.00079	.00001	.0150	84 577.8	66.614	66.288	4 415.74	480

	Single Payment		Uniform Payment Series				Arithmetic Gradient		
	Compound Amount Factor	Present Worth Factor	Sinking Fund Factor	Capital Recovery Factor	Compound Amount Factor	Present Worth Factor	Gradient Uniform Series	Gradient Present Worth	
n	Find F Given P F/P	Find P Given F P/F	Find A Given F A/F	Find A Given P A/P	Find F Given A F/A	Find P Given A P/A	Find A Given G A/G	Find P Given G P/G	n
1	1.018	.9828	1.0000	1.0175	1.000	0.983	0	0	1
2	1.035	.9659	.4957	.5132	2.018	1.949	0.496	0.966	2
3	1.053	.9493	.3276	.3451	3.053	2.898	0.989	2.865	3
4	1.072	.9330	.2435	.2610	4.106	3.831	1.478	5.664	4
5	1.091	.9169	.1931	.2106	5.178	4.748	1.965	9.332	5
6	1.110	.9011	.1595	.1770	6.269	5.649	2.450	13.837	6
7	1.129	.8856	.1355	.1530	7.378	6.535	2.931	19.152	7
8	1.149	.8704	.1175	.1350	8.508	7.405	3.409	25.245	8
9	1.169	.8554	.1036	.1211	9.656	8.261	3.885	32.088	9
10	1.189	.8407	.0924	.1099	10.825	9.101	4.357	39.655	10
11	1.210	.8263	.0832	.1007	12.015	9.928	4.827	47.918	11
12	1.231	.8121	.0756	.0931	13.225	10.740	5.294	56.851	12
13	1.253	.7981	.0692	.0867	14.457	11.538	5.758	66.428	13
14	1.275	.7844	.0637	.0812	15.710	12.322	6.219	76.625	14
15	1.297	.7709	.0589	.0764	16.985	13.093	6.677	87.417	15
16	1.320	.7576	.0547	.0722	18.282	13.851	7.132	98.782	16
17	1.343	.7446	.0510	.0685	19.602	14.595	7.584	110.695	17
18	1.367	.7318	.0477	.0652	20.945	15.327	8.034	123.136	18
19	1.390	.7192	.0448	.0623	22.311	16.046	8.481	136.081	19
20	1.415	.7068	.0422	.0597	23.702	16.753	8.924	149.511	20
21	1.440	.6947	.0398	.0573	25.116	17.448	9.365	163.405	21
22	1.465	.6827	.0377	.0552	26.556	18.130	9.804	177.742	22
23	1.490	.6710	.0357	.0532	28.021	18.801	10.239	192.503	23
24	1.516	.6594	.0339	.0514	29.511	19.461	10.671	207.671	24
25	1.543	.6481	.0322	.0497	31.028	20.109	11.101	223.225	25
26	1.570	.6369	.0307	.0482	32.571	20.746	11.528	239.149	26
27	1.597	.6260	.0293	.0468	34.141	21.372	11.952	255.425	27
28	1.625	.6152	.0280	.0455	35.738	21.987	12.373	272.036	28
29	1.654	.6046	.0268	.0443	37.363	22.592	12.791	288.967	29
30	1.683	.5942	.0256	.0431	39.017	23.186	13.206	306.200	30
36	1.867	.5355	.0202	.0377	49.566	26.543	15.640	415.130	36
40	2.002	.4996	.0175	.0350	57.234	28.594	17.207	492.017	40
48	2.300	.4349	.0135	.0310	74.263	32.294	20.209	652.612	48
50	2.381	.4200	.0127	.0302	78.903	33.141	20.932	693.708	50
52	2.465	.4057	.0119	.0294	83.706	33.960	21.644	735.039	52
60	2.832	.3531	.00955	.0271	104.676	36.964	24.389	901.503	60
70	3.368	.2969	.00739	.0249	135.331	40.178	27.586	1 108.34	70
72	3.487	.2868	.00704	.0245	142.127	40.757	28.195	1 149.12	72
80	4.006	.2496	.00582	.0233	171.795	42.880	30.533	1 309.25	80
84	4.294	.2329	.00531	.0228	188.246	43.836	31.644	1 387.16	84
90	4.765	.2098	.00465	.0221	215.166	45.152	33.241	1 500.88	90
96	5.288	.1891	.00408	.0216	245.039	46.337	34.756	1 610.48	96
100	5.668	.1764	.00375	.0212	266.753	47.062	35.721	1 681.09	100
104	6.075	.1646	.00345	.0209	290.028	47.737	36.652	1 749.68	104
120	8.019	.1247	.00249	.0200	401.099	50.017	40.047	2 003.03	120
240	64.308	.0156	.00028	.0178	3 617.6	56.254	53.352	3 001.27	240
360	515.702	.00194	.00003	.0175	29 411.5	57.032	56.443	3 219.08	360
480	4 135.5	.00024		.0175	236 259.0	57.129	57.027	3 257.88	480

2% — Compound Interest Factors — 2%

	Single Payment		Uniform Payment Series				Arithmetic Gradient		
	Compound Amount Factor	Present Worth Factor	Sinking Fund Factor	Capital Recovery Factor	Compound Amount Factor	Present Worth Factor	Gradient Uniform Series	Gradient Present Worth	
n	Find F Given P F/P	Find P Given F P/F	Find A Given F A/F	Find A Given P A/P	Find F Given A F/A	Find P Given A P/A	Find A Given G A/G	Find P Given G P/G	n
1	1.020	.9804	1.0000	1.0200	1.000	0.980	0	0	1
2	1.040	.9612	.4951	.5151	2.020	1.942	0.495	0.961	2
3	1.061	.9423	.3268	.3468	3.060	2.884	0.987	2.846	3
4	1.082	.9238	.2426	.2626	4.122	3.808	1.475	5.617	4
5	1.104	.9057	.1922	.2122	5.204	4.713	1.960	9.240	5
6	1.126	.8880	.1585	.1785	6.308	5.601	2.442	13.679	6
7	1.149	.8706	.1345	.1545	7.434	6.472	2.921	18.903	7
8	1.172	.8535	.1165	.1365	8.583	7.325	3.396	24.877	8
9	1.195	.8368	.1025	.1225	9.755	8.162	3.868	31.571	9
10	1.219	.8203	.0913	.1113	10.950	8.983	4.337	38.954	10
11	1.243	.8043	.0822	.1022	12.169	9.787	4.802	46.996	11
12	1.268	.7885	.0746	.0946	13.412	10.575	5.264	55.669	12
13	1.294	.7730	.0681	.0881	14.680	11.348	5.723	64.946	13
14	1.319	.7579	.0626	.0826	15.974	12.106	6.178	74.798	14
15	1.346	.7430	.0578	.0778	17.293	12.849	6.631	85.200	15
16	1.373	.7284	.0537	.0737	18.639	13.578	7.080	96.127	16
17	1.400	.7142	.0500	.0700	20.012	14.292	7.526	107.553	17
18	1.428	.7002	.0467	.0667	21.412	14.992	7.968	119.456	18
19	1.457	.6864	.0438	.0638	22.840	15.678	8.407	131.812	19
20	1.486	.6730	.0412	.0612	24.297	16.351	8.843	144.598	20
21	1.516	.6598	.0388	.0588	25.783	17.011	9.276	157.793	21
22	1.546	.6468	.0366	.0566	27.299	17.658	9.705	171.377	22
23	1.577	.6342	.0347	.0547	28.845	18.292	10.132	185.328	23
24	1.608	.6217	.0329	.0529	30.422	18.914	10.555	199.628	24
25	1.641	.6095	.0312	.0512	32.030	19.523	10.974	214.256	25
26	1.673	.5976	.0297	.0497	33.671	20.121	11.391	229.196	26
27	1.707	.5859	.0283	.0483	35.344	20.707	11.804	244.428	27
28	1.741	.5744	.0270	.0470	37.051	21.281	12.214	259.936	28
29	1.776	.5631	.0258	.0458	38.792	21.844	12.621	275.703	29
30	1.811	.5521	.0247	.0447	40.568	22.396	13.025	291.713	30
36	2.040	.4902	.0192	.0392	51.994	25.489	15.381	392.036	36
40	2.208	.4529	.0166	.0366	60.402	27.355	16.888	461.989	40
48	2.587	.3865	.0126	.0326	79.353	30.673	19.755	605.961	48
50	2.692	.3715	.0118	.0318	84.579	31.424	20.442	642.355	50
52	2.800	.3571	.0111	.0311	90.016	32.145	21.116	678.779	52
60	3.281	.3048	.00877	.0288	114.051	34.761	23.696	823.692	60
70	4.000	.2500	.00667	.0267	149.977	37.499	26.663	999.829	70
72	4.161	.2403	.00633	.0263	158.056	37.984	27.223	1 034.050	72
80	4.875	.2051	.00516	.0252	193.771	39.744	29.357	1 166.781	80
84	5.277	.1895	.00468	.0247	213.865	40.525	30.361	1 230.413	84
90	5.943	.1683	.00405	.0240	247.155	41.587	31.793	1 322.164	90
96	6.693	.1494	.00351	.0235	284.645	42.529	33.137	1 409.291	96
100	7.245	.1380	.00320	.0232	312.230	43.098	33.986	1 464.747	100
104	7.842	.1275	.00292	.0229	342.090	43.624	34.799	1 518.082	104
120	10.765	.0929	.00205	.0220	488.255	45.355	37.711	1 710.411	120
240	115.887	.00863	.00017	.0202	5 744.4	49.569	47.911	2 374.878	240
360	1 247.5	.00080	.00002	.0200	62 326.8	49.960	49.711	2 483.567	360
480	13 429.8	.00007		.0200	671 442.0	49.996	49.964	2 498.027	480

	Single Payment		Uniform Payment Series				Arithmetic Gradient		
	Compound Amount Factor	Present Worth Factor	Sinking Fund Factor	Capital Recovery Factor	Compound Amount Factor	Present Worth Factor	Gradient Uniform Series	Gradient Present Worth	
n	Find F Given P F/P	Find P Given F P/F	Find A Given F A/F	Find A Given P A/P	Find F Given A F/A	Find P Given A P/A	Find A Given G A/G	Find P Given G P/G	n
1	1.025	.9756	1.0000	1.0250	1.000	0.976	0	0	1
2	1.051	.9518	.4938	.5188	2.025	1.927	0.494	0.952	2
3	1.077	.9286	.3251	.3501	3.076	2.856	0.984	2.809	3
4	1.104	.9060	.2408	.2658	4.153	3.762	1.469	5.527	4
5	1.131	.8839	.1902	.2152	5.256	4.646	1.951	9.062	5
6	1.160	.8623	.1566	.1816	6.388	5.508	2.428	13.374	6
7	1.189	.8413	.1325	.1575	7.547	6.349	2.901	18.421	7
8	1.218	.8207	.1145	.1395	8.736	7.170	3.370	24.166	8
9	1.249	.8007	.1005	.1255	9.955	7.971	3.835	30.572	9
10	1.280	.7812	.0893	.1143	11.203	8.752	4.296	37.603	10
11	1.312	.7621	.0801	.1051	12.483	9.514	4.753	45.224	11
12	1.345	.7436	.0725	.0975	13.796	10.258	5.206	53.403	12
13	1.379	.7254	.0660	.0910	15.140	10.983	5.655	62.108	13
14	1.413	.7077	.0605	.0855	16.519	11.691	6.100	71.309	14
15	1.448	.6905	.0558	.0808	17.932	12.381	6.540	80.975	15
16	1.485	.6736	.0516	.0766	19.380	13.055	6.977	91.080	16
17	1.522	.6572	.0479	.0729	20.865	13.712	7.409	101.595	17
18	1.560	.6412	.0447	.0697	22.386	14.353	7.838	112.495	18
19	1.599	.6255	.0418	.0668	23.946	14.979	8.262	123.754	19
20	1.639	.6103	.0391	.0641	25.545	15.589	8.682	135.349	20
21	1.680	.5954	.0368	.0618	27.183	16.185	9.099	147.257	21
22	1.722	.5809	.0346	.0596	28.863	16.765	9.511	159.455	22
23	1.765	.5667	.0327	.0577	30.584	17.332	9.919	171.922	23
24	1.809	.5529	.0309	.0559	32.349	17.885	10.324	184.638	24
25	1.854	.5394	.0293	.0543	34.158	18.424	10.724	197.584	25
26	1.900	.5262	.0278	.0528	36.012	18.951	11.120	210.740	26
27	1.948	.5134	.0264	.0514	37.912	19.464	11.513	224.088	27
28	1.996	.5009	.0251	.0501	39.860	19.965	11.901	237.612	28
29	2.046	.4887	.0239	.0489	41.856	20.454	12.286	251.294	29
30	2.098	.4767	.0228	.0478	43.903	20.930	12.667	265.120	30
31	2.150	.4651	.0217	.0467	46.000	21.395	13.044	279.073	31
32	2.204	.4538	.0208	.0458	48.150	21.849	13.417	293.140	32
33	2.259	.4427	.0199	.0449	50.354	22.292	13.786	307.306	33
34	2.315	.4319	.0190	.0440	52.613	22.724	14.151	321.559	34
35	2.373	.4214	.0182	.0432	54.928	23.145	14.512	335.886	35
40	2.685	.3724	.0148	.0398	67.402	25.103	16.262	408.221	40
45	3.038	.3292	.0123	.0373	81.516	26.833	17.918	480.806	45
50	3.437	.2909	.0103	.0353	97.484	28.362	19.484	552.607	50
55	3.889	.2572	.00865	.0337	115.551	29.714	20.961	622.827	55
60	4.400	.2273	.00735	.0324	135.991	30.909	22.352	690.865	60
65	4.978	.2009	.00628	.0313	159.118	31.965	23.660	756.280	65
70	5.632	.1776	.00540	.0304	185.284	32.898	24.888	818.763	70
75	6.372	.1569	.00465	.0297	214.888	33.723	26.039	878.114	75
80	7.210	.1387	.00403	.0290	248.382	34.452	27.117	934.217	80
85	8.157	.1226	.00349	.0285	286.278	35.096	28.123	987.026	85
90	9.229	.1084	.00304	.0280	329.154	35.666	29.063	1 036.54	90
95	10.442	.0958	.00265	.0276	377.663	36.169	29.938	1 082.83	95
100	11.814	.0846	.00231	.0273	432.548	36.614	30.752	1 125.97	100

	Single Payment		Uniform Payment Series				Arithmetic Gradient		
	Compound Amount Factor	Present Worth Factor	Sinking Fund Factor	Capital Recovery Factor	Compound Amount Factor	Present Worth Factor	Gradient Uniform Series	Gradient Present Worth	
n	Find F Given P F/P	Find P Given F P/F	Find A Given F A/F	Find A Given P A/P	Find F Given A F/A	Find P Given A P/A	Find A Given G A/G	Find P Given G P/G	n
1	1.030	.9709	1.0000	1.0300	1.000	0.971	0	0	1
2	1.061	.9426	.4926	.5226	2.030	1.913	0.493	0.943	2
3	1.093	.9151	.3235	.3535	3.091	2.829	0.980	2.773	3
4	1.126	.8885	.2390	.2690	4.184	3.717	1.463	5.438	4
5	1.159	.8626	.1884	.2184	5.309	4.580	1.941	8.889	5
6	1.194	.8375	.1546	.1846	6.468	5.417	2.414	13.076	6
7	1.230	.8131	.1305	.1605	7.662	6.230	2.882	17.955	7
8	1.267	.7894	.1125	.1425	8.892	7.020	3.345	23.481	8
9	1.305	.7664	.0984	.1284	10.159	7.786	3.803	29.612	9
10	1.344	.7441	.0872	.1172	11.464	8.530	4.256	36.309	10
11	1.384	.7224	.0781	.1081	12.808	9.253	4.705	43.533	11
12	1.426	.7014	.0705	.1005	14.192	9.954	5.148	51.248	12
13	1.469	.6810	.0640	.0940	15.618	10.635	5.587	59.419	13
14	1.513	.6611	.0585	.0885	17.086	11.296	6.021	68.014	14
15	1.558	.6419	.0538	.0838	18.599	11.938	6.450	77.000	15
16	1.605	.6232	.0496	.0796	20.157	12.561	6.874	86.348	16
17	1.653	.6050	.0460	.0760	21.762	13.166	7.294	96.028	17
18	1.702	.5874	.0427	.0727	23.414	13.754	7.708	106.014	18
19	1.754	.5703	.0398	.0698	25.117	14.324	8.118	116.279	19
20	1.806	.5537	.0372	.0672	26.870	14.877	8.523	126.799	20
21	1.860	.5375	.0349	.0649	28.676	15.415	8.923	137.549	21
22	1.916	.5219	.0327	.0627	30.537	15.937	9.319	148.509	22
23	1.974	.5067	.0308	.0608	32.453	16.444	9.709	159.656	23
24	2.033	.4919	.0290	.0590	34.426	16.936	10.095	170.971	24
25	2.094	.4776	.0274	.0574	36.459	17.413	10.477	182.433	25
26	2.157	.4637	.0259	.0559	38.553	17.877	10.853	194.026	26
27	2.221	.4502	.0246	.0546	40.710	18.327	11.226	205.731	27
28	2.288	.4371	.0233	.0533	42.931	18.764	11.593	217.532	28
29	2.357	.4243	.0221	.0521	45.219	19.188	11.956	229.413	29
30	2.427	.4120	.0210	.0510	47.575	19.600	12.314	241.361	30
31	2.500	.4000	.0200	.0500	50.003	20.000	12.668	253.361	31
32	2.575	.3883	.0190	.0490	52.503	20.389	13.017	265.399	32
33	2.652	.3770	.0182	.0482	55.078	20.766	13.362	277.464	33
34	2.732	.3660	.0173	.0473	57.730	21.132	13.702	289.544	34
35	2.814	.3554	.0165	.0465	60.462	21.487	14.037	301.627	35
40	3.262	.3066	.0133	.0433	75.401	23.115	15.650	361.750	40
45	3.782	.2644	.0108	.0408	92.720	24.519	17.156	420.632	45
50	4.384	.2281	.00887	.0389	112.797	25.730	18.558	477.480	50
55	5.082	.1968	.00735	.0373	136.072	26.774	19.860	531.741	55
60	5.892	.1697	.00613	.0361	163.053	27.676	21.067	583.052	60
65	6.830	.1464	.00515	.0351	194.333	28.453	22.184	631.201	65
70	7.918	.1263	.00434	.0343	230.594	29.123	23.215	676.087	70
75	9.179	.1089	.00367	.0337	272.631	29.702	24.163	717.698	75
80	10.641	.0940	.00311	.0331	321.363	30.201	25.035	756.086	80
85	12.336	.0811	.00265	.0326	377.857	30.631	25.835	791.353	85
90	14.300	.0699	.00226	.0323	443.349	31.002	26.567	823.630	90
95	16.578	.0603	.00193	.0319	519.272	31.323	27.235	853.074	95
100	19.219	.0520	.00165	.0316	607.287	31.599	27.844	879.854	100

	Single Payment		Uniform Payment Series				Arithmetic Gradient		
	Compound Amount Factor	Present Worth Factor	Sinking Fund Factor	Capital Recovery Factor	Compound Amount Factor	Present Worth Factor	Gradient Uniform Series	Gradient Present Worth	
n	Find F Given P F/P	Find P Given F P/F	Find A Given F A/F	Find A Given P A/P	Find F Given A F/A	Find P Given A P/A	Find A Given G A/G	Find P Given G P/G	n
1	1.035	.9662	1.0000	1.0350	1.000	0.966	0	0	1
2	1.071	.9335	.4914	.5264	2.035	1.900	0.491	0.933	2
3	1.109	.9019	.3219	.3569	3.106	2.802	0.977	2.737	3
4	1.148	.8714	.2373	.2723	4.215	3.673	1.457	5.352	4
5	1.188	.8420	.1865	.2215	5.362	4.515	1.931	8.719	5
6	1.229	.8135	.1527	.1877	6.550	5.329	2.400	12.787	6
7	1.272	.7860	.1285	.1635	7.779	6.115	2.862	17.503	7
8	1.317	.7594	.1105	.1455	9.052	6.874	3.320	22.819	8
9	1.363	.7337	.0964	.1314	10.368	7.608	3.771	28.688	9
10	1.411	.7089	.0852	.1202	11.731	8.317	4.217	35.069	10
11	1.460	.6849	.0761	.1111	13.142	9.002	4.657	41.918	11
12	1.511	.6618	.0685	.1035	14.602	9.663	5.091	49.198	12
13	1.564	.6394	.0621	.0971	16.113	10.303	5.520	56.871	13
14	1.619	.6178	.0566	.0916	17.677	10.921	5.943	64.902	14
15	1.675	.5969	.0518	.0868	19.296	11.517	6.361	73.258	15
16	1.734	.5767	.0477	.0827	20.971	12.094	6.773	81.909	16
17	1.795	.5572	.0440	.0790	22.705	12.651	7.179	90.824	17
18	1.857	.5384	.0408	.0758	24.500	13.190	7.580	99.976	18
19	1.922	.5202	.0379	.0729	26.357	13.710	7.975	109.339	19
20	1.990	.5026	.0354	.0704	28.280	14.212	8.365	118.888	20
21	2.059	.4856	.0330	.0680	30.269	14.698	8.749	128.599	21
22	2.132	.4692	.0309	.0659	32.329	15.167	9.128	138.451	22
23	2.206	.4533	.0290	.0640	34.460	15.620	9.502	148.423	23
24	2.283	.4380	.0273	.0623	36.666	16.058	9.870	158.496	24
25	2.363	.4231	.0257	.0607	38.950	16.482	10.233	168.652	25
26	2.446	.4088	.0242	.0592	41.313	16.890	10.590	178.873	26
27	2.532	.3950	.0229	.0579	43.759	17.285	10.942	189.143	27
28	2.620	.3817	.0216	.0566	46.291	17.667	11.289	199.448	28
29	2.712	.3687	.0204	.0554	48.911	18.036	11.631	209.773	29
30	2.807	.3563	.0194	.0544	51.623	18.392	11.967	220.105	30
31	2.905	.3442	.0184	.0534	54.429	18.736	12.299	230.432	31
32	3.007	.3326	.0174	.0524	57.334	19.069	12.625	240.742	32
33	3.112	.3213	.0166	.0516	60.341	19.390	12.946	251.025	33
34	3.221	.3105	.0158	.0508	63.453	19.701	13.262	261.271	34
35	3.334	.3000	.0150	.0500	66.674	20.001	13.573	271.470	35
40	3.959	.2526	.0118	.0468	84.550	21.355	15.055	321.490	40
45	4.702	.2127	.00945	.0445	105.781	22.495	16.417	369.307	45
50	5.585	.1791	.00763	.0426	130.998	23.456	17.666	414.369	50
55	6.633	.1508	.00621	.0412	160.946	24.264	18.808	456.352	55
60	7.878	.1269	.00509	.0401	196.516	24.945	19.848	495.104	60
65	9.357	.1069	.00419	.0392	238.762	25.518	20.793	530.598	65
70	11.113	.0900	.00346	.0385	288.937	26.000	21.650	562.895	70
75	13.199	.0758	.00287	.0379	348.529	26.407	22.423	592.121	75
80	15.676	.0638	.00238	.0374	419.305	26.749	23.120	618.438	80
85	18.618	.0537	.00199	.0370	503.365	27.037	23.747	642.036	85
90	22.112	.0452	.00166	.0367	603.202	27.279	24.308	663.118	90
95	26.262	.0381	.00139	.0364	721.778	27.483	24.811	681.890	95
100	31.191	.0321	.00116	.0362	862.608	27.655	25.259	698.554	100

	Single Payment		Uniform Payment Series				Arithmetic Gradient		
	Compound Amount Factor	Present Worth Factor	Sinking Fund Factor	Capital Recovery Factor	Compound Amount Factor	Present Worth Factor	Gradient Uniform Series	Gradient Present Worth	
n	Find F Given P F/P	Find P Given F P/F	Find A Given F A/F	Find A Given P A/P	Find F Given A F/A	Find P Given A P/A	Find A Given G A/G	Find P Given G P/G	n
1	1.040	.9615	1.0000	1.0400	1.000	0.962	0	0	1
2	1.082	.9246	.4902	.5302	2.040	1.886	0.490	0.925	2
3	1.125	.8890	.3203	.3603	3.122	2.775	0.974	2.702	3
4	1.170	.8548	.2355	.2755	4.246	3.630	1.451	5.267	4
5	1.217	.8219	.1846	.2246	5.416	4.452	1.922	8.555	5
6	1.265	.7903	.1508	.1908	6.633	5.242	2.386	12.506	6
7	1.316	.7599	.1266	.1666	7.898	6.002	2.843	17.066	7
8	1.369	.7307	.1085	.1485	9.214	6.733	3.294	22.180	8
9	1.423	.7026	.0945	.1345	10.583	7.435	3.739	27.801	9
10	1.480	.6756	.0833	.1233	12.006	8.111	4.177	33.881	10
11	1.539	.6496	.0741	.1141	13.486	8.760	4.609	40.377	11
12	1.601	.6246	.0666	.1066	15.026	9.385	5.034	47.248	12
13	1.665	.6006	.0601	.1001	16.627	9.986	5.453	54.454	13
14	1.732	.5775	.0547	.0947	18.292	10.563	5.866	61.962	14
15	1.801	.5553	.0499	.0899	20.024	11.118	6.272	69.735	15
16	1.873	.5339	.0458	.0858	21.825	11.652	6.672	77.744	16
17	1.948	.5134	.0422	.0822	23.697	12.166	7.066	85.958	17
18	2.026	.4936	.0390	.0790	25.645	12.659	7.453	94.350	18
19	2.107	.4746	.0361	.0761	27.671	13.134	7.834	102.893	19
20	2.191	.4564	.0336	.0736	29.778	13.590	8.209	111.564	20
21	2.279	.4388	.0313	.0713	31.969	14.029	8.578	120.341	21
22	2.370	.4220	.0292	.0692	34.248	14.451	8.941	129.202	22
23	2.465	.4057	.0273	.0673	36.618	14.857	9.297	138.128	23
24	2.563	.3901	.0256	.0656	39.083	15.247	9.648	147.101	24
25	2.666	.3751	.0240	.0640	41.646	15.622	9.993	156.104	25
26	2.772	.3607	.0226	.0626	44.312	15.983	10.331	165.121	26
27	2.883	.3468	.0212	.0612	47.084	16.330	10.664	174.138	27
28	2.999	.3335	.0200	.0600	49.968	16.663	10.991	183.142	28
29	3.119	.3207	.0189	.0589	52.966	16.984	11.312	192.120	29
30	3.243	.3083	.0178	.0578	56.085	17.292	11.627	201.062	30
31	3.373	.2965	.0169	.0569	59.328	17.588	11.937	209.955	31
32	3.508	.2851	.0159	.0559	62.701	17.874	12.241	218.792	32
33	3.648	.2741	.0151	.0551	66.209	18.148	12.540	227.563	33
34	3.794	.2636	.0143	.0543	69.858	18.411	12.832	236.260	34
35	3.946	.2534	.0136	.0536	73.652	18.665	13.120	244.876	35
40	4.801	.2083	.0105	.0505	95.025	19.793	14.476	286.530	40
45	5.841	.1712	.00826	.0483	121.029	20.720	15.705	325.402	45
50	7.107	.1407	.00655	.0466	152.667	21.482	16.812	361.163	50
55	8.646	.1157	.00523	.0452	191.159	22.109	17.807	393.689	55
60	10.520	.0951	.00420	.0442	237.990	22.623	18.697	422.996	60
65	12.799	.0781	.00339	.0434	294.968	23.047	19.491	449.201	65
70	15.572	.0642	.00275	.0427	364.290	23.395	20.196	472.479	70
75	18.945	.0528	.00223	.0422	448.630	23.680	20.821	493.041	75
80	23.050	.0434	.00181	.0418	551.244	23.915	21.372	511.116	80
85	28.044	.0357	.00148	.0415	676.089	24.109	21.857	526.938	85
90	34.119	.0293	.00121	.0412	827.981	24.267	22.283	540.737	90
95	41.511	.0241	.00099	.0410	1 012.8	24.398	22.655	552.730	95
100	50.505	.0198	.00081	.0408	1 237.6	24.505	22.980	563.125	100

	Single Payment		Uniform Payment Series				Arithmetic Gradient		
	Compound Amount Factor	Present Worth Factor	Sinking Fund Factor	Capital Recovery Factor	Compound Amount Factor	Present Worth Factor	Gradient Uniform Series	Gradient Present Worth	
n	Find F Given P F/P	Find P Given F P/F	Find A Given F A/F	Find A Given P A/P	Find F Given A F/A	Find P Given A P/A	Find A Given G A/G	Find P Given G P/G	n
1	1.045	.9569	1.0000	1.0450	1.000	0.957	0	0	1
2	1.092	.9157	.4890	.5340	2.045	1.873	0.489	0.916	2
3	1.141	.8763	.3188	.3638	3.137	2.749	0.971	2.668	3
4	1.193	.8386	.2337	.2787	4.278	3.588	1.445	5.184	4
5	1.246	.8025	.1828	.2278	5.471	4.390	1.912	8.394	5
6	1.302	.7679	.1489	.1939	6.717	5.158	2.372	12.233	6
7	1.361	.7348	.1247	.1697	8.019	5.893	2.824	16.642	7
8	1.422	.7032	.1066	.1516	9.380	6.596	3.269	21.564	8
9	1.486	.6729	.0926	.1376	10.802	7.269	3.707	26.948	9
10	1.553	.6439	.0814	.1264	12.288	7.913	4.138	32.743	10
11	1.623	.6162	.0722	.1172	13.841	8.529	4.562	38.905	11
12	1.696	.5897	.0647	.1097	15.464	9.119	4.978	45.391	12
13	1.772	.5643	.0583	.1033	17.160	9.683	5.387	52.163	13
14	1.852	.5400	.0528	.0978	18.932	10.223	5.789	59.182	14
15	1.935	.5167	.0481	.0931	20.784	10.740	6.184	66.416	15
16	2.022	.4945	.0440	.0890	22.719	11.234	6.572	73.833	16
17	2.113	.4732	.0404	.0854	24.742	11.707	6.953	81.404	17
18	2.208	.4528	.0372	.0822	26.855	12.160	7.327	89.102	18
19	2.308	.4333	.0344	.0794	29.064	12.593	7.695	96.901	19
20	2.412	.4146	.0319	.0769	31.371	13.008	8.055	104.779	20
21	2.520	.3968	.0296	.0746	33.783	13.405	8.409	112.715	21
22	2.634	.3797	.0275	.0725	36.303	13.784	8.755	120.689	22
23	2.752	.3634	.0257	.0707	38.937	14.148	9.096	128.682	23
24	2.876	.3477	.0240	.0690	41.689	14.495	9.429	136.680	24
25	3.005	.3327	.0224	.0674	44.565	14.828	9.756	144.665	25
26	3.141	.3184	.0210	.0660	47.571	15.147	10.077	152.625	26
27	3.282	.3047	.0197	.0647	50.711	15.451	10.391	160.547	27
28	3.430	.2916	.0185	.0635	53.993	15.743	10.698	168.420	28
29	3.584	.2790	.0174	.0624	57.423	16.022	10.999	176.232	29
30	3.745	.2670	.0164	.0614	61.007	16.289	11.295	183.975	30
31	3.914	.2555	.0154	.0604	64.752	16.544	11.583	191.640	31
32	4.090	.2445	.0146	.0596	68.666	16.789	11.866	199.220	32
33	4.274	.2340	.0137	.0587	72.756	17.023	12.143	206.707	33
34	4.466	.2239	.0130	.0580	77.030	17.247	12.414	214.095	34
35	4.667	.2143	.0123	.0573	81.497	17.461	12.679	221.380	35
40	5.816	.1719	.00934	.0543	107.030	18.402	13.917	256.098	40
45	7.248	.1380	.00720	.0522	138.850	19.156	15.020	287.732	45
50	9.033	.1107	.00560	.0506	178.503	19.762	15.998	316.145	50
55	11.256	.0888	.00439	.0494	227.918	20.248	16.860	341.375	55
60	14.027	.0713	.00345	.0485	289.497	20.638	17.617	363.571	60
65	17.481	.0572	.00273	.0477	366.237	20.951	18.278	382.946	65
70	21.784	.0459	.00217	.0472	461.869	21.202	18.854	399.750	70
75	27.147	.0368	.00172	.0467	581.043	21.404	19.354	414.242	75
80	33.830	.0296	.00137	.0464	729.556	21.565	19.785	426.680	80
85	42.158	.0237	.00109	.0461	914.630	21.695	20.157	437.309	85
90	52.537	.0190	.00087	.0459	1 145.3	21.799	20.476	446.359	90
95	65.471	.0153	.00070	.0457	1 432.7	21.883	20.749	454.039	95
100	81.588	.0123	.00056	.0456	1 790.9	21.950	20.981	460.537	100

	Single Payment		Uniform Payment Series				Arithmetic Gradient		
	Compound Amount Factor	Present Worth Factor	Sinking Fund Factor	Capital Recovery Factor	Compound Amount Factor	Present Worth Factor	Gradient Uniform Series	Gradient Present Worth	
	Find F Given P F/P	Find P Given F P/F	Find A Given F A/F	Find A Given P A/P	Find F Given A F/A	Find P Given A P/A	Find A Given G A/G	Find P Given G P/G	
n									n
1	1.050	.9524	1.0000	1.0500	1.000	0.952	0	0	1
2	1.102	.9070	.4878	.5378	2.050	1.859	0.488	0.907	2
3	1.158	.8638	.3172	.3672	3.152	2.723	0.967	2.635	3
4	1.216	.8227	.2320	.2820	4.310	3.546	1.439	5.103	4
5	1.276	.7835	.1810	.2310	5.526	4.329	1.902	8.237	5
6	1.340	.7462	.1470	.1970	6.802	5.076	2.358	11.968	6
7	1.407	.7107	.1228	.1728	8.142	5.786	2.805	16.232	7
8	1.477	.6768	.1047	.1547	9.549	6.463	3.244	20.970	8
9	1.551	.6446	.0907	.1407	11.027	7.108	3.676	26.127	9
10	1.629	.6139	.0795	.1295	12.578	7.722	4.099	31.652	10
11	1.710	.5847	.0704	.1204	14.207	8.306	4.514	37.499	11
12	1.796	.5568	.0628	.1128	15.917	8.863	4.922	43.624	12
13	1.886	.5303	.0565	.1065	17.713	9.394	5.321	49.988	13
14	1.980	.5051	.0510	.1010	19.599	9.899	5.713	56.553	14
15	2.079	.4810	.0463	.0963	21.579	10.380	6.097	63.288	15
16	2.183	.4581	.0423	.0923	23.657	10.838	6.474	70.159	16
17	2.292	.4363	.0387	.0887	25.840	11.274	6.842	77.140	17
18	2.407	.4155	.0355	.0855	28.132	11.690	7.203	84.204	18
19	2.527	.3957	.0327	.0827	30.539	12.085	7.557	91.327	19
20	2.653	.3769	.0302	.0802	33.066	12.462	7.903	98.488	20
21	2.786	.3589	.0280	.0780	35.719	12.821	8.242	105.667	21
22	2.925	.3419	.0260	.0760	38.505	13.163	8.573	112.846	22
23	3.072	.3256	.0241	.0741	41.430	13.489	8.897	120.008	23
24	3.225	.3101	.0225	.0725	44.502	13.799	9.214	127.140	24
25	3.386	.2953	.0210	.0710	47.727	14.094	9.524	134.227	25
26	3.556	.2812	.0196	.0696	51.113	14.375	9.827	141.258	26
27	3.733	.2678	.0183	.0683	54.669	14.643	10.122	148.222	27
28	3.920	.2551	.0171	.0671	58.402	14.898	10.411	155.110	28
29	4.116	.2429	.0160	.0660	62.323	15.141	10.694	161.912	29
30	4.322	.2314	.0151	.0651	66.439	15.372	10.969	168.622	30
31	4.538	.2204	.0141	.0641	70.761	15.593	11.238	175.233	31
32	4.765	.2099	.0133	.0633	75.299	15.803	11.501	181.739	32
33	5.003	.1999	.0125	.0625	80.063	16.003	11.757	188.135	33
34	5.253	.1904	.0118	.0618	85.067	16.193	12.006	194.416	34
35	5.516	.1813	.0111	.0611	90.320	16.374	12.250	200.580	35
40	7.040	.1420	.00828	.0583	120.799	17.159	13.377	229.545	40
45	8.985	.1113	.00626	.0563	159.699	17.774	14.364	255.314	45
50	11.467	.0872	.00478	.0548	209.347	18.256	15.223	277.914	50
55	14.636	.0683	.00367	.0537	272.711	18.633	15.966	297.510	55
60	18.679	.0535	.00283	.0528	353.582	18.929	16.606	314.343	60
65	23.840	.0419	.00219	.0522	456.795	19.161	17.154	328.691	65
70	30.426	.0329	.00170	.0517	588.525	19.343	17.621	340.841	70
75	38.832	.0258	.00132	.0513	756.649	19.485	18.018	351.072	75
80	49.561	.0202	.00103	.0510	971.222	19.596	18.353	359.646	80
85	63.254	.0158	.00080	.0508	1 245.1	19.684	18.635	366.800	85
90	80.730	.0124	.00063	.0506	1 594.6	19.752	18.871	372.749	90
95	103.034	.00971	.00049	.0505	2 040.7	19.806	19.069	377.677	95
100	131.500	.00760	.00038	.0504	2 610.0	19.848	19.234	381.749	100

	Single Payment		Uniform Payment Series				Arithmetic Gradient		
	Compound Amount Factor	Present Worth Factor	Sinking Fund Factor	Capital Recovery Factor	Compound Amount Factor	Present Worth Factor	Gradient Uniform Series	Gradient Present Worth	
n	Find F Given P F/P	Find P Given F P/F	Find A Given F A/F	Find A Given P A/P	Find F Given A F/A	Find P Given A P/A	Find A Given G A/G	Find P Given G P/G	n
1	1.060	.9434	1.0000	1.0600	1.000	0.943	0	0	1
2	1.124	.8900	.4854	.5454	2.060	1.833	0.485	0.890	2
3	1.191	.8396	.3141	.3741	3.184	2.673	0.961	2.569	3
4	1.262	.7921	.2286	.2886	4.375	3.465	1.427	4.945	4
5	1.338	.7473	.1774	.2374	5.637	4.212	1.884	7.934	5
6	1.419	.7050	.1434	.2034	6.975	4.917	2.330	11.459	6
7	1.504	.6651	.1191	.1791	8.394	5.582	2.768	15.450	7
8	1.594	.6274	.1010	.1610	9.897	6.210	3.195	19.841	8
9	1.689	.5919	.0870	.1470	11.491	6.802	3.613	24.577	9
10	1.791	.5584	.0759	.1359	13.181	7.360	4.022	29.602	10
11	1.898	.5268	.0668	.1268	14.972	7.887	4.421	34.870	11
12	2.012	.4970	.0593	.1193	16.870	8.384	4.811	40.337	12
13	2.133	.4688	.0530	.1130	18.882	8.853	5.192	45.963	13
14	2.261	.4423	.0476	.1076	21.015	9.295	5.564	51.713	14
15	2.397	.4173	.0430	.1030	23.276	9.712	5.926	57.554	15
16	2.540	.3936	.0390	.0990	25.672	10.106	6.279	63.459	16
17	2.693	.3714	.0354	.0954	28.213	10.477	6.624	69.401	17
18	2.854	.3503	.0324	.0924	30.906	10.828	6.960	75.357	18
19	3.026	.3305	.0296	.0896	33.760	11.158	7.287	81.306	19
20	3.207	.3118	.0272	.0872	36.786	11.470	7.605	87.230	20
21	3.400	.2942	.0250	.0850	39.993	11.764	7.915	93.113	21
22	3.604	.2775	.0230	.0830	43.392	12.042	8.217	98.941	22
23	3.820	.2618	.0213	.0813	46.996	12.303	8.510	104.700	23
24	4.049	.2470	.0197	.0797	50.815	12.550	8.795	110.381	24
25	4.292	.2330	.0182	.0782	54.864	12.783	9.072	115.973	25
26	4.549	.2198	.0169	.0769	59.156	13.003	9.341	121.468	26
27	4.822	.2074	.0157	.0757	63.706	13.211	9.603	126.860	27
28	5.112	.1956	.0146	.0746	68.528	13.406	9.857	132.142	28
29	5.418	.1846	.0136	.0736	73.640	13.591	10.103	137.309	29
30	5.743	.1741	.0126	.0726	79.058	13.765	10.342	142.359	30
31	6.088	.1643	.0118	.0718	84.801	13.929	10.574	147.286	31
32	6.453	.1550	.0110	.0710	90.890	14.084	10.799	152.090	32
33	6.841	.1462	.0103	.0703	97.343	14.230	11.017	156.768	33
34	7.251	.1379	.00960	.0696	104.184	14.368	11.228	161.319	34
35	7.686	.1301	.00897	.0690	111.435	14.498	11.432	165.743	35
40	10.286	.0972	.00646	.0665	154.762	15.046	12.359	185.957	40
45	13.765	.0727	.00470	.0647	212.743	15.456	13.141	203.109	45
50	18.420	.0543	.00344	.0634	290.335	15.762	13.796	217.457	50
55	24.650	.0406	.00254	.0625	394.171	15.991	14.341	229.322	55
60	32.988	.0303	.00188	.0619	533.126	16.161	14.791	239.043	60
65	44.145	.0227	.00139	.0614	719.080	16.289	15.160	246.945	65
70	59.076	.0169	.00103	.0610	967.928	16.385	15.461	253.327	70
75	79.057	.0126	.00077	.0608	1 300.9	16.456	15.706	258.453	75
80	105.796	.00945	.00057	.0606	1 746.6	16.509	15.903	262.549	80
85	141.578	.00706	.00043	.0604	2 343.0	16.549	16.062	265.810	85
90	189.464	.00528	.00032	.0603	3 141.1	16.579	16.189	268.395	90
95	253.545	.00394	.00024	.0602	4 209.1	16.601	16.290	270.437	95
100	339.300	.00295	.00018	.0602	5 638.3	16.618	16.371	272.047	100

Compound Interest Factors

	Single Payment		Uniform Payment Series				Arithmetic Gradient		
	Compound Amount Factor	Present Worth Factor	Sinking Fund Factor	Capital Recovery Factor	Compound Amount Factor	Present Worth Factor	Gradient Uniform Series	Gradient Present Worth	
n	Find F Given P F/P	Find P Given F P/F	Find A Given F A/F	Find A Given P A/P	Find F Given A F/A	Find P Given A P/A	Find A Given G A/G	Find P Given G P/G	n
1	1.070	.9346	1.0000	1.0700	1.000	0.935	0	0	1
2	1.145	.8734	.4831	.5531	2.070	1.808	0.483	0.873	2
3	1.225	.8163	.3111	.3811	3.215	2.624	0.955	2.506	3
4	1.311	.7629	.2252	.2952	4.440	3.387	1.416	4.795	4
5	1.403	.7130	.1739	.2439	5.751	4.100	1.865	7.647	5
6	1.501	.6663	.1398	.2098	7.153	4.767	2.303	10.978	6
7	1.606	.6227	.1156	.1856	8.654	5.389	2.730	14.715	7
8	1.718	.5820	.0975	.1675	10.260	5.971	3.147	18.789	8
9	1.838	.5439	.0835	.1535	11.978	6.515	3.552	23.140	9
10	1.967	.5083	.0724	.1424	13.816	7.024	3.946	27.716	10
11	2.105	.4751	.0634	.1334	15.784	7.499	4.330	32.467	11
12	2.252	.4440	.0559	.1259	17.888	7.943	4.703	37.351	12
13	2.410	.4150	.0497	.1197	20.141	8.358	5.065	42.330	13
14	2.579	.3878	.0443	.1143	22.551	8.745	5.417	47.372	14
15	2.759	.3624	.0398	.1098	25.129	9.108	5.758	52.446	15
16	2.952	.3387	.0359	.1059	27.888	9.447	6.090	57.527	16
17	3.159	.3166	.0324	.1024	30.840	9.763	6.411	62.592	17
18	3.380	.2959	.0294	.0994	33.999	10.059	6.722	67.622	18
19	3.617	.2765	.0268	.0968	37.379	10.336	7.024	72.599	19
20	3.870	.2584	.0244	.0944	40.996	10.594	7.316	77.509	20
21	4.141	.2415	.0223	.0923	44.865	10.836	7.599	82.339	21
22	4.430	.2257	.0204	.0904	49.006	11.061	7.872	87.079	22
23	4.741	.2109	.0187	.0887	53.436	11.272	8.137	91.720	23
24	5.072	.1971	.0172	.0872	58.177	11.469	8.392	96.255	24
25	5.427	.1842	.0158	.0858	63.249	11.654	8.639	100.677	25
26	5.807	.1722	.0146	.0846	68.677	11.826	8.877	104.981	26
27	6.214	.1609	.0134	.0834	74.484	11.987	9.107	109.166	27
28	6.649	.1504	.0124	.0824	80.698	12.137	9.329	113.227	28
29	7.114	.1406	.0114	.0814	87.347	12.278	9.543	117.162	29
30	7.612	.1314	.0106	.0806	94.461	12.409	9.749	120.972	30
31	8.145	.1228	.00980	.0798	102.073	12.532	9.947	124.655	31
32	8.715	.1147	.00907	.0791	110.218	12.647	10.138	128.212	32
33	9.325	.1072	.00841	.0784	118.934	12.754	10.322	131.644	33
34	9.978	.1002	.00780	.0778	128.259	12.854	10.499	134.951	34
35	10.677	.0937	.00723	.0772	138.237	12.948	10.669	138.135	35
40	14.974	.0668	.00501	.0750	199.636	13.332	11.423	152.293	40
45	21.002	.0476	.00350	.0735	285.750	13.606	12.036	163.756	45
50	29.457	.0339	.00246	.0725	406.530	13.801	12.529	172.905	50
55	41.315	.0242	.00174	.0717	575.930	13.940	12.921	180.124	55
60	57.947	.0173	.00123	.0712	813.523	14.039	13.232	185.768	60
65	81.273	.0123	.00087	.0709	1 146.8	14.110	13.476	190.145	65
70	113.990	.00877	.00062	.0706	1 614.1	14.160	13.666	193.519	70
75	159.877	.00625	.00044	.0704	2 269.7	14.196	13.814	196.104	75
80	224.235	.00446	.00031	.0703	3 189.1	14.222	13.927	198.075	80
85	314.502	.00318	.00022	.0702	4 478.6	14.240	14.015	199.572	85
90	441.105	.00227	.00016	.0702	6 287.2	14.253	14.081	200.704	90
95	618.673	.00162	.00011	.0701	8 823.9	14.263	14.132	201.558	95
100	867.720	.00115	.00008	.0701	12 381.7	14.269	14.170	202.200	100

	Single Payment		Uniform Payment Series				Arithmetic Gradient		
	Compound Amount Factor	Present Worth Factor	Sinking Fund Factor	Capital Recovery Factor	Compound Amount Factor	Present Worth Factor	Gradient Uniform Series	Gradient Present Worth	
n	Find F Given P F/P	Find P Given F P/F	Find A Given F A/F	Find A Given P A/P	Find F Given A F/A	Find P Given A P/A	Find A Given G A/G	Find P Given G P/G	n
1	1.080	.9259	1.0000	1.0800	1.000	0.926	0	0	1
2	1.166	.8573	.4808	.5608	2.080	1.783	0.481	0.857	2
3	1.260	.7938	.3080	.3880	3.246	2.577	0.949	2.445	3
4	1.360	.7350	.2219	.3019	4.506	3.312	1.404	4.650	4
5	1.469	.6806	.1705	.2505	5.867	3.993	1.846	7.372	5
6	1.587	.6302	.1363	.2163	7.336	4.623	2.276	10.523	6
7	1.714	.5835	.1121	.1921	8.923	5.206	2.694	14.024	7
8	1.851	.5403	.0940	.1740	10.637	5.747	3.099	17.806	8
9	1.999	.5002	.0801	.1601	12.488	6.247	3.491	21.808	9
10	2.159	.4632	.0690	.1490	14.487	6.710	3.871	25.977	10
11	2.332	.4289	.0601	.1401	16.645	7.139	4.240	30.266	11
12	2.518	.3971	.0527	.1327	18.977	7.536	4.596	34.634	12
13	2.720	.3677	.0465	.1265	21.495	7.904	4.940	39.046	13
14	2.937	.3405	.0413	.1213	24.215	8.244	5.273	43.472	14
15	3.172	.3152	.0368	.1168	27.152	8.559	5.594	47.886	15
16	3.426	.2919	.0330	.1130	30.324	8.851	5.905	52.264	16
17	3.700	.2703	.0296	.1096	33.750	9.122	6.204	56.588	17
18	3.996	.2502	.0267	.1067	37.450	9.372	6.492	60.843	18
19	4.316	.2317	.0241	.1041	41.446	9.604	6.770	65.013	19
20	4.661	.2145	.0219	.1019	45.762	9.818	7.037	69.090	20
21	5.034	.1987	.0198	.0998	50.423	10.017	7.294	73.063	21
22	5.437	.1839	.0180	.0980	55.457	10.201	7.541	76.926	22
23	5.871	.1703	.0164	.0964	60.893	10.371	7.779	80.673	23
24	6.341	.1577	.0150	.0950	66.765	10.529	8.007	84.300	24
25	6.848	.1460	.0137	.0937	73.!06	10.675	8.225	87.804	25
26	7.396	.1352	.0125	.0925	79.954	10.810	8.435	91.184	26
27	7.988	.1252	.0114	.0914	87.351	10.935	8.636	94.439	27
28	8.627	.1159	.0105	.0905	95.339	11.051	8.829	97.569	28
29	9.317	.1073	.00962	.0896	103.966	11.158	9.013	100.574	29
30	10.063	.0994	.00883	.0888	113.283	11.258	9.190	103.456	30
31	10.868	.0920	.00811	.0881	123.346	11.350	9.358	106.216	31
32	11.737	.0852	.00745	.0875	134.214	11.435	9.520	108.858	32
33	12.676	.0789	.00685	.0869	145.951	11.514	9.674	111.382	33
34	13.690	.0730	.00630	.0863	158.627	11.587	9.821	113.792	34
35	14.785	.0676	.00580	.0858	172.317	11.655	9.961	116.092	35
40	21.725	.0460	.00386	.0839	259.057	11.925	10.570	126.042	40
45	31.920	.0313	.00259	.0826	386.506	12.108	11.045	133.733	45
50	46.902	.0213	.00174	.0817	573.771	12.233	11.411	139.593	50
55	68.914	.0145	.00118	.0812	848.925	12.319	11.690	144.006	55
60	101.257	.00988	.00080	.0808	1 253.2	12.377	11.902	147.300	60
65	148.780	.00672	.00054	.0805	1 847.3	12.416	12.060	149.739	65
70	218.607	.00457	.00037	.0804	2 720.1	12.443	12.178	151.533	70
75	321.205	.00311	.00025	.0802	4 002.6	12.461	12.266	152.845	75
80	471.956	.00212	.00017	.0802	5 887.0	12.474	12.330	153.800	80
85	693.458	.00144	.00012	.0801	8 655.7	12.482	12.377	154.492	85
90	1 018.9	.00098	.00008	.0801	12 724.0	12.488	12.412	154.993	90
95	1 497.1	.00067	.00005	.0801	18 701.6	12.492	12.437	155.352	95
100	2 199.8	.00045	.00004	.0800	27 484.6	12.494	12.455	155.611	100

	Single Payment		Uniform Payment Series				Arithmetic Gradient		
	Compound Amount Factor	Present Worth Factor	Sinking Fund Factor	Capital Recovery Factor	Compound Amount Factor	Present Worth Factor	Gradient Uniform Series	Gradient Present Worth	
n	Find *F* Given *P* *F/P*	Find *P* Given *F* *P/F*	Find *A* Given *F* *A/F*	Find *A* Given *P* *A/P*	Find *F* Given *A* *F/A*	Find *P* Given *A* *P/A*	Find *A* Given *G* *A/G*	Find *P* Given *G* *P/G*	*n*
1	1.090	.9174	1.0000	1.0900	1.000	0.917	0	0	1
2	1.188	.8417	.4785	.5685	2.090	1.759	0.478	0.842	2
3	1.295	.7722	.3051	.3951	3.278	2.531	0.943	2.386	3
4	1.412	.7084	.2187	.3087	4.573	3.240	1.393	4.511	4
5	1.539	.6499	.1671	.2571	5.985	·3.890	1.828	7.111	5
6	1.677	.5963	.1329	.2229	7.523	4.486	2.250	10.092	6
7	1.828	.5470	.1087	.1987	9.200	5.033	2.657	13.375	7
8	1.993	.5019	.0907	.1807	11.028	5.535	3.051	16.888	8
9	2.172	.4604	.0768	.1668	13.021	5.995	3.431	20.571	9
10	2.367	.4224	.0658	.1558	15.193	6.418	3.798	24.373	10
11	2.580	.3875	.0569	.1469	17.560	6.805	4.151	28.248	11
12	2.813	.3555	.0497	.1397	20.141	7.161	4.491	32.159	12
13	3.066	.3262	.0436	.1336	22.953	7.487	4.818	36.073	13
14	3.342	.2992	.0384	.1284	26.019	7.786	5.133	39.963	14
15	3.642	.2745	.0341	.1241	29.361	8.061	5.435	43.807	15
16	3.970	.2519	.0303	.1203	33.003	8.313	5.724	47.585	16
17	4.328	.2311	.0270	.1170	36.974	8.544	6.002	51.282	17
18	4.717	.2120	.0242	.1142	41.301	8.756	6.269	54.886	18
19	5.142	.1945	.0217	.1117	46.019	8.950	6.524	58.387	19
20	5.604	.1784	.0195	.1095	51.160	9.129	6.767	61.777	20
21	6.109	.1637	.0176	.1076	56.765	9.292	7.001	65.051	21
22	6.659	.1502	.0159	.1059	62.873	9.442	7.223	68.205	22
23	7.258	.1378	.0144	.1044	69.532	9.580	7.436	71.236	23
24	7.911	.1264	.0130	.1030	76.790	9.707	7.638	74.143	24
25	8.623	.1160	.0118	.1018	84.701	9.823	7.832	76.927	25
26	9.399	.1064	.0107	.1007	93.324	9.929	8.016	79.586	26
27	10.245	.0976	.00973	.0997	102.723	10.027	8.191	82.124	27
28	11.167	.0895	.00885	.0989	112.968	10.116	8.357	84.542	28
29	12.172	.0822	.00806	.0981	124.136	10.198	8.515	86.842	29
30	13.268	.0754	.00734	.0973	136.308	10.274	8.666	89.028	30
31	14.462	.0691	.00669	.0967	149.575	10.343	8.808	91.102	31
32	15.763	.0634	.00610	.0961	164.037	10.406	8.944	93.069	32
33	17.182	.0582	.00556	.0956	179.801	10.464	9.072	94.931	33
34	18.728	.0534	.00508	.0951	196.983	10.518	9.193	96.693	34
35	20.414	.0490	.00464	.0946	215.711	10.567	9.308	98.359	35
40	31.409	.0318	.00296	.0930	337.883	10.757	9.796	105.376	40
45	48.327	.0207	.00190	.0919	525.860	10.881	10.160	110.556	45
50	74.358	.0134	.00123	.0912	815.085	10.962	10.430	114.325	50
55	114.409	.00874	.00079	.0908	1 260.1	11.014	10.626	117.036	55
60	176.032	.00568	.00051	.0905	1 944.8	11.048	10.768	118.968	60
65	270.847	.00369	.00033	.0903	2 998.3	11.070	10.870	120.334	65
70	416.731	.00240	.00022	.0902	4 619.2	11.084	10.943	121.294	70
75	641.193	.00156	.00014	.0901	7 113.3	11.094	10.994	121.965	75
80	986.555	.00101	.00009	.0901	10 950.6	11.100	11.030	122.431	80
85	1 517.9	.00066	.00006	.0901	16 854.9	11.104	11.055	122.753	85
90	2 335.5	.00043	.00004	.0900	25 939.3	11.106	11.073	122.976	90
95	3 593.5	.00028	.00003	.0900	39 916.8	11.108	11.085	123.129	95
100	5 529.1	.00018	.00002	.0900	61 422.9	11.109	11.093	123.233	100

10% Compound Interest Factors 10%

	Single Payment		Uniform Payment Series				Arithmetic Gradient		
	Compound Amount Factor	Present Worth Factor	Sinking Fund Factor	Capital Recovery Factor	Compound Amount Factor	Present Worth Factor	Gradient Uniform Series	Gradient Present Worth	
n	Find F Given P F/P	Find P Given F P/F	Find A Given F A/F	Find A Given P A/P	Find F Given A F/A	Find P Given A P/A	Find A Given G A/G	Find P Given G P/G	n
1	1.100	.9091	1.0000	1.1000	1.000	0.909	0	0	1
2	1.210	.8264	.4762	.5762	2.100	1.736	0.476	0.826	2
3	1.331	.7513	.3021	.4021	3.310	2.487	0.937	2.329	3
4	1.464	.6830	.2155	.3155	4.641	3.170	1.381	4.378	4
5	1.611	.6209	.1638	.2638	6.105	3.791	1.810	6.862	5
6	1.772	.5645	.1296	.2296	7.716	4.355	2.224	9.684	6
7	1.949	.5132	.1054	.2054	9.487	4.868	2.622	12.763	7
8	2.144	.4665	.0874	.1874	11.436	5.335	3.004	16.029	8
9	2.358	.4241	.0736	.1736	13.579	5.759	3.372	19.421	9
10	2.594	.3855	.0627	.1627	15.937	6.145	3.725	22.891	10
11	2.853	.3505	.0540	.1540	18.531	6.495	4.064	26.396	11
12	3.138	.3186	.0468	.1468	21.384	6.814	4.388	29.901	12
13	3.452	.2897	.0408	.1408	24.523	7.103	4.699	33.377	13
14	3.797	.2633	.0357	.1357	27.975	7.367	4.996	36.801	14
15	4.177	.2394	.0315	.1315	31.772	7.606	5.279	40.152	15
16	4.595	.2176	.0278	.1278	35.950	7.824	5.549	43.416	16
17	5.054	.1978	.0247	.1247	40.545	8.022	5.807	46.582	17
18	5.560	.1799	.0219	.1219	45.599	8.201	6.053	49.640	18
19	6.116	.1635	.0195	.1195	51.159	8.365	6.286	52.583	19
20	6.728	.1486	.0175	.1175	57.275	8.514	6.508	55.407	20
21	7.400	.1351	.0156	.1156	64.003	8.649	6.719	58.110	21
22	8.140	.1228	.0140	.1140	71.403	8.772	6.919	60.689	22
23	8.954	.1117	.0126	.1126	79.543	8.883	7.108	63.146	23
24	9.850	.1015	.0113	.1113	88.497	8.985	7.288	65.481	24
25	10.835	.0923	.0102	.1102	98.347	9.077	7.458	67.696	25
26	11.918	.0839	.00916	.1092	109.182	9.161	7.619	69.794	26
27	13.110	.0763	.00826	.1083	121.100	9.237	7.770	71.777	27
28	14.421	.0693	.00745	.1075	134.210	9.307	7.914	73.650	28
29	15.863	.0630	.00673	.1067	148.631	9.370	8.049	75.415	29
30	17.449	.0573	.00608	.1061	164.494	9.427	8.176	77.077	30
31	19.194	.0521	.00550	.1055	181.944	9.479	8.296	78.640	31
32	21.114	.0474	.00497	.1050	201.138	9.526	8.409	80.108	32
33	23.225	.0431	.00450	.1045	222.252	9.569	8.515	81.486	33
34	25.548	.0391	.00407	.1041	245.477	9.609	8.615	82.777	34
35	28.102	.0356	.00369	.1037	271.025	9.644	8.709	83.987	35
40	45.259	.0221	.00226	.1023	442.593	9.779	9.096	88.953	40
45	72.891	.0137	.00139	.1014	718.905	9.863	9.374	92.454	45
50	117.391	.00852	.00086	.1009	1 163.9	9.915	9.570	94.889	50
55	189.059	.00529	.00053	.1005	1 880.6	9.947	9.708	96.562	55
60	304.482	.00328	.00033	.1003	3 034.8	9.967	9.802	97.701	60
65	490.371	.00204	.00020	.1002	4 893.7	9.980	9.867	98.471	65
70	789.748	.00127	.00013	.1001	7 887.5	9.987	9.911	98.987	70
75	1 271.9	.00079	.00008	.1001	12 709 0	9.992	9.941	99.332	75
80	2 048.4	.00049	.00005	.1000	20 474.0	9.995	9.961	99.561	80
85	3 299.0	.00030	.00003	.1000	32 979.7	9.997	9.974	99.712	85
90	5 313.0	.00019	.00002	.1000	53 120.3	9.998	9.983	99.812	90
95	8 556.7	.00012	.00001	.1000	85 556.9	9.999	9.989	99.877	95
100	13 780.6	.00007	.00001	.1000	137 796.3	9.999	9.993	99.920	100

	Single Payment		Uniform Payment Series				Arithmetic Gradient		
	Compound Amount Factor	Present Worth Factor	Sinking Fund Factor	Capital Recovery Factor	Compound Amount Factor	Present Worth Factor	Gradient Uniform Series	Gradient Present Worth	
	Find F Given P F/P	Find P Given F P/F	Find A Given F A/F	Find A Given P A/P	Find F Given A F/A	Find P Given A P/A	Find A Given G A/G	Find P Given G P/G	
n									n
1	1.120	.8929	1.0000	1.1200	1.000	0.893	0	0	1
2	1.254	.7972	.4717	.5917	2.120	1.690	0.472	0.797	2
3	1.405	.7118	.2963	.4163	3.374	2.402	0.925	2.221	3
4	1.574	.6355	.2092	.3292	4.779	3.037	1.359	4.127	4
5	1.762	.5674	.1574	.2774	6.353	3.605	1.775	6.397	5
6	1.974	.5066	.1232	.2432	8.115	4.111	2.172	8.930	6
7	2.211	.4523	.0991	.2191	10.089	4.564	2.551	11.644	7
8	2.476	.4039	.0813	.2013	12.300	4.968	2.913	14.471	8
9	2.773	.3606	.0677	.1877	14.776	5.328	3.257	17.356	9
10	3.106	.3220	.0570	.1770	17.549	5.650	3.585	20.254	10
11	3.479	.2875	.0484	.1684	20.655	5.938	3.895	23.129	11
12	3.896	.2567	.0414	.1614	24.133	6.194	4.190	25.952	12
13	4.363	.2292	.0357	.1557	28.029	6.424	4.468	28.702	13
14	4.887	.2046	.0309	.1509	32.393	6.628	4.732	31.362	14
15	5.474	.1827	.0268	.1468	37.280	6.811	4.980	33.920	15
16	6.130	.1631	.0234	.1434	42.753	6.974	5.215	36.367	16
17	6.866	.1456	.0205	.1405	48.884	7.120	5.435	38.697	17
18	7.690	.1300	.0179	.1379	55.750	7.250	5.643	40.908	18
19	8.613	.1161	.0158	.1358	63.440	7.366	5.838	42.998	19
20	9.646	.1037	.0139	.1339	72.052	7.469	6.020	44.968	20
21	10.804	.0926	.0122	.1322	81.699	7.562	6.191	46.819	21
22	12.100	.0826	.0108	.1308	92.503	7.645	6.351	48.554	22
23	13.552	.0738	.00956	.1296	104.603	7.718	6.501	50.178	23
24	15.179	.0659	.00846	.1285	118.155	7.784	6.641	51.693	24
25	17.000	.0588	.00750	.1275	133.334	7.843	6.771	53.105	25
26	19.040	.0525	.00665	.1267	150.334	7.896	6.892	54.418	26
27	21.325	.0469	.00590	.1259	169.374	7.943	7.005	55.637	27
28	23.884	.0419	.00524	.1252	190.699	7.984	7.110	56.767	28
29	26.750	.0374	.00466	.1247	214.583	8.022	7.207	57.814	29
30	29.960	.0334	.00414	.1241	241.333	8.055	7.297	58.782	30
31	33.555	.0298	.00369	.1237	271.293	8.085	7.381	59.676	31
32	37.582	.0266	.00328	.1233	304.848	8.112	7.459	60.501	32
33	42.092	.0238	.00292	.1229	342.429	8.135	7.530	61.261	33
34	47.143	.0212	.00260	.1226	384.521	8.157	7.596	61.961	34
35	52.800	.0189	.00232	.1223	431.663	8.176	7.658	62.605	35
40	93.051	.0107	.00130	.1213	767.091	8.244	7.899	65.116	40
45	163.988	.00610	.00074	.1207	1 358.2	8.283	8.057	66.734	45
50	289.002	.00346	.00042	.1204	2 400.0	8.304	8.160	67.762	50
55	509.321	.00196	.00024	.1202	4 236.0	8.317	8.225	68.408	55
60	897.597	.00111	.00013	.1201	7 471.6	8.324	8.266	68.810	60
65	1 581.9	.00063	.00008	.1201	13 173.9	8.328	8.292	69.058	65
70	2 787.8	.00036	.00004	.1200	23 223.3	8.330	8.308	69.210	70
75	4 913.1	.00020	.00002	.1200	40 933.8	8.332	8.318	69.303	75
80	8 658.5	.00012	.00001	.1200	72 145.7	8.332	8.324	69.359	80
85	15 259.2	.00007	.00001	.1200	127 151.7	8.333	8.328	69.393	85
90	26 891.9	.00004		.1200	224 091.1	8.333	8.330	69.414	90
95	47 392.8	.00002		.1200	394 931.4	8.333	8.331	69.426	95
100	83 522.3	.00001		.1200	696 010.5	8.333	8.332	69.434	100

	Single Payment		Uniform Payment Series				Arithmetic Gradient		
	Compound Amount Factor	Present Worth Factor	Sinking Fund Factor	Capital Recovery Factor	Compound Amount Factor	Present Worth Factor	Gradient Uniform Series	Gradient Present Worth	
n	Find F Given P F/P	Find P Given F P/F	Find A Given F A/F	Find A Given P A/P	Find F Given A F/A	Find P Given A P/A	Find A Given G A/G	Find P Given G P/G	n
1	1.150	.8696	1.0000	1.1500	1.000	0.870	0	0	1
2	1.322	.7561	.4651	.6151	2.150	1.626	0.465	0.756	2
3	1.521	.6575	.2880	.4380	3.472	2.283	0.907	2.071	3
4	1.749	.5718	.2003	.3503	4.993	2.855	1.326	3.786	4
5	2.011	.4972	.1483	.2983	6.742	3.352	1.723	5.775	5
6	2.313	.4323	.1142	.2642	8.754	3.784	2.097	7.937	6
7	2.660	.3759	.0904	.2404	11.067	4.160	2.450	10.192	7
8	3.059	.3269	.0729	.2229	13.727	4.487	2.781	12.481	8
9	3.518	.2843	.0596	.2096	16.786	4.772	3.092	14.755	9
10	4.046	.2472	.0493	.1993	20.304	5.019	3.383	16.979	10
11	4.652	.2149	.0411	.1911	24.349	5.234	3.655	19.129	11
12	5.350	.1869	.0345	.1845	29.002	5.421	3.908	21.185	12
13	6.153	.1625	.0291	.1791	34.352	5.583	4.144	23.135	13
14	7.076	.1413	.0247	.1747	40.505	5.724	4.362	24.972	14
15	8.137	.1229	.0210	.1710	47.580	5.847	4.565	26.693	15
16	9.358	.1069	.0179	.1679	55.717	5.954	4.752	28.296	16
17	10.761	.0929	.0154	.1654	65.075	6.047	4.925	29.783	17
18	12.375	.0808	.0132	.1632	75.836	6.128	5.084	31.156	18
19	14.232	.0703	.0113	.1613	88.212	6.198	5.231	32.421	19
20	16.367	.0611	.00976	.1598	102.444	6.259	5.365	33.582	20
21	18.822	.0531	.00842	.1584	118.810	6.312	5.488	34.645	21
22	21.645	.0462	.00727	.1573	137.632	6.359	5.601	35.615	22
23	24.891	.0402	.00628	.1563	159.276	6.399	5.704	36.499	23
24	28.625	.0349	.00543	.1554	184.168	6.434	5.798	37.302	24
25	32.919	.0304	.00470	.1547	212.793	6.464	5.883	38.031	25
26	37.857	.0264	.00407	.1541	245.712	6.491	5.961	38.692	26
27	43.535	.0230	.00353	.1535	283.569	6.514	6.032	39.289	27
28	50.066	.0200	.00306	.1531	327.104	6.534	6.096	39.828	28
29	57.575	.0174	.00265	.1527	377.170	6.551	6.154	40.315	29
30	66.212	.0151	.00230	.1523	434.745	6.566	6.207	40.753	30
31	76.144	.0131	.00200	.1520	500.957	6.579	6.254	41.147	31
32	87.565	.0114	.00173	.1517	577.100	6.591	6.297	41.501	32
33	100.700	.00993	.00150	.1515	664.666	6.600	6.336	41.818	33
34	115.805	.00864	.00131	.1513	765.365	6.609	6.371	42.103	34
35	133.176	.00751	.00113	.1511	881.170	6.617	6.402	42.359	35
40	267.864	.00373	.00056	.1506	1 779.1	6.642	6.517	43.283	40
45	538.769	.00186	.00028	.1503	3 585.1	6.654	6.583	43.805	45
50	1 083.7	.00092	.00014	.1501	7 217.7	6.661	6.620	44.096	50
55	2 179.6	.00046	.00007	.1501	14 524.1	6.664	6.641	44.256	55
60	4 384.0	.00023	.00003	.1500	29 220.0	6.665	6.653	44.343	60
65	8 817.8	.00011	.00002	.1500	58 778.6	6.666	6.659	44.390	65
70	17 735.7	.00006	.00001	.1500	118 231.5	6.666	6.663	44.416	70
75	35 672.9	.00003		.1500	237 812.5	6.666	6.665	44.429	75
80	71 750.9	.00001		.1500	478 332.6	6.667	6.666	44.436	80
85	144 316.7	.00001		.1500	962 104.4	6.667	6.666	44.440	85

18% Compound Interest Factors 18%

	Single Payment		Uniform Payment Series				Arithmetic Gradient		
	Compound Amount Factor	Present Worth Factor	Sinking Fund Factor	Capital Recovery Factor	Compound Amount Factor	Present Worth Factor	Gradient Uniform Series	Gradient Present Worth	
n	Find F Given P F/P	Find P Given F P/F	Find A Given F A/F	Find A Given P A/P	Find F Given A F/A	Find P Given A P/A	Find A Given G A/G	Find P Given G P/G	n
1	1.180	.8475	1.0000	1.1800	1.000	0.847	0	0	1
2	1.392	.7182	.4587	.6387	2.180	1.566	0.459	0.718	2
3	1.643	.6086	.2799	.4599	3.572	2.174	0.890	1.935	3
4	1.939	.5158	.1917	.3717	5.215	2.690	1.295	3.483	4
5	2.288	.4371	.1398	.3198	7.154	3.127	1.673	5.231	5
6	2.700	.3704	.1059	.2859	9.442	3.498	2.025	7.083	6
7	3.185	.3139	.0824	.2624	12.142	3.812	2.353	8.967	7
8	3.759	.2660	.0652	.2452	15.327	4.078	2.656	10.829	8
9	4.435	.2255	.0524	.2324	19.086	4.303	2.936	12.633	9
10	5.234	.1911	.0425	.2225	23.521	4.494	3.194	14.352	10
11	6.176	.1619	.0348	.2148	28.755	4.656	3.430	15.972	11
12	7.288	.1372	.0286	.2086	34.931	4.793	3.647	17.481	12
13	8.599	.1163	.0237	.2037	42.219	4.910	3.845	18.877	13
14	10.147	.0985	.0197	.1997	50.818	5.008	4.025	20.158	14
15	11.974	.0835	.0164	.1964	60.965	5.092	4.189	21.327	15
16	14.129	.0708	.0137	.1937	72.939	5.162	4.337	22.389	16
17	16.672	.0600	.0115	.1915	87.068	5.222	4.471	23.348	17
18	19.673	.0508	.00964	.1896	103.740	5.273	4.592	24.212	18
19	23.214	.0431	.00810	.1881	123.413	5.316	4.700	24.988	19
20	27.393	.0365	.00682	.1868	146.628	5.353	4.798	25.681	20
21	32.324	.0309	.00575	.1857	174.021	5.384	4.885	26.300	21
22	38.142	.0262	.00485	.1848	206.345	5.410	4.963	26.851	22
23	45.008	.0222	.00409	.1841	244.487	5.432	5.033	27.339	23
24	53.109	.0188	.00345	.1835	289.494	5.451	5.095	27.772	24
25	62.669	.0160	.00292	.1829	342.603	5.467	5.150	28.155	25
26	73.949	.0135	.00247	.1825	405.272	5.480	5.199	28.494	26
27	87.260	.0115	.00209	.1821	479.221	5.492	5.243	28.791	27
28	102.966	.00971	.00177	.1818	566.480	5.502	5.281	29.054	28
29	121.500	.00823	.00149	.1815	669.447	5.510	5.315	29.284	29
30	143.370	.00697	.00126	.1813	790.947	5.517	5.345	29.486	30
31	169.177	.00591	.00107	.1811	934.317	5.523	5.371	29.664	31
32	199.629	.00501	.00091	.1809	1 103.5	5.528	5.394	29.819	32
33	235.562	.00425	.00077	.1808	1 303.1	5.532	5.415	29.955	33
34	277.963	.00360	.00065	.1806	1 538.7	5.536	5.433	30.074	34
35	327.997	.00305	.00055	.1806	1 816.6	5.539	5.449	30.177	35
40	750.377	.00133	.00024	.1802	4 163.2	5.548	5.502	30.527	40
45	1 716.7	.00058	.00010	.1801	9 531.6	5.552	5.529	30.701	45
50	3 927.3	.00025	.00005	.1800	21 813.0	5.554	5.543	30.786	50
55	8 984.8	.00011	.00002	.1800	49 910.1	5.555	5.549	30.827	55
60	20 555.1	.00005	.00001	.1800	114 189.4	5.555	5.553	30.846	60
65	47 025.1	.00002		.1800	261 244.7	5.555	5.554	30.856	65
70	107 581.9	.00001		.1800	597.671.7	5.556	5.555	30.860	70

	Single Payment		Uniform Payment Series				Arithmetic Gradient		
	Compound Amount Factor	Present Worth Factor	Sinking Fund Factor	Capital Recovery Factor	Compound Amount Factor	Present Worth Factor	Gradient Uniform Series	Gradient Present Worth	
n	Find F Given P F/P	Find P Given F P/F	Find A Given F A/F	Find A Given P A/P	Find F Given A F/A	Find P Given A P/A	Find A Given G A/G	Find P Given G P/G	n
1	1.200	.8333	1.0000	1.2000	1.000	0.833	0	0	1
2	1.440	.6944	.4545	.6545	2.200	1.528	0.455	0.694	2
3	1.728	.5787	.2747	.4747	3.640	2.106	0.879	1.852	3
4	2.074	.4823	.1863	.3863	5.368	2.589	1.274	3.299	4
5	2.488	.4019	.1344	.3344	7.442	2.991	1.641	4.906	5
6	2.986	.3349	.1007	.3007	9.930	3.326	1.979	6.581	6
7	3.583	.2791	.0774	.2774	12.916	3.605	2.290	8.255	7
8	4.300	.2326	.0606	.2606	16.499	3.837	2.576	9.883	8
9	5.160	.1938	.0481	.2481	20.799	4.031	2.836	11.434	9
10	6.192	.1615	.0385	.2385	25.959	4.192	3.074	12.887	10
11	7.430	.1346	.0311	.2311	32.150	4.327	3.289	14.233	11
12	8.916	.1122	.0253	.2253	39.581	4.439	3.484	15.467	12
13	10.699	.0935	.0206	.2206	48.497	4.533	3.660	16.588	13
14	12.839	.0779	.0169	.2169	59.196	4.611	3.817	17.601	14
15	15.407	.0649	.0139	.2139	72.035	4.675	3.959	18.509	15
16	18.488	.0541	.0114	.2114	87.442	4.730	4.085	19.321	16
17	22.186	.0451	.00944	.2094	105.931	4.775	4.198	20.042	17
18	26.623	.0376	.00781	.2078	128.117	4.812	4.298	20.680	18
19	31.948	.0313	.00646	.2065	154.740	4.843	4.386	21.244	19
20	38.338	.0261	.00536	.2054	186.688	4.870	4.464	21.739	20
21	46.005	.0217	.00444	.2044	225.026	4.891	4.533	22.174	21
22	55.206	.0181	.00369	.2037	271.031	4.909	4.594	22.555	22
23	66.247	.0151	.00307	.2031	326.237	4.925	4.647	22.887	23
24	79.497	.0126	.00255	.2025	392.484	4.937	4.694	23.176	24
25	95.396	.0105	.00212	.2021	471.981	4.948	4.735	23.428	25
26	114.475	.00874	.00176	.2018	567.377	4.956	4.771	23.646	26
27	137.371	.00728	.00147	.2015	681.853	4.964	4.802	23.835	27
28	164.845	.00607	.00122	.2012	819.223	4.970	4.829	23.999	28
29	197.814	.00506	.00102	.2010	984.068	4.975	4.853	24.141	29
30	237.376	.00421	.00085	.2008	1 181.9	4.979	4.873	24.263	30
31	284.852	.00351	.00070	.2007	1 419.3	4.982	4.891	24.368	31
32	341.822	.00293	.00059	.2006	1 704.1	4.985	4.906	24.459	32
33	410.186	.00244	.00049	.2005	2 045.9	4.988	4.919	24.537	33
34	492.224	.00203	.00041	.2004	2 456.1	4.990	4.931	24.604	34
35	590.668	.00169	.00034	.2003	2 948.3	4.992	4.941	24.661	35
40	1 469.8	.00068	.00014	.2001	7 343.9	4.997	4.973	24.847	40
45	3 657.3	.00027	.00005	.2001	18 281.3	4.999	4.988	24.932	45
50	9 100.4	.00011	.00002	.2000	45 497.2	4.999	4.995	24.970	50
55	22 644.8	.00004	.00001	.2000	113 219.0	5.000	4.998	24.987	55
60	56 347.5	.00002		.2000	281 732.6	5.000	4.999	24.994	60

	Single Payment		Uniform Payment Series				Arithmetic Gradient		
	Compound Amount Factor	Present Worth Factor	Sinking Fund Factor	Capital Recovery Factor	Compound Amount Factor	Present Worth Factor	Gradient Uniform Series	Gradient Present Worth	
n	Find F Given P F/P	Find P Given F P/F	Find A Given F A/F	Find A Given P A/P	Find F Given A F/A	Find P Given A P/A	Find A Given G A/G	Find P Given G P/G	n
1	1.250	.8000	1.0000	1.2500	1.000	0.800	0	0	1
2	1.563	.6400	.4444	.6944	2.250	1.440	0.444	0.640	2
3	1.953	.5120	.2623	.5123	3.813	1.952	0.852	1.664	3
4	2.441	.4096	.1734	.4234	5.766	2.362	1.225	2.893	4
5	3.052	.3277	.1218	.3718	8.207	2.689	1.563	4.204	5
6	3.815	.2621	.0888	.3388	11.259	2.951	1.868	5.514	6
7	4.768	.2097	.0663	.3163	15.073	3.161	2.142	6.773	7
8	5.960	.1678	.0504	.3004	19.842	3.329	2.387	7.947	8
9	7.451	.1342	.0388	.2888	25.802	3.463	2.605	9.021	9
10	9.313	.1074	.0301	.2801	33.253	3.571	2.797	9.987	10
11	11.642	.0859	.0235	.2735	42.566	3.656	2.966	10.846	11
12	14.552	.0687	.0184	.2684	54.208	3.725	3.115	11.602	12
13	18.190	.0550	.0145	.2645	68.760	3.780	3.244	12.262	13
14	22.737	.0440	.0115	.2615	86.949	3.824	3.356	12.833	14
15	28.422	.0352	.00912	.2591	109.687	3.859	3.453	13.326	15
16	35.527	.0281	.00724	.2572	138.109	3.887	3.537	13.748	16
17	44.409	.0225	.00576	.2558	173.636	3.910	3.608	14.108	17
18	55.511	.0180	.00459	.2546	218.045	3.928	3.670	14.415	18
19	69.389	.0144	.00366	.2537	273.556	3.942	3.722	14.674	19
20	86.736	.0115	.00292	.2529	342.945	3.954	3.767	14.893	20
21	108.420	.00922	.00233	.2523	429.681	3.963	3.805	15.078	21
22	135.525	.00738	.00186	.2519	538.101	3.970	3.836	15.233	22
23	169.407	.00590	.00148	.2515	673.626	3.976	3.863	15.362	23
24	211.758	.00472	.00119	.2512	843.033	3.981	3.886	15.471	24
25	264.698	.00378	.00095	.2509	1 054.8	3.985	3.905	15.562	25
26	330.872	.00302	.00076	.2508	1 319.5	3.988	3.921	15.637	26
27	413.590	.00242	.00061	.2506	1 650.4	3.990	3.935	15.700	27
28	516.988	.00193	.00048	.2505	2 064.0	3.992	3.946	15.752	28
29	646.235	.00155	.00039	.2504	2 580.9	3.994	3.955	15.796	29
30	807.794	.00124	.00031	.2503	3 227.2	3.995	3.963	15.832	30
31	1 009.7	.00099	.00025	.2502	4 035.0	3.996	3.969	15.861	31
32	1 262.2	.00079	.00020	.2502	5 044.7	3.997	3.975	15.886	32
33	1 577.7	.00063	.00016	.2502	6 306.9	3.997	3.979	15.906	33
34	1 972.2	.00051	.00013	.2501	7 884.6	3.998	3.983	15.923	34
35	2 465.2	.00041	.00010	.2501	9 856.8	3.998	3.986	15.937	35
40	7 523.2	.00013	.00003	.2500	30 088.7	3.999	3.995	15.977	40
45	22 958.9	.00004	.00001	.2500	91 831.5	4.000	3.998	15.991	45
50	70 064.9	.00001		.2500	280 255.7	4.000	3.999	15.997	50
55	213 821.2			.2500	855 280.7	4.000	4.000	15.999	55

	Single Payment		Uniform Payment Series				Arithmetic Gradient		
	Compound Amount Factor	Present Worth Factor	Sinking Fund Factor	Capital Recovery Factor	Compound Amount Factor	Present Worth Factor	Gradient Uniform Series	Gradient Present Worth	
n	Find F Given P F/P	Find P Given F P/F	Find A Given F A/F	Find A Given P A/P	Find F Given A F/A	Find P Given A P/A	Find A Given G A/G	Find P Given G P/G	n
1	1.300	.7692	1.0000	1.3000	1.000	0.769	0	0	1
2	1.690	.5917	.4348	.7348	2.300	1.361	0.435	0.592	2
3	2.197	.4552	.2506	.5506	3.990	1.816	0.827	1.502	3
4	2.856	.3501	.1616	.4616	6.187	2.166	1.178	2.552	4
5	3.713	.2693	.1106	.4106	9.043	2.436	1.490	3.630	5
6	4.827	.2072	.0784	.3784	12.756	2.643	1.765	4.666	6
7	6.275	.1594	.0569	.3569	17.583	2.802	2.006	5.622	7
8	8.157	.1226	.0419	.3419	23.858	2.925	2.216	6.480	8
9	10.604	.0943	.0312	.3312	32.015	3.019	2.396	7.234	9
10	13.786	.0725	.0235	.3235	42.619	3.092	2.551	7.887	10
11	17.922	.0558	.0177	.3177	56.405	3.147	2.683	8.445	11
12	23.298	.0429	.0135	.3135	74.327	3.190	2.795	8.917	12
13	30.287	.0330	.0102	.3102	97.625	3.223	2.889	9.314	13
14	39.374	.0254	.00782	.3078	127.912	3.249	2.969	9.644	14
15	51.186	.0195	.00598	.3060	167.286	3.268	3.034	9.917	15
16	66.542	.0150	.00458	.3046	218.472	3.283	3.089	10.143	16
17	86.504	.0116	.00351	.3035	285.014	3.295	3.135	10.328	17
18	112.455	.00889	.00269	.3027	371.518	3.304	3.172	10.479	18
19	146.192	.00684	.00207	.3021	483.973	3.311	3.202	10.602	19
20	190.049	.00526	.00159	.3016	630.165	3.316	3.228	10.702	20
21	247.064	.00405	.00122	.3012	820.214	3.320	3.248	10.783	21
22	321.184	.00311	.00094	.3009	1 067.3	3.323	3.265	10.848	22
23	417.539	.00239	.00072	.3007	1 388.5	3.325	3.278	10.901	23
24	542.800	.00184	.00055	.3006	1 806.0	3.327	3.289	10.943	24
25	705.640	.00142	.00043	.3004	2 348.8	3.329	3.298	10.977	25
26	917.332	.00109	.00033	.3003	3 054.4	3.330	3.305	11.005	26
27	1 192.5	.00084	.00025	.3003	3 971.8	3.331	3.311	11.026	27
28	1 550.3	.00065	.00019	.3002	5 164.3	3.331	3.315	11.044	28
29	2 015.4	.00050	.00015	.3001	6 714.6	3.332	3.319	11.058	29
30	2 620.0	.00038	.00011	.3001	8 730.0	3.332	3.322	11.069	30
31	3 406.0	.00029	.00009	.3001	11 350.0	3.332	3.324	11.078	31
32	4 427.8	.00023	.00007	.3001	14 756.0	3.333	3.326	11.085	32
33	5 756.1	.00017	.00005	.3001	19 183.7	3.333	3.328	11.090	33
34	7 483.0	.00013	.00004	.3000	24 939.9	3.333	3.329	11.094	34
35	9 727.8	.00010	.00003	.3000	32 422.8	3.333	3.330	11.098	35
40	36 118.8	.00003	.00001	.3000	120 392.6	3.333	3.332	11.107	40
45	134 106.5	.00001		.3000	447 018.3	3.333	3.333	11.110	45

Compound Interest Factors

	Single Payment		Uniform Payment Series				Arithmetic Gradient		
	Compound Amount Factor	Present Worth Factor	Sinking Fund Factor	Capital Recovery Factor	Compound Amount Factor	Present Worth Factor	Gradient Uniform Series	Gradient Present Worth	
n	Find F Given P F/P	Find P Given F P/F	Find A Given F A/F	Find A Given P A/P	Find F Given A F/A	Find P Given A P/A	Find A Given G A/G	Find P Given G P/G	n
1	1.350	.7407	1.0000	1.3500	1.000	0.741	0	0	1
2	1.822	.5487	.4255	.7755	2.350	1.289	0.426	0.549	2
3	2.460	.4064	.2397	.5897	4.173	1.696	0.803	1.362	3
4	3.322	.3011	.1508	.5008	6.633	1.997	1.134	2.265	4
5	4.484	.2230	.1005	.4505	9.954	2.220	1.422	3.157	5
6	6.053	.1652	.0693	.4193	14.438	2.385	1.670	3.983	6
7	8.172	.1224	.0488	.3988	20.492	2.508	1.881	4.717	7
8	11.032	.0906	.0349	.3849	28.664	2.598	2.060	5.352	8
9	14.894	.0671	.0252	.3752	39.696	2.665	2.209	5.889	9
10	20.107	.0497	.0183	.3683	54.590	2.715	2.334	6.336	10
11	27.144	.0368	.0134	.3634	74.697	2.752	2.436	6.705	11
12	36.644	.0273	.00982	.3598	101.841	2.779	2.520	7.005	12
13	49.470	.0202	.00722	.3572	138.485	2.799	2.589	7.247	13
14	66.784	.0150	.00532	.3553	187.954	2.814	2.644	7.442	14
15	90.158	.0111	.00393	.3539	254.739	2.825	2.689	7.597	15
16	121.714	.00822	.00290	.3529	344.897	2.834	2.725	7.721	16
17	164.314	.00609	.00214	.3521	466.611	2.840	2.753	7.818	17
18	221.824	.00451	.00158	.3516	630.925	2.844	2.776	7.895	18
19	299.462	.00334	.00117	.3512	852.748	2.848	2.793	7.955	19
20	404.274	.00247	.00087	.3509	1 152.2	2.850	2.808	8.002	20
21	545.769	.00183	.00064	.3506	1 556.5	2.852	2.819	8.038	21
22	736.789	.00136	.00048	.3505	2 102.3	2.853	2.827	8.067	22
23	994.665	.00101	.00035	.3504	2 839.0	2.854	2.834	8.089	23
24	1 342.8	.00074	.00026	.3503	3 833.7	2.855	2.839	8.106	24
25	1 812.8	.00055	.00019	.3502	5 176.5	2.856	2.843	8.119	25
26	2 447.2	.00041	.00014	.3501	6 989.3	2.856	2.847	8.130	26
27	3 303.8	.00030	.00011	.3501	9 436.5	2.856	2.849	8.137	27
28	4 460.1	.00022	.00008	.3501	12 740.3	2.857	2.851	8.143	28
29	6 021.1	.00017	.00006	.3501	17 200.4	2.857	2.852	8.148	29
30	8 128.5	.00012	.00004	.3500	23 221.6	2.857	2.853	8.152	30
31	10 973.5	.00009	.00003	.3500	31 350.1	2.857	2.854	8.154	31
32	14 814.3	.00007	.00002	.3500	42 323.7	2.857	2.855	8.157	32
33	19 999.3	.00005	.00002	.3500	57 137.9	2.857	2.855	8.158	33
34	26 999.0	.00004	.00001	.3500	77 137.2	2.857	2.856	8.159	34
35	36 448.7	.00003	.00001	.3500	104 136.3	2.857	2.856	8.160	35

| | Single Payment | | Uniform Payment Series | | | | Arithmetic Gradient | | |
| | Compound Amount Factor | Present Worth Factor | Sinking Fund Factor | Capital Recovery Factor | Compound Amount Factor | Present Worth Factor | Gradient Uniform Series | Gradient Present Worth | |
n	Find *F* Given *P* F/P	Find *P* Given *F* P/F	Find *A* Given *F* A/F	Find *A* Given *P* A/P	Find *F* Given *A* F/A	Find *P* Given *A* P/A	Find *A* Given *G* A/G	Find *P* Given *G* P/G	*n*
1	1.400	.7143	1.0000	1.4000	1.000	0.714	0	0	1
2	1.960	.5102	.4167	.8167	2.400	1.224	0.417	0.510	2
3	2.744	.3644	.2294	.6294	4.360	1.589	0.780	1.239	3
4	3.842	.2603	.1408	.5408	7.104	1.849	1.092	2.020	4
5	5.378	.1859	.0914	.4914	10.946	2.035	1.358	2.764	5
6	7.530	.1328	.0613	.4613	16.324	2.168	1.581	3.428	6
7	10.541	.0949	.0419	.4419	23.853	2.263	1.766	3.997	7
8	14.758	.0678	.0291	.4291	34.395	2.331	1.919	4.471	8
9	20.661	.0484	.0203	.4203	49.153	2.379	2.042	4.858	9
10	28.925	.0346	.0143	.4143	69.814	2.414	2.142	5.170	10
11	40.496	.0247	.0101	.4101	98.739	2.438	2.221	5.417	11
12	56.694	.0176	.00718	.4072	139.235	2.456	2.285	5.611	12
13	79.371	.0126	.00510	.4051	195.929	2.469	2.334	5.762	13
14	111.120	.00900	.00363	.4036	275.300	2.478	2.373	5.879	14
15	155.568	.00643	.00259	.4026	386.420	2.484	2.403	5.969	15
16	217.795	.00459	.00185	.4018	541.988	2.489	2.426	6.038	16
17	304.913	.00328	.00132	.4013	759.783	2.492	2.444	6.090	17
18	426.879	.00234	.00094	.4009	1 064.7	2.494	2.458	6.130	18
19	597.630	.00167	.00067	.4007	1 419.6	2.496	2.468	6.160	19
20	836.682	.00120	.00048	.4005	2 089.2	2.497	2.476	6.183	20
21	1 171.4	.00085	.00034	.4003	2 925.9	2.498	2.482	6.200	21
22	1 639.9	.00061	.00024	.4002	4 097.2	2.498	2.487	6.213	22
23	2 295.9	.00044	.00017	.4002	5 737.1	2.499	2.490	6.222	23
24	3 214.2	.00031	.00012	.4001	8 033.0	2.499	2.493	6.229	24
25	4 499.9	.00022	.00009	.4001	11 247 2	2.499	2.494	6.235	25
26	6 299.8	.00016	.00006	.4001	15 747.1	2.500	2.496	6.239	26
27	8 819.8	.00011	.00005	.4000	22 046.9	2.500	2.497	6.242	27
28	12 347.7	.00008	.00003	.4000	30 866.7	2.500	2.498	6.244	28
29	17 286.7	.00006	.00002	.4000	43 214.3	2.500	2.498	6.245	29
30	24 201.4	.00004	.00002	.4000	60 501.0	2.500	2.499	6.247	30
31	33 882.0	.00003	.00001	.4000	84 702.5	2.500	2.499	6.248	31
32	47 434.8	.00002	.00001	.4000	118 584.4	2.500	2.499	6.248	32
33	66 408.7	.00002	.00001	.4000	166 019.2	2.500	2.500	6.249	33
34	92 972.1	.00001		.4000	232 427.9	2.500	2.500	6.249	34
35	130 161.0	.00001		.4000	325 400.0	2.500	2.500	6.249	35

	Single Payment		Uniform Payment Series				Arithmetic Gradient		
	Compound Amount Factor	Present Worth Factor	Sinking Fund Factor	Capital Recovery Factor	Compound Amount Factor	Present Worth Factor	Gradient Uniform Series	Gradient Present Worth	
n	Find F Given P F/P	Find P Given F P/F	Find A Given F A/F	Find A Given P A/P	Find F Given A F/A	Find P Given A P/A	Find A Given G A/G	Find P Given G P/G	n
1	1.450	.6897	1.0000	1.4500	1.000	0.690	0	0	1
2	2.103	.4756	.4082	.8582	2.450	1.165	0.408	0.476	2
3	3.049	.3280	.2197	.6697	4.553	1.493	0.758	1.132	3
4	4.421	.2262	.1316	.5816	7.601	1.720	1.053	1.810	4
5	6.410	.1560	.0832	.5332	12.022	1.876	1.298	2.434	5
6	9.294	.1076	.0543	.5043	18.431	1.983	1.499	2.972	6
7	13.476	.0742	.0361	.4861	27.725	2.057	1.661	3.418	7
8	19.541	.0512	.0243	.4743	41.202	2.109	1.791	3.776	8
9	28.334	.0353	.0165	.4665	60.743	2.144	1.893	4.058	9
10	41.085	.0243	.0112	.4612	89.077	2.168	1.973	4.277	10
11	59.573	.0168	.00768	.4577	130.162	2.185	2.034	4.445	11
12	86.381	.0116	.00527	.4553	189.735	2.196	2.082	4.572	12
13	125.252	.00798	.00362	.4536	276.115	2.204	2.118	4.668	13
14	181.615	.00551	.00249	.4525	401.367	2.210	2.145	4.740	14
15	263.342	.00380	.00172	.4517	582.982	2.214	2.165	4.793	15
16	381.846	.00262	.00118	.4512	846.325	2.216	2.180	4.832	16
17	553.677	.00181	.00081	.4508	1 228.2	2.218	2.191	4.861	17
18	802.831	.00125	.00056	.4506	1 781.8	2.219	2.200	4.882	18
19	1 164.1	.00086	.00039	.4504	2 584.7	2.220	2.206	4.898	19
20	1 688.0	.00059	.00027	.4503	3 748.8	2.221	2.210	4.909	20
21	2 447.5	.00041	.00018	.4502	5 436.7	2.221	2.214	4.917	21
22	3 548.9	.00028	.00013	.4501	7 884.3	2.222	2.216	4.923	22
23	5 145.9	.00019	.00009	.4501	11 433.2	2.222	2.218	4.927	23
24	7 461.6	.00013	.00006	.4501	16 579.1	2.222	2.219	4.930	24
25	10 819.3	.00009	.00004	.4500	24 040.7	2.222	2.220	4.933	25
26	15 688.0	.00006	.00003	.4500	34 860.1	2.222	2.221	4.934	26
27	22 747.7	.00004	.00002	.4500	50 548.1	2.222	2.221	4.935	27
28	32 984.1	.00003	.00001	.4500	73 295.8	2.222	2.221	4.936	28
29	47 826.9	.00002	.00001	.4500	106 279.9	2.222	2.222	4.937	29
30	69 349.1	.00001	.00001	.4500	154 106.8	2.222	2.222	4.937	30
31	100 556.1	.00001		.4500	223 455.9	2.222	2.222	4.938	31
32	145 806.4	.00001		.4500	324 012.0	2.222	2.222	4.938	32
33	211 419.3			.4500	469 818.5	2.222	2.222	4.938	33
34	306 558.0			.4500	681 237.8	2.222	2.222	4.938	34
35	444 509.2			.4500	987 795.9	2.222	2.222	4.938	35

	Single Payment		Uniform Payment Series				Arithmetic Gradient		
	Compound Amount Factor	Present Worth Factor	Sinking Fund Factor	Capital Recovery Factor	Compound Amount Factor	Present Worth Factor	Gradient Uniform Series	Gradient Present Worth	
	Find F Given P F/P	Find P Given F P/F	Find A Given F A/F	Find A Given P A/P	Find F Given A F/A	Find P Given A P/A	Find A Given G A/G	Find P Given G P/G	
n									n
1	1.500	.6667	1.0000	1.5000	1.000	0.667	0	0	1
2	2.250	.4444	.4000	.9000	2.500	1.111	0.400	0.444	2
3	3.375	.2963	.2105	.7105	4.750	1.407	0.737	1.037	3
4	5.063	.1975	.1231	.6231	8.125	1.605	1.015	1.630	4
5	7.594	.1317	.0758	.5758	13.188	1.737	1.242	2.156	5
6	11.391	.0878	.0481	.5481	20.781	1.824	1.423	2.595	6
7	17.086	.0585	.0311	.5311	32.172	1.883	1.565	2.947	7
8	25.629	.0390	.0203	.5203	49.258	1.922	1.675	3.220	8
9	38.443	.0260	.0134	.5134	74.887	1.948	1.760	3.428	9
10	57.665	.0173	.00882	.5088	113.330	1.965	1.824	3.584	10
11	86.498	.0116	.00585	.5058	170.995	1.977	1.871	3.699	11
12	129.746	.00771	.00388	.5039	257.493	1.985	1.907	3.784	12
13	194.620	.00514	.00258	.5026	387.239	1.990	1.933	3.846	13
14	291.929	.00343	.00172	.5017	581.859	1.993	1.952	3.890	14
15	437.894	.00228	.00114	.5011	873.788	1.995	1.966	3.922	15
16	656.841	.00152	.00076	.5008	1 311.7	1.997	1.976	3.945	16
17	985.261	.00101	.00051	.5005	1 968.5	1.998	1.983	3.961	17
18	1 477.9	.00068	.00034	.5003	2 953.8	1.999	1.988	3.973	18
19	2 216.8	.00045	.00023	.5002	4 431.7	1.999	1.991	3.981	19
20	3 325.3	.00030	.00015	.5002	6 648.5	1.999	1.994	3.987	20
21	4 987.9	.00020	.00010	.5001	9 973.8	2.000	1.996	3.991	21
22	7 481.8	.00013	.00007	.5001	14 961.7	2.000	1.997	3.994	22
23	11 222.7	.00009	.00004	.5000	22 443.5	2.000	1.998	3.996	23
24	16 834.1	.00006	.00003	.5000	33 666.2	2.000	1.999	3.997	24
25	25 251.2	.00004	.00002	.5000	50 500.3	2.000	1.999	3.998	25
26	37 876.8	.00003	.00001	.5000	75 751.5	2.000	1.999	3.999	26
27	56 815.1	.00002	.00001	.5000	113 628.3	2.000	2.000	3.999	27
28	85 222.7	.00001	.00001	.5000	170 443.4	2.000	2.000	3.999	28
29	127 834.0	.00001		.5000	255 666.1	2.000	2.000	4.000	29
30	191 751.1	.00001		.5000	383 500.1	2.000	2.000	4.000	30
31	287 626.6			.5000	575 251.2	2.000	2.000	4.000	31
32	431 439.9			.5000	862 877.8	2.000	2.000	4.000	32

	Single Payment		Uniform Payment Series				Arithmetic Gradient		
	Compound Amount Factor	Present Worth Factor	Sinking Fund Factor	Capital Recovery Factor	Compound Amount Factor	Present Worth Factor	Gradient Uniform Series	Gradient Present Worth	
n	Find F Given P F/P	Find P Given F P/F	Find A Given F A/F	Find A Given P A/P	Find F Given A F/A	Find P Given A P/A	Find A Given G A/G	Find P Given G P/G	n
1	1.600	.6250	1.0000	1.6000	1.000	0.625	0	0	1
2	2.560	.3906	.3846	.9846	2.600	1.016	0.385	0.391	2
3	4.096	.2441	.1938	.7938	5.160	1.260	0.698	0.879	3
4	6.554	.1526	.1080	.7080	9.256	1.412	0.946	1.337	4
5	10.486	.0954	.0633	.6633	15.810	1.508	1.140	1.718	5
6	16.777	.0596	.0380	.6380	26.295	1.567	1.286	2.016	6
7	26.844	.0373	.0232	.6232	43.073	1.605	1.396	2.240	7
8	42.950	.0233	.0143	.6143	69.916	1.628	1.476	2.403	8
9	68.719	.0146	.00886	.6089	112.866	1.642	1.534	2.519	9
10	109.951	.00909	.00551	.6055	181.585	1.652	1.575	2.601	10
11	175.922	.00568	.00343	.6034	291.536	1.657	1.604	2.658	11
12	281.475	.00355	.00214	.6021	467.458	1.661	1.624	2.697	12
13	450.360	.00222	.00134	.6013	748.933	1.663	1.638	2.724	13
14	720.576	.00139	.00083	.6008	1 199.3	1.664	1.647	2.742	14
15	1 152.9	.00087	.00052	.6005	1 919.9	1.665	1.654	2.754	15
16	1 844.7	.00054	.00033	.6003	3 072.8	1.666	1.658	2.762	16
17	2 951.5	.00034	.00020	.6002	4 917.5	1.666	1.661	2.767	17
18	4 722.4	.00021	.00013	.6001	7 868.9	1.666	1.663	2.771	18
19	7 555.8	.00013	.00008	.6011	12 591.3	1.666	1.664	2.773	19
20	12 089.3	.00008	.00005	.6000	20 147.1	1.667	1.665	2.775	20
21	19 342.8	.00005	.00003	.6000	32 236.3	1.667	1.666	2.776	21
22	30 948.5	.00003	.00002	.6000	51 579.2	1.667	1.666	2.777	22
23	49 517.6	.00002	.00001	.6000	82 527.6	1.667	1.666	2.777	23
24	79 228.1	.00001	.00001	.6000	132 045.2	1.667	1.666	2.777	24
25	126 765.0	.00001		.6000	211 273.4	1.667	1.666	2.777	25
26	202 824.0			.6000	338 038.4	1.667	1.667	2.778	26
27	324 518.4			.6000	540 862.4	1.667	1.667	2.778	27
28	519 229.5			.6000	865 380.9	1.667	1.667	2.778	28

Continuous Compounding—Single Payment Factors

rn	Compound Amount Factor e^{rn} — Find F Given P F/P	Present Worth Factor e^{-rn} — Find P Given F P/F	rn	Compound Amount Factor e^{rn} — Find F Given P F/P	Present Worth Factor e^{-rn} — Find P Given F P/F
.01	1.0101	.9900	.51	1.6653	.6005
.02	1.0202	.9802	.52	1.6820	.5945
.03	1.0305	.9704	.53	1.6989	.5886
.04	1.0408	.9608	.54	1.7160	.5827
.05	1.0513	.9512	.55	1.7333	.5769
.06	1.0618	.9418	.56	1.7507	.5712
.07	1.0725	.9324	.57	1.7683	.5655
.08	1.0833	.9231	.58	1.7860	.5599
.09	1.0942	.9139	.59	1.8040	.5543
.10	1.1052	.9048	.60	1.8221	.5488
.11	1.1163	.8958	.61	1.8404	.5434
.12	1.1275	.8869	.62	1.8589	.5379
.13	1.1388	.8781	.63	1.8776	.5326
.14	1.1503	.8694	.64	1.8965	.5273
.15	1.1618	.8607	.65	1.9155	.5220
.16	1.1735	.8521	.66	1.9348	.5169
.17	1.1853	.8437	.67	1.9542	.5117
.18	1.1972	.8353	.68	1.9739	.5066
.19	1.2092	.8270	.69	1.9937	.5016
.20	1.2214	.8187	.70	2.0138	.4966
.21	1.2337	.8106	.71	2.0340	.4916
.22	1.2461	.8025	.72	2.0544	.4868
.23	1.2586	.7945	.73	2.0751	.4819
.24	1.2712	.7866	.74	2.0959	.4771
.25	1.2840	.7788	.75	2.1170	.4724
.26	1.2969	.7711	.76	2.1383	.4677
.27	1.3100	.7634	.77	2.1598	.4630
.28	1.3231	.7558	.78	2.1815	.4584
.29	1.3364	.7483	.79	2.2034	.4538
.30	1.3499	.7408	.80	2.2255	.4493
.31	1.3634	.7334	.81	2.2479	.4449
.32	1.3771	.7261	.82	2.2705	.4404
.33	1.3910	.7189	.83	2.2933	.4360
.34	1.4049	.7118	.84	2.3164	.4317
.35	1.4191	.7047	.85	2.3396	.4274
.36	1.4333	.6977	.86	2.3632	.4232
.37	1.4477	.6907	.87	2.3869	.4190
.38	1.4623	.6839	.88	2.4109	.4148
.39	1.4770	.6771	.89	2.4351	.4107
.40	1.4918	.6703	.90	2.4596	.4066
.41	1.5068	.6637	.91	2.4843	.4025
.42	1.5220	.6570	.92	2.5093	.3985
.43	1.5373	.6505	.93	2.5345	.3946
.44	1.5527	.6440	.94	2.5600	.3906
.45	1.5683	.6376	.95	2.5857	.3867
.46	1.5841	.6313	.96	2.6117	.3829
.47	1.6000	.6250	.97	2.6379	.3791
.48	1.6161	.6188	.98	2.6645	.3753
.49	1.6323	.6126	.99	2.6912	.3716
.50	1.6487	.6065	1.00	2.7183	.3679

694

Economic Criteria

Method of Analysis	Fixed Input	Fixed Output	Neither Input Nor Output Fixed
PRESENT WORTH	Maximize PW of Benefits	Minimize PW of Costs	Maximize (PW of Benefits − PW of Costs), or Maximize Net Present Worth
ANNUAL CASH FLOW	Maximize Equivalent Uniform Annual Benefits (EUAB)	Minimize Equivalent Uniform Annual Cost (EUAC)	Maximize (EUAB − EUAC)
FUTURE WORTH	Maximize FW of Benefits	Minimize FW of Costs	Maximize (FW of Benefits − FW of Costs), or Maximize Net Future Worth
BENEFIT–COST RATIO	Maximize Benefit–Cost Ratio	Maximize Benefit–Cost Ratio	*Two Alternatives:* Compute the incremental Benefit–Cost ratio ($\Delta B/\Delta C$) on the increment of *investment* between the alternatives. If $\Delta B/\Delta C \geq 1$, choose higher-cost alternative; if not, choose lower-cost alternative. *Three or more Alternatives:* Incremental analysis is required (see Ch. 9).
RATE OF RETURN			

Two Alternatives: Compute the incremental rate of return (ΔROR) on the increment of *investment* between the alternatives. If $\Delta ROR \geq$ minimum attractive rate of return, choose the higher-cost alternative; if not, choose lower-cost alternative.
Three or more Alternatives: Incremental analysis is required (see Ch. 8).